Climate Change and Disadvantaged Communities

Challenges and Opportunities

Edited by

Sk Ajim Ali, Quoc Bao Pham, and Yunqing Xuan

CRC Press
Taylor & Francis Group
Boca Raton London New York

CRC Press is an imprint of the
Taylor & Francis Group, an **informa** business

Designed cover image: © Shutterstock

First edition published 2026
by CRC Press
2385 NW Executive Center Drive, Suite 320, Boca Raton FL 33431

and by CRC Press
4 Park Square, Milton Park, Abingdon, Oxon, OX14 4RN

CRC Press is an imprint of Taylor & Francis Group, LLC

ISBN: 9781041027201 (hbk)
ISBN: 9781041027218 (pbk)
ISBN: 9781003620747 (ebk)

DOI: 10.1201/9781003620747

Typeset in Times
by codeMantra

Dedicated to

Our Parents Who Made It Possible to Reach Where We Are Today...

Contents

PART 2 Urbanization, Public Health, and Microclimate Challenges in a Changing Climate

PART 3 Natural Resources, Ecosystems, and Climate Impact Assessments: Modeling, Mapping, and Risk Assessment

Preface

According to the World Health Organization (WHO) report, between 2030 and 2050, climate change is expected to result in an extra 2,50,000 fatalities per year. The most vulnerable populations are expected to be children and the underprivileged people of developing nations. In this regard, this book is to draw academics' and researchers' attention to the critical issue of how climate change disproportionately impacts vulnerable and disadvantaged communities by examining the socioeconomic, environmental, and infrastructure factors that contribute to these disproportions and offering a thorough understanding of why some communities are more affected than others.

The recent increase in temperature and alteration of rainfall patterns are undoubtedly signs of climate change, which will have an adverse impact on communities and society as a whole. This book explores how climate change affects the lives and livelihoods of people in general and climate-disadvantaged communities in particular by highlighting the specific challenges they face, such as increased risk of heatwaves, droughts, flooding, health issues, and economic losses, and explains why vulnerable populations are more at risk due to factors like lack of infrastructure, poverty, and limited access to resources. It also offers empirical insights about the unequal impacts of climate change and a conceptual framework to help vulnerable communities adapt and become more resilient to climate impacts. By emphasizing real-life examples, this book also aims to inform policymakers and the public about the urgent need for targeted initiatives to support disadvantaged communities in the face of climate change.

This book presents a vision for integrated solutions and a path forward for climate-disadvantaged communities, recognizing that climate change is intensifying extreme weather events such as heatwaves, droughts, and floods which disproportionately affect vulnerable populations, leading to death, illness, hunger, displacement, and the disruption of lives and livelihoods. This book helps readers gain a better understanding of the multiple impacts of climate change and the disproportionate effects on vulnerable communities by outlining the particular challenges that these people confront, along with practical responses and strategies to support and sustain them.

This book is organized into three thematic parts, each addressing a crucial dimension of the climate vulnerability-resilience nexus. Part I: Climate Change Vulnerability, Livelihoods, and Resilience of Disadvantaged Communities delves into the structural and place-based vulnerabilities that shape exposure and adaptive capacity in rural and peri-urban regions. The chapters in this section examine issues ranging from agrarian distress and gendered vulnerabilities to indigenous coping strategies and the role of social capital in resilience-building. Part II: Urbanization, Public Health, and Microclimate Challenges in a Changing Climate shifts focus to the rapidly urbanizing landscapes of the Global South, where informal settlements, limited infrastructure, and public health crises intersect with intensifying climatic events. This section presents empirical analyses and case studies on urban heat stress, air quality degradation, disease burdens, and governance deficits in urban climate adaptation. Part III: Natural Resources, Ecosystems, and Climate Impact Assessments: Modeling, Mapping, and Risk Assessment explores the role of data-driven tools in understanding climate impacts on ecosystems and resource-dependent communities. Employing a mix of spatial mapping, vulnerability modelling, and ecological assessments, these chapters provide methodological and policy insights into managing climate risks at various scales.

Taken together, the 19 chapters and one introductory chapter in this book offer diverse perspectives and critical reflections that emphasize the importance of inclusive, participatory, and context-specific approaches to climate action. By foregrounding the voices and experiences of disadvantaged communities, we hope this volume contributes meaningfully to ongoing debates on climate justice, sustainable development, and equitable adaptation pathways.

We extend our sincere gratitude to all the contributors who submitted their chapters with dedication and on time. Their insightful and rigorous scholarship has been instrumental in shaping this volume into a meaningful contribution to the discourse on climate change and disadvantaged communities. We believe this book will serve as a valuable resource for a wide audience, including climate scientists, geographers, ecologists, geologists, urban planners, environmentalists, research scholars, and policymakers engaged in the fields of environmental management, climate justice, and sustainable development. We are also deeply thankful to CRC Press – Taylor & Francis Group, especially Irma Shagla Britton, Senior Editor for GIS & Remote Sensing, Environmental Engineering & Science, and her team, for their professional guidance and support throughout the publication process.

Dr. Sk Ajim Ali

Dr. Quoc Bao Pham

Dr. Yunqing Xuan

About the Editors

Sk Ajim Ali is an environmental geographer who studies environmental concerns with a focus on places, politics, and the safe management of human and built environments. He received his PhD from Aligarh Muslim University, Aligarh. Currently, he is remotely working as a Research Associate at the Global Development Institute, School of Environment, Education and Development, University of Manchester, Manchester, UK. His principal areas of research include disaster management, climate change, and public health. Last year, he worked in a SEED-funded project on socio-spatial heat wave vulnerability and public health at the University of Manchester. Additionally, he studied geohazards vulnerability assessment through international collaboration with different countries of the world to understand environmental hazards, risk, and attendant issues of sustainable development. He was also employed as a guest lecturer at the Faculty of Life Sciences, Aligarh Muslim University, where he taught Environmental Studies to undergraduate students. He has contributed 57 scientific research papers in international journals on issues pertaining to environmental degradation, waste management, and geohazards vulnerabilities. He is a reviewer of many journals published by a variety of reputable academic publishers. He is the author of two books on remote sensing and quantum GIS. He is also contributing as an associate editor, editorial board member, and guest editor of different journals.

Quoc Bao Pham is currently working as an Assistant Professor at the Faculty of Natural Science, Institute of Earth Science, University of Silesia in Katowice, Katowice, Poland. He has obtained his PhD degree from the Department of Hydraulic and Ocean Engineering, National Cheng Kung University, Taiwan. He has more than eight years of research experience in hydrology, GIS, and remote sensing. His research interests are water resources engineering, applied machine learning, climate change, hydrology, remote sensing, GIS, and spatial analysis. He has collaborated actively with many international researchers in several other disciplines of engineering. He has published more than 80 SCI/SCIE research papers and is a reviewer of more than 20 SCI/SCIE journals. He is also a guest editor of some SCI/SCIE journals in the fields of water resources, remote sensing, and machine learning. He is also a reviewer of world-leading high-impact journals such as *Journal of Hydrology*, *Science of the Total Environment*, *Remote Sensing of Environment*, *Hydrological Sciences Journal*, *Journal of Environmental Management*, *Geocarto International*, and *Remote Sensing Applications*.

Yunqing Xuan, FRMetS, FIMA, FIHE, is a civil engineer by training but at the same time a hydrometeorologist with a range of interdisciplinary research interests in modelling hydrological processes interfacing with atmosphere and societies. He is currently working as an Associate Professor of Civil Engineering, Swansea University, Swansea, Wales, UK. In practical words, he has been working as both a practitioner and a researcher in several fields: hydrology and hydrometeorology using advanced computer modelling, studying extreme events; weather radars and rainfall nowcasting; and flood risk management and climate change impact/adaptation in the water sector. Dr. Xuan strives to contribute to the learner and professional communities. This includes his other undertakings as the Vice Chair of the Welsh Local Centre of the Royal Meteorological Society, Review Editor of the *Weather* journal, and Board Member of the Conference Committee of the Institute of Mathematics and Its Applications.

Contributors

Ateeque Ahmad
Department of Geography
Faculty of Science
Aligarh Muslim University
Aligarh, India

Syed Naushad Ahmad
Department of Geography
Aligarh Muslim University
Aligarh, UP, India

Snehashis Alam
Department of Humanities and Social Sciences
BITS Pilani
K. K. Birla Goa Campus
Goa, India

Rayees Ali
Department of Geography
Faculty of Sciences
Jamia Millia Islamia
New Delhi, India

Rukhsar Anjum
Aligarh Muslim University
Aligarh, India

Mohammad Sirazuddin Ansari
Department of Botany
Institute of Science
Banaras Hindu University
Varanasi, UP, India

Rohit Kumar Azad
Department of Geography
Centre for Disaster Management
Jamia Millia Islamia
New Delhi, India

Nasrin Banu
Department of Geography
Aliah University
Kolkata, India

Mohan Kumar Bera
Department of Humanities and Social Sciences
BITS Pilani
K. K. Birla Goa Campus
Goa, India

Jaysikha Bhattacharyya
Department of Geography
University of Kalyani
Kalyani, West Bengal, India

Nirsobha Bhuyan
Department of Geography
Faculty of Sciences
Jamia Millia Islamia
New Delhi, India

Ajay Dashora
Department of Civil Engineering
Indian Institute of Technology Guwahati
Guwahati, India

Daawar Bashir Ganaie
Department of Geography
Faculty of Sciences
Jamia Millia Islamia
New Delhi, India

Manotosh Gayen
Department of Geography
Aliah University
Kolkata, India

Mouparna Ghosh
Department of Geography
University of Kalyani
Kalyani, West Bengal, India

Milan Ghosh
Department of Geography
University of Kalyani
Kalyani, West Bengal, India

Mariya Hasnat
Department of Environmental Science
Integral University
Lucknow, India

Tran Thi Thu Hien
Department of Tourism Studies
School of Hospitality and Tourism
Hue University
Hue, Vietnam

Mujahid Husain
ARX, Smart Minds, One by Omniyat
Business Bay, Dubai

Anshul Jaiswal
Department of Botany
Institute of Science
Banaras Hindu University
Varanasi, UP, India

Mohit Joshi
Department of Civil Engineering
Galgotias College of Engineering and Technology
Greater Noida, India

Mohd Kashif Khan
Department of Environmental Science
Integral University
Lucknow, India

Zainab Khan
Geomatics Engineering Group
Civil Engineering Department
Indian Institute of Technology Roorkee
Uttarakhand, India

Zeba Khanam
Department of Geography
Jawaharlal Nehru University
New Delhi, India

Sania Kulsum
Department of Education
Integral University
Lucknow, India

Amit Kumar
Department of Geography
Faculty of Science
Aligarh Muslim University
Aligarh, UP, India

Pranay Rajendra Lanjekar
Extension Division
Saradar Patel Renewable Energy Research Institute (SPRERI)
Anand, Gujarat, India

Selim Molla
Department of Geography
University of Kalyani
Kalyani, West Bengal, India

Usman Mohseni
Department of Civil Engineering
Indian Institute of Technology Roorkee
Uttarakhand, India

Md Mohsin
Department of Geography
Aligarh Muslim University
Aligarh, UP, India

Mohammed Numan
Department of Geography
Rajendra University
Balangir, India

Aakansha Pandey
Department of Botany
Institute of Science
Banaras Hindu University
Varanasi, UP, India

Narayan Lal Panwar
Department of Renewable Energy Engineering
Maharana Pratap University of Agriculture and Technology
Udaipur, Rajasthan, India

Farhana Parvin
Department of Geography
Faculty of Science
Aligarh Muslim University
Aligarh, UP, India

Tapas Ranjan Patra
Department of Geography
Rajendra University
Balangir, India

Vinnarasi Rajendran
Department of Civil Engineering
Indian Institute of Technology Roorkee
Uttarakhand, India

Richa Rawat
Department of Botany
Institute of Science
Banaras Hindu University
Varanasi, UP, India

Arindam Roy
Department of Geography
Aliah University
Kolkata, India

Sadaf
Department of Geography
Veer Kunwar Singh University
Ara, Bihar, India

R. Sagar
Department of Botany
Institute of Science
Banaras Hindu University
Varanasi, UP, India

Haroon Sajjad
Department of Geography
Faculty of Sciences
Jamia Millia Islamia
New Delhi, India

Abhay Sankar Sahu
Department of Geography
University of Kalyani
Kalyani, Nadia, West Bengal, India

Amit Sardar
Department of Geography
Rajendra University
Balangir, India

Trilochan Sethy
Department of Geography
Rajendra University
Balangir, India

Aastha Sharma
Department of Geography
Faculty of Sciences
Jamia Millia Islamia
New Delhi, India

Md Mainul Sk
Department of Geography
Rajendra University
Balangir, India

Khemlata Soni
Department of Renewable Energy Engineering
Maharana Pratap University of Agriculture and Technology
Udaipur, Rajasthan, India

Deepika Vashishtha
Department of General and Applied Geography
Dr. Harisingh Gour Central University
Sagar, Madhya Pradesh, India

Deepankar Yadav
Department of Botany
Institute of Science
Banaras Hindu University
Varanasi, UP, India

1 Climate Change and Vulnerability of Disadvantaged Communities

Introduction and Overview

Sk Ajim Ali, Quoc Bao Pham, and Yunqing Xuan

INTRODUCTION

Climate change is one of the crucial global challenges of the 21st century, unequally affecting disadvantaged communities across the world (Lomborg, 2020; Parsons et al., 2024). These communities, often marginalized due to socio-economic, political, and geographical factors, bear the brunt of climate-induced disasters despite contributing the least to greenhouse gas emissions. The adverse impacts of climate change, ranging from extreme weather events to gradual environmental degradation, exacerbate existing vulnerabilities, deepening inequalities and threatening livelihoods, food security, and overall well-being (Abbass et al., 2022; Yang et al., 2024). The intersection of climate change and social inequality has gained significant attention in recent years, highlighting the need for a more just and equitable approach to climate resilience and adaptation.

According to a World Bank analysis, the 74 lowest-income nations create just a tenth of the world's greenhouse emissions, yet they will bear the brunt of climate change's repercussions (Dellinger, 2022). They have already seen around eight times as many natural catastrophes in the last ten years as they did in the 1980s. Over 200 million people may be forced to relocate within their own nations by 2050 due to unmanaged climate change, which may push up to 130 million people into poverty and undo decades of hard-won development gains (Tubridy et al., 2022; Yadav and Upadhyay, 2023). The effects of climate change are unevenly distributed across populations, influenced by a combination of economic disparities, geographic exposure, institutional capacity, and social stratification. Low-income communities, indigenous groups, and other marginalized populations often inhabit areas prone to environmental hazards such as coastal regions, floodplains, and arid lands. These locations, while rich in cultural and ecological significance, are increasingly vulnerable to rising sea levels, desertification, and extreme weather patterns. Furthermore, these communities frequently lack access to adequate infrastructure, healthcare, and emergency response mechanisms, making it difficult to cope with and recover from climate-related shocks (Jobaer et al., 2025; Hadley et al., 2023).

Indigenous people, small-scale food producers, and low-income households face malnourishment as a result, with children, elderly people, and pregnant women hit particularly hard (Gadhoke et al., 2023; Roy et al., 2023). Women are traveling further in drought-prone regions to obtain clean water, increasing their risks of encountering gender-based violence. Rising temperatures and humidity will allow mosquitoes and other vectors of disease to conquer new territory, spreading diseases in people (for example, dengue fever) and in crops, livestock, and wildlife (de Souza and Weaver, 2024; Prasad et al., 2024; Yadav and Upadhyay, 2023). This will mean more death and suffering in poorer countries, which lack health infrastructure, data collection, early warning systems, and vaccines. One in three people are already exposed to potentially deadly heat stress, and this number is projected to increase to as much as three in four by the end of the century (Klein and

DOI: 10.1201/9781003620747-1

Anderegg, 2021; Saeed et al., 2021). Meanwhile, a billion people living on coasts will be exposed to serious flooding every year by 2050 (Rentschler et al., 2022). Sea-level rise poses an "existential threat" for some small islands, according to the IPCC. Already, people living in these places face longer droughts, deeper floods, fish spoiled by ocean heating, and fresh water contaminated by encroaching seas. Storms increasingly flatten homes, schools, hospitals, shops, and crops. When floods or storms hit, people living with disabilities, older and younger people, and the poor often lack the necessary means to evacuate safely. Marginalized groups – including people who identify as lesbian, gay, bisexual, transgender, queer, and intersex – are more likely to live in insecure homes, which are more vulnerable to strong winds, for example.

The disproportionate effects of climate change are not limited to underdeveloped or developing nations; they are also becoming more noticeable in developed countries. According to a study by the U.S. Environmental Protection Agency (EPA), underprivileged areas are disproportionately affected by climate change, as they are less equipped to prepare for and recover from heat waves, poor air quality, flooding, and other effects. Racial and ethnic minority populations are most susceptible to the worst effects of climate change, according to EPA's research (EPA, 2021). According to a World Bank research, climate change might force an extra 68–135 million people into poverty by 2030 (World Bank, 2020). Global economic expansion has lifted millions out of severe poverty and decreased international inequality in recent decades. But unmanaged climate change threatens to set back that progress by damaging poverty eradication efforts worldwide and disproportionately affecting the poorest regions and people. Historical and structural inequalities further compound the vulnerabilities of disadvantaged populations. Colonial legacies, economic exploitation, and land dispossession have left many communities in precarious conditions with limited adaptive capacities. In both urban and rural settings, the absence of secure land tenure, limited access to financial resources, and political marginalization hinder effective climate adaptation efforts. Additionally, gender disparities often play a crucial role in determining the extent of climate vulnerability, as women in many societies face systemic barriers in accessing education, economic opportunities, and decision-making processes related to climate policies (Aziz and Anjum, 2024; Humayra et al., 2024; Lau et al., 2021).

The concept of climate justice emphasizes the moral and ethical responsibility to address these disproportionate impacts (Newell et al., 2021; Sardo, 2023). It calls for policies that recognize and rectify historical injustices while ensuring equitable distribution of resources and opportunities for climate adaptation. However, many climate adaptation and mitigation strategies continue to overlook the specific needs and priorities of disadvantaged communities. Large-scale infrastructure projects, for instance, may displace vulnerable populations, further exacerbating their exposure to environmental risks. Similarly, market-driven solutions such as carbon trading often fail to address localized vulnerabilities, leaving marginalized groups without tangible benefits (Perkiss, 2024). The global response to climate change has been marked by ambitious international agreements, including the Paris Agreement and the United Nations Sustainable Development Goals (SDGs). While these frameworks acknowledge the importance of equity in climate action, their implementation at national and local levels remains uneven (Fitzgerald, 2022). Developing countries, despite facing the most severe consequences of climate change, often struggle with inadequate funding, technological constraints, and political challenges in executing meaningful climate adaptation strategies (Biswas and Rahman, 2023; Garschagen and Doshi, 2022). Moreover, international climate finance mechanisms, designed to support vulnerable nations, have faced criticisms for bureaucratic inefficiencies and inequitable distribution of resources.

Community-based adaptation (CBA) approaches have emerged as effective alternatives to top-down climate policies (Masud-All-Kamal and Nursey-Bray, 2021). These initiatives prioritize local knowledge, participatory decision-making, and grassroots leadership in designing climate resilience strategies. Indigenous knowledge systems, traditional ecological practices, and collective resource management have historically enabled communities to adapt to environmental changes (Varghese and Crawford, 2021). Recognizing and integrating these approaches into national and global climate policies can enhance resilience while fostering social inclusion and environmental sustainability (Adebayo, 2025; Tamasiga et al., 2024). Despite the growing recognition of the need for inclusive climate action, significant gaps remain in research and policy implementation. Existing climate models and risk assessments often fail to capture the socio-economic dimensions of vulnerability, leading to inadequate planning and resource allocation. Moreover, there is a lack of representation of marginalized voices in climate governance structures, further alienating those most affected by climate change. Strengthening interdisciplinary research, promoting participatory governance, and ensuring equitable access to climate adaptation resources are critical steps toward addressing these challenges.

This study aims to explore the unequal impacts of climate change on disadvantaged communities, examining the socio-economic, political, and environmental factors that shape their vulnerabilities. By analyzing several literatures and case studies from different geographical contexts, the research seeks to highlight both the systemic barriers and the resilience strategies employed by these communities. Additionally, it highlights the importance of integrating climate justice principles into policy frameworks to create more inclusive and effective climate adaptation strategies. Ultimately, addressing climate vulnerability through an equity lens is not only a moral imperative but also a necessary step toward achieving sustainable and resilient societies in the face of an increasingly unpredictable climate.

EQUITY OR VULNERABILITY: WHAT ARE THE SOCIAL DIMENSIONS OF CLIMATE CHANGE?

Climate change is not just an environmental issue; it is deeply intertwined with social dimensions, influencing and being influenced by structures of inequality, power, and vulnerability. The question of whether climate change should be viewed primarily through the lens of equity or vulnerability is complex, as both perspectives highlight crucial aspects of the social consequences of climate change. Equity focuses on justice, fairness, and the disproportionate burden on marginalized communities, while vulnerability emphasizes the exposure, sensitivity, and adaptive capacity of different social groups. The disproportionate impacts of climate change on marginalized communities emphasize the issue of equity. Historically, disadvantaged groups, including the poor, women, indigenous populations, and ethnic minorities, contribute the least to greenhouse gas emissions but bear the brunt of climate-related disasters such as floods, droughts, and extreme heat (Abi Deivanayagam et al., 2023; Nielsen et al., 2021). Socio-economic disparities shape people's access to resources, information, and decision-making processes, which in turn affect their ability to respond to climate risks. Wealthier nations and communities have greater financial and technological resources to mitigate and adapt to climate change (Sovacool et al., 2022), whereas low-income communities often lack adequate infrastructure, social safety nets, and institutional support, deepening existing inequalities (Feng et al., 2023).

Climate justice movements have highlighted the need for fair and inclusive climate policies that consider historical responsibilities and current disparities. The principle of common but differentiated responsibilities (CBDR), established in international climate negotiations, acknowledges that industrialized nations have historically emitted more greenhouse gases and therefore bear a greater responsibility for mitigation and financial support to vulnerable regions (Stoddard et al., 2021). However, achieving climate justice remains a challenge due to geopolitical power imbalances, economic interests, and the slow pace of climate financing for adaptation and loss and damage in vulnerable regions (Chapman and Ahmed, 2021).

Vulnerability to climate change is shaped by multiple intersecting factors, including geography, economic status, gender, caste, and access to political power. Rural and agrarian communities that rely on climate-sensitive livelihoods, such as farming and fishing, are among the most vulnerable to shifting climate patterns. Smallholder farmers in South Asia, particularly Bangladesh and India, for instance, face erratic monsoons and prolonged droughts that threaten food security, leading to cycles of poverty and forced migration (Aziz, 2024). Similarly, indigenous communities that depend on forests and rivers for sustenance and cultural identity are increasingly facing displacement due to climate-induced environmental changes, such as rising sea levels and deforestation (Datta et al., 2025; Whyte, 2021). Urban areas, despite being hubs of economic opportunity, also reveal significant vulnerabilities. Informal settlements, or slums, in rapidly growing cities often lack basic services such as clean water, sanitation, and healthcare, making their residents particularly susceptible to climate-related health crises, including heat stress, vector-borne diseases, and respiratory illnesses from air pollution (Haque et al., 2022). The social and economic marginalization of these populations limits their ability to relocate or invest in climate-resilient infrastructure, leaving them in a cycle of perpetual risk (Friedman, 2023).

Gender is another critical dimension of climate vulnerability. Women, particularly in developing regions, often face greater exposure to climate risks due to their roles in household water collection, food production, and caregiving (Dickin et al., 2021; Tantoh et al., 2021). During climate disasters, they experience heightened vulnerabilities, including increased risks of gender-based violence, loss of livelihood opportunities, and barriers to accessing relief resources. However, women also play a crucial role in climate adaptation, with local women-led initiatives demonstrating effective strategies for sustainable resource management and disaster preparedness. The intersection of climate change and migration further complicates the social landscape. Climate-induced displacement is becoming a pressing global concern as rising sea levels, desertification, and extreme weather events force communities to relocate (Poddar, 2024). The social, economic, and psychological toll of displacement is immense, often leading to loss of cultural identity, economic instability, and tensions in receiving areas. Climate refugees, particularly those crossing international borders, face legal uncertainties due to the absence of specific international protections, exacerbating their vulnerability (Tahir and Tahir, 2024).

Social responses to climate change, including community-led adaptation strategies, policy interventions, and resilience-building efforts, play a crucial role in addressing both equity and vulnerability. Bottom-up approaches that integrate local and indigenous knowledge into climate governance have shown promise in strengthening community resilience (Ali et al., 2021). Participatory decision-making, equitable access to climate finance, and inclusive urban planning can help bridge the gap between vulnerability and adaptive capacity. Furthermore, intersectional climate policies that consider gender, caste, and socio-economic status can lead to more just and effective climate action. Ultimately, the social dimensions of climate change cannot be reduced to a binary choice between equity and vulnerability. Instead, they must be understood as interconnected forces that shape the lived experiences of communities affected by climate change. Addressing climate change requires a holistic approach that combines scientific, economic, and social perspectives to create policies that are both just and effective. The recognition of social inequalities and vulnerabilities in climate discourse is essential for developing sustainable solutions that prioritize the needs of those most affected, ensuring that climate action does not reinforce existing disparities but instead promotes a more equitable and resilient future.

Reducing poverty and tackling the root causes of vulnerability, such as social exclusion, are major priorities of the World Bank's work on the social aspects of climate change (Costella et al., 2023; Millard and Fucci, 2023). In order to increase social resilience at the local level, where the effects of climate change are usually felt the most, and to encourage meaningful engagement of communities and other social groups in climate change decision-making and action, the Bank takes a whole-of-society approach, collaborating with grassroots communities, national and local governments, civil society, and other partners. Taking action on climate change in a broad and immediate social context, acknowledging the urgency of current requirements, and charting an ambitious path to decarbonization are all necessary for green recovery from the recent global pandemic and the transition to low-carbon, climate-resilient growth. The World Bank assists in three main areas of action for achieving these goals: (i) Encourage locally led climate action: encouraging more transparency and accountability on climate finance; encouraging devolved climate finance and community and local development approaches that enable communities to drive a climate agenda in support of their development goals; connecting locally led climate action to national climate change priorities and strategies; encouraging efforts to improve resilience and adaptation M&E; (ii) Recognizing and monitoring the social effects and co-benefits of green development and mitigation policies and initiatives: supporting research and discussion on what constitutes a "just transition" in various industries and situations; recognizing and resolving the effects that various decarbonization strategies have on distribution and social inclusion; and (iii) Facilitating the procedures required to support important transitions: Building awareness and political will among governments and partners on the necessity of comprehending and addressing the social dimensions of climate change and green growth; involving communities and citizens in climate decision-making; and improving social learning as a form of regulatory feedback (e.g., citizen engagement, national climate dialogs, and improved governance).

In order to empower marginalized social groups and impoverished communities and lower the risk of future crises, the World Bank promotes strong collaboration across various practice areas through these areas of action. Additionally, it helps to close the gap between the local, sub-national, and national levels for effective climate change support.

WHY CLIMATE CHANGE DISPROPORTIONATELY IMPACTS THE DISADVANTAGED COMMUNITIES?

Climate change disproportionately impacts disadvantaged communities due to a combination of environmental, political, and socio-economic factors. These communities often face higher exposure to climate hazards, greater sensitivity to climate impacts, and limited adaptive capacity. Disadvantaged communities often live in high-risk areas such as floodplains, coastal zones, drought-prone regions, and urban slums. Inadequate housing, lack of drainage systems, and insufficient early warning systems increase exposure to extreme weather events (Dlamini et al., 2024). Many marginalized groups rely on agriculture, fishing, or informal labor, which are highly vulnerable to climate variability. Pre-existing health conditions and inadequate healthcare access make disadvantaged populations more susceptible to climate-related diseases, including heat stress (Lindsay et al., 2023). Recovery programs often favor wealthier areas, leaving marginalized communities struggling with long-term displacement and economic losses (Tan et al., 2022; Wilson et al., 2021). The specific areas in which climate change disproportionately impacts the disadvantaged groups are as follows:

Increased Food Insecurity and Hunger – Climate change disproportionately affects disadvantaged communities by exacerbating existing social and economic inequalities. One of the most severe consequences is hunger, as climate-induced disruptions in agriculture lead to declining crop yields, food shortages, and rising prices. Droughts, floods, and extreme weather events destroy crops and reduce soil fertility, making food less accessible to low-income populations (Farooq et al., 2022). Malnutrition becomes widespread, especially among children and pregnant women, weakening their immunity and overall health (Morales et al., 2023). As a result of climate change impacts on agricultural productivity and prices driven up by either gradual changes in long-term climate trends or more frequent and severe natural disasters, poor people will be affected by changes in food production, higher consumption prices, and changes in rural incomes. In global breadbaskets, the likelihood of crop yield failures is expected to increase by up to 4.5 times by 2030 and by up to 25 times by 2050 (Hasegawa et al., 2022; Kompas et al., 2024).

Worsening Public Health Crises – The implementation of universal health coverage (UHC) is seriously threatened by climate change in a number of ways, such as by increasing the burden of disease outbreaks and making it more difficult to obtain healthcare, frequently when it is most needed. Health conditions deteriorate further as climate change increases the prevalence of heat-related illnesses, respiratory diseases, and vector-borne infections such as malaria and dengue. Poor air quality from wildfires and pollution exacerbates respiratory issues, while extreme heatwaves disproportionately impact those without access to cooling or proper housing (Bayram et al., 2023). Disadvantaged communities, with limited access to healthcare facilities and essential medicines, face greater difficulties in coping with these health crises. An additional 150 million cases of malaria, or a 5% increase in the number of people at risk, might arise from global warming of 2°C–3°C. Increased water scarcity would affect the purity and quality of the water and increase the prevalence of diarrhea (Li and Managi, 2022). It is estimated that 48,000 more children under the age of 15 will die from diarrhea-related causes by 2030 (Hubbard et al., 2025).

Education: Disruptions and Inequalities – Education is also severely impacted, as climate-induced disasters damage school infrastructure and disrupt academic schedules. Children from poor families are often forced to drop out to support their households during climate-induced economic hardships (Bakti et al., 2025; de Carvalho, 2025). Girls are particularly vulnerable, as families struggling with food insecurity may prioritize early marriages over education. Limited access to schools in disaster-prone areas further reduces opportunities for quality education, reinforcing cycles of poverty and vulnerability (Pal et al., 2023). Many families will use counterproductive strategies to get through economic downturns caused by destroyed crops, such as putting their kids to work to make up for lost revenue or removing them from school to save money on tuition.

Water Scarcity: Limited Access to Safe Water – Water scarcity is another major issue, with prolonged droughts, desertification, and groundwater depletion reducing access to safe drinking water. Contaminated water sources increase the spread of waterborne diseases, disproportionately affecting already vulnerable populations (Lee et al., 2023). Women and children often bear the burden of walking long distances to fetch water, reducing their time for education and inc ome-generating activities (Ho et al., 2021). Meanwhile, in coastal regions, rising sea levels lead to saltwater intrusion, making freshwater resources unusable. Climate change is posing a risk to water access. Water availability is especially affected in regions like the sub-tropics, which already face water shortages, by frequent droughts, increased evaporation, and changes in rainfall patterns and runoff. The number of people impacted by water scarcity will increase from the current 1.7 billion to 5 billion by 2025 (du Plessis, 2023; Wang et al., 2025).

Work-Related Hazards: Increased Occupational Risks – Work-related hazards increase as extreme heat and unpredictable weather conditions make outdoor labor more dangerous. Agricultural and construction workers face severe health risks, including heat stress and dehydration, while declining crop yields threaten the livelihoods of subsistence farmers (El Khayat et al., 2022). Heating up at work reduces productivity and limits mental and physical capacities, which poses a danger to occupational health (Ioannou et al., 2022). Fisheries and coastal industries are also impacted by rising sea levels and stronger storms, displacing entire communities that depend on these sectors for survival. Informal and low-wage workers are hit hardest, as they often lack access to social safety nets and labor protections (Theodore, 2023). In 2030, heat stress is expected to cut global GDP by $2.4 billion and working hours by 2.2% globally (Borg et al., 2021).

Creating Refugees: Climate-Induced Displacement – The combined effects of these climate stressors lead to mass displacement, creating climate refugees who are forced to migrate due to environmental degradation. Because of climate change impacts on people's livelihoods and the loss of livability in areas that are particularly vulnerable, climate change is a significant driver of internal migration (Sarker, 2022). Rising sea levels, desertification, and severe weather events push communities to relocate, often to overcrowded urban slums with poor living conditions (Hussainzad and Gou, 2024). The loss of traditional lands and livelihoods not only causes economic instability but also erodes cultural and social identities, leaving displaced populations vulnerable to exploitation and marginalization. Six global areas would have seen the displacement of 216 million climate refugees by 2050, with sub-Saharan Africa accounting for the largest number at 86 million, East Asia and the Pacific at 49 million, and South Asia at 40 million (Almulhim et al., 2024).

In general, underprivileged groups are disproportionately affected by climate change, which increases their risk of displacement and restricts their access to safe livelihoods, food, healthcare, education, and water. Climate policies that prioritize vulnerable groups and increase resilience to environmental shocks are urgently needed to address these issues.

HOW INEQUALITY MAKES CLIMATE CHANGE IMPACTS WORSE?

Inequality intensifies the impacts of climate change by disproportionately exposing vulnerable populations to environmental hazards while limiting their ability to respond and adapt. The poorest communities, which contribute the least to global carbon emissions, often suffer the most from climate disasters. These inequalities manifest in access to resources, adaptive capacity, and policy influence, leading to severe economic and social consequences. Developing countries face greater risks from climate change due to their geographical exposure, reliance on climate-sensitive industries, and lack of infrastructure (Nath and Behera, 2011). For example, Bangladesh, a country highly vulnerable to sea-level rise, experiences recurrent floods and cyclones that displace millions (Sammonds et al., 2021). The 2020 Cyclone Amphan caused over $13 billion in damages, affecting over 2.4 million people, with the poorest households suffering the most due to their reliance on agriculture and informal labor (Crawford et al., 2023). In contrast, the Netherlands, which also faces flood risks, invests heavily in advanced flood defense systems, reducing its vulnerability despite its geographic exposure (Chan et al., 2022).

In sub-Saharan Africa, climate change is exacerbating drought conditions, threatening food security and livelihoods. Ethiopia, Kenya, and Somalia experienced one of the worst droughts in 2022, leaving over 22 million people facing severe hunger (John, 2024). Subsistence farmers and pastoralists in these regions lack financial reserves to recover from repeated climate shocks, forcing them to migrate or seek alternative livelihoods. In contrast, wealthier nations like Australia experience droughts but have government-funded relief programs and financial safety nets that protect farmers from total economic collapse (Reis et al., 2022).

Latin America also faces severe inequalities in climate change impacts. Guatemala, where agriculture is a primary livelihood, has seen crop failures due to erratic rainfall and rising temperatures (Zuñiga et al., 2021). The decline in maize and coffee production has pushed many families into poverty and forced migration toward the United States (Huber et al., 2023). In contrast, the United States, despite facing climate-related disasters like wildfires and hurricanes, has strong insurance and recovery programs that cushion the economic impact on affected populations (Hornstein, 2022). The disparity between rich and poor countries in contributing to climate change and facing its consequences is evident in carbon emissions data. The wealthiest 10% of the global population is responsible for over 50% of global emissions, while the poorest 50% contribute less than 10% (Chancel, 2022). The United States and China alone contribute nearly 45% of global carbon emissions, yet developing countries like Mozambique, Madagascar, and the Philippines are ranked among the most climate-vulnerable nations. In 2019, Mozambique faced two devastating cyclones – Idai and Kenneth – that caused over $3 billion in damages, a figure that dwarfs the country's GDP. Limited international aid and slow recovery efforts left millions in prolonged humanitarian crises.

Economic inequality also influences climate adaptation and resilience. Wealthier nations invest in green technology, resilient infrastructure, and disaster preparedness, while poorer nations struggle to secure climate finance (Bazbauers, 2022). The Paris Agreement's climate finance goal of $100 billion annually for developing nations remains largely unmet, limiting the capacity of low-income countries to build resilience. Countries like India and Indonesia, despite being emerging economies, still face funding gaps for adaptation projects, leaving millions exposed to climate risks

(Garschagen and Doshi, 2022). In urban areas, inequality within countries also determines climate impact. In cities like Mumbai and Jakarta, informal settlements and slums are highly vulnerable to flooding and heat stress. In 2021, Jakarta experienced severe flooding that displaced over 400,000 people, mostly from low-income communities living in poorly constructed housing (Widyaningsih and Van den Broeck, 2021). Meanwhile, wealthier neighborhoods were better protected due to elevated infrastructure and access to emergency services. Similar patterns exist in New Orleans, USA, where Hurricane Katrina disproportionately affected low-income and African American communities due to inadequate disaster preparedness and slower response efforts (Bivens and Miller, 2023).

Gender and social inequality further compound climate vulnerability. Women, particularly in developing countries, are more affected by climate change due to their roles in agriculture and household water collection. In Niger, where women make up the majority of subsistence farmers, climate-induced droughts have intensified food insecurity, leading to higher rates of malnutrition and maternal health risks (Olunusi, 2024). Additionally, in regions affected by climate disasters, women face increased risks of displacement, gender-based violence, and loss of livelihood opportunities (Desai and Mandal, 2021).

Climate-induced migration is another consequence of inequality. In Pacific Island nations like Tuvalu and Kiribati, rising sea levels threaten entire communities, yet limited resources hinder relocation efforts (McMichael et al., 2021). By contrast, wealthy coastal cities in the United States and Europe have begun implementing multi-billion-dollar coastal defense projects. The Maldives, another island nation facing existential climate threats, has invested in land reclamation and artificial island construction, but similar solutions remain unattainable for poorer nations. The economic cost of climate inequality is significant. Studies suggest that by 2050, climate change could reduce global GDP by over 10%, with developing countries suffering the largest declines (Farajzadeh et al., 2022). The World Bank estimates that climate change could push over 130 million people into extreme poverty by 2030 if inequality is not addressed through targeted climate adaptation and economic policies (Wilkinson, 2021).

Climate Risk and Livelihood: Developing nations, especially those in the Global South, are more susceptible to climate risks due to their geographical locations and reliance on climate-sensitive sectors like agriculture and fishing. For instance, the indigenous Wayuu community in northern Colombia faces worsening droughts and floods, leading to displacement and loss of traditional — livelihoods. These environmental changes force them into informal settlements lacking basic services, thereby increasing their vulnerability to health issues and economic instability. In Guatemala, climate change has intensified extreme weather events, adversely affecting agriculture – a primary livelihood for many. The recent reduction in international aid, such as the cuts to U.S. development funds, has further strained local efforts to build resilience against these climate-induced challenges. This situation highlights how diminished support exacerbates the vulnerabilities of communities already facing environmental hardships.

Rich Countries vs. Poor Countries: The disparity between rich and poor countries in contributing to and suffering from climate change is stark. The wealthiest 1% of the global population have been responsible for more than double the carbon emissions of the poorest 50% over a 25-year period (Saraswati et al., 2024). Despite their minimal contribution to global emissions, developing countries bear the brunt of climate-related damages. Economic projections indicate that climate change could reduce the GDP of wealthy nations like the United States by approximately 7%, while developing nations might experience losses of up to 20% or even 40% of their economic output (Farajzadeh et al., 2023). This economic strain hampers the ability of poorer countries to invest in necessary adaptation and mitigation strategies, perpetuating a cycle of vulnerability.

Furthermore, the recent cuts in U.S. overseas aid have significantly impacted global climate finance, exposing efforts in developing countries to combat climate change. This reduction in funds threatens to abandon many communities suffering the most from climate change, making it increasingly challenging to meet global climate finance targets. The inequality intensifies the adverse effects of climate change by placing a heavier burden on poorer nations and vulnerable communities

(Parsons et al., 2024). These groups often lack the resources needed to effectively respond to environmental challenges, leading to greater economic losses and threats to their livelihoods.

HOW COULD CLIMATE INEQUALITY REDUCE?

Reducing climate inequality requires a combination of financial commitments, policy reforms, and CBA strategies to ensure that vulnerable populations can effectively respond to climate risks. Wealthier nations, which contribute the most to global carbon emissions, must take greater responsibility in providing financial and technological support to developing countries. The Paris Agreement's climate finance goal of $100 billion annually for developing nations remains largely unmet, with only around $83.3 billion provided in 2020 (Gordon, 2023). Meeting and exceeding this target are crucial for enhancing climate resilience in poorer countries. Investment in climate adaptation is essential to protect vulnerable communities. According to the Global Commission on Adaptation, every $1 invested in climate resilience could yield $4 in economic benefits. Infrastructure projects such as flood defenses, drought-resistant agriculture, and early warning systems can significantly reduce disaster impacts. For example, Bangladesh's cyclone shelters and improved forecasting systems have drastically reduced cyclone-related fatalities over the past few decades (Akon and Mia, 2024).

Access to clean energy and sustainable technologies must be expanded in developing countries to reduce reliance on fossil fuels. While Africa has 60% of the world's best solar resources, it receives only 2% of global investments in renewable energy (Abdelrazik et al., 2022). Bridging this gap could create millions of jobs and reduce carbon emissions. Similarly, small-scale renewable energy projects in rural areas can provide affordable electricity to communities, improving health, education, and economic opportunities. Social protection programs, including climate insurance and direct financial aid, can help vulnerable populations recover from climate shocks. In Ethiopia, the Productive Safety Net Program has assisted millions of farmers in coping with droughts, reducing food insecurity (Kassaw and Worku, 2024). Strengthening such initiatives globally can prevent climate-induced poverty and migration. By prioritizing financial support, technological investment, and equitable policies, the global community can reduce climate inequality and build a more resilient future for all.

HOW CAN WE PROTECT DISADVANTAGED COMMUNITIES FROM CLIMATE IMPACTS?

Protecting disadvantaged communities from climate impacts requires a combination of policy interventions, financial support, infrastructure development, and CBA strategies. These communities, often the most vulnerable due to limited resources and weak institutional support, need targeted measures to enhance their resilience and adaptive capacity. Fighting climate change and ending poverty cannot be done separately; addressing both issues simultaneously in the world's poorest nations will make achieving both goals much easier. Through inclusive and climate-informed development strategies, developing nations have a window of opportunity to increase resilience and lessen the immediate effects of climate change on poverty. This is especially true for investments, like those in energy infrastructure, urban transportation, or deforestation, that have significant upfront costs but are desperately needed to avoid lock-ins into carbon-intensive patterns. There are solutions. It is also important to accelerate the structural transformation of rural economies so that less physical labor is required in hot weather and fewer agricultural workers are exposed to high temperatures. The development of skills, the encouragement of an environment that supports sustainable businesses, public infrastructure investment, and better integration of emerging nations into international commerce are other significant policy initiatives that might be beneficial.

One of the most effective ways to protect disadvantaged communities is through climate-resilient infrastructure. Investing in flood defenses, cyclone shelters, and improved housing can significantly reduce disaster-related fatalities and economic losses. For example, Bangladesh has built over

12,000 cyclone shelters and an advanced early warning system, reducing cyclone-related deaths from hundreds of thousands in the 1970s to just a few dozen in recent years (Islam, 2023). Similarly, in Jakarta, Indonesia, efforts to improve drainage systems and elevate roads in flood-prone areas have helped reduce the frequency and severity of urban flooding (Pamungkas et al., 2025).

Access to clean water and sustainable agriculture is critical for communities dependent on climate-sensitive livelihoods. Drought-resistant crops and improved irrigation systems can help farmers maintain food production despite changing rainfall patterns. In Kenya, the introduction of drought-tolerant maize has improved food security for thousands of smallholder farmers (Simtowe et al., 2021). In India, the government's National Adaptation Fund for Climate Change (NAFCC) supports projects that help rural communities adapt to climate variability through watershed management and sustainable farming techniques (Pathak, 2023).

Financial support, including climate insurance and social safety nets, plays a crucial role in helping communities recover from climate-related disasters. In Ethiopia, the Productive Safety Net Program (PSNP) provides food or cash assistance to vulnerable populations affected by droughts, preventing them from falling deeper into poverty (Kassaw and Worku, 2024). In the Caribbean, the Caribbean Catastrophe Risk Insurance Facility (CCRIF) offers parametric insurance, which provides quick financial relief to governments after hurricanes and extreme weather events, ensuring faster recovery efforts.

Enhancing community participation and indigenous knowledge in climate adaptation can lead to more effective and sustainable solutions. Indigenous communities have long practiced sustainable land and water management techniques that can be integrated into modern climate policies. In the Philippines, indigenous communities in the Cordillera region have used centuries-old rice terrace farming techniques that naturally conserve water and reduce soil erosion (Sumarwati, 2022). Recognizing and incorporating such practices into national climate strategies can strengthen local resilience. Renewable energy access is another crucial factor in protecting disadvantaged communities. Many rural areas in developing countries still rely on expensive and polluting energy sources, increasing their vulnerability to climate shocks. Expanding access to solar, wind, and microhydropower can provide reliable energy while reducing dependency on fossil fuels. In sub-Saharan Africa, the M-KOPA solar initiative has provided affordable, pay-as-you-go solar home systems to millions of low-income households, improving electricity access and reducing indoor air pollution from kerosene lamps (Mukisa et al., 2022).

Education and awareness programs also play a vital role in preparing communities for climate risks. Disaster preparedness training, climate literacy programs, and local adaptation planning ensure that communities understand the risks they face and know how to respond effectively. In Japan, community-based disaster preparedness programs have significantly reduced casualties during earthquakes and tsunamis by ensuring that local populations are well-trained in evacuation and emergency response (Nakai and Nakano, 2023). Ultimately, protecting disadvantaged communities from climate impacts requires a multi-faceted approach that combines infrastructure investments, financial mechanisms, community engagement, and sustainable resource management. By implementing these strategies, governments and organizations can help build climate resilience and ensure that vulnerable populations are not left behind in the fight against climate change.

The vulnerability of disadvantaged communities to climate change is a growing global concern that requires urgent and sustained action. Without targeted interventions, climate change will continue to deepen existing social and economic inequalities, leaving marginalized populations at greater risk of displacement, food insecurity, and livelihood loss. This study underscores the importance of integrating climate adaptation strategies into development policies, focusing on infrastructure resilience, sustainable resource management, and inclusive governance. Case studies demonstrate that investments in early warning systems, financial assistance, and local adaptation strategies can significantly enhance community resilience. However, addressing climate vulnerability requires long-term commitment, adequate funding, and collaborative efforts between governments, international organizations, and local communities. By prioritizing equity and sustainability, global efforts can ensure that vulnerable populations are better protected against the escalating threats of climate change.

ORGANIZATION OF THIS BOOK

This book has been organized into three thematic parts, each addressing a critical dimension of the complex relationship between climate change and disadvantaged communities. This book comprises **20 chapters**, beginning with an introductory overview by the editors and followed by in-depth empirical, theoretical, and methodological explorations across various geographies and disciplines. The introductory chapter sets the stage by offering an overview of the key themes, challenges, and opportunities in understanding the climate vulnerability of disadvantaged communities. It introduces the rationale, conceptual framework, and structure of the book (Chapter 1).

The **first part** of this book focuses on how climate change disproportionately affects vulnerable populations, particularly indigenous groups, women, and small-scale resource-dependent communities. Six chapters in this part explore livelihood disruptions, traditional knowledge, and the evolving nature of resistance, adaptation, and resilience to climate change. For example, Sk et al. (Chapter 2) examine indigenous resistance and environmental justice through a case study of the Dongria Kondh tribe in the Niyamgiri Hills of Odisha, India. Gayen et al. (Chapter 3) investigate the gendered impacts of climate change on natural resource-based livelihoods among women in the Sundarbans, India. Ali (Chapter 4) explores climate-induced hazards and their role in deepening agrarian distress among marginal farmers. Hien (Chapter 5) analyzes seasonal vulnerability and livelihood uncertainty in the ecotourism sector in Thua Thien Hue, Vietnam. Sk and Patra (Chapter 6) offer a scientometric analysis of global research trends related to climate change and indigenous livelihoods. Sahu et al. (Chapter 7) assess the socio-ecological vulnerability of Mousuni Island in the Indian Sundarbans and explore sustainable, tourism-based adaptation strategies.

The **second part** of this book shifts the focus to urban and peri-urban contexts, where rapid urbanization, inadequate infrastructure, and environmental degradation exacerbate climate-related risks. Seven chapters in this section examine public health, urban heat islands, heat index, air quality, and community resilience in cities. For example, Soni et al. provide a conceptual understanding of the multi-dimensional factors shaping climate change impacts in urban areas (Chapter 8). Mohseni and Rajendran present a comprehensive analysis of the Heat Index across different regions of India, offering insights into how rising temperatures, humidity levels, and regional climatic variations are intensifying heat stress. Using climatological data, the study develops spatially explicit maps and trend analyses to characterize heat stress patterns and identify the most affected regions. The findings highlight critical zones of concern and provide a scientific basis for targeted heat action planning and adaptive public health responses under climate change (Chapter 9). Khanam presents a case study of urban environmental vulnerability in the National Capital Region of India (Chapter 10). Sadaf utilizes remote sensing to study urban microclimates, with a focus on heat stress in Ghaziabad city, India (Chapter 11). Hasnat et al. explore the intersections of climate change, public health, and community empowerment in vulnerable urban settings (Chapter 12). Mohsin and Ahmad analyze the impacts of indoor air pollution on the health of older adults in India using nationally representative survey data (Chapter 13). Alam and Bera assess climate-related health vulnerabilities in flood-affected marginalized communities in Nirmal, highlighting gaps in adaptation and response (Chapter 14).

The **third part** or final section of this book brings together chapters that employ spatial analysis, modeling, and empirical assessments to understand climate impacts on natural resources and ecosystems. It emphasizes the role of technology and data in vulnerability mapping, risk assessment, and ecosystem monitoring. Parvin discusses the geopolitical and energy transitions driven by climate change and their implications for global power structures (Chapter 15). Joshi et al. compare glacier volume changes in two major Himalayan glaciers to assess climate-induced cryospheric transformations (Chapter 16). Sharma et al. investigate climate change and household-level adaptation and mitigation strategies to landslides in Shimla district, Himachal Pradesh, India (Chapter 17). Khan et al. use species distribution models to assess the impact of climate change on phytoplankton habitat suitability in Nainital Lake (Chapter 18). Husain et al. employ Google Earth Engine to map persistent climate-driven flooding and its socio-economic impacts in the Indo-Gangetic Basin (Chapter 19). Ansari et al. examine the vulnerability of dry tropical grasslands to climate change using an asymmetrical warming experiment (Chapter 20).

REFERENCES

Abbass, K., Qasim, M. Z., Song, H., Murshed, M., Mahmood, H., & Younis, I. (2022). A review of the global climate change impacts, adaptation, and sustainable mitigation measures. *Environmental Science and Pollution Research*, 29(28), 42539–42559.

Abdelrazik, M. K., Abdelaziz, S. E., Hassan, M. F., & Hatem, T. M. (2022). Climate action: Prospects of solar energy in Africa. *Energy Reports*, 8, 11363–11377.

Abi Deivanayagam, T., English, S., Hickel, J., Bonifacio, J., Guinto, R. R., Hill, K. X., et al. (2023). Envisioning environmental equity: Climate change, health, and racial justice. *The Lancet*, 402(10395), 64–78.

Adebayo, W. G. (2025). Resilience in the face of ecological challenges: Strategies for integrating environmental considerations into social policy planning in Africa. *Sustainable Development*, 33(1), 203–220.

Akon, M. S., & Mia, M. J. (2024). Performance of JICA funded cyclone shelter in the coastal areas of Bangladesh: An empirical analysis from Pirojpur district. *Progress in Disaster Science*, 23, 100342.

Ali, T., Buergelt, P. T., Paton, D., Smith, J. A., Maypilama, E. L., Yuŋgirrŋa, D., et al. (2021). Facilitating sustainable disaster risk reduction in Indigenous communities: Reviving Indigenous worldviews, knowledge and practices through two-way partnering. *International Journal of Environmental Research and Public Health*, 18(3), 855.

Almulhim, A. I., Alverio, G. N., Sharifi, A., Shaw, R., Huq, S., Mahmud, M. J., et al. (2024). Climate-induced migration in the Global South: An in-depth analysis. *NPJ Climate Action*, 3(1), 47.

Aziz, A. (2024). Navigating climate change and food security in South Asia: Challenges and strategic approaches. *Journal of Climate and Community Development*, 3(2), 171–185.

Aziz, M., & Anjum, G. (2024). Transformative strategies for enhancing women's resilience to climate change: A policy perspective for low-and middle-income countries. *Women's Health*, 20, 17455057241302032.

Bakti, I., Tursunov, D., & Pong, M. (2025). The effect of the climate crisis on social mobility and economic well-being. *Journal of Social Science Utilizing Technology*, 3(1), 32–41.

Bayram, H., Rice, M. B., Abdalati, W., Akpinar Elci, M., Mirsaeidi, M., Annesi-Maesano, et al. (2023). Impact of global climate change on pulmonary health: Susceptible and vulnerable populations. *Annals of the American Thoracic Society*, 20(8), 1088–1095.

Bazbauers, A. R. (2022). Sustainable, green, and climate-resilient cities: An analysis of multilateral development banks. *Climate and Development*, 14(8), 689–704.

Biswas, R. R., & Rahman, A. (2023). Adaptation to climate change: A study on regional climate change adaptation policy and practice framework. *Journal of Environmental Management*, 336, 117666.

Bivens, N. D., & Miller, D. S. (2023). Lessons learned and moving forward from hurricane Katrina: Emergency response planning to build an inclusive response framework. In R. J. Thorpe Jr., & D. S. Miller (Eds.), *Health Disparities, Disasters, and Crises Approaches for a Culture of Preparedness* (pp. 177–186). Routledge.

Borg, M. A., Xiang, J., Anikeeva, O., Pisaniello, D., Hansen, A., Zander, K., et al. (2021). Occupational heat stress and economic burden: A review of global evidence. *Environmental Research*, 195, 110781.

Chan, F. K. S., Yang, L. E., Mitchell, G., Wright, N., Guan, M., Lu, X., et al. (2022). Comparison of sustainable flood risk management by four countries–the United Kingdom, the Netherlands, the United States, and Japan–and the implications for Asian coastal megacities. *Natural Hazards and Earth System Sciences*, 22(8), 2567–2588.

Chancel, L. (2022). Global carbon inequality over 1990–2019. *Nature Sustainability*, 5(11), 931–938.

Chapman, A. R., & Ahmed, A. K. (2021). Climate justice, human's rights, and the case for reparations. *Health and Human Rights*, 23(2), 81.

Costella, C., van Aalst, M., Georgiadou, Y., Slater, R., Reilly, R., McCord, A., et al. (2023). Can social protection tackle emerging risks from climate change, and how? A framework and a critical review. *Climate Risk Management*, 40, 100501.

Crawford, N. J., Rahman, S., Nazia, T., Mattar, S. D., & Kalu, U. U. (2023). 'I was poor before, but Cyclone Amphan left me destitute': Disaster displacement and support in Bangladesh. In N. J. W. Crawford, K. Michael, M. Mikulewicz (Eds.), *Climate Justice in the Majority World Vulnerability, Resistance, and Diverse Knowledges* (pp. 82–104). Routledge.

Datta, R., Hurlbert, M., Kibria, A., Singha, R., Kairy, B., & Chattapadhya, S. (2025). *Rethinking and Relearning Disaster Adaptations from and within Indigenous Land-Based Perspectives.* Taylor & Francis.

de Carvalho, T. (2025). The cascading effects of climate change on children: Extreme floods, family mobility and child well-being in Amazonia. *Climate and Development, 17*(3), 191–199.

Dellinger, M. (2022). Carbon taxation for climate change mitigation. *LSU Journal of Energy Law & Resources*, 11, 141.

Desai, B. H., & Mandal, M. (2021). Role of climate change in exacerbating sexual and gender-based violence against women: A new challenge for international law. *Environmental Policy and Law*, 51(3), 137–157.

de Souza, W. M., & Weaver, S. C. (2024). Effects of climate change and human activities on vector-borne diseases. *Nature Reviews Microbiology*, 22(8), 476–491.

Dickin, S., Segnestam, L., & Sou Dakouré, M. (2021). Women's vulnerability to climate-related risks to household water security in Centre-East, Burkina Faso. *Climate and Development*, 13(5), 443–453.

Dlamini, S., Nhleko, B., & Ubisi, N. (2024). Understanding socioeconomic risk and vulnerability to climate change–induced disasters: The case of informal settlements in KwaZulu-Natal, South Africa. *Journal of Asian and African Studies*, 00219096241275398. https://doi.org/10.1177/00219096241275398.

du Plessis, A. (2023). Water resources from a global perspective. In A. du Plessis (Ed.), *South Africa's Water Predicament: Freshwater's Unceasing Decline* (pp. 1–25). Cham: Springer International Publishing.

El Khayat, M., Halwani, D. A., Hneiny, L., Alameddine, I., Haidar, M. A., & Habib, R. R. (2022). Impacts of climate change and heat stress on farmworkers' health: A scoping review. *Frontiers in Public Health*, 10, 782811.

EPA (2021). EPA Report Shows Disproportionate Impacts of Climate Change on Socially Vulnerable Populations in the United States. Accessed on 3rd March, 2025. Accessed from https://www.epa.gov/newsreleases/epa-report-shows-disproportionate-impacts-climate-change-socially-vulnerable

Farajzadeh, Z., Ghorbanian, E., & Tarazkar, M. H. (2022). The shocks of climate change on economic growth in developing economies: evidence from Iran. *Journal of Cleaner Production*, 372, 133687.

Farajzadeh, Z., Ghorbanian, E., & Tarazkar, M. H. (2023). The impact of climate change on economic growth: Evidence from a panel of Asian countries. *Environmental Development*, 47, 100898.

Farooq, M. S., Uzair, M., Raza, A., Habib, M., Xu, Y., Yousuf, M., et al. (2022). Uncovering the research gaps to alleviate the negative impacts of climate change on food security: A review. *Frontiers in Plant Science*, 13, 927535.

Feng, Y., Hu, J., Afshan, S., Irfan, M., Hu, M., & Abbas, S. (2023). Bridging resource disparities for sustainable development: A comparative analysis of resource-rich and resource-scarce countries. *Resources Policy*, 85, 103981.

Fitzgerald, J. (2022). Transitioning from urban climate action to climate equity. *Journal of the American Planning Association*, 88(4), 508–523.

Friedman, E. (2023). Constructing the adaptation economy: Climate resilient development and the economization of vulnerability. *Global Environmental Change*, 80, 102673.

Gadhoke, P., Brenton, B., & Katz, S. H. (Eds.). (2023). *Transformations of Global Food Systems for Climate Change Resilience: Addressing Food Security, Nutrition, and Health.* CRC Press.

Garschagen, M., & Doshi, D. (2022). Does funds-based adaptation finance reach the most vulnerable countries? *Global Environmental Change*, 73, 102450.

Gordon, N. J. (2023). Climate finance: An overview. *Environment: Science and Policy for Sustainable Development*, 65(4), 18–26.

Hadley, K., Wheat, S., Rogers, H. H., Balakumar, A., Gonzales-Pacheco, D., Davis, et al. (2023). Mechanisms underlying food insecurity in the aftermath of climate-related shocks: a systematic review. *The Lancet Planetary Health*, 7(3), e242–e250.

Haque, M. N., Saroar, M., Fattah, M. A., Morshed, S. R., & Fatema, N. (2022). Access to basic services during the transition from MDGs to SDGs: More rhetoric than reality in a Bangladesh slum. *Journal of Humanities and Applied Social Sciences*, 4(1), 57–75.

Hasegawa, T., Wakatsuki, H., & Nelson, G. C. (2022). Evidence for and projection of multi-breadbasket failure caused by climate change. *Current Opinion in Environmental Sustainability*, 58, 101217.

Ho, E. W., Strohmeier-Breuning, S., Rossanese, M., Charron, D., Pennise, D., & Graham, J. P. (2021). Diverse health, gender and economic impacts from domestic transport of water and solid fuel: A systematic review. *International Journal of Environmental Research and Public Health*, 18(19), 10355.

Hornstein, D. T. (2022). Public investment in climate resiliency: Lessons from the law and economics of natural disasters. *Ecology LQ*, 49, 137.

Hubbard, S., Wolf, J., Oza, H. H., Arnold, B. F., Freeman, M. C., & Levy, K. (2025). Differential effectiveness of water, sanitation, and handwashing interventions to reduce child diarrhea in dry and rainy seasons: A systematic review and meta-analysis of intervention trials. *Environmental Health Perspectives*, 133(2), 026001.

Huber, J., Madurga-Lopez, I., Murray, U., McKeown, P. C., Pacillo, G., Laderach, P., & Spillane, C. (2023). Climate-related migration and the climate-security-migration nexus in the Central American Dry Corridor. *Climatic Change*, 176(6), 79.

Humayra, J., Uddin, M. K., & Pushpo, N. Y. (2024). Deficiencies of women's participation in climate governance and sustainable development challenges in Bangladesh. *Sustainable Development*, 32(6), 7096–7113.

Hussainzad, E. A., & Gou, Z. (2024). Climate risk and vulnerability assessment in informal settlements of the global south: A critical review. *Land*, 13(9), 1357.

Ioannou, L. G., Foster, J., Morris, N. B., Piil, J. F., Havenith, G., Mekjavic, I. B., et al. (2022). Occupational heat strain in outdoor workers: A comprehensive review and meta-analysis. *Temperature*, 9(1), 67–102.

Islam, M. T. (2023). Role of Bangladesh government in disaster management. In M. T. Islam (Ed.), *Disaster, Governance and Development: Perspectives from Bangladesh* (pp. 235–279). Singapore: Springer Nature Singapore.

Jobaer, M. A., Hossain, M. Z., Ha-Mim, N. M., Haque, S. F., & Rahaman, K. R. (2025). Examining the link between marginality and differential climate resilience among disaster-affected communities in southwestern Bangladesh. *Climate and Development*, 17(2), 91–106.

John, M. (2024). Climate change, food insecurity, peace and sustainable development in East Africa: Case study of South Sudan, Sudan, Ethiopia and Kenya. In J. C. K. Kiyala & N. Chivasa (Eds.), *Climate Change and Socio-political Violence in Sub-Saharan Africa in the Anthropocene: Perspectives from Peace Ecology and Sustainable Development* (pp. 141–165). Cham: Springer Nature Switzerland.

Kassaw, K., & Worku, M. (2024). The contribution of productive safety net program for households food security and asset building in drought prone woredas of northeast Ethiopia. *Heliyon*, 10(3), 1–13.

Klein, T., & Anderegg, W. R. (2ßœœ¯021). A vast increase in heat exposure in the 21st century is driven by global warming and urban population growth. *Sustainable Cities and Society*, 73, 103098.

Kompas, T., Che, T. N., & Grafton, R. Q. (2024). Global impacts of heat and water stress on food production and severe food insecurity. *Scientific Reports*, 14(1), 14398.

Lau, J. D., Kleiber, D., Lawless, S., & Cohen, P. J. (2021). Gender equality in climate policy and practice hindered by assumptions. *Nature Climate Change*, 11(3), 186–192.

Lee, D., Gibson, J. M., Brown, J., Habtewold, J., & Murphy, H. M. (2023). Burden of disease from contaminated drinking water in countries with high access to safely managed water: A systematic review. *Water Research*, 242, 120244.

Li, C., & Managi, S. (2022). Global malaria infection risk from climate change. *Environmental Research*, 214, 114028.

Lindsay, S., Hsu, S., Ragunathan, S., & Lindsay, J. (2023). The impact of climate change related extreme weather events on people with pre-existing disabilities and chronic conditions: A scoping review. *Disability and Rehabilitation*, 45(25), 4338–4358.

Lomborg, B. (2020). Welfare in the 21st century: Increasing development, reducing inequality, the impact of climate change, and the cost of climate policies. *Technological Forecasting and Social Change*, 156, 119981.

Masud-All-Kamal, M., & Nursey-Bray, M. (2021). Socially just community-based climate change adaptation? Insights from Bangladesh. *Local Environment*, 26(9), 1092–1108.

McMichael, C., Farbotko, C., Piggott-McKellar, A., Powell, T., & Kitara, M. (2021). Rising seas, immobilities, and translocality in small island states: Case studies from Fiji and Tuvalu. *Population and Environment*, 43, 82–107.

Millard, J., & Fucci, V. (2023). The role of social innovation in tackling global poverty and vulnerability. *Frontiers in Sociology*, 8, 966918.

Morales, F., Montserrat-De la Paz, S., Leon, M. J., & Rivero-Pino, F. (2023). Effects of malnutrition on the immune system and infection and the role of nutritional strategies regarding improvements in children's health status: A literature review. *Nutrients*, 16(1), 1.

Mukisa, N., Manitisa, M. S., Nduhuura, P., Tugume, E., & Chalwe, C. K. (2022). Solar home systems adoption in Sub-Saharan African countries: Household economic and environmental benefits assessment. *Renewable Energy*, 189, 836–852.

Nakai, F., & Nakano, G. (2023). Community-mediated individual disaster preparedness practices: A case study in Kochi, Japan. *International Journal of Disaster Risk Reduction*, 86, 103532.

Nath, P. K., & Behera, B. (2011). A critical review of impact of and adaptation to climate change in developed and developing economies. *Environment, Development and Sustainability*, 13, 141–162.

Newell, P., Srivastava, S., Naess, L. O., Torres Contreras, G. A., & Price, R. (2021). Toward transformative climate justice: An emerging research agenda. *Wiley Interdisciplinary Reviews: Climate Change*, 12(6), e733.

Nielsen, K. S., Nicholas, K. A., Creutzig, F., Dietz, T., & Stern, P. C. (2021). The role of high-socioeconomic-status people in locking in or rapidly reducing energy-driven greenhouse gas emissions. *Nature Energy*, 6(11), 1011–1016.

Olunusi, B. O. (2024). Overview of climate-induced food insecurity in Nigeria. *African Journal of Food Science*, 18, 69–76.

Pal, A., Tsusaka, T. W., Nguyen, T. P. L., & Ahmad, M. M. (2023). Assessment of vulnerability and resilience of school education to climate-induced hazards: A review. *Development Studies Research*, 10(1), 2202826.

Pamungkas, P. J., Kristianto, A., Prianto, M. D., Rahman, R. Z., & Firmansyah, R. T. (2025). Revitalizing flood-prone areas through socio-environmental strategies: A qualitative approach to urban flood management. *Calamity: A Journal of Disaster Technology and Engineering*, 2(2), 87–99.

Parsons, E. S., Jowell, A., Veidis, E., Barry, M., & Israni, S. T. (2024). Climate change and inequality. *Pediatric Research*, 1–8. https://doi.org/10.1038/s41390-024-03153-z.

Pathak, H. (2023). Impact, adaptation, and mitigation of climate change in Indian agriculture. *Environmental Monitoring and Assessment*, 195(1), 52.

Perkiss, S. (2024). Climate apartheid: the failures of accountability and climate justice. *Accounting, Auditing & Accountability Journal*, 37(7/8), 1761–1794.

Poddar, A. K. (2024). Climate change and migration: Developing policies to address the growing challenge of climate-induced displacement. *International Journal of Climate Change: Impacts & Responses*, 16(1), 149–170. https://doi.org/10.18848/1835-7156/CGP/v16i01/149-170.

Prasad, P., Gupta, S. K., Mahto, K. K., Kumar, G., Rani, A., Velan, I. (2024). Influence of climatic factors on the life stages of Aedes mosquitoes and vectorial transmission: A review. *Journal of Vector Borne Diseases*, 61(2), 158–166.

Reis, K., Desha, C., Campbell, S., & Liddy, P. (2022). Working through disaster risk management to support regional food resilience: A case study in north-eastern Australia. *Sustainability*, 14(4), 2466.

Rentschler, J., Salhab, M., & Jafino, B. A. (2022). Flood exposure and poverty in 188 countries. *Nature Communications*, 13(1), 3527.

Roy, A. S., Mazaniello-Chézol, M., Rueda-Martinez, M., Shafique, S., & Adams, A. M. (2023). Food systems determinants of nutritional health and wellbeing in urban informal settlements: A scoping review in LMICs. *Social Science & Medicine*, 322, 115804.

Saeed, F., Schleussner, C. F., & Ashfaq, M. (2021). Deadly heat stress to become commonplace across South Asia already at 1.5C of global warming. *Geophysical Research Letters*, 48(7), e2020GL091191.

Sammonds, P., Shamsudduha, M., & Ahmed, B. (2021). Climate change driven disaster risks in Bangladesh and its journey towards resilience. *Journal of the British Academy*, 9(s8), 55–77.

Saraswati, C. M., Judge, M. A., Weeda, L. J., Bassat, Q., Prata, N., Le Souëf, P. N., & Bradshaw, C. J. (2024). Net benefit of smaller human populations to environmental integrity and individual health and wellbeing. *Frontiers in Public Health*, 12, 1339933.

Sardo, M. C. (2023). Responsibility for climate justice: Political not moral. *European Journal of Political Theory*, 22(1), 26–50.

Sarker, M. S. H. (2022). Assessing levels of migrant-friendliness in the context of vulnerability to climate variability, change and environmental hazard: A comparison of two different-sized cities. *International Journal of Disaster Risk Reduction*, 68, 102525.

Simtowe, F., Makumbi, D., Worku, M., Mawia, H., & Rahut, D. B. (2021). Scalability of adaptation strategies to drought stress: The case of drought tolerant maize varieties in Kenya. *International Journal of Agricultural Sustainability*, 19(1), 91–105.

Sovacool, B. K., Newell, P., Carley, S., & Fanzo, J. (2022). Equity, technological innovation and sustainable behaviour in a low-carbon future. *Nature Human Behaviour*, 6(3), 326–337.

Stoddard, I., Anderson, K., Capstick, S., Carton, W., Depledge, J., Facer, K., et al. (2021). Three decades of climate mitigation: Why haven't we bent the global emissions curve? *Annual Review of Environment and Resources*, 46(1), 653–689.

Sumarwati, S. (2022). Traditional ecological knowledge on the slope of Mount Lawu, Indonesia: All about non-rice food security. *Journal of Ethnic Foods*, 9(1), 9.

Tahir, S., & Tahir, W. (2024). Climate-related migration and displacement: Legal protections for climate refugees. *Mayo Journal of Communication World*, 1(2), 1–11.

Tamasiga, P., Mogomotsi, P. K., Onyeaka, H., & Mogomotsi, G. E. (2024). Amplifying climate resilience: The impact of social protection, social cohesion, and social capital on public support for climate change action. *Sustainable Environment*, 10(1), 2361568.

Tan, S. B., Waters, M. C., & Arcaya, M. C. (2022). Analyzing the long-term impact of post-disaster relocation and implications for disaster recovery policy. *International Journal of Disaster Risk Reduction*, 70, 102765.

Tantoh, H. B., McKay, T. T., Donkor, F. E., & Simatele, M. D. (2021). Gender roles, implications for water, land, and food security in a changing climate: A systematic review. *Frontiers in Sustainable Food Systems*, 5, 707835.

Theodore, N. (2023). Day-labor worker centers: Advancing new models of equity and inclusion in the informal economy. *Economic Development Quarterly*, 37(4), 363–374.

Tubridy, F., Lennon, M., & Scott, M. (2022). Managed retreat and coastal climate change adaptation: The environmental justice implications and value of a coproduction approach. *Land Use Policy*, 114, 105960.

Varghese, J., & Crawford, S. S. (2021). A cultural framework for indigenous, local, and science knowledge systems in ecology and natural resource management. *Ecological Monographs*, 91(1), e01431.

Wang, C., Shuai, C., Chen, X., Huang, W., Hou, W., Zhao, B., & Sun, J. (2025). Analyzing the potential local and distant economic loss of global construction sector due to water scarcity. *Environmental Impact Assessment Review*, 110, 107667.

Whyte, K. P. (2021). Indigenous peoples, climate change loss and damage, and the responsibilities of states. In M. Doelle & S. L. Seck (Eds.), *Research Handbook on Climate Change Law and Loss & Damage* (pp. 224–244). Edward Elgar Publishing.

Widyaningsih, A., & Van den Broeck, P. (2021). Social innovation in times of flood and eviction crisis: The making and unmaking of homes in the Ciliwung riverbank, Jakarta. *Singapore Journal of Tropical Geography*, 42(2), 325–345.

Wilkinson, J. (2021). Building more common wealth in a climate changed world. *The Round Table*, 110(5), 546–563.

Wilson, B., Tate, E., & Emrich, C. T. (2021). Flood recovery outcomes and disaster assistance barriers for vulnerable populations. *Frontiers in Water*, 3, 752307.

World Bank (2020). Global Action Urgently Needed to Halt Historic Threats to Poverty Reduction. Accessed on 1st March, 2025. Accessed from https://www.worldbank.org/en/news/feature/2020/10/07/global-action-urgently-needed-to-halt-historic-threats-to-poverty-reduction

Yadav, N., & Upadhyay, R. K. (2023). Global effect of climate change on seasonal cycles, vector population and rising challenges of communicable diseases: A review. *Journal of Atmospheric Science Research*, 6(1), 21–59.

Yang, Y., Tilman, D., Jin, Z., Smith, P., Barrett, C. B., Zhu, Y. G., et al., (2024). Climate change exacerbates the environmental impacts of agriculture. *Science*, 385(6713), eadn3747.

Zuñiga, R. A. A., Lima, G. N., & Villoria, A. M. G. (2021). Impact of slow-onset events related to climate change on food security in Latin America and the Caribbean. *Current Opinion in Environmental Sustainability*, 50, 215–224.

Part 1

Climate Change Vulnerability, Livelihoods, and Resilience of Disadvantaged Communities

2 Indigenous Resistance and Environmental Justice in the Face of Climate Change

A Study of the Dongria Kondh Tribe in Niyamgiri Hills, Odisha

Md Mainul Sk, Amit Sardar, Tapas Ranjan Patra, Mohammed Numan, and Trilochan Sethy

INTRODUCTION

The Niyamgiri Hills in Odisha are home to the Dongria Kondh, an indigenous tribal community recognized as particularly vulnerable tribal group (PVTG) in India, enriched by enormous ecological and cultural importance (Tatpati et al., 2016). The Dongria Kondh are known for their symbiotic relation with the environment, by which they sustain their livelihood through sustainable practices like shifting cultivation, forest products use and water resource management (Pandey, 2018). They regard the Niyamgiri Hills as the sacred abode of their deity, Niyam Raja, and their intimate connection to nature is not only spiritual but something beyond the utilitarian (Temper & Martinez-Alier, 2013). This spiritual bond forms the core of their identity and exhibits how they live in harmony with nature.

The discovery of rich bauxite reserves in Niyamgiri Hills during the early 2000s sparked the attention of Vedanta Resources, a multinational organization looking for means to tap into its valued natural resources (Jammulamadaka & Bhattacharjee, 2017). Kaushal (2017) claims that the proposed extraction was an existential threat to the community as their sacred sites, livelihoods and cultural heritage were threatened. The subsequent proposal for mining activities posed a dual threat to the Dongria Kondh: the destruction of their sacred land and the destruction of their way of life (Mohapatra, 2017). However, this impending crisis catalysed a grassroots resistance movement against such projects and drew on the support of not only human rights activists and civil society organizations, but even international human rights organizations as the Dongria Kondh was leading this resistance (Guha et al., 2012). Soon, the tribe's resistance to this project caught international attention and became a symbol of the broader struggle of indigenous communities defending their rights and ecosystems (Borde, 2017). This conflict draws a distinction on the essential contradiction between development and environmental justice, where economic imperatives are always imposed at the expense of marginalized communities (Campbell, 2013). The Dongria Kondh's resistance, thus, necessitates the urgent need for development paradigms that are inclusive and prioritize ecological integrity and social equity over unsustainable industrial expansion (Dey & Arunachalam, 2024). In addition, the case of Niyamgiri reflects the larger impacts of extractive activities and the question of whose voices get heard within policy making and who is burdened by development.

Worldwide, Indigenous people are key players in the management of the natural resources of indigenous communities, who account for 6% of the global population and protect around 80% of the world's biodiversity (Garnett et al., 2018; Toledo, 2001). Nevertheless, colonial histories and post-colonial policies marginalized them into a position that makes them vulnerable to resource

DOI: 10.1201/9781003620747-3

exploitation and socio-environmental injustices (Sahu et al., 2023; Domínguez & Luoma, 2020). For instance, the Sami people of Scandinavia struggle with the conservation of their traditional reindeer herding in the face of climate change, industrial encroachment and political exclusion (Kelman & Næss, 2019; Riseth, 2000). Similarly, indigenous people in Brazil's Amazon persist in opposing deforestation and extractive activities by stressing that indigenous knowledge is essential for global environmental management (Hecht et al., 2021; Nepstad et al., 2006). Indigenous people of the Pacific Islands are today also grappling with threats to their very existence stemming from climate change and the subsequent loss of their land through flooding and sea-level rise that also threatens their cultural and social existence (Kelman, 2018). Also, like many other indigenous communities, the Dongria Kondh are threatened with cultural survival and traditional knowledge systems, and the environmental stewardship that the Dongria Kondh possess are under threat from extractive industry pressures (Survival International, 2013). These examples reveal that indigenous people are at the cutting edge of climate change mitigation and biological diversity, but they remain germane targets of disadvantage (Ford et al., 2020).

Although there is a wealth of information regarding the environmental roles of indigenous people, there is still a lack of research on how cultural identity, indigenous knowledge and political activism affect environmental justice in the face of industrialization (Redvers et al., 2023; Whyte, 2018). Less attention, however, has been paid to how these factors interact and co-evolve in order to inform resistance strategies, or how these insights might be relevant for other global environmental governance initiatives (Mekonnen et al., 2021; Hill et al., 2020). To fill these gaps, this study presents a complex picture of the Dongria Kondh's resistance and enriches the global debate on indigenous rights, environmental justice and sustainable development. Therefore, this study aims to (i) investigate the resistance of the Dongria against the encroaching mining activity, (ii) evaluate the resistance strategies the Dongria have used to maintain their culture/environment and (iii) examine the broader implications of the struggle for global environmental justice. Specifically, it aims to answer questions such as: How did the Dongria Kondh mobilize to resist the mining? How did their cultural, social and ecological strategies assist in sustaining their identity? And what can we learn from their struggle for other similar conflicts occurring throughout the world?

The significance of the research lies in contributing to the study of indigenous knowledge systems concerning environmental conservation, social justice and all forms of human cultural expression. This work documents the Dongria Kondh's struggle against exploitative development paradigms and for global biodiversity and emphasizes the important voice of local communities in sustainable development. The findings also have broader relevance, as they indicate how human social systems can use their strengths to resist and to be resilient in the face of socio-environmental crises in a variety of contexts (Brown, 2014). Understanding this is indispensable for achieving sustainable development compatible with cultural heritage and ecological stewardship.

THEORETICAL FRAMEWORK

The study is anchored in two complementary theoretical perspectives: the Resistance and Resilience theories. Resistance theory is a critical framework for understanding and explaining how marginalized groups resist, confront and oppose structures of domination and exploitation to defend their autonomy, identity and resources (Rubin, 1995). This theory draws heavily from the works of Scott (2012) and Guha (1999) in its promotion of the beliefs that subaltern communities have the ability to resist externally imposed agency and pressures. With regard to Dongria Kondh, resistance has come to be attached to the cultural identity and the spiritual relation with Niyamgiri Hills. Seen as an act of cultural resistance, the internal drive behind their struggle against mining is using traditional practices, rituals and beliefs to assert their opposition to industrial encroachment (Pandey, 2017). The tribes used their gram sabhas (village councils), consistent with the unanimous rejection of mining proposals as empowered by the Supreme Court of India, which shows their strategic mobilization within democratic frameworks (Woodman, 2014). In addition, the Dongria Kondh's stance is part of a larger critique of the neoliberal development paradigm which puts the rights of indigenous peoples subordinate to economic growth in the industry of resource extraction. This critique resonates with the environmentalism of the poor, where the poor resist the project of economic growth by defending the ecological commons from predatory economic activities (Anguelovski & Alier, 2014).

In contrast, resilience theory builds on resistance theory but emphasizes community adaptive capacities to adapt and cope with and recover from socio-environmental disruptions (Lv et al., 2024; Norris et al., 2008). Resilience theory was originally developed in ecological studies (Gunderson, 2000) and has been more recently adapted for use in social-ecological systems to study how communities engage in and avoid external pressures (Folke, 2006; Folke et al., 2003). The Dongria Kondh's sustainable livelihood practices, cultural preservation and adaptive governance mechanisms are evidence of resilience within the communities (Bocha & Srinivas, 2021). The use of shifting cultivation, forest and water resources management and indigenous knowledge systems shows their deep understanding of ecological cycles and the sustainability of resources. These practices not only keep them alive, but they also reassert their group identity, helping them hang on when external pressure is coming in. It is important to discuss 'cultural resilience,' in order to understand how the Dongria Kondh have been able to continue with their traditions and values in the face of globalization and industrialization. As Berkes and Ross (2013) and Folke et al. (2003) pointed out, cultural resilience is essential for the long-term sustainability of social-ecological systems because learning to belong to a place can encourage community-based conservation. The Dongria Kondh's rituals and community gatherings serve as reservoirs of resilience because of which their way of life continues to sustain despite constant threats (Chowdhry, 2019).

Although resistance and resilience are separate concepts, they are very closely associated with Dongria Kondh's resistance. In fact, resistance often draws its strength from its resilience by building upon cultural and ecological stability as the ground upon which one can be organized (Adger, 2003). However, acts of resistance which protect the community from an existential threat reinforce resilience (Corry, 2014). This dynamic interplay brings the Dongria Kondh's ability to simultaneously refute oppressive systems and accommodate socio-environmental changes. The Dongria Kondh, in the case of Niyamgiri, have managed to establish a robust defence against rights and their environment with their dual reliance on resistance and resilience. By integrating these theoretical perspectives, we show that they have proven to be effective in understanding indigenous struggles against the extractive industry.

METHODOLOGY OF THE STUDY

To explore the resistance as well as the resilience of the Dongria Kondh community against mining in Niyamgiri Hills, a qualitative research paradigm was adopted in the study (Figure 2.1). Participant observation and semi-structured interviews of both men and women of the community were used to collect primary data, ensuring inclusion of diverse opinions. Table 2.1 lists the main questions used in the semi-structured interviews. The following questions were prepared in order to consider different aspects of the Dongria Kondh's resistance and resilience in fighting against the proposed mining project. Fieldwork was conducted in selected villages within the Dongria Kondh-inhabited Niyamgiri Hills. Participant observation included witnessing ceremonies, day-to-day activities and rituals so as to understand the cultural and spiritual linkages between the Niyamgiri Hills and the individuals and groups living there. Interviews were conducted in the local *Kui* language, with translation support as needed to record community member voices accurately. Historical and contextual insights were obtained through secondary data such as reports, legal documents, academic papers and media coverage. The collected data was then thematically analysed to understand and look for key patterns regarding resistance, resilience, cultural identity and environmental justice. The findings were juxtaposed to a global set of case studies of indigenous resistance to extractive industries in a comparative analysis, putting the Dongria Kondh's struggle in a larger context of environmental sustainability and indigenous rights.

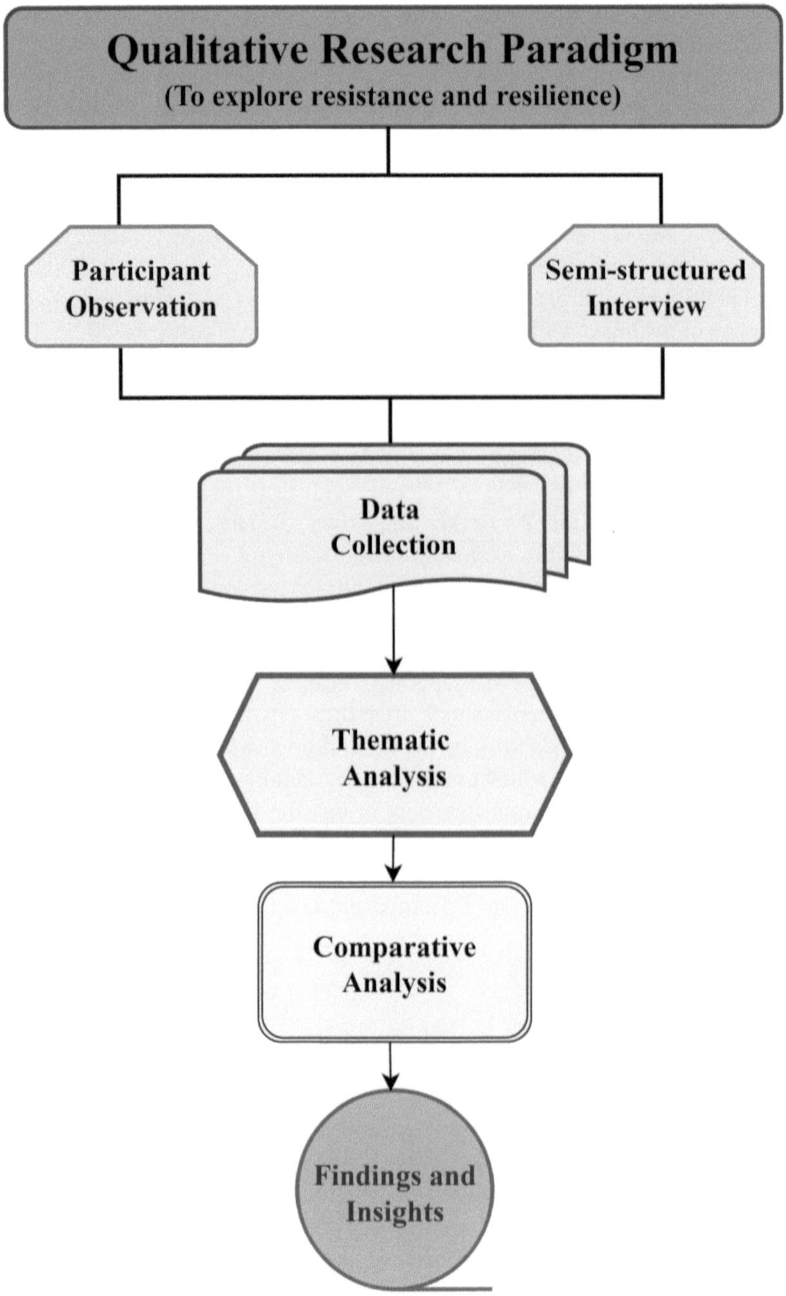

FIGURE 2.1 The research design and methodology of workflow.

RESULTS

The analysis of the findings in the study focuses on the Dongria Kondh's mobilization strategies and their implications for indigenous people's rights and environmental sustainability. Table 2.2 synthesizes these strategies by cultural, social, legal and ecological dimensions. These strategies were not only helpful in protecting the Niyamgiri Hills but also created a paradigm of resistance against the neo-imperialist development agenda. Table 2.2 provides a background for the subsequent analysis of the results, which is addressed in the following sections.

TABLE 2.1
Main Questions Used in Semi-Structured Interviews with the Dongria Kondh Community.

Themes	Questions
Perception of the environment	1. In what way do you perceive your connection to the Niyamgiri Hills?
	2. What makes you believe that the Niyamgiri Hills are important to you and your community for their identity and survival?
	3. In your opinion, how does Niyam Raja help in the management of your ecosystem?
Cultural and spiritual values	4. Could you please describe the cultural and spiritual importance of the Niyamgiri Hills in your normal lifestyle?
	5. In what ways do the practices related to the conservation of sacred groves support you and your family's livelihood?
Impact of mining	6. Since mining activities were proposed, what changes have you observed in the environment and the society?
	7. What strategies has the community of Niyamgiri Hills adopted to prevent the mining operations?
	8. How do the indigenous people use rituals, songs and other forms of traditional practices in order to prevent mining?
Governance and rights	9. In what ways has the Supreme Court ruling of 2013 impacted the community?
	10. What challenges does the community encounter in the struggle to assert the Niyamgiri Hills' sovereignty rights?
Role of external alliances	11. To what extent have alliances with the NGOs and activists helped your movement?
	12. How did the support from the international community plus media campaigns influence your struggle?
Legal strategies	13. How did the people go about opposing mining proposals through the gram sabhas and the courts?
Sustainable practices	14. In what way does the shifting cultivation contribute to the sustainability of the community that you come from?
	15. How can you ensure the sustainable use of non-timber forest products (NTFPs)?
Future and broader implications	16. What would be the fate of the Niyamgiri Hills if mining were allowed to be carried out?
	17. In your view, what is the potential role of indigenous governance in addressing global climate change?

TABLE 2.2
Insights from the Dongria Kondh's Resistance Movement

Research Question	Findings from the Case Study
How did the Dongria Kondh mobilize to resist the mining?	Through grassroots activism (palli sabhas and gram sabhas), the Dongria Kondh mobilized to resist the mining operation. They also formed alliances with NGOs, international advocacy groups and the media to highlight to the rest of the world what they were doing. Their means of resistance were public demonstrations, petitions and legal challenges.
How did their cultural, social and ecological strategies assist in sustaining their identity?	The Niyamgiri Hills were at their cultural identity, which they considered the sacred home of their deity, Niyam Raja. Social cohesion, traditional knowledge of sustainable living and forest conservation and agroforestry all helped to nurture their identity. They reinforced their ties with their land by means of song, ritual and oral tradition.
What can we learn from their struggle for other similar conflicts occurring throughout the world?	The case of the Dongria Kondh emphasizes the need to give recognition to indigenous land and cultural rights. The successful resistance reinforces the benefits of community-led initiatives, alliances with external organizations and the role that legal frameworks play in resolving such conflicts. This case provides us with insights into how to combat similar struggles, the value of local narratives and global solidarity power.

THE DONGRIA KONDH'S RESISTANCE

The Dongria Kondh's resistance to mining the Niyamgiri Hills is an important milestone in grassroots activism, marked by resistance to huge-scale corporate aggressions, as well as mobilization of cultural identity, legal strategies and alliance of international groups against mining (Figure 2.2). The act of this movement is the importance of indigenous agency in the struggle to defend their environmental and cultural resources from exploitative industrial practices. Their solid resistance was based on the community's undying caution of their sacred lands and their smart use of assets and shares. While on fieldwork in the village of Khambesi, elder Suresh Wadaka said, "I am actually from Niyamgiri; Niyamgiri is not only just our home but it is our soul, and without these hills areas our identity will no longer exist." A woman leader, Kalpana Sikoka, in Khambesi echoed this sentiment: "The forests give us food, the streams give us water, the hills protect us. How can we let anyone destroy this?" Dongria Kondh's defence of their land is one that has endured because this is how they feel about the land—they experienced these heartfelt statements—but their bond to the environment was so strong that their struggle became a moral and a spiritual struggle.

Indeed, their resistance was built on a strong community feeling and a cultural identity. As it made a stand about protecting the Niyamgiri Hills, the tribe mobilized the traditional with more contemporary strategies to push a powerful movement (Chakraborty, 2022). Through traditional songs, dances and rituals, these became symbolic tools of resistance and made the sacredness both of their land and reframing the conflict from land to the existential and spiritual. As Borde (2017) described it, this cultural resistance fits with global discourses around indigenous heritage as a medium for political resistance. Hence, such expressions carried broader resonances and translated regional conflict relative to the global environmental and cultural preservation narratives. In Kurli village, community members of Dongria Kondh gathered under the shade of an ancient huge banyan tree to perform rituals to seek Niyam Raja, their deity (Figure 2.3). Their amalgamated

FIGURE 2.2 Dongria Kondh's resistance.

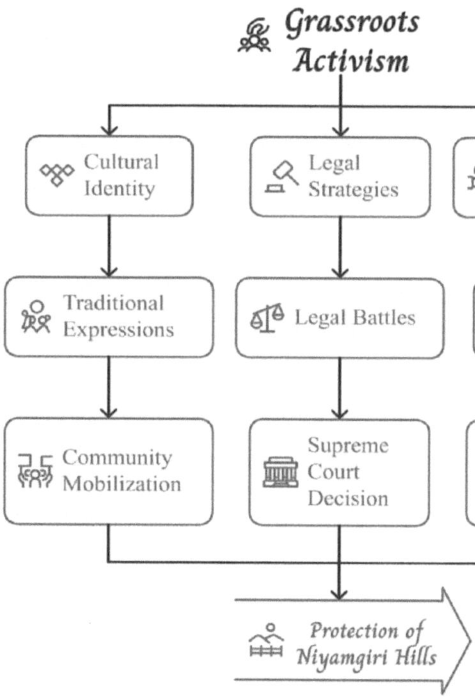

FIGURE 2.3 Dongria Kondhs performing rituals to seek Niyam Raja, their deity.

rhythmical drumming and chants echoed through the forest, representing their collective resolve. "These songs are our strength," says one young village man. "These make remind us that the hills are live, and they're watching over us." These cultural displays were not just acts of defiance, but they also provided a way to communicate to the world the sacredness of Niyamgiri.

The Dongria Kondh's efforts were further strengthened by forming strategic alliances. The tribe had little power to face Vedanta Resources, so it worked with local activists and national non-governmental organizations (NGOs) like Amnesty International and Survival International. The role of these alliances was to amplify their voices on global platforms and define the mining conflict as a large indigenous rights versus industrial exploitation story. Information was disseminated and protests mobilized through local networks, including youth groups and village elders that bridged critical resource gaps and helped the community to move across a complex legal and political landscape. In an interview, local activists in Lamba said that they worked alongside NGOs such as Amnesty International to document the impact of mining on the community. "Without them we could never have taken our case outside these hills, to the courts, to the media, to the international forums," said a local activist. In other villages I visited, villagers explained how alliances with national organizations allowed them to organize protests in Bhubaneswar and Delhi that brought the matter to the attention of the policymaker and the public.

One of the foundations of the Dongria Kondh's resistance was their fighting legal battles (Amnesty International, 2013). The community used India's constitutional provisions for tribal protection and environmental conservation to challenge the legality of the mining project. This was big news in India, where the fate of mining in Niyamgiri was decided in 2013 by the Supreme Court, with local gram sabhas having the final call (Boehm & Excell, 2017). Underpinned by assumptions of participatory governance, this decision localized the rights of indigenous communities. Following the judgement of the forest rights tribunal, Niyamgiri gram sabhas of all 12 communities in the region came to a unanimous decision, voting equally against the mining proposals on the grounds of cultural and environmental sustainability.

The Dongria Kondh's battle became a legal process, which was a very important strategy. The legality of the mining project was challenged by the community through the provision of the Indian Constitution on the protection of the tribes and environmental conservation. The most cherishable fact is that the Supreme Court of India, way back in 2013, gave local gram sabhas total power to decide the future of mining in Niyamgiri. This decision was made based on principles of participatory governance and was meant to defend indigenous communities' rights. But after the judgement in favour of the petitioners, all 12 gram sabhas, or village councils, along the length and breadth of the Niyamgiri region, had rejected the mining proposals on cultural preservation and ecological sustainability. This legal victory came not only as a victory for the protection of the Dongria Kondh's heritage but also set a global precedent for a form of indigenous resistance to industrial encroachment. Speaking to elders in Khambesi-II, they recalled how gram sabha meetings were held to decide the future of mining in Niyamgiri. "We were nervous but we realised this was our chance to tell the government we don't want mining," says a gram sabha participant. The unanimous rejection of mining proposals by all 12 gram sabhas embodied the community's collective will which demonstrated a level of collective will as a means to participatory governance, which sets a new precedent.

It was the Dongria Kondh's ability to bring their movement to international attention that helped keep it afloat. Strategic media campaigns alongside advocacy efforts shrewdly redefined what had once been a local fight for environmental justice as a global battle for the same. However, this mining conflict was portrayed through documentaries, articles and social media campaigns as a case of corporate reach surpassing the boundaries of indigenous survival (Survival International, 2013). Strategic narratives on the ethical and environmental stakes of the conflict were crafted by organizations from Survival International to Amnesty International (Borde & Bluemling, 2021). The Dongria's resistance could not have been brought forward without an alignment of their cause with global frameworks, such as the UN Declaration on the Rights of Indigenous Peoples (UNDRIP) and Sustainable Development Goals (SDGs), which gained them the solidarity of environmentalists, academics and policymakers from all around the world (Yap & Watene, 2019). Khambesi villagers remembered how filmmakers and journalists had come to film the Dongria Kondh, filming how they lived sustainably, while mining would have wrought destruction in the place where they lived. Mani Kutruk, a local storyteller, said they demonstrated to the world "we are not backward people but protectors of the environment."

The Dongria Kondh's resistance represents a potent example of how the power of cultural unity and strategic collaboration can be transformational. While battling for the Niyamgiri Hills, they didn't only fight for it; they also saw the need to protect the environment and Indigenous knowledge. "This fight was never just about us," one elder from Kurli said. "It's about protecting life itself for everyone, everywhere." It was a symbol of the indigenous communities rising up against industrial exploitation and still finding a way to win. Voices from Khambesi and Kurli remind us that the defence of cultural and environmental heritage is a universal struggle, with bearings in other movements for sustainability and justice all around the globe.

THE DONGRIA KONDH'S RESILIENCE STRATEGIES

The Dongria Kondh tribe's ability to resist the Niyamgiri Hills from mining activities flows from the symbiosis of sustainable livelihood, cultural preservation and adaptive governance. Their resilient strategies, integrated with the indigenous knowledge and socio-ecological systems, have made the community successful in their endeavours to defend the environment and culture from the diktats of the external world. All these approaches portray the interconnections between environmental conservation, cultural sustainability and indigenous resistance.

The Dongria Kondh's practices for their survival are clear evidence of the wisdom on the natural resources of the world acquired over the years. Shifting cultivation, locally named as *podu chasa*, is quite rampant and is one of the biggest agricultural practices of the community. In the village of

FIGURE 2.4 Dongria Kondhs shifting cultivation practices.

Khambesi-II, a local elder described their traditional agricultural practice, podu chasa, with pride: "We farm in harmony with the forest. Each year, we leave parts of the land to rest so it can heal, and the forest gives back to us." Shifting cultivation is the mainstay of the Dongria Kondh's subsistence strategies, and they cultivate millets, pulses, turmeric and ginger on rotational ploughing system (Figure 2.4). Such nutrient-rich and environmentally friendly crops provide food security with sustainable soil management, preserving fertility for crop production. Even though shifting cultivation has been criticized for being inefficient, it is now understood through rigorous research to be a sustainable agricultural practice that conserves soil fertility and maintains minimal use of chemical fertilizers (Pretty & Bharucha, 2014). While visiting Khambesi village, young farmer Sanjit Jakesika shared his views: "We do not require chemical fertilizers. The forest nourishes our soil and the rains bless our crops." These practices seem to contradict notions of inefficiency; rather, these practices reflect the deep engagements with sustainable agriculture that minimizes environmental degradation and supports biodiversity.

Dongria Kondh's sustainable forest resources management practices are also equally noteworthy. They collect non-timber forest products (NTFPs), including honey, herbs and fruits from the forests in a way that does not affect the stock for future uses. Sacred groves that are considered inviolable zones also show their concern for the conservation of biochemical diversity. These practices also ensure their income and at the same time reduce deforestation and promote conservation of the region's biodiversity. In Kurli village, women Balo Jakesi explained their meticulous methods of collecting NTFPs: "We only hunt for food; we do not kill more than we need, we take honey, berries and herbs for medicine. If we exploit it to the extent that we can, the forest is harmed, and, in turn, we are harmed as well." Through sustainable utilization of the resources and respecting the sacred groves, the ecosystem is well preserved. Sacred groves in Niyamgiri are also demarcated areas that are home to many old trees and special kind of plants and are protected by the Dongria Kondh as inviolable proofs of their nature conservation.

The resiliency of the Dongria Kondh is profoundly rooted in their cultural preservation (Figure 2.5). The rituals and festivals they practice are deeply entwined with the ecological significance of the

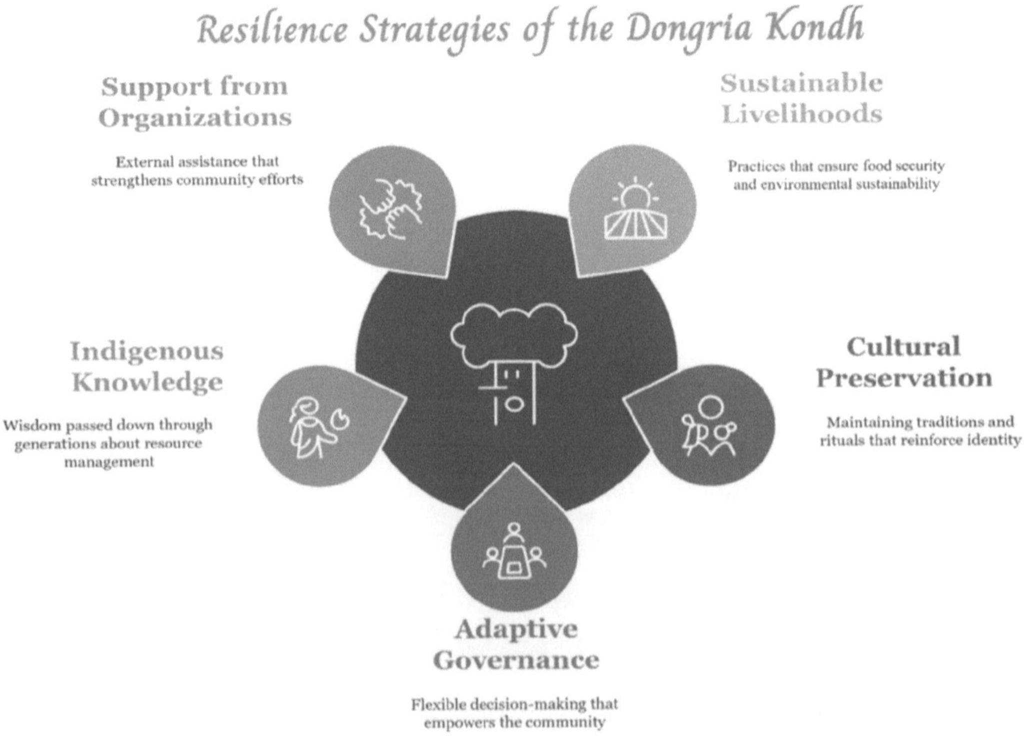

FIGURE 2.5 Resilience strategies of the Dongria Kondh.

Niyamgiri Hills, and the religious aspect of the hills represents the ecological importance of the hills to the community, where they worship Niyam Raja as their deity. In Khambesi village, a local social activist emphasized the spiritual connection of the Dongria Kondh to Niyam Raja: "This is not just land. It is the place of our God, the fountain of life and blessings." Niyam Raja Puja and Meriah are the two prime examples of the festivals that celebrate this particular bond (Figure 2.6). Such events not only help to enhance the spiritual connection of the people to the hills but also become a means of countering external extrusion. Additionally, oral traditions and songs that are shared around communal fires in Kurli-Khambesi village weave narratives of harmony with nature and stories of resistance. "The songs are our history of how Niyam Raja defends us and how we must defend him in return," narrated by Siba Jakesika, an elder storyteller. These narratives convey environmental wisdom and strategies of opposition and strengthen community unity.

The tribe's involvement in adaptive governance further added to their capacity in withstanding the changes. In Khambesi-II, a hamlet of Dongria Kondh emerged as a platform of unity and became a symbol of collective decision-making stance for all the gram sabhas. Gram sabha member Hikoka recalled, "I was a part of it. We spoke as one, and we said no to mining." The 2013 Supreme Court judgement, which gave the authority to the local gram sabhas regarding the mining activities, was a historic judgement for indigenous governance in India. Using this provision, all 12 gram sabhas in the Niyamgiri Hills rejected mining proposals, expressing the strength of decentralized governance in empowering the vulnerable groups. The fight of the Dongria Kondh is a good example of the participatory decision-making systems in which traditional leaders and outside supporters are involved (Karak, 2018). In a meeting held at Khambesi village, Sikaka, the activist leader said, "The old people steered us with knowledge while the young people introduce fresh ideas. Together we made the decisions that are fair and powerful."

Also, the tribe's engagements with national and international organizations have further strengthened their voices and proliferated on global platforms. Support from environmental NGOs

FIGURE 2.6 Dongria Kondhs celebrating a festival.

and human rights organizations facilitated technical and legal assistance and framed their fight within the broader discourse of environmental justice and indigenous people's rights. Local activists explained how they accomplished this with NGOs such as Survival International and Amnesty International in Bissamcuttack. "They helped take our voices to places where we cannot reach," said activist Radha Kadraka. It has not only strengthened the resists but also called on the public to notice the significance of the policies that aimed at the development of industries as well as the protection of the indigenous people and their environments.

The Dongria Kondh's multiple approaches show that indigenous knowledge and cultural persistence are effective levers to counter the penetration of extractive imperialism. Their sustenance economics, sustainable livelihood practices, cultural restoration and governance flexibility provide valuable insights for other struggles for environmental justice and sustainable development. Their struggle highlights the need for the enhancement of the rights of indigenous people as the managers of biological diversity and culture against the backdrop of a world that is fast being stripped of these resources. Community leader, Kuni Hikaka of Kurli village, articulated, "Niyamgiri is not just ours alone; it belongs to the world. If we are to protect it then we are protecting the future."

CHALLENGES TO RESISTANCE AND RESILIENCE

While well known for their resistance against mining in the Niyamgiri Hills, the Dongria Kondh community continues to struggle with multiple challenges, which put their lives on the test. These challenges come from socio-economic exclusion, environmental constraints and unrelenting industrialization and underpin the vulnerability of indigenous struggles in a global environment of increasing industrialism. Even after their historic win in 2013, when the Supreme Court allowed them to stop mining projects through the gram sabha, the community remains yet under relentless threats.

Mining companies, especially Vedanta Resources, continue to seek mining approvals and sometimes make the excuse of promoting economic development and job creation. When we were in Kurli village recently, a local elder, Daitri Kadraka, put it this way: "They speak of creating employment, but we understand that the employment they provide will destroy our hills and gods, how

FIGURE 2.7 Dongria Kondh settlements, undernourished children and women harvesting broom—capturing aspects of their livelihood, health and traditional practices.

can we accept this?" With the help of state policies that encourage industrial development, these corporations have been trying to challenge the 2013 ruling over and over. The people of Khambesi said that government officials have started paying surprise visits to conduct some forms of assessment that they believe will lead to the revival of mining activities. A young activist, Rebati Sikaka said that "they just come without our permission and try to survey our land, but we stand together to defend our own land."

The Dongria Kondh are socially and economically excluded and vulnerable which only add to their problems (Figure 2.7). The women of the community in Khambesi complained about the absence of health facilities. "When our children fall sick, we have to walk for hours to reach a health center," said a mother of three children. Poor maternal and infant mortality rates in these villages point to poor health care facilities, while traditional medicine cannot cope with the new health problems such as malnutrition and vector-borne diseases (Bhoi & Acharya, 2024). Also, poor literacy levels limit the ability of individuals to engage in legal and administrative environments that exist outside the organization. During the community meeting in Kurli, the youth complained about their poor opportunities to attend formal learning institutions. "We want to learn and to defend our rights," *added Muna Jakesi, a teenager,* "but schools are far away and many of us cannot go there." Economic problems also affect the stability of the community in the same way that they weaken the ability of the people in the community. Living in Khambesi, people described how exploitative middlemen unfairly charge a low price for their forest goods. "We pick fruits and herbs in the forest but they only offer us very little money," said Sarita Sikoka, another gatherer. This economic marginalization, together with reductions to the amount of land and the use of shifting cultivation, leads to dependency and poverty. The traditional and still employed system of shifting cultivation is a way of life and is threatened by the increasing pressure of environmental changes.

Climate change and environmental pressure are another threat to the Dongria Kondh people (Figure 2.8). Climate change–induced phenomena such as erratic rainfall, drought and other unusual weather changes are eroding age-old agriculture techniques that form the basis of their food security. Shifting cultivation, an important means of livelihood, is now threatened by climate change vulnerabilities. As we discussed, in Khambesi-II village, farmers shared how climate change effects

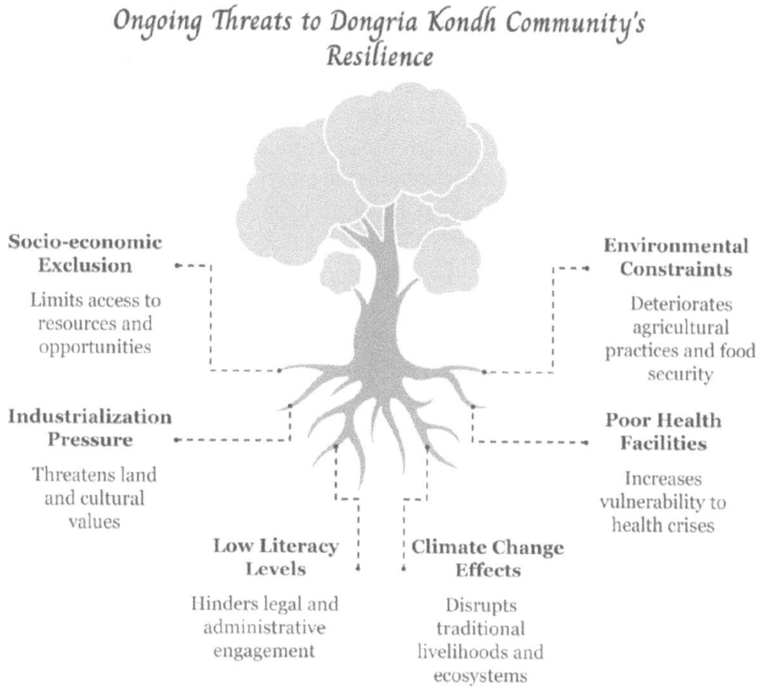

FIGURE 2.8 Challenges to the Dongria Kondh's resistance.

had affected their crop yields. "The rains come too late and when they come, they wash everything away," said Jakesika, a farmer. Besides, the agents of anthropogenic and climatic-induced shifts are causing the deterioration of the forest ecosystems which puts at risk their cultural and ecological values. The Niyamgiri Hills do not only hold economic value but are also home to spiritual value for the indigenous. Resource degradation, either through the illicit cutting down of trees or climate change, reduces the carrying capacity of the environment that the Dongria Kondh have always relied on for their sustenance. The villagers of Denguni disclosed the effects of the lea-dearth of major forest products that are invaluable to the sustenance of villagers. "The trees on which we depend are being lost," *said the elder, Mani Kutruk.* "Without them we are not only deprived of food but our connection to the hills."

Nonetheless, the Dongria Kondh still stands its ground. Residents of the Niyamgiri Hills vowed to defend Niyamgiri at all costs based on the discussions they had with the community. "We are prepared once again to defend our hills if the enemy attempts to regain what we have wrestled from them," said Kuni Sikoka, a community leader. This unity, however, has faced escalating pressure from forces outside and within that has made their fight a symbol of the fight of indigenous peoples and a call for better protections for indigeneity all over the world.

DISCUSSIONS

The Dongria Kondh's case against mining in the Niyamgiri Hills is one of the most historic examples of the fight in the discourse on environmental justice and has many layers of learning that are relevant across the world for sustainable development and participatory governance. Of course, the recognition, protection and assurance of the indigenous people's rights are central to this struggle, often most of the time violated by the neoliberal development agenda (Bargh, 2007). The Dongria Kondh people's perception of the environment, practices for preserving biodiversity and cultural heritage through implications of traditional ecological knowledge and their relation to the Niyamgiri Hills is a clear example of why indigenous peoples should be allowed to govern their environment.

Their respect for Niyam Raja, the guardian deity of the hills, captures how virtually all native cultural practices are anchored in environmental conservation. However, the community has been under threat from industrialization, especially mining. The celebrated ruling of the Supreme Court of India in 2013 which gave the right to gram sabhas to decide the future of mining at Niyamgiri, created a new normal for participatory decision-making processes globally and reinforced the principle that indigenous people have a right to defend their lands. Such decisions have an international significance, for example, the Dakota Access Pipeline protests in the United States, which also testify to the struggle for indigenous people's rights (Estes, 2024).

The Dongria Kondh's case also illustrates how community-based conservation is most likely to succeed (Figure 2.9). This shows that the endangered knowledge system of an indigenous community of the Niyamgiri Hills supports sustainable interactions with their environment which include the preservation of sacred groves and practising shifting cultivation. Several studies show that forests that are being managed by indigenous people are more diverse and more stable than other forests that are managed by the state or private companies (Watson et al., 2018; Agrawal et al., 2008). The Dongria Kondh depend on NTFPs for their food, medicine and other household needs; hence, the argument that indigenous people are often more effective at natural resource management than commercial operators. The Niyamgiri case counters the conventional, elite-based top-down approach to conservation by promoting the sharing of authority over natural resources with local peoples.

Besides, the Dongria Kondh's struggle reveals that sustainable development models cannot be limited to the extractive and neoliberal developmentalism model. The Niyamgiri Hills are equally important for bauxite reserves, although they have a greater cultural and environmental significance, as they are home to a variety of wildlife and are the only source of income for the Dongria Kondh tribe. Harmful effects of mining in the environment show that extractive industries are unsustainable and have a negative impact on deforestation, greenhouse gas emissions and socio-economic inequality (Odisha Sun Times, 2014; Adelphi, 2009). The Dongria Kondh community is fighting against a development model rooted in short-term economic returns and not ecological integrity and social equity.

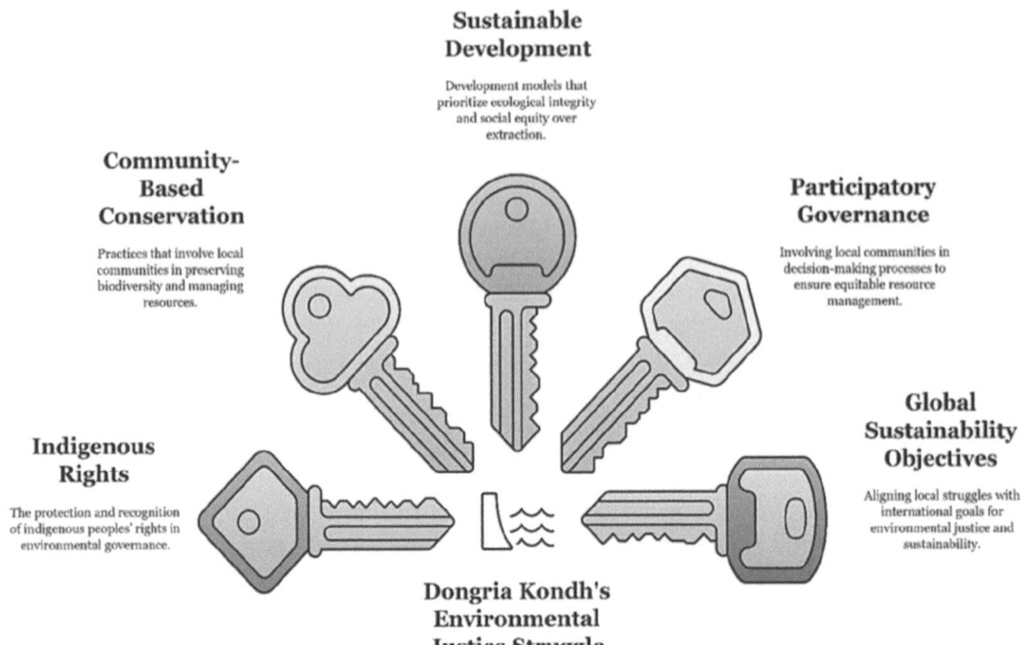

FIGURE 2.9 Indigenous empowerment and global sustainability.

Like the globally recognized sustainability goals, the United Nations' SDGs, the Dongria Kondh's fighting is aligned with SDG 15: Life on Land and SDG 16: Peace, Justice and Strong Institutions. The Dongria Kondh's model of participatory governance as a model for the conservation of ecosystems and environmental justice is demonstrated in this study. Furthermore, their case underlines the importance of indigenous peoples in climate change mitigation in the global context and underscores their part in sustainable futures. Through their fight against mining and their struggle to protect their culture and natural values, the Dongria Kondh have come to represent local communities' struggle for environmental justice on the world stage.

The account of the Dongria Kondh's struggle as presented in this research therefore enriches the existing knowledge on environmental justice and indigenous management of natural resources. It shows that the implementation of indigenous people's rights is not just the right thing to do, but that it is the right thing to do in order to support the sustainable and adaptive development of indigenous communities. This work presents the ideals of participatory governance to create a model that may be used for incorporating indigenous knowledge into international environmental management. The case of the Dongria Kondh lays down a precedent for the way that locally driven action can contribute to promoting change in global scenarios and highlights the critical role of indigenous peoples in determining the future of sustainable development.

CONCLUSION AND RECOMMENDATIONS

The study evidences how indigenous communities play a key role in conserving their ecosystems and cultural heritage against mine exploitation in the Niyamgiri Hills. As a paradigm of grassroots environmentalism, this enduring struggle against extractive industries is deeply enmeshed in what they understand to be the interconnectedness between their socio-cultural worlds and Niyam Raja and ecology. This is not an isolated or localized struggle; it is part of a global agenda of environmental justice of the marginalized communities taking their right to defend territories and life forms.

This study finds that the resilience strategies employed by the Dongria Kondh are an example of how indigenous knowledge systems can cope with the modern world. Their sustainable agriculture, including crop diversification such as millets cultivation and forest dwelling practices, presents an opposite model of coexistence with the environment than the extractivist models of growth. Sacred groves and customary governance also show how they can sustain ecological balance even under much pressure from the outside. The complete non-adoption of mining projects by the gram sabhas under the 2013 Supreme Court judgement is the assertion of indigenous people's self-determination in India's legal system. This decision not only benefited the Dongria Kondh but also made a legal basis for the local people's participation in environmental protection (Amnesty International, 2013). However, the Niyamgiri ecosystem and the socio-economic marginalization of the Dongria Kondh remain under threat (Nath, 2024).

The Dongria Kondh's resistance has important global consequences. Examples of indigenous struggles similar to theirs question the growth-oriented development models that are considered hegemonic, demanding the recognition of cultural and ecological differences. These movements make a huge impact in the discourse for sustainable development, emphasizing that both economic objectives and equitable social and ecological interests must be harmonized. Thus, the Dongria Kondh's victory over the mining in Niyamgiri proves that indigenous people's struggle against persecution of their rights is a strong tool of environmental human rights and can influence the policies, opinions and even the conservation of the whole world. Such movements should be emphasized, and through legislative support, systematic development programmes and global advocacy, we need to recognize and strengthen them for the sake of sustainable and equal future, where environmental integrity and cultural heritage are equally important for progress.

Further to strengthen their resilience, it is imperative to have targeted recommendations for climate change adaptation as shown in Figure 2.10. This involves encouraging the use of local wisdom such as the cultivation of farm crops such as inter-cropping and millet farming and documenting

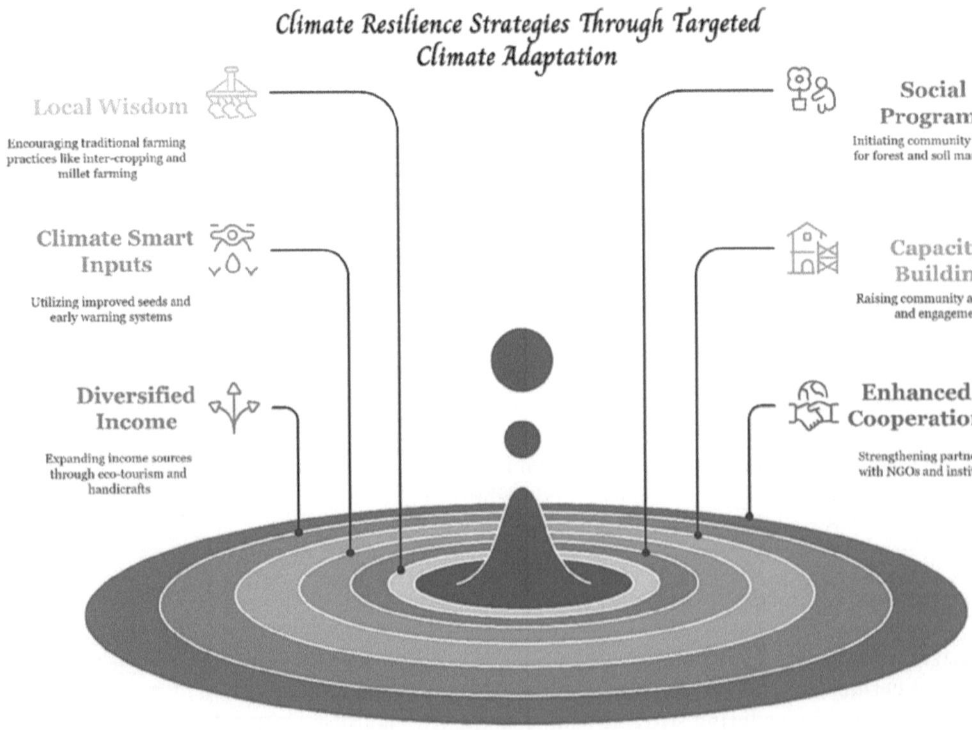

FIGURE 2.10 Climate resilience strategies through targeted climate adaptation.

the same. Other social programmes for forest and soil management and climate-smart agriculture should be initiated in the community in conjunction with the tribes. Climate-smart inputs like improved seeds resistant to drought, early warning systems on extreme climatic conditions, climate education and information are crucial. This chapter has argued that capacity-building interventions may raise awareness of the rights of the community and enable them to engage in climate change decisions. As shown above, there is a need to diversify income sources through eco-tourism, handicrafts and other income-generating activities that are less sensitive to climate variability and change. By enhancing the cooperation with NGOs, academic institutions and international organizations their advocacy will be more powerful and the strategies for adaptation will be more efficient. They can also help in preserving their culture and the environment despite the challenges posed by climate change and thus set that pace for other countries to follow in as far as development is concerned.

REFERENCES

Adelphi. (2009). Mining conflict in the Niyamgiri Hills, India. Retrieved from https://climate-diplomacy.org/case-studies/mining-conflict-niyamgiri-hills-india

Adger, W. N. (2003). Building resilience to promote sustainability. *IHDP Update*, *2*(2003), 1–3. https://www.researchgate.net/profile/John-Mcmanus-15/publication/259482152

Agrawal, A., Chhatre, A., & Hardin, R. (2008). Changing governance of the world's forests. *Science*, *320*(5882), 1460–1462. https://doi.org/10.1126/science.1155369

Amnesty International. (2013, April 18). India: Landmark Supreme Court ruling a great victory for indigenous rights. *Refworld.* Retrieved May 30, 2023, from https://www.refworld.org/docid/517662e24.html

Anguelovski, I., & Alier, J. M. (2014). The environmentalism of the poor revisited: Territory and place in disconnected global struggles. *Ecological Economics*, *102*, 167–176. https://dx.doi.org/10.1016/j.ecolecon.2014.04.005

Bargh, M. (Ed.). (2007). *Resistance: An Indigenous Response to Neoliberalism*. Huia Publishers. https://books.google.co.in/books?hl=en&lr=&id=6wKMe_kLgWoC&oi=fnd&pg=PP9&dq

Berkes, F., & Ross, H. (2013). Community resilience: Toward an integrated approach. *Society & Natural Resources*, *26*(1), 5–20. https://doi.org/10.1080/08941920.2012.736605

Bhoi, N., & Acharya, S. K. (2024). Health status of particularly vulnerable tribal groups (PVTGs) of Odisha: A narrative review. *Journal of Health, Population and Nutrition*, *43*(1), 176. https://doi.org/10.1186/s41043-024-00671-8

Bocha, K., & Srinivas, B. K. (2021). Livelihood practices and sustenance among Dongria Kondhs of Odisha. *IJRDO – Journal of Social Science and Humanities Research*, *6*(4), 11–17. https://www.researchgate.net/publication/351915580

Boehm, S., & Excell, C. (2017, April 24). *The man who stopped the mine: Q&A with Prafulla Samantara, 2017 Goldman Environmental Prize winner.* World Resources Institute. Retrieved January 10, 2025, from https://www.wri.org/insights/man-who-stopped-mine-qa-prafulla-samantara-2017-goldman-environmental-prize-winner

Borde, R. (2017). Protecting indigenous land from mining: A study of activist representations of indigenous people, in the context of anti-mining movements, with a focus on an Indian case (Doctoral dissertation, Wageningen University and Research). Retrieved from https://www.proquest.com/openview/e45bf4d8eedc7c42b0c9e3f90336c617/1?

Borde, R., & Bluemling, B. (2021). Representing indigenous sacred land: The case of the Niyamgiri movement in India. *Capitalism Nature Socialism*, *32*(1), 68–87.

Brown, K. (2014). Global environmental change I: A social turn for resilience? *Progress in Human Geography*, *38*(1), 107–117. https://doi.org/10.1177/0309132513498837

Campbell, S. D. (2013). Sustainable development and social justice: Conflicting urgencies and the search for common ground in urban and regional planning. *Michigan Journal of Sustainability*, *1*. https://dx.doi.org/10.3998/mjs.12333712.0001.007

Chakraborty, S. L. (2022, April 29). *The Niyamgiri movement and the failure to implement alternative visions of development*. Australian Institute of International Affairs. Retrieved from https://www.internationalaffairs.org.au/australianoutlook/the-niyamgiri-movement-and-the-failure-to-implement-alternative-visions-of-development/

Chowdhry, K. (2019, July 30). Kondh ingenuity: An unsentimental reverence to nature. *Goya Journal*. Retrieved from https://www.goya.in/blog/an-unsentimental-reverence-to-nature-desia-kondh-niyamgiri.

Corry, O. (2014). From defense to resilience: Environmental security beyond neo-liberalism. *International Political Sociology*, *8*(3), 256–274. https://doi.org/10.1111/ips.12057

Dey, D., & Arunachalam, A. (2024). Asian conservation consciousness. In *Sustainable and Traditional Environmental Conservation in Asia: Concept, Approach and Case Studies of Satoyama* (pp. 85–112). *Singapore*: Springer Nature Singapore. https://doi.org/10.1007/978-981-97-1806-1_5

Domínguez, L., & Luoma, C. (2020). Decolonising conservation policy: How colonial land and conservation ideologies persist and perpetuate indigenous injustices at the expense of the environment. *Land*, *9*(3), 65. https://doi.org/10.3390/land9030065

Estes, N. (2024). *Our History Is the Future: Standing Rock versus the Dakota Access Pipeline, and the Long Tradition of Indigenous Resistance*. Haymarket Books. https://books.google.co.in/books?hl=en&lr=&id=RV_eEAAAQBAJ&oi=fnd&pg

Folke, C. (2006). Resilience: The emergence of a perspective for social–ecological systems analyses. *Global Environmental Change*, *16*(3), 253–267. https://doi.org/10.1016/j.gloenvcha.2006.04.002

Folke, C., Colding, J., & Berkes, F. (2003). Building resilience and adaptive capacity in social-ecological systems. In *Navigating Social-Ecological Systems*. Cambridge University Press, *Cambridge*, 352–387. https://www.remineo.org/repositorio/ciao/xviii/talleres/t1/Mirada7.Folke-et-al-2016.pdf

Ford, J. D., King, N., Galappaththi, E. K., Pearce, T., McDowell, G., & Harper, S. L. (2020). The resilience of indigenous peoples to environmental change. *One Earth*, *2*(6), 532–543. https://www.cell.com/one-earth/fulltext/S2590-3322(20)30250-5

Garnett, S. T., Burgess, N. D., Fa, J. E., Fernández-Llamazares, Á., Molnár, Z., Robinson, C. J., & Leiper, I. (2018). A spatial overview of the global importance of Indigenous lands for conservation. *Nature Sustainability*, *1*(7), 369–374. https://doi.org/10.1038/s41893-018-0100-6

Guha, R. (1999). *The Unquiet Woods: Ecological Change and Peasant Resistance in the Himalaya*. University of California Press.

Guha, R., Sundar, N., Baviskar, A., Kothari, A., Pathak, N., Saxena, N. C., Lélé, S., et al. (2012). *Deeper Roots of Historical Injustice*. Rights and Resources Initiative. https://www.rightsandresources.org/wp-content/uploads/2014/01/doc_5589.pdf

Gunderson, L. H. (2000). Ecological resilience—in theory and application. *Annual Review of Ecology and Systematics*, *31*(1), 425–439. https://doi.org/10.1146/annurev.ecolsys.31.1.425

Hecht, S., Schmink, M., Abers, R., Assad, E., Humphreys Bebbington, D., Eduaro, B., & Pinedo-Vazquez, M. (2021). The Amazon in motion: Changing politics, development strategies, peoples, landscapes, and livelihoods. *Amazon Assessment Report 2021, Part II*, 14–2. https://doi.org/10.3929/ethz-b-000526184

Hill, R., Walsh, F. J., Davies, J., Sparrow, A., Mooney, M., Council, C. L.,... & Tengö, M. (2020). Knowledge co-production for indigenous adaptation pathways: Transform post-colonial articulation complexes to empower local decision-making. *Global Environmental Change*, *65*, 102161. https://doi.org/10.1016/j.gloenvcha.2020.102161

Jammulamadaka, N., & Bhattacharjee, S. (2017). Of gods and demons 1: The sacred hills of Niyamgiri and Vedanta Aluminium Ltd (VAL). In *The Dark Side 2* (pp. 23–41). Routledge. https://www.taylorfrancis.com/chapters/edit/10.4324/9781351277129-3

Karak, M. (2018). Insurgent difference: An ethnography of an Indian resource frontier (Doctoral dissertation, The Graduate Center, City University of New York). CUNY Academic Works. https://academicworks.cuny.edu/gc_etds/2856

Kaushal, L. A. (2017). A case study on Vedanta Alumina Ltd (VAL) Orissa India: State and FDI versus democracy? *Valahian Journal of Economic Studies*, *8*(1), 107. https://pdfs.semanticscholar.org/21ee/c15f37c2ef72b35699ce19c8e8da5267dd1f.pdf

Kelman, I. (2018). Islandness within climate change narratives of small island developing states (SIDS). *Island Studies Journal*, *13*(1), 149–166. https://doi.org/10.24043/isj.52

Kelman, I., & Næss, M. W. (2019). Climate change and migration for Scandinavian Saami: A review of possible impacts. *Climate*, *7*(4), 47. https://doi.org/10.3390/cli7040047

Lv, Y., Sarker, M. N. I., & Firdaus, R. R. (2024). Disaster resilience in climate-vulnerable community context: Conceptual analysis. *Ecological Indicators*, *158*, 111527. https://doi.org/10.1016/j.ecolind.2023.111527

Mekonnen, Z., Kidemu, M., Abebe, H., Semere, M., Gebreyesus, M., Worku, A.,... & Chernet, A. (2021). Traditional knowledge and institutions for sustainable climate change adaptation in Ethiopia. *Current Research in Environmental Sustainability*, *3*, 100080. https://doi.org/10.1016/j.crsust.2021.100080

Mohapatra, T. (2017). Adivasi's struggle for Niyamgiri: A study. *Afro Asian Journal of Anthropology and Social Policy*, *8*(1), 35–56. https://dx.doi.org/10.5958/2229-4414.2017.00004.7

Nath, S. R. (2024). Defending the sacred: Discourses of development, identity and everyday resistance among the Dongria Kondh (Doctoral dissertation, University of Sussex). Sussex Figshare. https://sussex.figshare.com/articles/thesis/Defending_the_sacred_discourses_of_development_identity_and_everyday_resistance_among_the_Dongria_Kondh/25966813?file=46807669

Nepstad, D., Schwartzman, S., Bamberger, B., Santilli, M., Ray, D., Schlesinger, P.,... & Rolla, A. (2006). Inhibition of Amazon deforestation and fire by parks and indigenous lands. *Conservation Biology*, *20*(1), 65–73. https://doi.org/10.1111/j.1523-1739.2006.00351.x

Norris, F. H., Stevens, S. P., Pfefferbaum, B., Wyche, K. F., & Pfefferbaum, R. L. (2008). Community resilience as a metaphor, theory, set of capacities, and strategy for disaster readiness. *American Journal of Community Psychology*, *41*, 127–150. https://dx.doi.org/10.1007/s10464-007-9156-6

Odisha Sun Times. (2014). *WIO for action against Vedanta for Encroaching Land in Odisha.* Retrieved from https://www.bluelotusmind.com/

Pandey, A. D. (2017). The challenges of neoliberal policies and the indigenous people's resistance movement in Odisha, India. *e-cadernos ces*, (28). https://doi.org/10.4000/eces.2340

Pandey, A. D. (2018). Kondhs' resistance movement to save sacred Niyamgiri, Odisha. In *Indigenous Perspectives on Sacred Natural Sites* (pp. 61–89). Routledge. https://www.taylorfrancis.com/chapters/edit/10.4324/9781351234900-5

Pretty, J., & Bharucha, Z. P. (2014). Sustainable intensification in agricultural systems. *Annals of Botany*, *114*(8), 1571–1596. https://doi.org/10.1093/aob/mcu205

Redvers, N., Aubrey, P., Celidwen, Y., & Hill, K. (2023). Indigenous peoples: Traditional knowledges, climate change, and health. *PLOS Global Public Health*, *3*(10), e0002474. https://doi.org/10.1371/journal.pgph.0002474

Riseth, J. Å. (2000). Sami reindeer management under technological change 1960–1990: Implications for common-pool resource use under various natural and institutional conditions: a comparative analysis of regional development paths in West Finnmark, North Trondelag and South Trondelag/Hedmark, Norway, Vol. 1. https://hdl.handle.net/10535/3618

Rubin, J. W. (1995). Defining resistance: Contested interpretations of everyday acts. *Studies in Law, Politics, and Society, 15*, 237.

Sahu, M., Chattopadhyay, B., Das, R., & Chaturvedi, S. (2023). Measuring impact of climate change on Indigenous health in the background of multiple disadvantages: A scoping review for equitable public health policy formulation. *Journal of Prevention*, *44*(4), 421–456. https://doi.org/10.1007/s10935-022-00718-8

Scott, J. C. (2012). *Decoding Subaltern Politics: Ideology, Disguise, and Resistance in Agrarian Politics* (1st ed.). Routledge. https://doi.org/10.4324/9780203095041

Survival International. (2013). *The Dongria Kondh*. Retrieved January 3, 2025, from https://www.survivalinternational.org/tribes/dongria

Tatpati, M., Kothari, A., Mishra, R., Tatpati, M., Kothari, A., & Mishra, R. (2016). The Niyamgiri story: Challenging the idea of growth without limits. *Kalpabriksh Pune*. Retrieved from https://vikalpsangam.org/wp-content/uploads/migrate/Resources/niyamgiricasestudyjan2018.pdf

Temper, L., & Martinez-Alier, J. (2013). The god of the mountain and Godavarman: Net present value, indigenous territorial rights and sacredness in a bauxite mining conflict in India. *Ecological Economics*, *96*, 79–87. https://doi.org/10.1016/j.ecolecon.2013.09.011

Toledo, V. M. (2001). Indigenous peoples and biodiversity. *Encyclopedia of Biodiversity*, *3*, 451–463. https://dx.doi.org/10.1016/B978-0-12-384719-5.00299-9

Watson, J. E., Evans, T., Venter, O., Williams, B., Tulloch, A., Stewart, C., & Lindenmayer, D. (2018). The exceptional value of intact forest ecosystems. *Nature Ecology & Evolution*, *2*(4), 599–610. https://doi.org/10.1038/s41559-018-0490-x

Whyte, K. (2018). Settler colonialism, ecology, and environmental injustice. *Environment and Society*, *9*(1), 125–144. https://doi.org/10.3167/ares.2018.090109

Woodman, J. (2014, January 14). India's rejection of Vedanta's bauxite mine is a victory for tribal rights. *The Guardian*. Retrieved from https://www.theguardian.com/global-development/poverty-matters/2014/jan/14/india-rejection-vedanta-mine-victory-tribal-rights

Yap, M. L. M., & Watene, K. (2019). The sustainable development goals (SDGs) and indigenous peoples: another missed opportunity? *Journal of Human Development and Capabilities*, *20*(4), 451–467. https://dx.doi.org/10.1080/19452829.2019.1574725

3 Fight an Uphill Battle

Current Challenges of Natural Resource–Based Livelihood of Women in Sundarbans (India) in the Wake of Climate Change

Manotosh Gayen, Arindam Roy, and Nasrin Banu

INTRODUCTION

Over the past few decades, climate change has emerged as one of the alarming global threats to the sustenance of lives and livelihood of the millions of the commons. During 1991–2005, nearly one million people have died worldwide due to increasing occurrences of climate change-induced hazards (ISDR, 2008), which have encouraged scholars to explore the nature, dynamics, and dimensions of climate change impacts on human sphere from global to local scale (Jain et al., 2013; Dastagir, 2015). As the "hotspot" of climate change-induced hazards, the impacts of ongoing climate changes on the physical and human environment of the tropical deltas have gained attention of the scholars worldwide and comprehensively been studied from diverse perspectives (Neal & Phillips, 2011; Ahmed et al., 2017). The Sundarbans, the largest mangrove-estuarine ecosystem of the globe with unique biota (Hazra et al., 2010) and a huge socio-economically backward population, has now been experienced severe threats of climate change. Nowadays, it has become an irrefutable truth that the climate variability in the Sundarbans region is noticeable higher than global average (Sahana et al., 2020). At present, the lives and livelihoods of the inhabitants of deltaic Sundarbans are under diverse challenges due of increasing temperature, rainfall variability, increasing numbers of extreme climatic events, sea level rise (amplifying the risk of salt water intrusion and frequent coastal flooding), coastal erosion, etc. Increasing frequencies and severity of coastal disasters, especially tropical cyclones and their associated storm surges, in the wake of recent climate change have significantly impacted the prosperity of the delta, by destroying the resource bases and livelihood capitals of the locals. Over the recent years, climate change has not only significantly impacted the physical landscape of the Sundarbans, but also deteriorated the social structure, environment, and livelihood in the area leading to social inequality (Roy & Guha, 2017).

Over the past century, the Indian Sundarbans has experienced 0.5°C increase in the average air temperature, which is anticipated to reach up to 1°C by 2050 with the continuation of the present rising rate (Hazra et al., 2002). The overall rainfall has increased in the region with the maximum increase rate in pre-monsoon, followed by monsoon, while rainfall has reported a declining trend during post-monsoon or winter (January–February). Increasing numbers of annual wet days (0.001 days/year), particularly heavy rainfall days (having >10 mm of rainfall), heavy rainfall during rainy season (due to the influence of southwest monsoon, proximity to the sea, etc.), delayed monsoon, and spatial differentiation in rainfall distribution (due to varied topography) have become common phenomena in the region (Nandargi & Barman, 2018). The region has become highly susceptible to tropical storms and severe cyclones, intensification of which leads to

DOI: 10.1201/9781003620747-4

disastrous storm surges, causing massive coastal flooding and extensive damage. Coastal erosion, intensified by sea level rise and cyclone impacts, jeopardizes the environmental stability of the region and has directly threatened the communities that thrive on its resources (Ghosh & Mistri, 2023). In addition, rising temperatures and rainfall variability disrupt the delicate ecological balance of the area and hamper the unique biological resources of the region by influencing breeding patterns, loss of habitat, and alteration in niche, contributing to the decline in biodiversity (Gopal & Chauhan, 2006). Such destruction of ecological resources directly hampers local livelihoods to a greater extent, as the local communities largely rely on nature-based primary activities such as fishing and agriculture supported by the delta's resources (Jamal et al., 2022). Moreover, with such environmental changes the number of "ecological refuges" started to increase under the constant risk of displacement and out-migration of communities due to the loss of habitable land and all other livelihood capitals (Mistri & Das, 2020).

The role of women in natural resource–based livelihoods is crucial and multifaceted. As key agents in the intersection between human societies and natural resources, women play an integral role in shaping sustainable and resilient livelihoods for present and future generations. Women play pivotal roles in various sectors, such as agriculture, forestry, fisheries, and other resource-dependent activities (Folasade, 2016). Their contributions often extend beyond traditional roles and encompass diverse responsibilities such as farming, gathering, processing, and marketing of natural resources (Brody et al., 2008). In many societies, women are primary stewards of household food security, actively participating in agricultural activities and ensuring the well-being of their families (Goodrich et al., 2017). In addition, women often engage in sustainable resource management practices, utilizing their knowledge of local ecosystems to balance conservation and livelihood needs (Uniyal et al., 2007). Despite their significant contributions, women in natural resource–based livelihoods often face gender-specific challenges, including limited access to resources, land, credit, and technology (Jensen et al., 2013). Recognizing and addressing these challenges is crucial for promoting gender equality and enhancing the overall sustainability of natural resource management. Empowering women in these livelihoods not only benefits them individually but also contributes to broader community development and environmental conservation (Ellis & Allison, 2004). The women-dependent livelihoods in the Sundarbans are often more vulnerable to environmental change for several reasons. Generally, women in the area are engaged in primary sectors such as agriculture, aquaculture, fishing, and forestry, which have directly been impacted by climate variations and environmental shifts. Additionally, societal norms may limit women's access to resources, information, and decision-making, exacerbating their vulnerability. Recognizing and addressing these specific challenges is crucial for building resilience and promoting sustainable development in the context of changing environmental conditions. Therefore, the recent changes in climatic conditions and their impact on the natural resource–based livelihood of women in the Indian Sundarbans need to be addressed to reduce the extent of vulnerability of the womenfolk of the region. Till date, very few literature studies (Mukherjee, 2023) have attempted to examine the extent of climate variability from gendered perspectives. The specific challenges that the womenfolk of the area faced and were compelled to deal with have not been discussed well. In such context, the present study sincerely attempted to address the variability of the natural resource–based livelihood of women through contemplative field-based observations and interactions with the stakeholders. The principal objectives of the study are as follows:

- To explore the current challenges faced by women of Sundarbans due to recent climate change;
- To evaluate the adaptability of women in sustaining natural resource–based livelihoods; and
- To explore potential strategies to promote sustainable livelihoods of women in the context of climate change.

BRIEF LITERATURE REVIEW: THEORETICAL FRAMEWORK LINKING GENDER, CLIMATE CHANGE, AND LIVELIHOODS

Intersectional studies offer valuable insights into understanding the complex dynamics shaping sustainable development in vulnerable regions. In Sundarbans, traditionally men and women have distinct roles tied to conventional occupations and responsibilities (Sen & Pattanaik, 2017). Generally, men have engaged in nature-based activities such as fishing, agriculture, and forestry, considering these jobs as physically demanding, tiresome, and requiring outdoor work (Mistri, 2013; Mistri, 2019). On the contrary, women are engaged in household chores, raising children, and small-scale agriculture and animal husbandry (Datta & Mete, 2022). Such typical labor divisions determine the role and authority of the genders in resource management practices in the society, where men play a dominant role in decision-making for larger-scale economic activities (such as selling of fish/agricultural produce), while women lag far behind (Jamal et al., 2022). Till today, the women in the area typically lack equal authority, decision-making capacity, or access to services compared to men. As the Sundarbans is an underdeveloped region, women constantly face new barriers to their empowerment, including illiteracy, a male-dominated society, lack of economic independence, a conservative social structure, and religiously obscurantist practices and beliefs (Ghosh, 2014; Chowdhury et al., 2016a).

However, the situation started to change with the increasing impact of climate change and associated consequences (such as cyclones, rising sea levels, and salinity intrusion) on the economic and livelihood structures of the region (Mondal et al., 2022). With diminishing incomes from agriculture, fishing, and forestry, men relocate elsewhere, leaving women to assume household leadership (Mistri, 2013). Under such circumstances, the women employ unique livelihood strategies and adaptation practices to deal with the specific challenges posed by the fragile and dynamic coastal ecosystem in the recent wake of climate change (Md et al., 2022). They have become engaged in cultivating salt-tolerant crops (such as rice varieties, vegetables, and fruit orchards) to manage soil salinity (Sardar, 2022); fishing and crab harvesting, utilizing traditional knowledge and innovative techniques to sustainably manage aquatic resources (Patel & Rajagopalan, 2009; Sen & Pattanaik, 2017). They often use organic farming methods and adaptive crop varieties to enhance resilience to climate change impacts (Dasgupta, 2018). Besides, women in the Sundarbans are participating in income-generating activities such as handicrafts production, micro-enterprises, and involvement in community-based organizations, which are reshaping traditional gender roles (Datta et al., 2011; Jamal et al., 2022). They also actively participate in mangrove conservation efforts through mangrove plantation, eco-friendly tourism initiatives, and sustainable livelihood options such as honeybee farming, ecotourism guides, and handicraft production using mangrove-derived materials (Ghosh, 2012; Chowdhury et al., 2016b). Women-led self-help groups and cooperatives play a crucial role in building community resilience by promoting savings, microfinance, disaster preparedness, and health education programs (Chowdhury & Maiti, 2014; Dhara, 2021; Hazra et al., 2021). Despite this fact, environmental challenges such as climate change, natural disasters, and resource depletion have severely impacted the livelihood dynamics of the area. Women, in particular, have to bear the brunt of these climate change-induced challenges as they are responsible for securing water, food, and fuel for their families amidst changing adverse conditions.

In addition to institutional arrangements, the endeavor tried to establish a framework for vulnerability analysis, encompassing a set of linked components such as women's background characteristics, biophysical characteristics, information, and technology. This framework also explains the functions of relations, structures, and agency. The relationship between women and climate change, as well as susceptibility and adaptation, is comprehensively explained by this paradigm. If women's vulnerabilities are not addressed on multiple levels, it will be impossible to increase their access to sustainable livelihoods. The sustainable livelihood approach (SLA), a comprehensive framework developed by department for international development (DFID) of the United Kingdom (Figure 3.1), can also be used to demonstrate how reducing vulnerabilities to the effects of disasters and climate change leads to sustainable livelihoods. The study seeks to explore how the SLA can serve as an

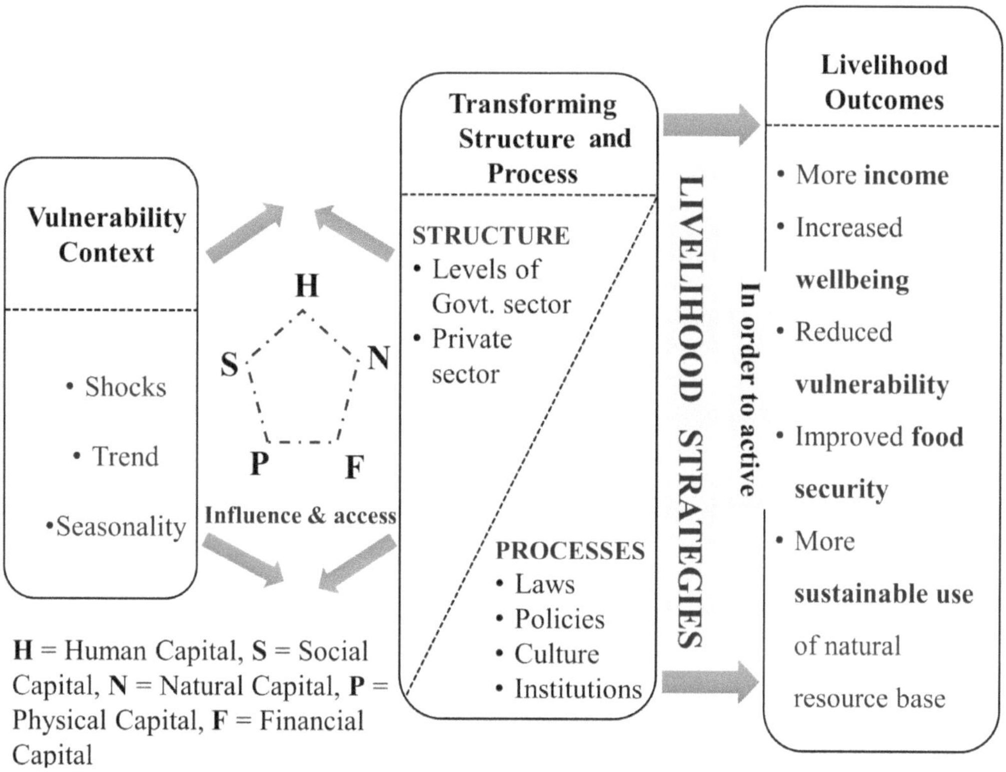

FIGURE 3.1 Sustainable livelihoods framework (Source: DFID, 1999).

effective tool for understanding, analyzing, and enhancing the resilience of women's livelihoods in the face of recent climate changes. Through a comprehensive examination of the socio-economic dynamics, resource availability, and adaptation strategies employed by women in the Sundarbans region, this research aims to contribute valuable insights toward the development of sustainable and gender-inclusive policies for climate resilience in the region.

The most vulnerable women are considered for the present study, who are financially poor and currently experience discrimination in terms of decision-making, access to resources (natural, human, and social capital) in farming areas that are highly susceptible to natural disasters and climate change. At present, the people in the Sundarbans in general and women in particular face serious challenges due to adversities of climate change, as most of them rely on natural resources for their livelihoods. Hence, a complete framework should be utilized to comprehend the multidimensional components of women's livelihoods in the present context of climate change in the Sundarbans region. Primarily, the framework should analyze the ecological shifts of the deltaic region due to climate change, affecting the availability of resources (such as fisheries, agriculture, and forest products), crucial for women's livelihoods. Understanding the interplay among temperature changes, rising sea levels, frequent cyclones, and their effects on biological resources is essential in this regard. The socio-economic dimensions must be considered in the second phrase, with special focus on the economic repercussions of climate change on different climate-sensitive primary sectors such as agriculture and fishing, on which women are more dependent for sustenance. Besides, income disparities, market access, and financial vulnerabilities faced by women should be delved simultaneously. Finally, the framework should incorporate the socio-cultural aspects, recognizing the gender-specific roles and responsibilities within the local communities to assess the climate-induced alterations in livelihood patterns affect women's social status, health, and education. Thus, by incorporating ecological, social, economic, and cultural dimensions, the framework provides a holistic understanding of the intricate challenges faced by women in the Sundarbans, enabling the development of targeted interventions and policies to promote resilience.

MATERIALS AND METHODS

Study Area

Located in the southwestern part of Bangladesh and the eastern part of India, the Sundarbans Delta has been formed by continuous alluvium deposition by several major rivers including the Ganga, Brahmaputra, and Meghna (Mistri & Das, 2020). This active delta region (as the deposition process is still going on) comprises the largest mangrove forest in the world with highest species diversity, covering an area of approximately 10,000 km^2, of which nearly 38% lies within the Indian territory (Sahana et al., 2020). It has been assumed that the region is named after the dominant mangrove tree species, the Sundari (*Heritiera fomes*) (Sen & Pattanaik, 2017). The region has been declared as UNESCO World Heritage Site in 1987. The delta has also been considered as a 'biodiversity hotspot,' as the unique estuarine ecosystem of the delta harbors a diverse range of flora and fauna, including various species of mangrove trees, amphibians, reptiles, and a variety of fish (Chowdhury et al., 2016a). Notably, it is the habitat of the Royal Bengal tiger (*Panthera tigristigris*), which has adapted to the unique mangrove environment (Jalais, 2010). The region is characterized by low-lying flat alluvial plains, mudflats, and small islands, dissected by complex and interconnected network of tidal waterways. At present, the Sundarbans faces various threats, including habitat loss, overexploitation of resources, pollution, and, above all, the consequences of climate change (Mahadevia & Vikas, 2012). Overall, the delta represents remarkable example of the interdependence of natural systems, human communities, and the need for sustainable conservation practices.

The Indian part of Sundarbans Delta extends between 21°32′N–22°40′N latitudes and 88°05′E–89°51′E longitudes, covering an area of 4,260 km^2 and administratively sub-divided into 19 CD. Among the Blocks, Gosaba Block of South 24 Parganas, located at the easternmost boundary of the Indian Sundarbans, has been considered for primary survey. Three geographically isolated riverine locations Satjelia, Lahiripur, and Dayapur in the Gosaba Block have been chosen as study sites for empirical investigation based on higher susceptibility to climate change-induced disasters (Figure 3.2). Considering the frequency and severity of climatic hazards and other vulnerabilities (in terms of economic and social factors), the study sites were categorized into three primary stressors: cyclones-impacting coastal regions, riverbank erosion or flood-prone areas, and heightened salinity in water-stressed zones. The study is encompassed a region situated between 22°05′15″N–22°10′03″N latitude and 88°49′54″E–88°55′26″E longitude, traversed by the Garal River and Datta River. These rivers posed significant challenges, leading to the loss of livelihood assets, crops, and agricultural land, particularly during instances of riverbank erosion.

METHODS OF ANALYSIS

Both quantitative and qualitative data as well as analyzing methods, literary sources, and empirical evidences have been used for the present study to examine women's vulnerability to climate change, considering the present socio-economic, political, health, and psychosocial challenges. Data for this review are drawn from diverse sources, including peer-reviewed literature, thematic books, government and non-government agency reports, and thematic discussions with individuals actively involved in climate change responses. Primarily, the present endeavor searches for available research works on the impact of climate change on natural resource–based livelihoods, food security, health, water, sanitation, etc., especially among women in the Sundarbans region. The second line of inquiry investigated the resilience and coping strategies of individuals. The focus was on climate-related

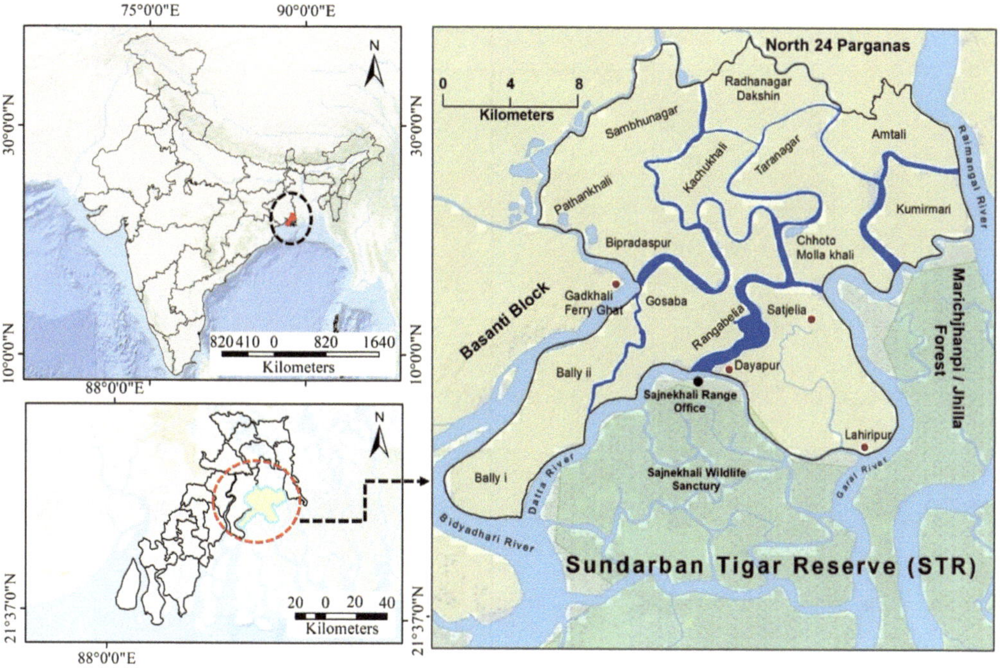

FIGURE 3.2 Location of the study area in the Indian Sundarbans (Source: Prepared by the Authors).

extreme weather events (temperature, rainfall, floods, and cyclones) natural disasters, livelihood, climate change and women, gender, impact, resilience, adaptation, coping mechanisms, policy, program, and strategy. Searches for both peer-reviewed and generic literature, were conducted through various platforms such as Google Scholar, PubMed, ScienceDirect, SpringerLink, Wiley, government websites, and other databases. For empirical investigation, field surveys were also conducted in different part of the selected sites (i.e., **Satjelia, Lahiripur**, and **Dayapur**) in different seasons of 2023–2024. Nearly 300 women (100 women from each village), having at least 40 years of age and actively engaged in economic activities, were selected deliberately for interview. Semi-structured questionnaire has been used for the purpose and emphasis was given on the shared experiences and views of the respondents regarding the issue. The study is basically an empirical study that has vividly described the current challenges, faced in natural resource–based livelihood of women in Sundarbans (Indian) due to recent climate change over the past.

RESULTS AND DISCUSSION

Major Sources of Livelihood of the Women in Sundarbans

The Sundarbans provides sustainable livelihood to millions of people, which has notably affected by recent climate change and environmental degradation, leading to income insecurity (Jamal et al., 2022). As an ecologically sensitive region, the Indian Sundarbans creates various challenges to its inhabitants. Especially, women in the area face interconnected challenges in livelihood, health, education, and socio-economic status. Over the past few decades, climate change appears as the biggest challenge in front of the women in the Sundarbans Delta, where the community are subjected to regular natural disasters such as cyclones, floods, salt water intrusion, coastal erosion, and sea level rising (Mondal et al., 2022). The acquired result of the primary survey reveals that majority of women respondents in the area belong to low socio-economic status and a significant proportion of the womenfolk remain unemployed, with no jobs to engage (Figure 3.3).

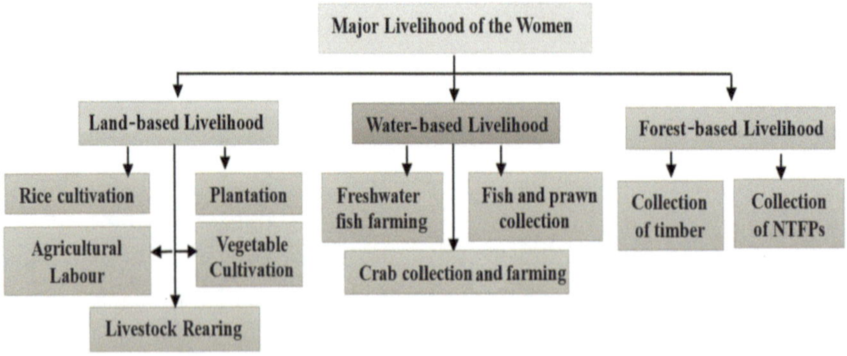

FIGURE 3.3 Major livelihoods of the women in the study area (Source: Field survey, 2023–2024).

FIGURE 3.4 Sowing of paddy by women agricultural labor.

Women in the Sundarbans are primarily engaged in traditional occupations or in primary sectors such as crop cultivation (Figure 3.4), mangrove plantation (Figure 3.5), fishing (fish, crab, and "meen" or prawn seed) (Figures 3.6–3.9), or collection of forest products (such as firewood, leaves, honey, fruit, seeds, and wax) to meet family needs (Field Survey, 2023–2024). As a whole, the livelihood sources of the women in the study area can be classified into three broad categories, that is, land-based, water–based, and forest-based livelihood (Figure 3.3). Despite the problems of soil water salinity, crop farming comprises the mainstay economy of the area, with the larger proportion of working population engaged with it.

The unique estuarine ecosystem of the region serves as an extensive breeding ground for numerous fish and crustacean species, supports a thriving fishing industry, providing livelihoods for local communities. It has been estimated that nearly three lakhs people in the region opt fishing as their source of livelihood, out of which approximately one-third proportion are women (Dasgupta et al., 2020; Tripathy & Paul, 2024). Collection of crabs and prawn seeds (locally known as "meen") is quite popular among the women worker in the area. Many fisher women everyday make a living by selling the crabs, prawns, and small fishes they manage to catch. Mrs. M. Mondal (42 years old: fishing worker) stated that, "As we don't have any land properties, fishing is the only option for me to manage livelihood. My husband moved to Andhra Pradesh as agricultural labor to sown rice. I catch meen for nearly three hours a day, mainly during low tide by using 'tempo jal' (local net) and sell it to the Aratdar (local shopkeeper) at the price of 300 Rs. for 1000 pieces." Most of the women

FIGURE 3.5 Women engaged in planting of mangrove saplings to protect embankments from cyclones and tidal surges.

FIGURE 3.6 Fishing by using tempo net in the muddy coastal areas.

FIGURE 3.7 Collection of prawn seed.

FIGURE 3.8 Woman collects crab by using "thopa" technique.

FIGURE 3.9 Woman put her hand in Crab's hole for crab catching.

fishing worker in the study area stand nearly three to six hours per day keeping almost half of their body under salt water, which lead to increasing prevalence of infectious diseases among the women. Besides, fishing in forest fringe areas increases the incidence of man-animal conflicts in the area. With increasing imposition of strict government regulations on fishing in forest fringe areas, the scope of fishing and crab collection gradually became dwindled.

Over the recent past, a tendency has been observed in the area, where female member of the households started to engage in non-primary activities for earning due to diminishing scope and income along with increasing challenges in primary sectors due to changes in prevailing geo-climatic situations. Nowadays, women often operate small shops (Figure 3.10a), worked as cultural performer in local cultural groups (to entertain tourists) (Figure 3.10b) or migrate to the nearby cities. Hence, besides male migration, female out-migration has gradually become prominent in the area. Moreover, they often receive training in handicrafts through various Govt. initiatives or by self-help groups and get engaged in artisan activities (Figure 3.10c). Figure 3.10 highlights the women's participation and main sources of different economic activities to earn livelihoods.

PRESENT CHALLENGES OF LIVELIHOOD FACED BY WOMEN

It has been affirmed by majority of the respondents (92.06%) that recent climate change has significant detrimental impact on their livelihood (Table 3.1). Bulk of the respondents have argued that recent changes in climatic attributes along with increasing numbers of extreme events, water scarcity, and increased salinity have disrupted the base of primary activities, especially the agricultural practices on which the local women largely depend.

Figure 3.11 demonstrates the impact of climate change in different aspects of livelihood as perceived by the respondents. Rising sea levels coupled with increasing frequency and severity of cyclones and accompanying storm surges leads to lack of access to forest resources, loss of crop land and hamper crop production (or crop failure), loss of home/homestead land, water crisis, livestock loss and destruction of other livelihood capitals of hundreds of women in different part of the

FIGURE 3.10 Newly emerged livelihood options: (a) A woman operates a tea and confectionary shop; (b) a group of women perform dance; and (c) women get trained in Kantha embroidery.

TABLE 3.1
Local Women's Perception on Impact of Climate Change on Livelihood.

Nature of Impact		Satjelia	Lahiripur	Dayapur	Average
Yes	Positive	2.37	3.55	4.39	3.44
	Negative	93.2	93.9	89.07	92.06
	Both	3.2	1.45	5.28	3.31
No		1.23	1.1	1.26	1.20

Source: Primary Survey, 2023–2024.

delta from short to long-term basis. Salinity intrusion and contamination of water sources make access to safe drinking water a challenge and the women, who are typically responsible for water collection are compelled to deal with it. They have to wake early, go out in groups of four to five and walk for a long distance to collect water from distant tube wells, and returned home around noon with heavy loads of water buckets or bottles. As the women have to spend a lot of time to fetch water from distance sources, they do not have much time for eating or sleeping. Thus, the health

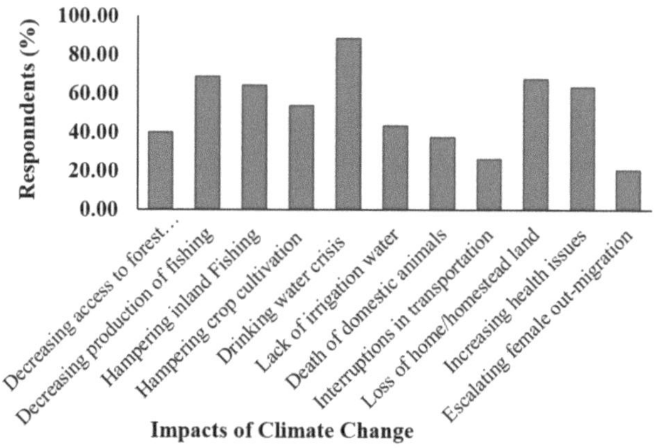

FIGURE 3.11 Climate change impact on livelihood: Rural women's perceptions (Source: Field survey, 2023–2024).

and well-being of the rural women of the study area are greatly affected by such scarcity of clean water (Figure 3.11). Furthermore, the quality of water they collect from ponds is not satisfactory in many cases.

After the cyclone Aila in 2009, efforts have been made to install water pipelines to provide drinking water to all villagers, although the implementation was not success at all. Primary survey reveals that in few parts of the region installation of water purification machines and other supportive equipment have been done, albeit in limited numbers. Therefore, women, resided along the coastal areas, often travel to the village center to collect water from these installed machines. Thus, in spite of resource allocation by the government, still a large segment of womenfolks is not in a position to get benefited evenly. The following Table 3.1 elaborately discusses the multifarious ways through which climate change hampers women's livelihoods in the study area (Table 3.2).

Remote locations and poor infrastructure in the delta region limit women's access to healthcare facilities. Along with cultural barriers, limited healthcare facilities impact women's health in general, and maternal health (both prenatal and postnatal) and child health in particular. Primary findings reveal that women often prefer to give birth at home to avoid nearly 8 km travel to reach to the nearest government hospital. However, in the Sundarbans, accessibility challenges go beyond mere distance, as short distances in some areas can pose significant difficulties due to damaged transport links or lack of available water transport when needed. For instance, a pregnant woman of Lahiripur village (Gosaba block) has to travel approximately 9 km by walking or van rickshaw, cross a river, and, then, cover another 5.5 km by van rickshaw to reach Gosaba BPHC (Block Primary Health Centre) for institutional delivery. This whole journey would take around 3–3.5 h. If she decides to avoid the break journey and travel in single mode, she will reach the destination after a journey of 6–7 h by river route (Lahiripur to Bali, then Bali to Gosaba BPHC). Alternatively, she might choose to take the risk of delivering at home.

Additionally, gender-based violence, including domestic violence and harassment, is a concern in the studied area. Isolation, lack of awareness, and limited access to support services exacerbate the problem of the womenfolk in this regard. Poor infrastructure, including transportation facilities hinders women's mobility and access to markets or essential services. Traditional gender roles often limit women's participation in community decision-making processes, which impacts their ability to voice concerns and influence policies that affect their lives. Efforts to address these challenges require a comprehensive approach that considers the intersectionality of gender, climate change, and socio-economic factors. Implementing sustainable development initiatives, promoting education, and empowering women in the Sundarbans Delta are essential steps to enhance their resilience and well-being.

TABLE 3.2
Local Women's Perception on Impact of Climate Change on Their Livelihoods

Climatic Events	Women' Perceptions on Climate Change	Effects on Lives and Livelihoods
Temperature	**Increase** • Summer temperature • Frequency of extremely warm days • Shortage of drinking water • Increased frequency of diseases • Rate of evaporation resulting in salinity enhancement	• Increased negative impact on health and hygiene • Reduced agricultural earnings and productivity • Lack of freshwater irrigation • Increased salinity in water and soil
Precipitation	**Decrease** • Rainfall during the rainy season • Total number of rainy • Growth and production of crops • Shorter rainy season and late arrival of monsoon	• Loss of food crops (mainly rice) and vegetables • Loss of freshwater fish
Cold wave	**Delay** • Hamper to winter crops (potatoes, cabbage, cauliflower, etc.) • Mango inflorescence • Disruption of rice production in the summer	• Reduced crop yield • Diseased elderly and children • Lack of access to crabs and honeycombs in cold climates • Seasonal diseases affecting cattle and birds
Tropical cyclone	**Increase** • Frequency and intensity of cyclones • Height of storm surges • More areas inundated by storm surges • Affected mangroves • Hamper fish collection	• Damage to the physical and organizational infrastructure of the local area • Harm to crops, trees, homesteads, gardens, and plants • Lack of opportunity for wage labor
Tidal surge and sea level rise	**Increase** • More land flooded during high tide • Regions affected by salinity • Hamper prawn collection	• Physically damaged homes, freshwater aquaculture, crops, and livestock • Agricultural land submerged • Submerged tube well and inadequate safe drinking water • Health risks for all family members and sanitation problems

Source: Field survey, 2023–2024.

The education rate and health status of women in this region are very poor. The influx of salt water through the frequent occurrences of tidal surges, cyclones, and coastal flooding, especially in the coastal areas, raise water salinity of water and land, which make water farming almost impossible. As a consequence, women have to be more dependent on fishing for livelihood. Due to prolonged contact with salt water by keeping almost half of their body inundated in salt water to catch shrimps, and bathing in polluted pond water, outbreaks of complicated gynecological diseases are more common in the area. Even so, they cannot use sanitary pads regularly due to poor financial affordability and use cloth during menstruation days. Nearly 69% of the women in the Indian

Sundarbans have suffered from skin diseases and white discharge problems. Thus, both health and livelihood of poor rural women in Sundarbans region have severely been affected by direct or indirect effects of recent climatic variability as well as climate change.

INSIGHTS INTO ADAPTIVE STRATEGIES EMPLOYED BY WOMEN TO COPE WITH CLIMATE CHANGE

The climate change-affected women in the Sundarbans Delta often demonstrate resilience and employ adaptive strategies to cope with the challenges posed by changing environmental conditions (Figure 3.12). It is important to recognize that adaptive strategies can vary based on local contexts, cultural norms, and the specific challenges faced in each region. Tailoring interventions to the unique circumstances of women in different communities is essential for effective climate change adaptation. Primary survey reveals that in many cases the women diversify their sources of income by engaging in multiple livelihood activities. They often combine traditional occupations (such as agriculture and fishing) with alternative income-generating activities such as handicrafts or small-scale entrepreneurship. They also actively participate in community-based adaptation initiatives, which encourage local communities working together to address climate-related challenges, such as building resilient infrastructure, implementing sustainable agricultural practices, and creating early warning systems for natural disasters. Climate-resilient agriculture practices, such as crop diversification and the use of climate-tolerant varieties, are often adopted by women, which help ensure food security in the face of changing weather patterns. In this context, the women may involve in the establishment and management of community-based seed banks, preserving and sharing traditional seeds that are suitable to local environmental conditions. This will further help in maintaining biodiversity and ensuring crop resilience. In the areas of water scarcity or salinity intrusion, women adapt innovative water management practices such as rainwater harvesting, using water-efficient technologies, or exploring alternative sources of freshwater for domestic and agricultural purposes. Empowering women to participate in decision-making processes at the community as well as household levels is crucial. Empowering women through education and skill development programs enables them to adapt to changing circumstances as educated women are better equipped to explore new opportunities, participate in decision-making

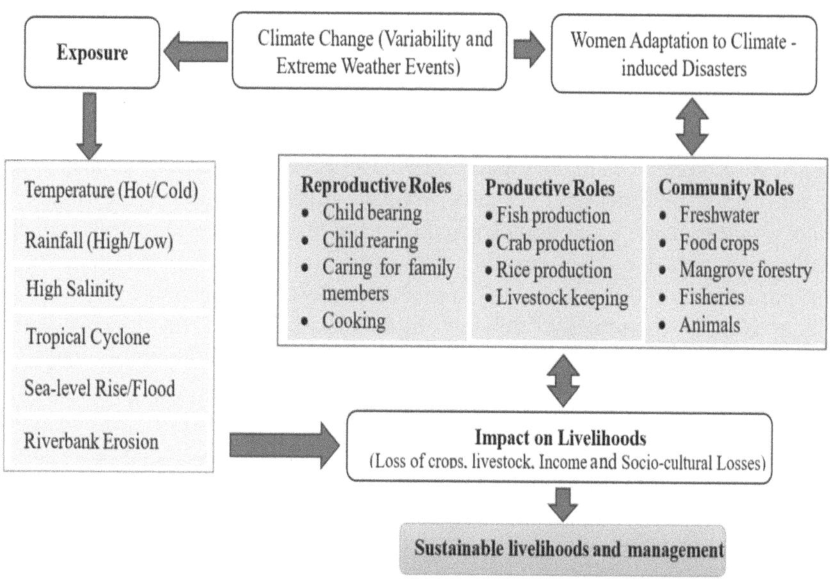

FIGURE 3.12 Conceptual framework on women's perspectives on climate change adaptation strategies for sustainable livelihoods (Source: Prepared by authors based on Nagasha et al., 2019).

processes, and contribute to sustainable development (Md et al., 2022). Women can also take the lead in implementing community health programs (such as creating awareness about vector-borne diseases, maternal health, and preventive healthcare measures) to address the health impacts of climate change (WHO, 2014). Building and strengthening social networks within communities will enhance women's resilience as women can share information, resources, and support, fostering a sense of solidarity and cooperation through these networks. Their involvement ensures that climate adaptation strategies are more inclusive and address the specific needs and priorities of women. Furthermore, transitioning to renewable energy sources, such as solar power, can be an effective adaptive strategy. Women may play a role in promoting and adopting these technologies for household energy needs, reducing reliance on non-renewable sources.

RECOMMENDATIONS FOR ENHANCING SUSTAINABILITY OF WOMEN-DEPENDENT LIVELIHOODS

Enhancing the sustainability of women-dependent livelihoods involves implementing measures that address the unique challenges faced by women and promote their economic resilience. Here are some recommendations:

Enhancing Climate Resilience: Promote ecosystem-based adaptation (EBA) strategies that integrate women's traditional knowledge and practices for sustainable resource management, such as mangrove conservation, sustainable fishing practices, and water harvesting techniques. Supporting women farmers with climate-resilient crop varieties, organic farming techniques, and access to weather information and advisory services to enhance agricultural productivity by mitigating climate risks. Facilitating educational opportunities for women, including vocational training programs will be helpful in this regard as education equips women with skills that enhance their employability and adaptability to changing economic conditions. Promoting climate-resilient women-dependent livelihoods, such as sustainable agriculture and alternative income sources will reduce the adversities of changing environmental conditions.

Empowering Women in Natural Resource Management: Ensure women's land rights will enhance their control over land which is crucial for women-dependent livelihoods, especially in agricultural communities. Besides, secure access to forest resources will also be helpful for sustainable livelihoods for women. Implementation of climate-resilient agricultural practices, providing training to women farmers for organic farming and crop diversification, and water management techniques will help escalate crop yield and sustain agricultural practices in the area. Improve infrastructure, including roads, transportation, and market facilities will enhance women's access to markets and reduce barriers to accessing resources. Providing targeted training programs on climate-smart practices, financial literacy, value chain development, and entrepreneurship will enhance women's skills and decision-making capacities in natural resource management.

Promoting Gender-Responsive Policies and Institutions: Conduct gender-sensitive research and allocate budgetary funds to address gendered vulnerability and disparities in access to agricultural inputs, microfinance, and social protection programs targeting women-headed households, which may help to mitigate the current challenges faced by women. Ensure active participation of women in local governance, social organizations, and disaster risk reduction committees to amplify their voices in decision-making processes and policy formulation. Relevant policies are needed to address gender disparities and to include women in economic development encompassing equal opportunities in employment, land rights, and participation in decision-making processes. Investing in healthcare infrastructure and services, especially in maternal care, is critical for sustaining women's livelihoods by ensuring their well-being.

Improving Access to Climate Information and Services: Strengthen community-based early warning systems is needed to ensure timely weather forecasts and climate-related hazards (such as cyclones, floods, and storm surges) and to prepare effective disaster preparedness measures including evacuation plans. Develop accessible and localized climate information services tailored to women's needs is required including climate-resilient farming practices, water management strategies, and livelihood diversification options.

Promoting Sustainable Livelihood Diversification: Adoption of appropriate technologies (such as training on the use of modern tools and digital platforms for marketing) will enhance productivity in women-dependent livelihoods. Introduction of sustainable fisheries practices will ensure women's access to fair market opportunities. Facilitate access to microcredit and market linkages for women engaged in alternative livelihoods such as ecotourism, handicrafts, sustainable aquaculture, and renewable energy initiatives will enhance their sustenance. Adoption of integrated coastal zone management approaches will balance economic development with ecosystem conservation, ensuring women's participation in sustainable fisheries management, coastal afforestation, and water resource governance. Facilitate the formation of microfinance groups and savings cooperatives among women can provide financial support, encourage savings, and foster a sense of economic independence. Promote financial inclusion by providing women with access to credit, savings, and insurance services will help women in income-generating activities, cope with emergencies, and build financial resilience. Besides, initiatives should be created that support women entrepreneurs, (including training programs, access to markets, and financial assistance) and empower women to establish and grow their businesses, contributing to sustainable livelihoods. Encouraging the creation of community-based support systems (such as women's cooperatives and networks) can facilitate resource-sharing, information exchange, and mutual support among women in the community.

Monitoring, Evaluation, and Learning: Collect and analyze gender-disaggregated data on climate impacts, livelihood strategies, and adaptation outcomes to inform evidence-based policy planning, monitoring, and evaluation processes. Foster knowledge sharing platforms, community-based learning networks, and participatory research initiatives involving women, academia, NGOs, and government agencies to co-create innovative solutions and best practices for sustainable development in the Sundarbans Delta. Measures should be implemented to address and prevent gender-based violence, as it can significantly impact women's ability to engage in sustaining livelihoods. This includes awareness campaigns, legal support, and safe spaces for women.

Implementing these recommendations requires collaboration between governments, non-governmental organizations, local communities, and other stakeholders. A holistic and gender-sensitive approach is essential for creating sustainable livelihoods that benefit women and contribute to overall community development. By contributing empirical evidence, theoretical frameworks, policy insights, and practical recommendations, research on gender, climate change, and livelihoods in the Indian Sundarbans can inform evidence-based interventions, policy dialogues, capacity-building initiatives, and community resilience strategies aimed at achieving Sustainable Development Goals, gender equality, and environmental sustainability in deltaic and coastal regions facing climate risks.

CONCLUSION

The study highlights the disproportionate impact of climate change on the natural resource–based livelihood of the womenfolk in the Sundarbans. Factors such as water and food scarcity, limited control on resources and loss of livelihood capitals exacerbate women's vulnerability at socio-economic sphere. Furthermore, climate change has impacted their health, especially concerning menstruation, pregnancy, and skin diseases. Changes in rainfall patterns and water scarcity strain women's livelihoods in agriculture, often leading to male migration for alternative occupations and increased household responsibilities for women. Despite these challenges, women in the region actively contribute to agriculture, fishing, and water harvesting, bolstering community livelihoods and fostering social resilience. The study emphasizes the positive impact of existing adaptation measures of local communities, highlighting the need for gender-inclusive climate-resilient approaches for disaster risk reduction. Empowering women through socio-economic freedoms and decision-making roles will help to minimize climate change impacts and fostering broader social transformation. Effective strategies must involve tailored policies, collaborative planning, good governance, and multi-stakeholder cooperation to mitigate disaster risks and improve socio-economic conditions. Educating and mobilizing women toward the Sustainable Development Goals is essential for long-term resilience and a brighter future for climate-affected vulnerable communities in the area.

ACKNOWLEDGMENTS

The work is a part of a major research project, sponsored by the **Indian Council of Social Science Research (ICSSR)**, New Delhi. The authors have gratefully acknowledged the financial assistance of ICSSR for conducting the research.

COMPLIANCE WITH ETHICAL STANDARDS

Conflict of interest: The authors have declared that there is no conflict of interest.

Ethical approval: This research does not involve individual participants in the study, and thus, ethical approval is not required.

Informed consent: There is consent to publish this article in this book.

REFERENCES

Ahmed, N., Cheung, W.W., Thompson, S., & Glaser, M. (2017). Solutions to blue carbon emissions: Shrimp cultivation, mangrove deforestation and climate change in coastal Bangladesh. *Marine Policy*, *82*, 68–75.

Brody, A., Demetriades, J., & Esplen, E. (2008). Gender and climate change: Mapping the linkages a scoping study on knowledge and gaps. BRIDGE, Institute of Development Studies, Brighton.

Chowdhury, A., & Maiti, S.K. (2014). Mangrove reforestation through participation of vulnerable population: Engineering a sustainable management solution for resource conservation. *International Journal of Environmental Research and Development, 4*(1), 1–8.

Chowdhury, A., Maiti, S.K., & Bhattacharyya, S. (2016b). How to communicate climate change 'impact and solutions' to vulnerable population of Indian Sundarbans? From theory to practice. *SpringerPlus*, *5*, 1219. https://doi.org/10.1186/s40064-016-2816-y.

Chowdhury, A.N., Mondal, R., Brahma, A., & Biswas, M.K. (2016a). Ecopsychosocial aspects of human–tiger conflict: An ethnographic study of tiger widows of sundarban delta, India. *Environmental Health Insights*. https://doi.org/10.4137/EHi.s24899.

Dasgupta, D. (2018). Organic farming gathers steam in the Sundarbans. Available from https://vikalpsangam.org/article/organic-rice-sundarbans/

Dasgupta, S., Wheeler, D., Sobhan, M.I., Bandyopadhyay, S., Nishat, A., & Paul, T. (2020). *Coping with Climate Change in the Sundarbans Lessons from Multidisciplinary Studies*. Washington DC: World Bank. https://doi.org/10.1596/978-1-4648-1587-4.

Dastagir, M.R. (2015). Modeling recent climate change induced extreme events in Bangladesh: A review. *Weather and Climate Extremes*, *7*, 49–60.

Datta, D., Chattopadhyay, R.N., & Deb, S. (2011). Prospective livelihood opportunities from the mangroves of the sunderbans, India. *Research Journal of Environmental Science*, *5*(6), 536–543. https://doi.org/10.3923/rjes.2011.536.543.

Datta, R., & Mete, J. (2022). The socio-economic status of females of Sundarban regions in West Bengal. *Journal of Applied Development Economics, 1*, 33–41.

Dhara, K.C. (2021). Empowerment of women farmers in sundarban. Unpublished Doctoral Thesis, West Bengal University of Animal and Fishery Sciences, Kolkata, India.

Ellis, F., & Allison, E.H. (2004). Livelihood diversification and natural resource areas. LSP Working Paper, Food and Agriculture Organization of the United Nations.

Folasade, O. (2016). The role of women in natural resources management. *Continental Jounal of Social Sciences*, *9*(1), 26–36.

Ghosh, A. (2012). *Living with Changing Climate–Impact, Vulnerability and Adaptation Challenges in Indian Sundarbans.* New Delhi: Centre for Science and Environment.

Ghosh, P. (2014). Subsistence and biodiversity conservation in the sundarban biosphere reserve, West Bengal India. Theses and Dissertations-Geography.26. https://uknowledge.uky.edu/geography_etds/26.

Ghosh, S., & Mistri, B. (2023). Cyclone induced coastal vulnerability, livelihood challenges and mitigation measures of Matla–Bidya interestuarine area, Indian Sundarban. *Natural Hazards.* https://doi.org/10.1007/s11069-023-05840-2.

Goodrich, C.G., Mehta, M., & Bisht, S. (2017). Status of gender, vulnerabilities and adaptation to climate change in the Hindu Kush Himalaya: Impacts and implications for livelihoods, and sustainable mountain development. Kathmandu: ICIMOD

Gopal, B., & Chauhan, M. (2006). Biodiversity and its conservation in the Sundarban mangrove ecosystem. *Aquatic Sciences*, *68*(3), 338–354. https://doi.org/10.1007/s00027-006-0868-8.

Hazra, S., Basu, O., Samanta, S., Das, I., Das, S., & Chanda, A. (2021). Regenerating mangrove biodiversity in the sundarbans biosphere reserve India – An assessment of scope and present interventions for ecological mangrove restoration in Sundarban Biosphere Reserve India. Caritus, India.

Hazra, S., Ghosh, T., DasGupta, R., & Sen, G. (2002). Sea level and associated changes in the Sundarbans. *Science and Culture*, *68*(9/12), 309–321.

Hazra, S., Samanta, K., Mukhopadhyay, A., & Akhand, A. (2010). *Temporal Change Detection (2001–2008) Study of Sundarban.* Kolkata: School of Oceanographic Studies, Jadavpur University.

ISDR. (2008). Climate change and disaster risk reduction. Geneva: Retrieved September 12, 2018, from https://www.wmo.int/pages /dra/vcp/docum ents/7607_Climate-Change-DRR.pdf.

Jain, A.K., Meiyappan, P., Song, Y., & House, J.I. (2013). CO_2 emissions from land-use change affected more by nitrogen cycle, than by the choice of land-cover data. *Global Change Biology*, *19*(9), 2893–2906.

Jalais, A. (2010). *Forest of Tigers: People, Politics and Environment in the Sundarbans.* New Delhi: Routledge.

Jamal, S., Ghosh, A., Hazarika, R., & Sen, A. (2022). Livelihood, conflict and tourism: An assessment of livelihood impact in Sundarbans, West Bengal. *International Journal of Geoheritage & Parks*, *10*, 383–399. https://doi.org/10.1016/j.ijgeop.2022.07.004

Jensen, D., Halle, S., Boyer, G., Stork, A., Specker, L., & Travis, C. (2013). *Women and Natural Resources: Unlocking the Peacebuilding Potential.* UNEP, UN Women, PBSO and UNDP. ISBN: 978-92-807-3361-7

Mahadevia, G.K., & Vikas, M. (2012). Climate change—Impact on the Sundarbans, a case study. *International Scientific Journal: Environmental Science*, *2*(1), 7–15.

Md, A., Gomes, C., Dias, J.M., &Cerdà, A. (2022). Exploring gender and climate change nexus, and empowering women in the south western coastal region of Bangladesh for adaptation and mitigation. *Climate, 10* (172). https://doi.org/10.3390/cli10110172.

Mistri, A. (2013). Migration and sustainable livelihoods: A study from sundarban biosphere reserve. ISAPS. *Asia Pacific Journal of Social Sciences*, *2*, 76–102.

Mistri, A. (2019). Is the migration from Indian Sundarban an environmental migration? Investigating through Sustainable Livelihood Approach (SLA). *Asian Profile*, *47*(3), 195–219.

Mistri, A., & Das, B. (2020). *Environmental Change, Livelihood Issues and Migration, Sundarban Biosphere Reserve, India.* Springer Nature Singapore Pte Ltd. https://doi.org/10.1007/978-981-13-8735-7.

Mondal, M., Biswas, A., Haldar, S., Mandal, S., Mandal, P., Bhattacharya, S., & Paul, S. (2022). Climate change, multi-hazards and society: An empirical study on the coastal community of Indian Sundarban. *Natural Hazards Research*, 2, 84–96. https://doi.org/10.1016/j.nhres.2022.04.002.

Mukherjee, N. (2023). Apprehension of climate change and its consequences on human society in southern part of the Indian Sundarbans region, West Bengal, Unpublished Doctoral Thesis, The University of Burdwan.

Nagasha, J.I., Ocaido, M., & Kaase-Bwanga, E. (2019). Theoretical and conceptual framework for gender analysis of attitudes and adaptation mechanisms to climate change for sustainable livelihoods in Uganda. *Journal of African Studies and Development*, *11*(4), 51–58.

Nandargi, S.S., & Barman, K. (2018). Evaluation of climate change impact on rainfall variation in West Bengal. *Acta Scientific Agriculture*, *2*(7), 74–82.

Neal, R.A., & Phillips, I.D. (2011). Winter daily precipitation variability over Cumbria, Northwest England. *Theoretical and Applied Climatology*, *106*, 245–262.

Patel, V., & Rajagopalan, R. (2009). *Fishing Community Issues in the Sundarban Tiger Reserve: A Case Study*. Chennai: International Collective in Support of Fishworker. Available from https://core.ac.uk/download/pdf/11017826.pdf.

Roy, C., & Guha, I. (2017). Economics of climate change in the Indian Sundarbans. *Global Business Review*, *18*(2), 493–508.

Sahana, M., Rehman, S., Ahmed, R., & Sajjad, H. (2020). Analyzing climate variability and its effects in Sundarban Biosphere Reserve, India: reaffirmation from local communities. *Environment, Development & Sustainability*. https://doi.org/10.1007/s10668-020-00682-5.

Sardar, A. (2022). Empowering women through alternative livelihoods in the coastal areas of the Sundarbans, west Bengal, India. IJCRT, *10*(7), 6–15.

Sen, A., & Pattanaik, S. (2017). How can traditional livelihoods find a place in contemporary conservation politics debates in India? Understanding community perspectives in Sundarban, West Bengal. *Journal of Political Ecology*, *24*(1), 861–880. https://doi.org/10.2458/v24i1.20971.

Tripathy, B., & Paul, A.K. (2024). A study on problems of fishing communities in SundarbanRegion of India. *Alochana Journal*, *13*(2), 128–132.

Uniyal, B., Uniyal, V.P., & Bhatt, V.K. (2007). Empowering rural women for sustainable livelihoods through natural resource management in Garhwal Himalaya, Uttarakhand, India. https://www.researchgate.net/publication/228706378.

WHO. (2014). Gender, climate change and health. Retrieved on February 12, 2005 from https://iris.who.int/bitstream/handle/10665/144781/9789241508186_eng.pdf.

4 Climate-Induced Hazards and Agrarian Distress among Marginal Farmers

An Evidence-Based Study

Sk Ajim Ali

INTRODUCTION

Climate change is a significant global environmental challenge being faced by humanity and is emerging as a potent threat to agriculture-based livelihoods (Das and Ansari, 2021). Any change in the climate over time, whether brought on by human action or natural variability, is referred to as climate change (Salinger, 2005). Rising temperatures, altered rainfall patterns, and a rise in the frequency of weather events like droughts, floods, and cyclones are all signs of climate change's effects being felt globally (Rawat et al., 2024; Singh et al., 2023). Because of its anticipated effects on food, water, the natural ecology, and human health, climate change is not just an environmental problem but also a major worry for countries worldwide (Abbass et al., 2022). Human life; property; infrastructure; productive systems; and broader social, economic, and ecological systems might all be lost or damaged as a result of climate change (Loucks, 2021; Patel et al., 2021). The agricultural industry in underdeveloped nations is predicted to be more severely impacted by climate change (Raihan, 2023). Although India's economy is developing throughout several geographical areas, it is the world's most vulnerable region to the effects of climate change (Yaduvanshi et al., 2021). Over half of India's population resides in rural regions, where people mostly rely on agriculture, fishing, and other industries that are susceptible to climate change for their livelihoods. Rural India is extremely susceptible because 70% of the population lives in rural areas and heavily depends on climate-dependent agricultural activities for their livelihoods (Das et al., 2024; Ghosh and Ghosal, 2021).

According to FAO (2019), about 26% of the economic effects of climate-related disasters in developing nations are borne by the agriculture sector. Climate change is predicted to exacerbate these losses by increasing the frequency and severity of weather events (such as unpredictable rainfall and rising temperatures), dispersing pests and diseases, deteriorating ecosystems, and increasing water scarcity, which will ultimately result in forced migration and food and livelihood insecurity (Codur, 2021). Because they depend largely on agriculture for their survival and means of subsistence and lack the resources and ability to cope with the negative effects of climate change, small and subsistence farmers are especially vulnerable to these effects (Verma and Sudan, 2021). In some states of India, particularly West Bengal, the average minimum and maximum temperature range increased 0.1°C–0.9°C across the state (Bal and Kirchner, 2023). Average rainfall has decreased, and monsoons have also been delayed; as a result, the effects of climate change are being felt largely in the agriculture sector (Yang et al., 2021).

Agrarian distress, a persistent issue in India's rural landscape, has increasingly become intertwined with the adverse impacts of climate-induced hazards (Matthan, 2023). In recent years, the intersection between climatic variability and the precarious livelihoods of marginal farmers has garnered critical scholarly and policy attention. In the state of West Bengal, particularly in

DOI: 10.1201/9781003620747-5

the alluvial plains of Bardhaman district, climatic fluctuations such as erratic rainfall, unseasonal storms, and extended dry spells have posed growing challenges to the sustainability of smallholder agriculture (Chakraborty and Chakraborty, 2021). Memari-I block, located in the eastern part of Bardhaman, serves as a representative site to examine the multifaceted vulnerabilities of marginal farmers who constitute a substantial segment of the rural economy yet remain largely underrepresented in macro-level agricultural discourse.

The relationship between climate variability and agrarian distress has been examined across multiple domains – ranging from yield reduction and crop failure (Vesco et al., 2021) to market volatility and indebtedness (Kandikuppa and Gray, 2022), and from food insecurity (Dasgupta and Robinson, 2022) to out-migration and livelihood shifts (Thapa and Mishra, 2022). In the Indian context, numerous studies have highlighted how prolonged exposure to extreme weather events exacerbates existing socio-economic inequalities, disproportionately affecting those with limited access to land, capital, and institutional support (Datta and Behera, 2022; Matthan, 2023; Ray et al., 2021). Yet, despite the growing body of literature on climate-change impacts on agriculture, there remains a critical gap in localized, micro-level studies that foreground the lived experiences of marginal farmers in specific agroecological contexts. In West Bengal, where agriculture is predominantly smallholder-based, marginal farmers – defined as those owning less than one hectare of land – constitute over 80% of the farming population. These farmers are often excluded from institutional credit systems, face inadequate access to irrigation infrastructure, and are rarely covered under formal insurance mechanisms. While state-level studies have acknowledged the vulnerability of these farmers, very few have provided a granular understanding of how recurrent climate-induced shocks are reshaping agrarian livelihoods at the block or village level.

Studies indicate a rising trend of distress-driven occupational shifts (Dhara, 2021; Rajak, 2025), but these are often generalized across districts without probing into the complex agrarian ecology that governs the vulnerability profiles of marginal farmers in specific locales such as Memari-I. The block located in the lower Gangetic plain experiences a monsoon-driven agricultural calendar. However, recent climatic aberrations – such as delayed monsoons, erratic rainfall, and unexpected temperature increases – have destabilized sowing and harvesting cycles, particularly for paddy and potato, the staple crops of the region. These climatic irregularities are not merely environmental issues but have cascading socio-economic consequences, triggering cycles of crop loss, rising debt burdens, distress sale of assets, and eventual withdrawal from farming. What remains insufficiently explored is how these farmers navigate these crises – what coping strategies they employ, what institutional mechanisms support or fail them, and how their perceptions of climate risk influence agricultural decision-making. Moreover, the concept of agrarian distress itself requires contextual unpacking. Much of the mainstream policy discourse defines distress in economic terms – primarily focusing on income loss or rural indebtedness. However, as argued by Huang et al. (2021) and Rivera-Ferre et al. (2021), agrarian distress must be understood within a broader framework that includes the degradation of agrarian institutions, erosion of community resilience, and the disintegration of traditional knowledge systems. In this context, a study of the Memari-I block offers an opportunity to delve deeper into how climate-induced hazards interact with structural vulnerabilities in the rural economy to produce differentiated experiences of agrarian distress.

This study is significant for several reasons. First, it addresses the paucity of block-level empirical data on climate vulnerability and agrarian responses. Second, it foregrounds the agency of marginal farmers by documenting their strategies of adaptation, resilience, and resistance in the face of recurring climate shocks. Third, it seeks to inform local governance and policy by identifying specific institutional gaps that amplify agrarian precarity. In doing so, the study contributes to both academic and policy-oriented understandings of rural livelihoods under climate stress. The objectives of the study are fourfold. Firstly, it aims to document the nature and frequency of climate-induced hazards affecting agricultural practices in the Memari-I block over the last decade. Secondly, it seeks to assess the economic and social dimensions of agrarian distress among marginal farmers, including patterns of crop loss, income instability, indebtedness, and food insecurity. Lastly, the research aims to evaluate the role of local governance, extension services, and climate adaptation schemes in mitigating agrarian distress.

A distinctive feature of this study lies in its methodological orientation. By combining focus group discussions (FGDs) and participatory rural appraisal (PRA) techniques, the research adopts a mixed-methods approach that enables both quantitative assessment and qualitative insight. Furthermore, the study incorporates farmer-led narratives to understand perceptions of climate change, thereby aligning with the growing body of literature that advocates for the integration of indigenous and experiential knowledge into climate adaptation frameworks (Ensor and de Bruin, 2022; Staub and Clarkson, 2021). This focus on farmers' voices not only democratizes the knowledge production process but also reveals the nuanced and often-overlooked dimensions of vulnerability and resilience in agrarian communities. The novelty of the study also stems from its spatial and temporal specificity. While many climate-agriculture studies operate at national or regional scales, this research prioritizes the micro-context of a single administrative block to unpack the interplay of climatic and socio-economic stressors. By doing so, it avoids the pitfall of overgeneralization and offers grounded insights that are directly relevant for local planning and policy intervention. Additionally, the study's temporal focus on the last decade allows for an understanding of how vulnerability trajectories have evolved in response to both climatic trends and shifting policy regimes.

CLIMATE CHANGE AND AGRARIAN UNCERTAINTY – GLOBAL TO LOCAL CASE

Climate change has emerged as a defining challenge for global agricultural systems, introducing profound uncertainties that jeopardize food security, rural livelihoods, and socio-ecological stability. Across the world, rising temperatures, erratic precipitation patterns, increasing frequency of extreme weather events, and shifting agroecological zones are altering the very foundation on which agriculture depends (Roy et al., 2023). These climatic disturbances have not only impacted crop productivity and livestock systems but have also introduced a layer of unpredictability in farming decisions, intensifying vulnerability among smallholder and marginal farmers, particularly in the Global South (Jones-Garcia and Krishna, 2021). Globally, the effects of climate change on agriculture are unevenly distributed. While temperate regions may benefit marginally from longer growing seasons in some areas, tropical and subtropical regions – home to a significant portion of the world's smallholder farmers – are disproportionately burdened. Sub-Saharan Africa, South Asia, and parts of Latin America have reported significant yield losses in climate-sensitive crops such as maize, rice, and wheat due to temperature extremes and rainfall variability (Omotoso et al., 2023). For instance, studies by the Intergovernmental Panel on Climate Change (IPCC, 2021) warn that global warming beyond 1.5°C could drastically reduce the productivity of rain-fed agriculture, on which more than 60% of farmers in these regions depend. This uncertainty compels farmers to make cultivation decisions with limited information, further compounding the risks associated with farming in vulnerable landscapes (Jagadale and Shaikh, 2025; Khatri et al., 2024).

In India, the manifestation of climate-induced agrarian uncertainty is particularly acute. As an agrarian economy with over half of its workforce directly or indirectly dependent on agriculture, India stands at the crossroads of climate vulnerability and agricultural sustainability (Singh et al., 2021). The country has already begun to experience the implications of climate change in the form of increased frequency of droughts, unseasonal rainfall, and extreme heatwaves (Charak et al., 2024). The kharif and rabi seasons – traditionally anchored in predictable monsoon cycles – are increasingly misaligned with changing rainfall patterns, leading to lower sowing and harvesting rates, pest outbreaks, and unpredictable market outcomes. For example, in recent years, states like Maharashtra, Karnataka, and Telangana have faced recurring droughts that not only decimated crop yields but also triggered a wave of agrarian distress, manifested in rising farmer suicides and migration (Deshpande, 2022). Agrarian uncertainty in India is not merely a product of climatic variables but is compounded by structural issues such as land fragmentation, lack of irrigation, inadequate crop insurance, and policy inconsistencies. These factors intersect with climate risks to create a deeply unstable agrarian environment, particularly for marginal and small farmers who constitute more than 85% of the agricultural population (Verma and Sudan, 2021). While state and central governments have launched adaptation initiatives such as the Pradhan Mantri Fasal Bima Yojana (PMFBY) and National Mission for Sustainable Agriculture (NMSA), the effectiveness of these interventions remains limited by gaps in implementation, awareness, and accessibility.

Within India, the state of West Bengal presents a unique case of climate-agriculture interaction. Located in the eastern part of the Indo-Gangetic plain and the lower Gangetic delta, the region is characterized by high climatic variability due to its geographic position and monsoon dependency. The state's agrarian system is largely smallholder-based and heavily reliant on seasonal rainfall, making it particularly vulnerable to climate shocks. In the last 5–10 years, districts such as Purba Bardhaman, Hooghly, and both Medinipurs have experienced increased frequency of untimely rains, rising temperature, and prolonged dry spells, all of which have disrupted traditional paddy and potato cycles and reduced productivity. For example, unseasonal rains during the flowering stage of paddy crops have been increasingly reported in Bardhaman and Hooghly districts, leading to substantial crop losses and heightened financial insecurity for marginal farmers. Furthermore, the coastal and deltaic regions of West Bengal, particularly the Sundarbans, face compound risks from rising sea levels, salinity intrusion, and tropical cyclones. These areas witness frequent crop failures and declining soil fertility, pushing communities toward alternative or distress-driven livelihoods. The recent cyclones Amphan (2020) and Yaas (2021) inflicted heavy damage on standing crops, irrigation structures, and rural infrastructure, further amplifying the uncertainty surrounding agricultural livelihoods (Biswas and Sen, 2024).

In blocks like Memari-I, which are neither coastal nor upland but situated in fertile alluvial zones, the challenge lies in the unpredictability of rainfall and temperature that disrupts multiple cropping cycles. Traditional knowledge systems, which once provided cues for weather forecasting and farming rhythms, are being rendered inadequate in the face of rapidly shifting climatic baselines. Marginal farmers, in particular, face compounded vulnerabilities due to small landholdings, debt cycles, and dependence on monocropping. As they grapple with the consequences of agrarian uncertainty, loss of income, food insecurity, increased reliance on non-farm employment, or even migration; their stories underline the urgent need for a localized understanding of climate-agriculture dynamics.

RESEARCH DATA AND METHODOLOGY

CASE STUDY

This study is based on cases from different gram panchayats and villages of the Memari-I block of Purba Bardhaman district, West Bengal. Deep observation on changes in rainfall and temperature patterns and their impacts on crops were recorded from 2015 to 2025 to understand how marginalized farmers faced losses and damages. Memari-I subdivision has a population of 2,18,425, out of which the urban population is 4,420 while the rural population is 2,14,005, consisting of 111 villages under 10 Gram panchayats. There are about 51,148 houses in the sub-district, including 1,072 urban houses and 50,076 rural houses. When it comes to literacy, 66.54% of the population of the Memari-I subdivision is literate, out of which 72.62% are males and 60.29% are females.

The area as a whole under Aman paddy constituted 64.32% of the total area under paddy cultivation, while the area under Boro and Aus paddy constituted 32.87% and 2.81%, respectively. The expansion of Boro paddy cultivation, with higher yield rates, was the result of the expansion of the irrigation system and intensive cropping. As per the District Statistical Handbook, Memari-I Block produced 3,342 tons of Aman paddy, the main winter crop, from 1,263 hectares, 6,682 tons of Aus paddy (summer crop) from 2,534 hectares, 13,255 tons of Boro paddy (spring crop) from 3,576 hectares, 1,063 tons of wheat from 397 hectares, 358,206 tons of potatoes from 10,836 hectares and

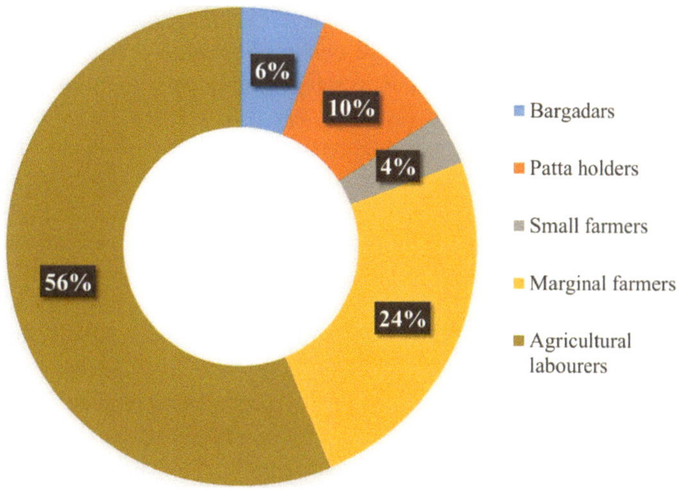

FIGURE 4.1 People engaged in agriculture and share of agriculture participation.

1,192 tons of sugarcane from 19 hectares. It also produced pulses and oilseeds (District Statistical Handbook Burdwan, 2014). About 80% of people are marginalized farmers and engaged in agricultural labor (Figure 4.1).

Data Collection

This investigation relies exclusively on primary evidence gathered over a decade-long field engagement designed to capture both the episodic shocks and the slow-burn stresses of climate change on marginal farming livelihoods. Three complementary methods – direct observation, FGDs, and PRA – were deployed in an iterative, overlapping manner to build a richly triangulated dataset. Continuous, immersive observation began in March 2015 and ran through March 2025, with the researcher spending extended periods each agricultural season in selected villages, recording phenological changes, sowing and harvesting calendars, visible crop stress, market arrivals, and household adjustments to weather aberrations. These longitudinal field notes produced a 10-year seasonal chronology of climate irregularities and their agronomic and socio-economic reverberations. To probe collective experience and shared interpretations, five FGDs were convened with marginal farmers (landholdings < 1 ha) in different places of the block timed to coincide with post-harvest windows when participants could reflect on recent seasons. Each FGD, averaging 12–15 participants balanced by gender and age, explored perceptions of rainfall volatility, temperature extremes, pest incursions, and the cascading effects on yield, debt, and household food security; sessions were audio-recorded, transcribed, and thematically coded. PRA techniques – seasonal calendars, impact ranking matrices, hazard mapping, and livelihood timelines – were integrated into both the observation cycles and the FGDs to elicit farmer-generated visual data, validate researcher observations, and foreground local knowledge of climatic differentiation. Combining these methods over ten years yielded a multi-layered corpus of field diaries, FGD transcripts, PRA charts, and photographic evidence that together underpin the study's analysis of how recurring climate-induced hazards translate into escalating agrarian distress among marginal farmers.

Ethical Consideration

Ethical integrity guided every stage of the research. In each village, community leaders were approached first to explain the study aims; only after their endorsement were individual participants invited. Informed consent was obtained verbally and, where literacy allowed, in writing using plain-language versions translated into the local language (Bengali) and reiterating that participation was entirely voluntary, could be paused or withdrawn at any time, and would have no bearing on access to government schemes or extension services. To protect confidentiality, all personal identifiers (e.g., name, gender, age, and caste) were removed from field notes and transcripts; unique codes replaced names, and any potentially identifying contextual details were aggregated or masked in publications. Audio recordings and PRA visuals were stored on encrypted drives accessible only to the author who had signed nondisclosure agreements. During FGDs, ground rules were established to discourage disclosure of others' private information, and participants were offered the option to speak privately with the researcher afterward if sensitive topics arose. Because discussions of crop losses and indebtedness can trigger distress, efforts were made to replace questions with encouragement. No cash incentives were provided. Findings were shared back with the communities in feedback meetings where preliminary results were presented in accessible formats (infographics and village-level climate calendars), allowing participants to verify interpretations and suggest revisions. Finally, all photographs used for dissemination were taken only with explicit permission, and no personal photos visualizing the faces of farmers were used. The study sought to uphold the principles of respect, beneficence, and justice while amplifying the voices of marginal farmers in climate-change discourse.

FINDINGS

From 2015 to 2025, systematic field observation was used to track how climate variability deepens distress among marginal farmers in the region. In 2025, this was supplemented by a survey of 150 respondents who completed a structured, choice-based questionnaire covering socio-economic status, perceptions of climate change, and the challenges they face; the responses were coded numerically for quantitative analysis. Five FGDs held in different villages provided additional qualitative insight into climate impacts and agrarian distress. The key findings are discussed below.

Socio-Economic Profiles of Marginal Farmers

Sociodemographic characteristics of the participants in the Memari-I block are presented in Figure 4.2. The 150 respondents included in the study comprised 98 males (65.33%) and 52 females (34.67%). The distribution of age across the sampled locations indicates that 52.76% of the population is between 36 and 60 years old, while 32.00% of the respondents are between 21 and 35 years old, and interviewees above 61 years old constitute 09.33% of the sampled population. Marital status of the respondents indicates that most of the farmers are married, about 71.33%. It is also reported that most people get married between the age groups of 20 and 25. During the interview, 32.00% indicated that they had no formal education at all, 50.67% of the respondents said they had primary schooling, and 10.00% indicated that they had secondary education. Therefore, it can be said that lack of education, early marriage, and family responsibilities are major barriers to their way of living and quality of life. Majority of the farmers (52.00%) earn <5,000/- monthly, whereas 37.33% of respondents earn Rs. 5,000–10,000/- monthly and only 1.33% earn a monthly income >20,000/-. Nevertheless, an assessment of the occupation of respondents indicates that 65.33% are permanent farmers, they have no other occupation. 28.00% are sessional farmers, which means they engaged in leased cultivation either in Rabi or Kharif or sessional vegetable cultivation. Only 6.67% are engaged in other types of activities including labor, fishing, and mason. From the above analysis, it may be concluded that the farmers community in this region is socioeconomically and educationally backward, which subsequently shapes their lives and livelihood patterns.

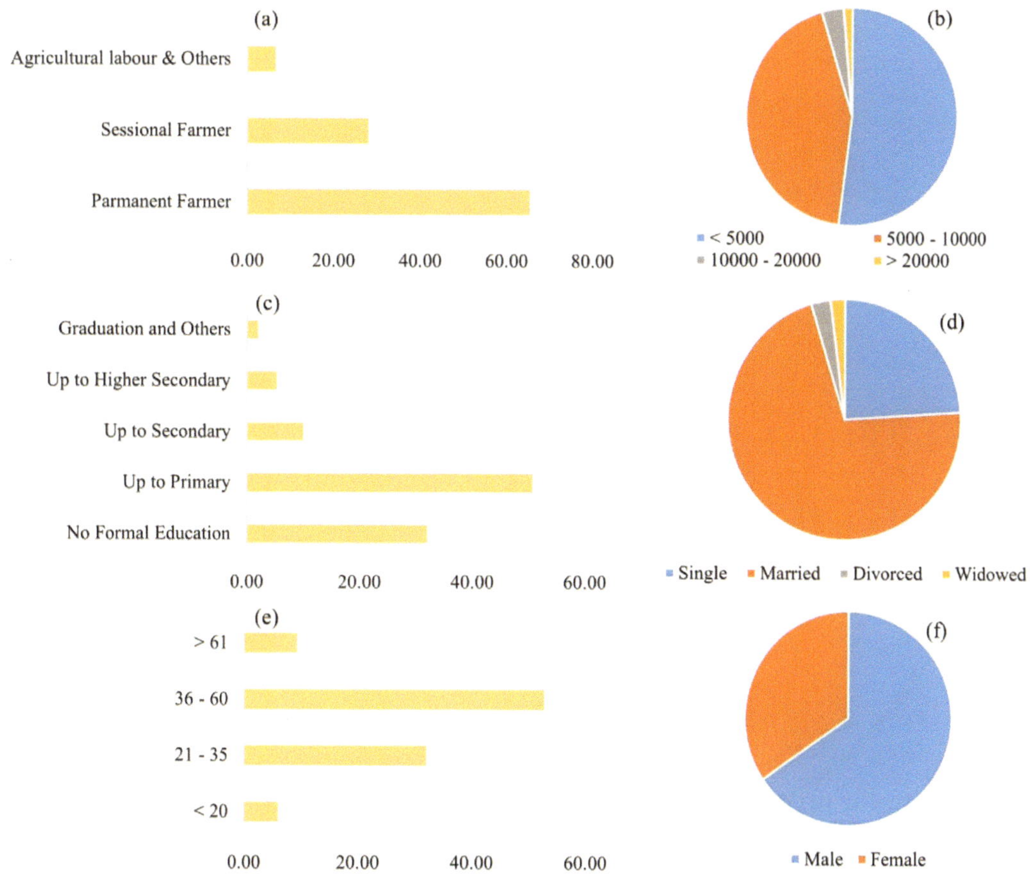

FIGURE 4.2 Socio-economic profiles of the farmers/respondents: (a) types of farmers, main occupations they engaged in; (b) monthly income of the farmers; (c) educational attainment; (d) marital status; (e) age structure of the respondents; and (f) gender consideration of the respondents.

Climatic Uncertainty

As per my perception based on longitudinal observation and FGDs, local-level climatic variation is highly observed in this area. Only between 2020 and 2024, devastating damages are seen due to unexpected rainfall and temperature increase. In normal conditions, for paddy, farmers manage two distinct ecotypes. Boro paddy, a high-yield winter-spring crop nourished by tube-well irrigation, starts with nursery sowing in late December or the first week of January. Transplanting into well-leveled, irrigated fields is carried out from late January through early February, and harvesting is finished by late April to the first half of May. Aman paddy, the backbone of household food security, begins with seedbed preparation and sowing from mid-June to early July, immediately after the onset of steady monsoon rains. Transplanting of 25- to 30-day-old seedlings peaks between mid-July and the first week of August, when puddled fields hold adequate standing water. Grain filling benefits from receding rainfall and lengthening sunshine in October, allowing harvest from early-November to mid-November. Potato, the district's chief cash crop, slots neatly into the cool-dry interlude that follows the Aman harvest. Certified seed tubers are planted from the first week of November through mid-November, once minimum nighttime temperatures drop below 20°C and residual soil moisture is still adequate for sprouting. Earthing-up and critical irrigation coincide with the crisp, fog-free mornings of December. Tuber bulking accelerates through January, and harvesting is undertaken from late February to early March, enabling timely field turnaround (Figure 4.3).

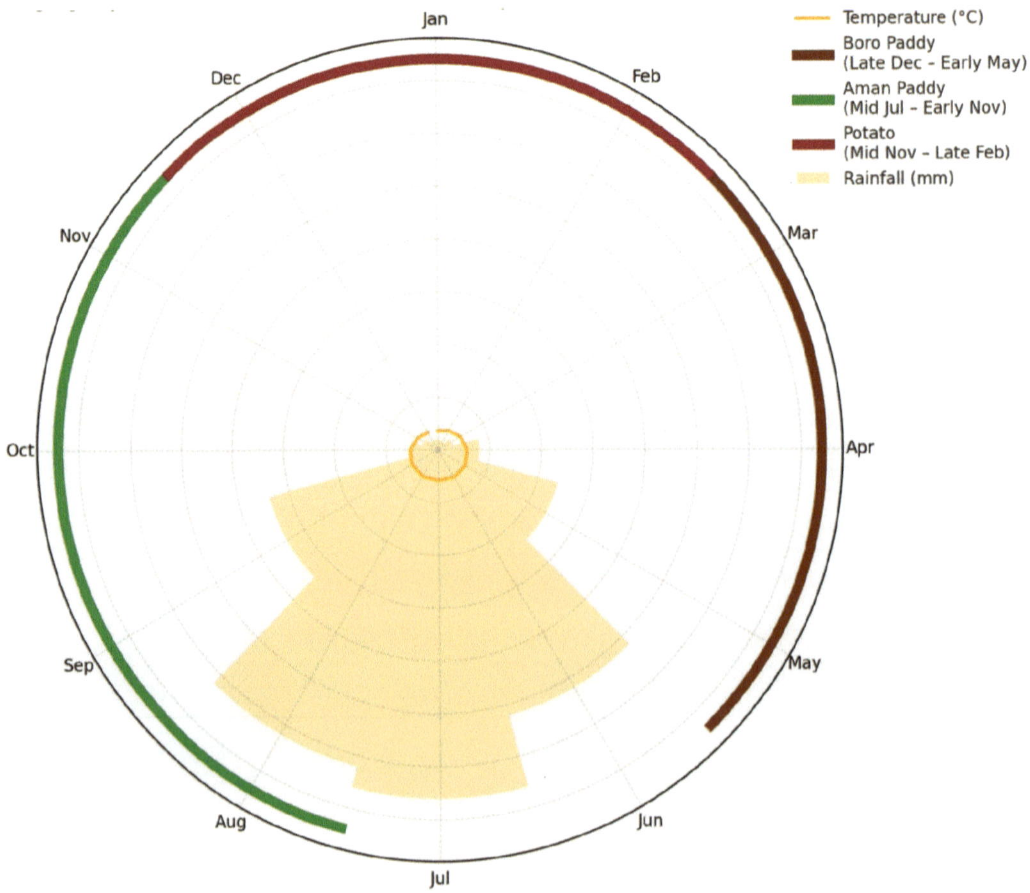

FIGURE 4.3 Climate and crop calendar in normal conditions as expected; showing average monthly rainfall (yellow bars), mean temperature (orange line), and the sowing-to-harvest windows for Boro paddy (brown ring), Aman paddy (green ring), and potato (red ring). It illustrates how crop calendars align with the normal monsoon-centered rains and the cooler, drier winter period.

However, farmers reported that unseasonal rainfall in November and December 2021 and again in early December 2023 devastated both paddy and potato crops. During the FGDs, farmers explained that these months are typically dry with clear skies, enabling them to harvest paddy and sow potato. The heavy downpours of 2021, however, ruined both crops. For marginal farmers, re-preparing a field for potatoes costs about ₹30,000 per bigha, and many could not afford to replant after such losses. Just two years later, the unexpected rains of December 2023 caused fresh havoc: thousands of hectares were inundated, many farmers lost their unharvested paddy, and extensive potato fields were destroyed (Figure 4.4).

Note: Yellow bars, with blue arrows marking the unseasonal rainfall in November and December, and an orange line, with a red arrow indicating the sudden March rise to ≈ 35°C, which normally did not occur before 20 years.

How Does Climate Change Cause Agrarian Distress?

Perception studies are inherently qualitative. In Memari-I, marginal farmers possess only limited knowledge of climate science, yet they are acutely aware of weather-related hazards. Drawing on respondents' accounts and local knowledge, it is evident that these communities have noticed

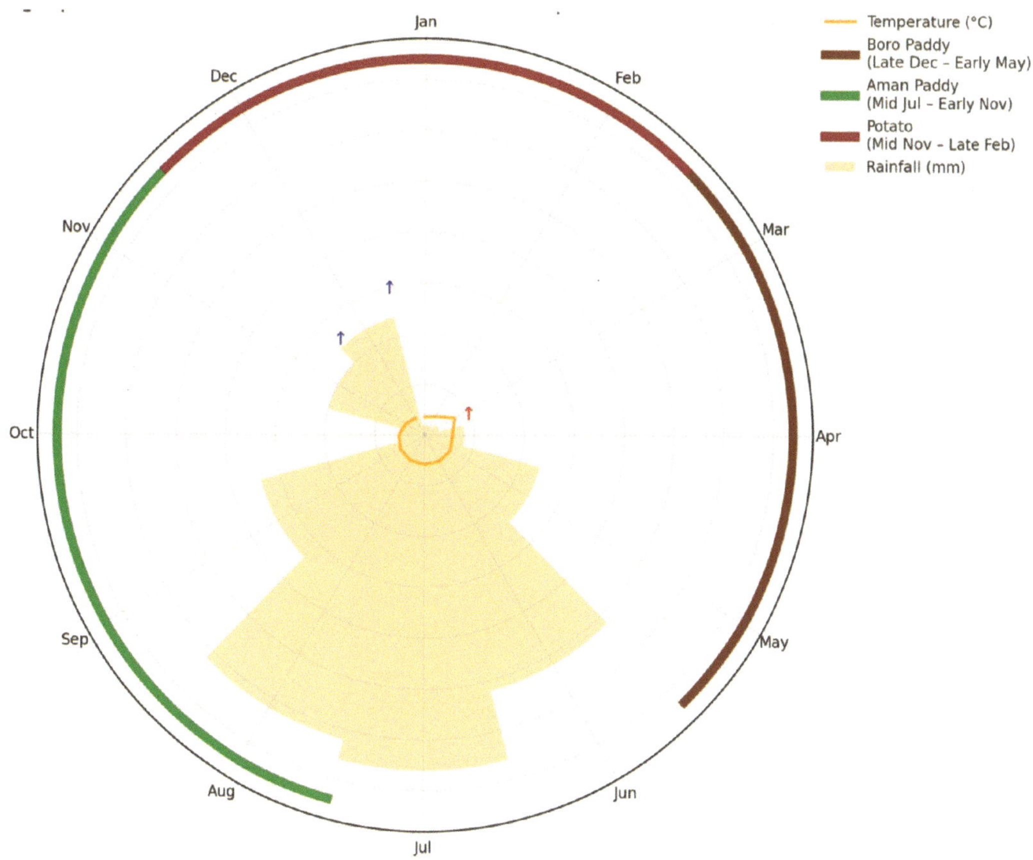

FIGURE 4.4 Anomalies and crop calendar in case of unexpected condition, illustrating the pathway from climate irregularities to agrarian distress. (Source: Author.)

marked shifts in climate over the past decade, even if they cannot name the underlying processes. Nearly 80% of farmers reported a sharp decline in rainfall during the critical paddy-sowing period, contrasted with an increase in unseasonal downpours from late November to early December. They have also observed notably higher temperatures in early March. Together, these shifts have repeatedly wrecked seasonal crops and inflicted heavy financial losses (Figure 4.5). During the FGDs, farmers voiced additional worries about groundwater. One cultivator from Keja village (Amadpur Gram Panchayat) spoke urgently about falling water tables. Because monsoon rains no longer supply adequate moisture, Aman paddy now depends on pumping from shallow wells and tubewells. Thus, altered rainfall and temperature patterns are driving not only agrarian distress but also serious pressure on environmental resources such as groundwater.

Farmers in these areas describe climate change not in scientific terms but through the sudden betrayals of a familiar sky. Their agricultural calendars, once guided by long-observed rhythms, have been upended by November and December downpours that arrive "like the monsoon has lost its way," as one grower put it. Paddy, ready for cutting and drying in the open sun, is drenched instead; stacked sheaves sprout on the threshing floor, and grain discolors, slashing both yield and market price. Potato land, freshly ridged and manured in anticipation of cool, rain-free nights, turns to slurry; seed tubers rot before they sprout, forcing farmers to rehire tractors, repay advance labor fees, and find another ₹30,000 per bigha that few can spare. When the rain retreats and replanting finally begins, a new shock follows. In early March, when potatoes need to be harvested in normal temperatures of 18°C–22°C weather, temperatures now leap above 30°C–33°C. The tubers stop

FIGURE 4.5 Unexpected rainfall during late November 2021 and 2023 caused damage to the paddy and potato plantations. (Source: Photos collected by the author during the field visit and observation.)

swelling, the haulms yellow weeks too soon, and yields tumble. Farmers speak of "boiled potatoes in the soil" (Figure 4.6). Insurance rarely covers these "out-of-season" losses because, on paper, March is still classed as winter. Creditors, however, remain punctual, and many households "fall back on costly informal loans or distress sales of cattle and jewellery." This is the ground reality reported during the FGDs in most of the villages settled by marginalized farmers. Asked about climate change, most cultivators confess only a hazy idea: "The seasons have shifted, that is all we know." Yet their ledgers tell the story precisely – rising input costs, shrinking harvests, and mounting debt tied to each bout of unseasonal rain or heat. What they experience as erratic weather translates into skipped school fees, postponed medical care, migration of younger relatives to labor markets, and the slow erosion of community seed banks and cooperative labor days. In everyday language, climate change is the moment the rain ruins a year's income overnight and the sun turns lethal before spring is done. Through these repeated shocks, agrarian distress ceases to be an abstract economic trend and becomes a lived reality etched onto each field, harvest, and household balance sheet in Memari-I.

ADAPTATION STRATEGIES OF THE MARGINALIZED FARMERS

It has been found that the adaptation strategies of marginalized farmers are very poor. They have the least knowledge about climatic uncertainty and no early warning system. Marginal farmers confront climate shocks with resourcefulness born of necessity, yet their adaptation repertoire remains narrow and largely unsupported. Few possess formal knowledge of climate change or probabilistic forecasts; decisions are still guided by observation of sky color, soil moisture, and elders'

FIGURE 4.6 Sudden temperature increase in early March 2025 caused huge damage to potatoes. (Source: Photos taken by the author during the field visit and observation.)

recollections of "normal" seasons. When unseasonal November rains drowned paddy windrows and potato ridges, most households responded with ad hoc measures. No respondent held an active crop-insurance policy. Government relief – typically a one-time cash transfer of ₹2,000–3,000 per affected bigha arrived months after the loss and covered only a fraction of replanting costs that often exceed ₹20,000 per bigha for potatoes. Lacking formal safety nets, farmers turned to informal ones: borrowing from relatives, pawning jewelry, or taking high-interest loans from local moneylenders. Community labor exchanges, once a robust fallback, have weakened as male migration rises, reducing the pool of able-bodied workers for urgent re-sowing.

Water management illustrates both ingenuity and limits. In years of deficient monsoon rainfall farmers deepen shallow tube wells or share electric pumps, but the fuel outlay erodes already thin margins and accelerates groundwater depletion, setting up longer-term vulnerability. Collective adaptations such as group purchase of solar pumps or coordinated crop diversification, remains rare, hampered by mistrust and the absence of facilitative institutions. Overall, the adaptation strategies are reactive, labor-intensive, and chronically undercapitalized. Without accessible insurance, timely climate information, and better-targeted state support, marginal farmers will continue to shoulder disproportionate risk, exhausting their meager buffers in the face of each new climatic surprise.

DISCUSSION

This study set out to trace the pathways through which climate-induced hazards translate into agrarian distress among marginal farmers using a decade of direct observation, mixed-method fieldwork, and farmer testimony. The findings establish three interlocking patterns. First, weather anomalies once regarded as rare outliers heavy post-monsoon downpours, early-season heat spikes, and erratic onset of the kharif rains now occur with sufficient frequency to erode the reliability of local crop calendars. Second, the impact of those anomalies is amplified by structural conditions that define marginal farming: tiny, fragmented holdings; high input-to-output ratios; and limited access to formal credit, crop insurance, or irrigation infrastructure capable of buffering climatic shocks. Third, repeated climate shocks trigger cumulative socio-economic consequences that push households beyond simple year-to-year coping into a chronic state of indebtedness and livelihood precarity. In combination, these patterns demonstrate that "climate-induced hazards" cannot be treated as discrete environmental events; they are embedded in, and magnified by, the political economy of smallholder agriculture.

Rainfall analysis and farmer recall converge on a marked temporal shift: a discernible decline in early-monsoon precipitation (critical for Aman nurseries) and a counterintuitive increase in late-November and December downpours, which coincide with paddy harvest and potato planting. Long-term meteorological records corroborate the observational dataset, showing that post-monsoon rainfall events above 80 mm per week have more than doubled in frequency since 2020, farmers are becoming panicked with this unseasonal rain. The agronomic repercussions are immediate. Paddy that would ordinarily dry in farmyards now ferments on the stalk; tubers rot in waterlogged furrows; and the window for Boro nursery preparation narrows, forcing hurried operations that compromise seedling vigor. Each of these effects registers directly in cost–benefit calculations: farmers incur higher labor and machinery expenses to salvage crops or prepare land a second time but rarely recoup commensurate returns at market.

Crucially, the study highlights how marginal farmers' vulnerability is not merely biophysical but institutional. Government schemes such as Pradhan Mantri Fasal Bima Yojana (PMFBY) remain under-enrolled in the block: fewer than 20% of the respondents held an active policy in 2024. Reasons include premium costs perceived as high relative to income, cumbersome claims procedures, and disbelief that climate-related losses will be reimbursed promptly. Meanwhile, public irrigation commands cover less than 15% of cultivable land, leaving the vast majority of marginal farms dependent on private pumps or uncertain rainfall. The synergy of inadequate risk-transfer mechanisms and dwindling common-pool water resources ensures that a single freak storm or heatwave can erase an entire season's profit. Farmers' cognitive framing of these changes adds a further dimension. Although most respondents do not use the term "climate change," they articulate a lived phenomenology of shifting seasons: the monsoon is misplaced, winter ends too quickly, or the sky cannot be trusted. Such experiential knowledge is neither naïve nor insignificant; it guides decisions to stagger sowing dates. Yet adaptation is constrained by capital, tenure insecurity, and market signals that still reward water-intensive paddy over less risky alternatives. The local seed market, for instance, offers limited access to drought-tolerant cultivars, and extension advice often arrives after the planting window has closed. Consequently, incremental on-farm adjustments remain insufficient to neutralize systemic risk.

At the local level, repeated climate shocks have triggered what sociologists' term "distress cascade effects." Survey data show a doubling of high-interest informal borrowing between 2018 and 2024, and temporary migration to construction sites outside the area is also increasing. Focus-group testimonies reveal a gendered dimension: women absorb additional care burdens when men migrate, while also assuming greater responsibility for water management and livestock feeding, tasks made more onerous by groundwater depletion and fodder scarcity. These shifts compound the social costs of climate stress, eroding community reciprocity networks that once functioned as informal insurance against crop failure. The evidence also complicates any simple narrative of "farmer fatalism." Many respondents exhibit proactive experimentation, trialing staggered transplanting dates, intercropping potatoes with short-duration pulses, or pooling resources to purchase mobile pumps. Yet the success of these measures is highly uneven and contingent on local social capital. Villages with active self-help groups or producer cooperatives weathered the 2023 rainfall anomaly with fewer losses, suggesting that collective agency can mediate climatic risk when supportive institutional scaffolding is present. That scaffolding, however, remains patchy.

Policy implications thus extend beyond climate-proofing hardware such as embankments or check dams. They point to a need for integrated interventions that mesh timely climate information services, affordable credit, crop-insurance reform, and community-level water-governance mechanisms. For example, climate-indexed insurance products triggered by specific rainfall or temperature thresholds could shorten claim cycles and reduce moral-hazard disputes. Likewise, subsidizing solar pumps conditioned on adherence to community groundwater-use plans could align energy access with resource conservation. Crucially, pluralizing the crop portfolio through procurement guarantees for less water-intensive cereals or high-value vegetables would reduce the current over-reliance on rice-potato rotations that amplify climate exposure.

Methodologically, the study demonstrates the value of marrying long-horizon ethnographic observation with periodic quantitative surveys. The ten-year immersion captured incremental environmental change that might evade short-term studies, while structured questionnaires provided the numerical ballast needed to examine correlations between specific weather events and economic outcomes. Nevertheless, limitations persist. The rainfall and temperature series are based on a single block-level weather station; micro-climatic variation within the block is therefore under-represented. Future research would benefit from installing participatory weather sensors at the panchayat level to corroborate farmer-reported anomalies. Similarly, the study's focus on marginal farmers leaves out the experiences of sharecroppers and landless laborers, whose exposure to climate risk may differ yet intertwine with marginal cultivators' coping strategies. Addressing these challenges requires moving beyond reactive disaster relief toward anticipatory governance that integrates climate science with locally grounded knowledge, equitable resource access, and institutional support structures capable of turning seasonal uncertainty from a source of perpetual crisis into a manageable risk.

CONCLUSION

The evidence from Memari-I paints a stark portrait of how intensifying climate irregularities intersect with structural marginality to propel smallholders into chronic distress. Unseasonal downpours during harvest and planting windows, early-season heat spikes, and increasingly unreliable monsoon onset have recast farming from a calculated risk into a gamble few can afford. Marginal farmers, though highly perceptive of these shifts, lack the institutional buffers, affordable credit, timely extension, and dependable irrigation – that might convert awareness into resilience. As a result, each climatic shock triggers a cascade of costs: crop loss, debt accumulation, asset liquidation, male migration, and the erosion of communal support systems. Over the past decade, these shocks have ceased to be aberrations; they now define the production environment. The study highlights that those technical fixes alone new varieties, pumps, or embankments will not suffice. What is required is an integrated adaptation architecture that couples localized climate-information services with index-based insurance, expands procurement incentives for less water-intensive crops, and democratizes control over groundwater. Strengthening women's self-help groups and farmer cooperatives emerges as equally vital, because collective action proved the most effective informal cushion against recent anomalies. Finally, research and policy must treat marginal farmers not as passive victims but as knowledge holders whose experiential insights can refine early-warning thresholds and adaptation planning. Climate-induced hazards have become a structural driver of agrarian precarity in Memari-I. Reversing this trajectory demands policies that are as systemic and sustained as the risks themselves policies that embed climate resilience within the everyday economics of smallholder agriculture and restore a measure of predictability, dignity, and hope to rural livelihoods.

REFERENCES

Abbass, K., Qasim, M. Z., Song, H., Murshed, M., Mahmood, H., & Younis, I. (2022). A review of the global climate change impacts, adaptation, and sustainable mitigation measures. *Environmental Science and Pollution Research*, 29(28), 42539–42559.

Bal, S., & Kirchner, I. (2023). Future changes in thermal bioclimate conditions over West Bengal, India, based on a climate model. *Atmosphere*, 14(3), 505.

Biswas, D., & Sen, A. (2024). An analysis of local political forces around the management of climate change-led disasters in the Indian Sundarbans. *Asian Journal of Social Science*, 52(2), 63–75.

Chakraborty, S. K., & Chakraborty, S. K. (2021). Land-use changes: floodplains, dams, and reservoirs–integrated river basins management. In S. K. Chakraborty (Ed.), *Riverine Ecology Volume 2: Biodiversity Conservation, Conflicts and Resolution* (pp. 531–607). Springer Nature, Print media in Cham, Switzerland. https://doi.org/10.1007/978-3-030-53941-2_6

Charak, A., Ravi, K., & Verma, A. (2024). Review of various climate change exacerbated natural hazards in India and consequential socioeconomic vulnerabilities. *IDRiM Journal*, 13(2), 142–177.

Codur, A. M. (2021). Agriculture, food, and environment. In T. Tietenberg & L. Lewis (Eds.), *Environmental and Natural Resource Economics* (pp. 467–500). Routledge.

Das, U., & Ansari, M. A. (2021). The nexus of climate change, sustainable agriculture and farm livelihood: Contextualizing climate smart agriculture. *Climate Research*, 84, 23–40.

Das, U., Ansari, M. A., & Ghosh, S. (2024). Measures of livelihoods and their effect on vulnerability of farmers to climate change: Evidence from coastal and non-coastal regions in India. *Environment, Development and Sustainability*, 26(2), 4801–4836.

Dasgupta, S., & Robinson, E. J. (2022). Attributing changes in food insecurity to a changing climate. *Scientific Reports*, 12(1), 4709.

Datta, P., & Behera, B. (2022). Climate change and Indian agriculture: A systematic review of farmers' perception, adaptation, and transformation. *Environmental Challenges*, 8, 100543.

Deshpande, R. S. (2022). Disaster management in India: Are we fully equipped? *Journal of Social and Economic Development*, 24(Suppl 1), 242–281.

Dhara, A. (2021). Quarter century of economic reforms: Transformation of the occupational structure in rural India. In S. N. Rajesh Raj & Komol Singha (Eds.), *The Routledge Handbook of Post-Reform Indian Economy* (pp. 498–511). Routledge India.

District Statistical Handbook Burdwan (2014). Department of Statistics and Programme Implementation, Government of West Bengal. Archived from the original on 10 March 2025. Accessed from https://purbabardhaman.nic.in/district-statistical-handbook/

Ensor, J., & de Bruin, A. (2022). The role of learning in farmer-led innovation. *Agricultural Systems*, 197, 103356.

FAO. (2019). Agriculture and climate change: Challenges and opportunities at the global and local level Collaboration on Climate-Smart Agriculture. Accessed https://openknowledge.fao.org/server/api/core/bitstreams/4b6ecf83-2641-4134-8af2-8eeb384178f2/content.

Ghosh, M., & Ghosal, S. (2021). Climate change vulnerability of rural households in flood-prone areas of Himalayan foothills, West Bengal, India. *Environment, Development and Sustainability*, 23, 2570–2595.

Huang, S. M., Che-Hao, H., & Chang, Y. H. (2021). The paradox of cultivating community resiliency: Re-agrarianisation and De-peasantisation of indigenous farmers in Taiwan. *Journal of Rural Studies*, 83, 96–105.

Jagadale, S. R., & Shaikh, J. M. (2025). Governmentality and marketing system failure: The case of stubble burning and climate change in neoliberal India. *Journal of Macromarketing*, 02761467251318608.

Jones-Garcia, E., & Krishna, V. V. (2021). Farmer adoption of sustainable intensification technologies in the maize systems of the Global South. A review. *Agronomy for Sustainable Development*, 41(1), 8.

Kandikuppa, S., & Gray, C. (2022). Climate change and household debt in rural India. *Climatic Change*, 173(3), 20.

Khatri, P., Kumar, P., Shakya, K. S., Kirlas, M. C., & Tiwari, K. K. (2024). Understanding the intertwined nature of rising multiple risks in modern agriculture and food system. *Environment, Development and Sustainability*, 26(9), 24107–24150.

Loucks, D. P. (2021). Impacts of climate change on economies, ecosystems, energy, environments, and human equity: A systems perspective. In T. Letcher (Ed.), *The Impacts of Climate Change* (pp. 19–50). Elsevier.

Matthan, T. (2023). Beyond bad weather: Climates of uncertainty in rural India. In I. Scoones, S. M. Borras Jr., A. Baviskar, M. Edelman, N. L. Peluso, & W. Wolford (Eds.), *Climate Change and Critical Agrarian Studies* (pp. 164–185). Routledge.

Omotoso, A. B., Letsoalo, S., Olagunju, K. O., Tshwene, C. S., & Omotayo, A. O. (2023). Climate change and variability in sub-Saharan Africa: A systematic review of trends and impacts on agriculture. *Journal of Cleaner Production*, 414, 137487.

Patel, S., Dey, A., Singh, S. K., Singh, R., & Singh, H. P. (2021). Socio-economic impacts of climate change. In P. Kumar, R. K. Singh, M. Kumar, M. Rani, & P. Sharma (Eds.), *Climate Impacts on Sustainable Natural Resource Management* (pp. 237–267). Wiley, New York City, United States. https://doi.org/10.1002/9781119793403.ch12

Raihan, A. (2023). A review of the global climate change impacts, adaptation strategies, and mitigation options in the socio-economic and environmental sectors. *Journal of Environmental Science and Economics*, 2(3), 36–58.

Rajak, D. (2025). Pattern of occupational diversification among rural agricultural households: A field view. *Artha Journal of Social Sciences*, 24(1), 1–30.

Rawat, A., Kumar, D., & Khati, B. S. (2024). A review on climate change impacts, models, and its consequences on different sectors: A systematic approach. *Journal of Water and Climate Change*, 15(1), 104–126.

Ray, K., Giri, R. K., Ray, S. S., Dimri, A. P., & Rajeevan, M. (2021). An assessment of long-term changes in mortalities due to extreme weather events in India: A study of 50 years' data, 1970–2019. *Weather and Climate Extremes*, 32, 100315.

Rivera-Ferre, M. G., Di Masso, M., Vara, I., Cuellar, M., López-i-Gelats, F., Bhatta, G. D., & Gallar, D. (2021). Traditional agricultural knowledge in land management: The potential contributions of ethnographic research to climate change adaptation in India, Bangladesh, Nepal, and Pakistan. *Climate and Development*, 13(7), 644–661.

Roy, T., Kalambukattu, J. G., Biswas, S. S., & Kumar, S. (2023). Agro-climatic variability in climate change scenario: Adaptive approach and sustainability. In U. Chatterjee, A. O. Akanwa, S. K., S. K. Singh, & A. D. Roy (Eds.), *Ecological Footprints of Climate Change: Adaptive Approaches and Sustainability* (pp. 313–348). Cham: Springer International Publishing.

Salinger, M. J. (2005). Climate variability and change: past, present and future–an overview. *Climatic Change*, 70(1), 9–29.

Singh, A. P., Kumar, J., Chauhan, I., Saxena, P. K., Gaur, P. K., Sharma, S. K., et al. (2023). Role of climate change in disasters occurrences: Forecasting and management options. In A. Srivastav, A. Dubey, & M. A. Khan (Eds.), *Visualization Techniques for Climate Change with Machine Learning and Artificial Intelligence* (pp. 149–180). Elsevier.

Singh, N. P., Anand, B., Singh, S., Srivastava, S. K., Rao, C. S., Rao, K. V., & Bal, S. K. (2021). Synergies and trade-offs for climate-resilient agriculture in India: an agro-climatic zone assessment. *Climatic Change*, 164, 1–26.

Staub, C. G., & Clarkson, G. (2021). Farmer-led participatory extension leads Haitian farmers to anticipate climate-related risks and adjust livelihood strategies. *Journal of Rural Studies*, 81, 235–245.

Thapa, R., & Mishra, D. K. (2022). Agrarian distress and seasonal out-migration: Insights from a field survey in North Bengal. *Sociological Bulletin*, 71(3), 396–420.

Verma, S., & Sudan, F. K. (2021). Impact of climate change on marginal and small farmers' livelihood and their adaptation strategies-A review. *Regional Economic Development Research*, 2(2), 96–112. https://doi.org/10.37256/redr.222021896.

Vesco, P., Kovacic, M., Mistry, M., & Croicu, M. (2021). Climate variability, crop and conflict: Exploring the impacts of spatial concentration in agricultural production. *Journal of Peace Research*, 58(1), 98–113.

Yaduvanshi, A., Nkemelang, T., Bendapudi, R., & New, M. (2021). Temperature and rainfall extremes change under current and future global warming levels across Indian climate zones. *Weather and Climate Extremes*, 31, 100291.

Yang, B., Qin, C., Bräuning, A., Osborn, T. J., Trouet, V., Ljungqvist, F. C., et al., (2021). Long-term decrease in Asian monsoon rainfall and abrupt climate change events over the past 6,700 years. *Proceedings of the National Academy of Sciences*, 118(30), e2102007118.

5 Climate Change, Seasonality, and Livelihood Challenges in Ecotourism

A Case Study from Thua Thien Hue, Vietnam

Tran Thi Thu Hien

INTRODUCTION

Climate change is now widely recognized as one of the major threats to the ecological and human well-being globally. Over the past few years, growing climate variability and more frequent and harsh storms, uncertain rainfall, and long periods of drought have significantly impacted the sustainability of ecotourism (Dogru et al., 2019). Destinations that rely on outdoor nature-based tourism, such as waterfalls, rivers, and forests, are becoming more susceptible to these climatic interruptions (Becken, 2013).

Ecotourism is an increasingly global concept in the world today; it is regarded as an environmentally friendly way of generating income since it merges the economic and gainful aspect of tourism with environmental preservation. There have been many clear definitions of ecotourism. It was introduced in 1983 by Hector Ceballos-Lascurain. Ecotourism is a form of environmentally responsible travel and visitation to relatively undisturbed natural areas, in order to enjoy and appreciate nature that promotes conservation, has low visitor impact, and provides for beneficially active socio-economic involvement of local populations (Ceballos-Lascuráin, 1996). According to Honey (1999), ecotourism is "travel to fragile, pristine, and usually protected areas that strives to be low impact and usually small scale." The concept goes on to cover topics including growing respect for cultural variety, local communities' economic development, and travel education. A simpler definition, adopted from the International Ecotourism Society (TIES), says the practice is: "responsible travel to natural areas that conserves the environment, sustains the well-being of the local people and involves interpretation and education" (TIES, 2015). The definition of ecotourism by the Global Ecotourism Network (GEN): "Ecotourism is responsible travel to natural areas that conserves the environment and sustains the well-being of local people, and that creates knowledge and understanding through interpretation and education of all involved (visitors, staff and the visited)" (GEN, 2016). Fennell (1999) has also defined ecotourism as a nature-based tourism activity that is relatively sustainable (consumptive, scale, carrying capacity considerations) and essentially non-consumptive, especially in the third world, and which is "primarily experiences, focus, and management" (control, benefits, scale). It is usually found in outdoor settings and should be related to the protection or preservation of the natural setting. Although definitions may differ, ecotourism shares common characteristics, such as: a form of tourism that is based on nature and indigenous culture; its relatedness to awareness-raising and stimulation of the environmental responsibility; creating conditions for sustainable development of natural, cultural and social values; and possibility for local communities to take part in it and derive economic benefits from it.

DOI: 10.1201/9781003620747-6

Located in Central Vietnam, Thua Thien Hue, is home to various ecotourism areas, including national parks, waterfalls, rivers, and marine ecosystems. The abundance of natural resources and the diversity of these resources make Thua Thien Hue attractive to the development of ecotourism, making ecotourism activity one of the two general objectives that Thua Thien Hue will focus on investing in tourism activities in the conservation of the ecological environment in Thua Thien Hue. Under Resolution No. 03-NQ/TU-UBND of Thua Thien Hue Provincial People's Committee dated November 8, 2016 on the planning to develop Thua Thien Hue into a centrally run city on the foundation of heritage, culture, ecology, landscape, and environmental friendliness, ecotourism is among the forms of tourist attractions which are priority for development as part of its traditional tourism species. In particular, the plan puts stress on "tapping Thua Thien Hue's ecological potential by developing some precedent key tourism products such as community-based rural tourism, forest, lake and lagoon, and marine ecotourism." Moreover, taking advantage of the strengths of resort ecotourism and marine ecotourism in Chan May, Can Duong, Lang Co, Hai Van, the two regions of the South-East are the central area of spatial structuring of tourism activities. Bach Ma is a national tourist area, and the Lang Co–Canh Duong region is established as a national key integrated management area.

However, Thua Thien Hue province is also one of Vietnam's most disaster-prone areas. This area has frequently suffered from natural hazards (Ministry of Natural Resources and Environment, 2020). Typhoons, flooding, and other extreme weather occurrences, as well as rising temperatures and rainfall patterns, pose serious threats to the province's ecotourism industry. It's not only that declining biodiversity and shrinking natural areas of interest make travel less attractive, but also affect the lives and livelihoods of people who rely on ecotourism. The combination of its rich ecotourism resources and a high level of climate vulnerability shows this province is a valuable place to examine the intersections of climate change, seasonality, and livelihood challenges.

Traditionally, the tourism season in Thua Thien Hue aligns with the dry months when streams and waterfalls are most accessible and visually appealing. However, with the shifting onset of rainy seasons with unexpected storms, the predictability of the tourist season has been compromised (Butler, 2001). Tourists are now more cautious when choosing destinations prone to extreme weather, which leads to fluctuating visitor numbers and reduced income for local businesses (Amelung et al., 2007). Seasonality under the influence of climate change not only reduces the activities of tourism-related businesses but also increases operational costs due to the need for maintenance, repairs, and infrastructure (Jopp et al., 2010). Waterfalls and streams suffer from both water scarcity during extended dry periods and dangerous flooding during intensified rainy seasons, deterring tourist visits and decreasing revenue streams (Kaján & Saarinen, 2013). Consequently, local communities, many of whom rely primarily on ecotourism for their livelihoods, face mounting challenges in sustaining their income throughout the year (López-i-Gelats et al., 2016). The dependence on climate-sensitive sectors is both necessary and difficult (Ellis, 2000).

To establish a scientific foundation for proposing solutions that enhance adaptive capacity and ensure the livelihoods of people in areas of Thua Thien Hue province, this study analyses the perceptions of relevant stakeholders regarding the causes and impacts of climate change, as well as the existing response strategies in the ecotourism businesses of local communities.

In a few qualitative studies in this area, researchers started from a bottom-up approach that allows communities to share their perceptions and experiences regarding climate-related changes. The results demonstrate that active community participation and diversified livelihoods are crucial to enhancing resilience to climate change and seasonality. Macleod and Russell (2010) used the SLF (Sustainable Livelihoods Framework) in the Caribbean to assess the vulnerability of the local communities in the ecotourism sectors to climate variability. Their results show that community members deal with seasonal tourism shifts and suffer from the consequences of climate change, including longer rainy seasons. This study emphasizes the importance of diversification strategies in building resilience. Zhou and Ap (2017) conducted a participatory study in Southeast Asia to explore how communities are coping with the impacts of climate change and seasonality. They found that local communities with active participation in decision-making processes and those that diversified their economic activities were more resilient to the challenges posed by climate shifts and fluctuating tourist seasons.

Other works used statistical and econometric approaches to quantify the direct and indirect impacts of climate change/seasonality on the ecotourism industry. Becken et al. (2014) estimated the effects of temperature and precipitation variations on ecotourism in New Zealand and Australia. They polled tourism businesses and found that climate change is shortening the duration of tourism seasons, especially in places that rely on particular types of weather. And this change is leading to falling income from tourism and hardship for ecotourism businesses. Scott et al. (2008) analyzed long-term data to estimate the implications of snowfall variability for ski tourism in North America. Their study showed that as snowfall decreases due to rising temperatures, the ski season becomes shorter, leading to reduced visitor numbers and lower revenue for local communities. Zhou et al. (2017) applied a mixed-methods approach, combining surveys and econometric models to assess the economic impact of changing seasons on coastal ecotourism communities. Their results indicated that communities with diversified income sources are more likely to withstand the economic shocks from fluctuating tourist numbers during the off-season.

Although awareness of the effects of climate change on tourism is increasing around the world, there is still a lack of community-level research. Therefore, this study addresses an important gap by exploring the ecotourism experience of community-based tourists in Thua Thien Hue. This chapter aims to study the current status of ecotourism operations and management in ecotourism sites in the Thua Thien Hue province, identify the challenges confronting local communities engaged in ecotourism livelihoods, explore how communities respond and adapt to climate variability, and suggest practical adaptation strategies that could enhance the sustainability of ecotourism-based livelihoods in climate-sensitive areas.

STUDY AREA

Thua Thien Hue is a province in the center of Vietnam, with Quang Tri to the north, Da Nang at the south, the west is Lao with a boundary about the Truong Son mountain range, and the east is more than 128 km of coastline (Figure 5.1). With an area of around 5,048 km^2, 70% of its land area is mountainous and forestland. As of 2024, the population of Thua Thien Hue was greater than 1.23 million.

FIGURE 5.1 Map of Thua Thien Hue province, Vietnam. (Source: Hien & Chu, 2024.)

Tropical monsoon climate prevails in the province with distinct wet seasons during September to December and dry season during March to August. Its variety of geographical terrain, consisting of littoral plains, lagoons, mountain ranges, rivers, and falls, facilitates the development of ecotourism. The area is home to various ethnic minority communities such as the Ta Oi, Co Tu, and Pa Co, with their rich cultural traditions associated with natural resources.

Nevertheless, the geographical and topographical features of Thua Hue make this province very sensitive to climate hazards. Climate shifts predict that the temperature may increase by 1.4°C by 2050 and 2.6°C by 2100 on average in the province. Simultaneously, the annual precipitation will also increase 3.7% in 2050 and 6.8% in 2100, with higher rainfall magnitude in the wet season and reduced amount in the dry season. Thua Thien Hue is becoming more vulnerable to natural disasters, including typhoons, floods, landslides, and droughts. Over the last 20 years, they have been increasing by about 15% and there are significant variations in seasonal rain patterns (MONRE, 2020). These changes have had profound implications on the socio-economics of local communities, especially where nature-based livelihoods, including farming, fishing, and ecotourism, have traditionally been practiced.

This study was conducted in 39 typical ecotourism sites in Thua Thien Hue province, including garden houses, national parks, mangroves, lagoon ecosystems, streams, and waterfalls (Table 5.1). Figure 5.2 shows some of the typical ecotourism sites in the province.

DATABASE AND METHODOLOGY

The secondary data includes data about climate patterns, seasonality, and tourism flows in Thua Thien Hue were collected from the Department of Tourism of Thua Thien Hue Province, and reports on the socio-economic situation of Thua Thien Hue province. The primary data collection focused on community involvement and perceptions of residents engaged in ecotourism activities. Primary data sources include:

- **Household Surveys:** A total of 150 households engaged in ecotourism were surveyed to gather information on the livelihoods of local communities using semi-structured interviews between January and August 2022. Households were randomly selected from the list of farm households provided by the commune people's committee in Thua Thien Hue province. Based on the study framework, semi-structured interviews are conducted with local people to explore in-depth information. The data collected includes information on livelihood challenges experienced by local communities (income, environmental and climate risks, skills and capacity, access to resources and support, participation in management and decision-making), community responses, and adaptive strategies to climate variability.
- **Interviews with Local Authorities and Travel Agencies:** In-depth interviews were conducted with 20 representatives from local authorities and 30 travel agencies to understand the institutional framework and challenges faced in ecotourism management. Secondary data on climate patterns, seasonality, and tourism flows in Thua Thien Hue were collected from the Department of Tourism of Thua Thien Hue Province.
- **Field Observations:** On-site observations were made to assess the condition of ecotourism infrastructure, natural attractions, and local biodiversity.

The research uses a mixed-method approach to examine the connections among seasonality, infrastructure, human resources, and livelihood dimensions in ecotourism. Both approaches complement each other and bring a more complete perspective to the problem, which will hopefully contribute to the development of adaptation policies in ecotourism destinations. Statistical analysis was computed using SPSS version 20 software: Descriptive statistics, frequency, percent, and mean were also applied.

TABLE 5.1
Management Models of Tourist Sites in Thua Thien Hue Province

No	Name of Area	Name of Ecotourism Site	Type of Resource	Management Model
1	Hue city	Hue Ecolodge	Garden house	Enterprise
		An Hien	Garden house	Individual
		Thuy Bieu Eco-village	Garden house	Individual
		Xuan Vien Tieu Cung	Garden house	Individual
		Phu Mong	Garden house	Individual
		Phuoc Tich	Garden house	Individual
		Hue Riverside Boutique	Garden house	Enterprise
2	Phu Loc district	Bach Ma	National park	Government
		Bach Ma village	Stream and waterfall	Enterprise
		Nhi Ho	Stream and waterfall	Cooperative
		Khe Su	Stream and waterfall	Enterprise
		Suoi Voi	Stream and waterfall	Enterprise
		Bo Ghe	Stream and waterfall	Community Management Board
		Mo	Stream and waterfall	Cooperative
		Tam Thac Do	Stream and waterfall	Community Management Board
		Ho Truoi	Stream and waterfall	Cooperative
		Tien	Stream and waterfall	Cooperative
		Mo–Hoi Mit	Stream and waterfall	Cooperative
		Khe Su	Stream and waterfall	Enterprise
		Cay Sen	Stream and waterfall	Unregulated operation
3	A Luoi district	A Nor	Stream and waterfall	Cooperative
		Parle	Stream and waterfall	Cooperative
		Can Te	Stream and waterfall	Unregulated operation
		A Lin	Stream and waterfall	Community Management Board
		A Rum	Stream and waterfall	Unregulated operation
4	Nam Dong district	YesHue Eco	Stream and waterfall	Enterprise
		Kazan	Stream and waterfall	Unregulated operation
		Phuon	Stream and waterfall	Unregulated operation
5	Huong Tra town	Ru Cha	Mangrove forest	Government
		Khe Day	Stream and waterfall	Community Management Board
		Mau	Stream and waterfall	Unregulated operation
		Khe Lanh	Stream and waterfall	Unregulated operation
		Khe Hung	Stream and waterfall	Unregulated operation
6	Huong Thuy town	Chin Chang	Stream and waterfall	Enterprise
		Da Dam	Stream and waterfall	Unregulated operation
7	Phong Dien district	A Don	Stream and waterfall	Community Management Board
		O Lau	Stream and waterfall	Community Management Board
8	Phu Vang district	Dam Chuon	Lagoon ecosystem	Unregulated operation
9	Quang Dien district	Tam Giang	Lagoon ecosystem	Unregulated operation

Source: Department of Tourism, Thua Thien Hue province, 2022.

FIGURE 5.2 Some of the typical ecotourism sites in Thua Thien Hue province. (a) An Hien Garden House. (b) Bach Ma National Park. (c) A Nor Waterfall. (d) Ru Cha Mangrove Forest. (e) Bach Ma Village. (f) Tam Thac Do waterfall. (g) Dam Chuon lagoon. (h) Mo Waterfall. (YesHue Eco Area.)

RESULTS

Current Ecotourism Tour Operations in Thua Thien Hue Province

Survey results from 30 travel agencies in Thua Thien Hue province indicate that only 33.3% of the surveyed units are engaged in operating ecotourism tours. Figure 5.3 shows the barriers to the travel development of ecotourism tours.

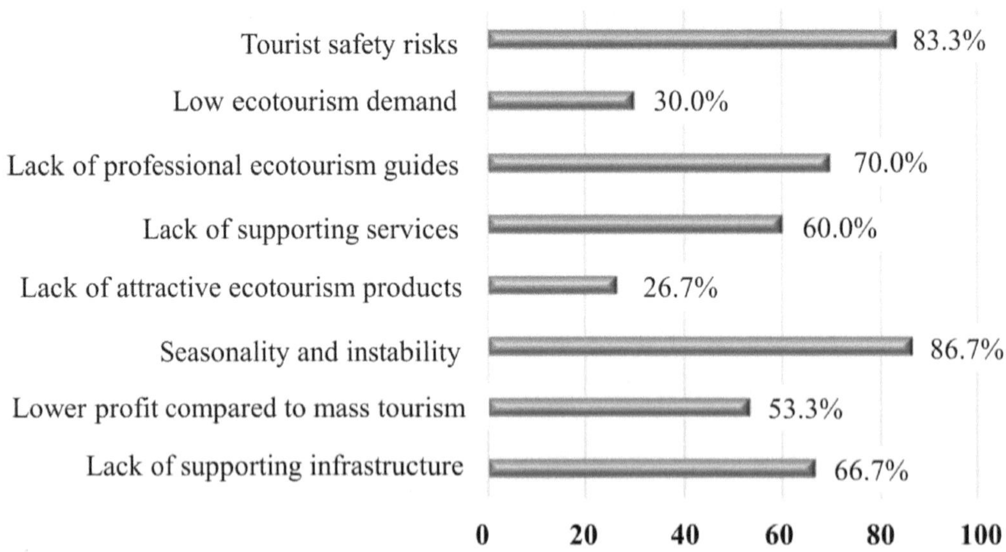

FIGURE 5.3 Barriers to travel companies' development of ecotourism tours. (Source: Survey results, 2022.)

The findings of the survey highlight a number of significant obstacles that restrict the growth of ecotourism among Thua Thien Hue tour companies. The survey results reflect many constraints that undermine the development of ecotourism in companies, local tours in Thua Thien Hue. Seasonality and unpredictability were named as the top two main points, with 86.7% of businesses flagging them up as barriers. Eighty-three percent of operators were worried about tourists' safety in the event of landslides or floods. Seventy percent reported that they did not have access to adequate ecotourism guides, which identified a human resource deficit. Other most often cited types of problems were lack of supporting infrastructure (66.7% of the companies) and the fact that ecotourism is less attractive for companies due to its lower profit comparative to the mass tourist trips (63.3% of the companies). Absence of supportive facilities (60%), such as food and refuge, represented additional important barriers. Low demand for ecotourism was noted by 30% of enterprises, and 26.7 % of companies reported the absence of attractive ecotourism products. This demonstrates that ecotourism is promising for development in Hue, but the majority of businesses are not willing to do so because of many limitations, including weak infrastructure, seasonality, insecurity, lack of service systems, and professional staff. These challenges will need to be dealt with in order to develop viable ecotourism in Thua Thien Hue province.

In Thua Thien Hue province, the ecotourism brand of the sea and Tam Giang–Cau Hai lagoon has been forming. Particularly, experiential tours "A Day on Tam Giang Lagoon" and "Morning in Thuy Bieu, Afternoon in Tam Giang" have overwhelmingly appealed to tourists from home and abroad. Furthermore, tourism at folk cultural festivals has become more and more professionalized, such as the Cau Ngu Festival, Sinh village wrestling, Thu Le wrestling, and regatta. It is also a part of the biennial Hue Cultural Festival including various themed programs such as "Thuan An -The sea calling," "Tam Giang – Huong Tra and the waves of water," and "Lang Co – The sea legend."

Coastal and lagoon fishing villages leave strong impressions on visitors, such as the Ngư Mỹ Thạnh mural village, Ru Cha mangrove forest, Dam Chuon lagoon, and Son Cha Island Sea area. Community-based tourism and ecotourism linked with traditional craft villages are also developing in the lagoon region, such as the Thuy Lap bamboo weaving village, along with connections to other craft villages outside the lagoon area, like the ancient village of Phuoc Tich and the Bao La bamboo weaving village.

The survey results also show that although Thua Thien Hue has many beautiful ecotourism sites, especially streams, waterfalls, and beaches, they are rarely included in ecotourism tours. Most tours focus on well-known destinations such as Lang Co, Truoi lake, Truc Lam Bach Ma Zen monastery, Bach Ma national park, Tam Giang—Cau Hai lagoon, Ru Cha mangrove forest, Voi stream, Mo waterfall, and A Nor waterfall. Despite the abundance of natural landscapes such as lagoons, coastal ecosystems, and traditional villages, especially streams, waterfalls, many of these resources have yet to be integrated into well-structured tourism products.

THE CURRENT ECOTOURISM MANAGEMENT IN THUA THIEN HUE PROVINCE

Traditional garden houses, streams and waterfalls, national parks, and lagoon–mangrove ecosystems are just a few of the natural and cultural resources that serve as the foundation for Thua Thien Hue ecotourism destinations. Table 5.1 illustrates how each resource type is linked to a particular management model, indicating the degree of participation in tourism development by private companies, governmental agencies, and local communities.

The most common management style for garden house tourism is individual ownership. This implies that small-scale, spontaneous tourism businesses are still operated by the majority of local households. More professional services are provided because only a few well-known establishments, such as Hue Ecolodge and Hue Riverside Boutique, are operated by private businesses. Individual ownership is the most popular management style for garden house tourism. This suggests that most local households still run small-scale, impromptu tourism businesses. The fact that only a small number of well-known locations—like Hue Ecolodge and Hue Riverside Boutique—are run by private companies results in more expert services.

On the other hand, sites that rely on streams, waterfalls, and forests present a rich diversity of management models (private enterprises, cooperatives, community management boards, government initiatives (e.g., national parks)). Of these, cooperatives and private businesses are important in formalizing services as well as sustaining livelihoods locally. However, a significant number of sites still lack official management models—such as Can Te stream, A Rum stream, Kazan waterfall, Phuon waterfall, Mau stream, Khe Lanh waterfall, Khe Hung waterfall, Da Dam waterfall, Dam Chuon lagoon, and Tam Giang lagoon. These sites operate informally, which may pose risks to environmental sustainability and tourist safety.

Sites with unique ecosystems like lagoons and mangrove forests also often lack clear management structures. This reflects a lack of sufficient attention despite their high ecological value. Therefore, these unregulated sites should be considered in future planning and development strategies.

THE LIVELIHOOD CHALLENGES EXPERIENCED BY LOCAL COMMUNITIES ENGAGED IN ECOTOURISM

Seasonal Income

Despite showing great promise, particularly in regions surrounded by rivers, lakes, streams, and waterfalls, Thua Thien Hue's ecotourism industry is still more susceptible than other types of travel because of its heavy reliance on seasonal variations and natural conditions. According to an analysis of 150 local households, 83.3% of ecotourism businesses are seasonal, with the busiest months being March through August. Just 16.7% of them continued to operate all year long. The livelihoods of workers and households involved in tourism are significantly impacted by this strong seasonality, especially in areas where ecotourism is the main or only source of income.

In terms of income, Figure 5.4 reflects generally modest earnings. Approximately 52.2% of employees earned between VND 3–5 million per month, 18.9% received VND 5–7 million, and only 15.6% earned over VND 7 million. Notably, 13.3% of workers earned less than VND 3 million per month. These limited income levels present significant barriers to livelihood sustainability and may undermine long-term local engagement in ecotourism development.

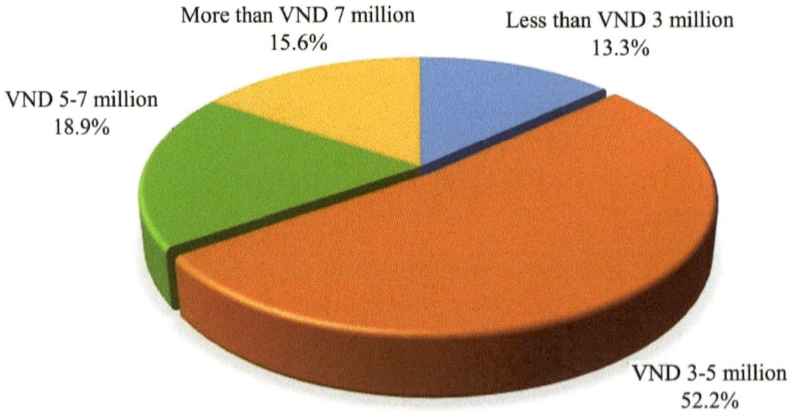

FIGURE 5.4 Seasonal income per month of local people surveyed in ecotourism sites. (Source: Survey results, 2022.)

Environmental and Climate Risks

Residents say they fear floods and storm damage most. Landslides have also hit some provincial ecotourism sites and garden houses, some neck-deep in water, in the province due to climate change-triggered extreme weather-related events such as heavy rains and rising sea level. Beaches have been eroded significantly, and some lodgings have been damaged by erosion. Serious submerged tides have caused some lagoons to bar ecotourism activities, as multiple mangrove ecotourism sites, such as Ru Cha or on the O Lau, can be wiped out.

Similarly, locations for ecotourism activities in the mountainous areas of Thua Thien Hue are much influenced by climate change, especially those located in the headwaters of streams and spring waterfalls. Challenging water conditions and sudden flash floods resulting from greater water volumes and unpredictable weather have posed significant risks to both visitors and local staff. For example, due to the risk of floods and landslides, even popular sites like Voi stream, Kazan waterfall, Khe Su stream, and A Rum stream were often temporarily closed during the rainy season. On certain occasions, tourism services are discontinued due to broken or washed away infrastructure (walkways, small bridges). Numerous access road sites leading to remote waterfalls and eco-sites are mediocre at best, and can be washed out by flood water and heavy rain. Known ecotourism attractions that are frequented by both domestic and international tourists are Bach Ma National Park and Elephant Spring, and Nhi Ho. However, the degradation of natural resources and unfavorable weather conditions brought on by climate change have had a significant impact on ecotourism activities at these sites in recent years, delaying and impeding travel.

In the dry season, these streams and waterfalls' water levels are considerably lowered by extended heat and less rainfall, which deters tourists. Due to dry conditions, locals reported a dramatic decline in the number of tourists visiting places like Phuon waterfall and Mau stream during the summer.

Animal behavior and biodiversity have changed as a result of changes in the natural environment. For instance, the ecosystems of the Tam Giang–Cau Hai wetland are at serious risk from rising sea levels and climate change, which are predicted to make lagoon waters more saline. The mangrove forest in Ru Cha and the distinctive bird habitat of O Lau are under threat. Numerous threatened species may be at increased risk of going extinct.

Skills and Capacity of the Local Community

Overall, the level of local community participation in tourism remains low. Most local residents rely on seasonal tourism-related activities, such as food and other products, including beaches, rivers, and waterfalls for swimming.

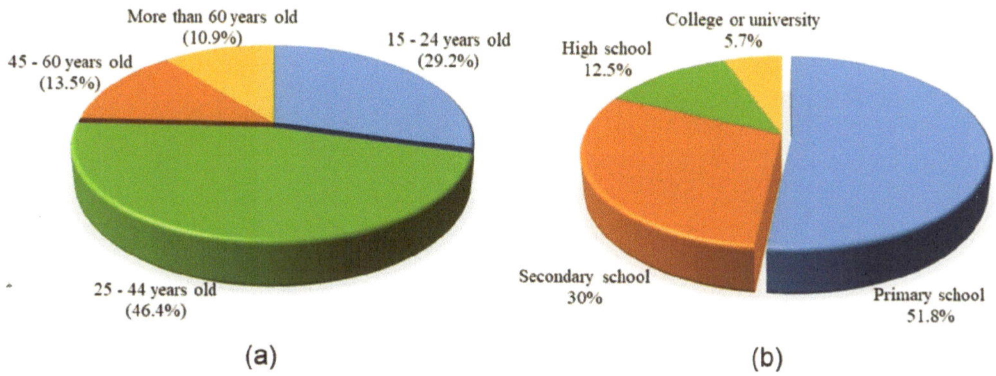

FIGURE 5.5 (a) Age and education background. (b) Local people surveyed.

Among 150 households surveyed, 850 people were employed in tourism, and 76.7% of the households had two to seven employees, reflecting the large number of small and home-based enterprises. The majority of women (56%) are relatively young, with 46.4% aged 25–44 years and 29.2% aged 15–24 years. Meanwhile, 13.5% are between 45 and 60 years old, and only 10.9% are 60 or older (Figure 5.5).

However, the overall skill level and educational background of the workforce remain limited. Over half of the people (51.8%) had completed only primary school, while just 5.7% held college or university degrees. This low level of education directly influences the ability of workers to access and benefit from professional training, adopt service standards, and effectively engage with tourism activities.

Local authorities and tourism departments have promoted training programs, but 22.2% of businesses had never taken any kind of training, according to the survey. Those that had joined had low membership, reducing the effectiveness of the programs. Another major obstacle to professional development is the prevalent low level of English language skills, greatly limiting communication with international visitors.

As local communities are the key actors in right-based tourism such as ecotourism and community-based tourism, the findings in this study stress the necessity of designing continued training programs to enhance knowledge, service skills, and professional attitude.

Access of the Local Community to Resources and Support

There are 76.7% of households reported that they face difficulties in accessing preferential loans to renovate their homes and develop infrastructure to serve tourists. Accommodation at ecotourism sites is highly attractive to tourists seeking authentic experiences. The survey results indicate that one of the main issues faced by households involved in ecotourism is the lack of financial resources to build accommodation facilities that meet tourism standards. The existing accommodation facilities are also greatly affected during the rainy seasons, and they have to bear maintenance costs despite not generating income during the low season. Many households believe that, despite the potential for ecotourism development in the area, the cost of investing in infrastructure, particularly upgrading accommodation facilities, exceeds their financial capabilities.

At certain natural streams and waterfalls in Hue, overnight camping in tents is available. However, camping at these remote and undeveloped sites poses significant safety risks. Security and safety concerns are paramount, with 83% of surveyed travel agencies emphasizing their importance, given the potential risks to tourists' health and lives. Therefore, better management and stricter regulations are urgently needed.

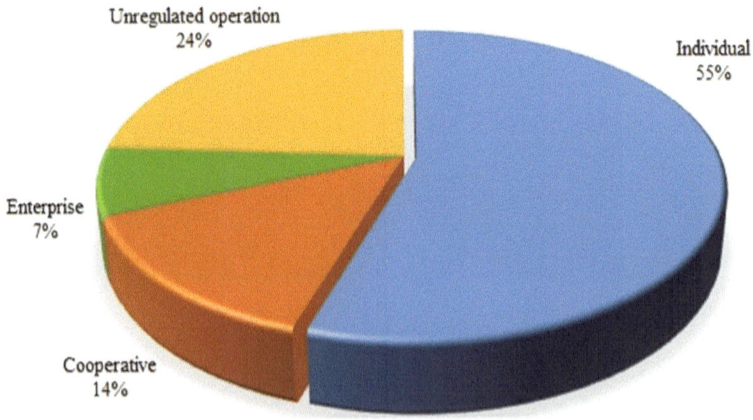

FIGURE 5.6 Management models adopted by surveyed households. (Source: Survey results, 2022.)

Participation of the Local Community in Management and Decision-Making

A survey of 150 households engaged in ecotourism in Thua Thien Hue province indicates that individual household businesses dominate the local management model, accounting for 55% of all cases. Informal, unregulated operations represent the second most common model at 24%, followed by cooperatives at 14%, and independent enterprises at just 7% (Figure 5.6).

Cooperative is an organization or business that is owned and operated by a group of individuals for their mutual benefit. In the ecotourism industry, a cooperative refers to local community members (e.g., households, workers, or farm holders) jointly managing a service rather than individually. Cooperatives help to strengthen bargaining power, minimize risk at the individual level, and promote sustainable, inclusive ecotourism development.

These figures reflect limited formal organization and suggest that local participation in management and decision-making processes remains weak. Nearly one-third of the surveyed households operate without formal management structures, and many report a lack of awareness or access to local tourism development plans. This limited engagement in governance and planning may increase the vulnerability of communities to the risks of uncoordinated tourism development.

ADAPTATION TO CLIMATE CHANGE BY LOCAL COMMUNITIES

According to the interviews, Thua Thien Hue ecotourism-related local communities have started implementing a number of coping mechanisms to deal with the effects of climate change. The most popular strategy, though, is to change the travel schedule to avoid the rainy season, particularly at locations with streams and waterfalls. Nowadays, a lot of households focus their tourism efforts on the dry season (March to August) and temporarily halt or scale back operations during storm-prone months. During the off-season, many households salvage iron implements and go back to traditional fishing or farming, that too is also increasingly affected by climate variability.

In some areas of Thua Thien Hue province, especially in lowland and flood-prone areas of the Tam Giang–Cau Hai Lagoon, people have long experience in adapting their lives and production to adapt to climate change. Local roads flatten in Quang Dien, Huong Tra, Phong Dien, and Phu Vang districts, a nearly everyone is equipped with a boat or boat to travel in rainy days. On the production side, lots of households have reformed the way they produce vegetables (using net trays), reintegrated mixed aquaculture, a well-suited process both for adaptation to the climate and for mitigating the impacts of the latter. In tourism, the households have not been able to develop effective responses and to ensure their income from tourism.

Local authorities in some districts have also intervened timely in the conservation of the Ru Cha forest with a "community model for coping with climate change." For example, the People's Committee of Huong Phong commune developed a mangrove forest development plan for the commune, with a total area of more than 30.8 hectares of land to be used for planting trees to combat climate change. The plan strictly forbids any forest destruction and calls for the participation of local residents. In order to improve the vegetation of the Ru Cha forest, the residents of Huong Phong commune, working with the Thua Thien Hue Environmental Protection Sub-Department, have planted more than 3,000 mangrove trees. The "Explore the Tam Giang–Cau Hai Lagoon System" tourism route includes the Ru Cha forest in Huong Phong.

Many adaptation strategies continue to operate unregulated and without technical assistance in spite of these efforts. There are very few available courses on early warning systems, climate resilience, and sustainable infrastructure. Another option would be to mainstream climate adaptation into tourism planning and strengthen cooperation with local authorities to enhance long-term resilience.

DISCUSSION

Due to existing characteristics of nature and culture, such as streams, waterfalls, and different living environments, ecotourism in Thua Thien Hue province has much potential. Results from the surveys of local tour operators, government, and communities show that the region has the opportunity to develop as an ecotourism destination; however, the sector struggles with problems such as seasonality, lack of infrastructure, and susceptibility to climate change. The discussion elaborates on the current situation of ecotourism in operation, the challenges facing local communities in terms of livelihoods, and the impacts of climate change on the Province's ecotourism sustainability. The issue of living sources for the livelihoods of the ecotourism-dependent population in Thua Thien Hue has been diversified.

Ecotourism in most ecotourism sites in Thua Thien Hue faces mounting challenges due to increased environmental instability and seasonal income variation. The impacts are seen as natural disasters such as storms, floods, landslides, flash floods, riverbank and coastal erosion, and forest fires have become more common and more frequent due to climate change. Ecotourism is one of the types of tourism most at risk. This seasonal variation can be made worse by climate change, limiting prospects for sustainable development, and lowering the dependability of tourism revenue. Moreover, the degradation of natural attractions, such as streams and waterfalls, during extreme weather events diminishes the appeal of these sites to tourists. This phenomenon has been documented in various ecotourism destinations, where climate-induced environmental changes directly impact tourist satisfaction and visitation rates (Nyaupane et al., 2004). Hien and Chu (2024) indicated that positive influences, such as vegetation, accessibility, restaurant, and hotel variables, demonstrate remarkable contributions to this ecotourism potential, with vegetation playing the most important role.

The ecotourism sites of Thua Thien Hue are more susceptible to the impact of the climate than some other destinations because of the insufficient public infrastructure there. Underdeveloped public transportation limits access to distant natural attractions such as streams, waterfalls, and access roads constructed in an unstable manner can be damaged by heavy rain or flooding. Infrastructure damaged by the impact of natural hazards poses a noticeable challenge to the travel industry sector. Both visitors and residents are at risk from inadequate infrastructure and safety precautions. Many locations lack official management, which encourages uncontrolled tourism and raises the risk of environmental damage. To support long-term development, enforce safety regulations, and upgrade infrastructure, coordinated efforts are required.

The community's ability to diversify revenue streams and enhance service quality is further constrained by human capital constraints, such as low educational attainment, a lack of training in sustainable tourism practices, and a lack of understanding of climate adaptation tactics (Adger, 2003; Jones & Boyd, 2011). Thua Thien Hue's local communities still have limited influence over the decisions and management of ecotourism. Long-term success and sustainability of ecotourism depend on these communities being empowered through inclusive governance frameworks and capacity-building programs. Research indicates that when local communities actively participate in the planning and management of tourism, they demonstrate a greater sense of ownership and commitment to environmental preservation (An & Dinh, 2023). Education programs that feature tourism management, environmentalism, and foreign languages can significantly enhance local engagement. A variety of research demonstrates the potential contribution ecotourism can make through its localized capacity-building through education, skills training, and language development (Ashley, 2000; Black & Crabtree, 2007).

In addition, short high seasons and irregular visitors flows discourages investors from developing tourism-related facilities such as accommodations, and local tour operators in the vicinity because it is what prevents them from making a return (Belle & Bramwell, 2005). A lot of small parties do not have enough capital reserves or credit lines to invest in or repair assets damaged by climate effects, which leads to a downward spiral of decay and of being priced out of the market (Simpson et al., 2008). It is necessary to have financial support tools, such as relief funds, subsidized credit, or community savings groups, which are designed to protect the local community. To this end, enhancing the resilience of the ecotourism economy and the supply of essential services to which local livelihoods are tied requires strengthening access to financial resources.

Climate change adaptation is most important for the future of ecotourism given these vulnerable conditions, notably in Thua Thien Hue province. Nonetheless, research suggests adaptation at the local level is limited and often tends to be avoidance and passive rather than proactive and strategic. In terms of organized disaster preparedness planning and insurance mechanisms, communities are left very vulnerable to extreme weather-related events, which magnify post-disaster recovery constraints (Becken & Hay, 2007; Moreno & Becken, 2009). It is essential to strengthen the community's resilience and adaptive capacity to climate change.

To build local adaptation capacity to the effects of climate change, it is important to promote community and resident engagement in the decision-making process, creating awareness on climate change among local communities and tourism managers, and business enterprises are crucial to create synergy on environmental protection and for disaster risk reduction. This is by far the most important aspect, highlighting the need to promote awareness and empowerment among the local communities for sustainable action on climate issues, especially with the fact that their level of information is still very low. There is a need to develop awareness among local authorities and the community of climate change issues, and to facilitate the integration of natural resource management at the local level, through the training of service providers and beneficiaries. We must also document and use the local wisdom for sustainable deployment and resource use.

Beyond these specific instances, it becomes clear that sustainable ecotourism development is not attained through natural attractions alone. It needs the holistic development of facilities for accommodation, governance, transport, and committed human capacity. Policymakers and tourism planners must adopt a holistic approach that addresses these interconnected dimensions simultaneously (Honey, 2008; Wearing & Neil, 2009).

CONCLUSION

In summary, despite an obvious existing potential for the development of ecotourism, to materialize this potential will require sustainable multi-sectoral facilitation to overcome the existing structural limitations of ecotourism development in Thua Thien Hue province. Although this study has documented several weaknesses of the livelihood impacts of local communities involving in ecotourism such as seasonal income, environmental and climate risks, the skill and ability of the local community and the accessibility of local community in resource and support, local community retention in management and decision-making in the process of developing ecotourism in Thua Thien Hue province. These restrictions limit the sustainable development of ecotourism in the area.

The study aimed to investigate factors influencing ecotourism as the success of environmental and socio-economic conditions. The findings of the study can contribute to the debate by showing that seasonality, climate variability, and constrained investments and institutional arrangements are important deterrents of a robust ecotourism sector. Reliable as tourism may be as a seasonal livelihood, it becomes slightly less reliable when a natural disaster can come around sporadically. In addition, lack of infrastructure, from poor roads to insufficient public transportation, makes the challenge even greater for visitors and local service providers. These results support the call for strong policy actions and integrated planning to respond to the multiple levels of vulnerability of ecotourism sites.

The study underscores the challenge for residents, especially those in rural regions, in accessing benefits from ecotourism without appropriate support. Low levels of education, limited exposure to sustainable tourism training, and low capacity to adapt to the impacts of climate change hinder communities from diversifying their sources of income and improving the quality of services. Their exclusion from tourism planning and control inhibits creativity and tends to sow dependence. Intentional policy responses are required to mitigate these issues, including financial mechanisms that buttress community-based operators, inclusive styles of governance, and capacity-building efforts. The research highlights the importance of developing urgently effective integrated management that builds local skills. In order to ensure that the development of tourism is, in fact, beneficial for local communities, the engagement of communities in decision-making as well as in service delivery is necessary.

The implication for this study is that local governments and national agencies should offer training programs, put in place a funding or credit support system for rural tourism businesses, and facilitate a platform on which local community members could express their concerns and suggestions. Additionally, the encouragement of community inclusion can boost the environmental protection and the financial sustainability of ecotourism areas.

Notwithstanding these contributions, this research is not void of limitations. First, they concentrate on Thua Thien Hue province, which may result in restricted representativeness for other locations with diverse ecological and socio-political features. Second is that it is predominantly an adaptation response at the community scale, and it does not delve into the roles of government agencies or intersectoral partnerships, which also shape the success of ecotourism.

In the future, it would be fruitful to have comparisons between provinces or countries, which would contribute to knowledge and the delegation of best practices. We also believe that it would be interesting to investigate other digital tools (e.g., virtual tours or remote sensing technology for environmental monitoring) to discover possible other ways to supplement sustainable tourism policies.

Overall, this research has provided new knowledge and contributions to the emerging literature on ecotourism, climate resilience, and community development in vulnerable regions. Policymakers, government leaders, and people in the tourism business need to work together to turn these findings into policies that preserve natural capital, empower local people, and help the sector grow sustainably in the face of increasingly stressed environmental conditions.

REFERENCES

Adger, W. N. (2003). Social capital, collective action, and adaptation to climate change. *Economic Geography*, 79(4), 387–404.

Amelung, B., Nicholls, S., & Viner, D. (2007). Implications of global climate change for tourism flows and seasonality. *Journal of Travel Research,* 45(3), 285–296.

An, L.T. & Dinh, N.C. (2023). Residents' perceptions towards nature-based tourism development in Bach Ma National Park, Thua Thien Hue. *Hue University Journal of Science: Agriculture and Rural Development*, 131(3D). https://doi.org/10.26459/hueunijard.v131i3D.6801

Ashley, C. (2000). The impacts of tourism on rural livelihoods: Namibia's experience. *Overseas Development Institute Working Paper.* London: Portland House Stag Place, SW1E 5DP.

Becken, S. (2013). A review of tourism and climate change as an evolving knowledge domain. *Tourism Management Perspectives*, 6, 53–62.

Becken, S., & Hay, J. E. (2007). *Tourism and Climate Change: Risks and Opportunities*. Channel View Publications.

Becken, S., McManus, P., & Harvey, M. (2014). Impacts of climate change on tourism in New Zealand and Australia: A cross-sectional study of tourism businesses. *Tourism Management*, 45, 224–234.

Belle, N., & Bramwell, B. (2005). Climate change and small island tourism: Policy-maker and industry perspectives in Barbados. *Journal of Travel Research*, 44(1), 32–41.

Black, R., & Crabtree, A. (2007). *Quality Assurance and Certification in Ecotourism*. CABI Publishing.

Butler, R. W. (2001). Seasonality in tourism: Issues and implications. In *Tourism: Critical Concepts in the Social Sciences* (pp. 5–21). Routledge.

Ceballos-Lascuráin, H. (1996). *Tourism, Ecotourism, and Protected Areas: The State of Nature-based Tourism Around the World and Guidelines for Its Development*. IUCN – The World Conservation Union.

Dogru, T., Marchio, E. A., Bulut, U., & Suess, C. (2019). Climate change: Vulnerability and resilience of tourism and hospitality industry. *Tourism Management*, 72, 292–305.

Ellis, F. (2000). *Rural Livelihoods and Diversity in Developing Countries*. Oxford: Oxford University Press.

Fennell, D. A. (1999) *Ecotourism: An Introduction*. London: Routledge.

Global Ecotourism Network (GEN). (2016). *What is ecotourism?* Retrieved from https://www.globalecotourismnetwork.org/definition-and-key-concepts/

Hien, T. T. T., & Chu, H. J. (2024). Revealing ecotourism potential of streams and waterfalls: Identification of influential factors in a case study from Thua Thien Hue Province, Vietnam. *Sustainability*, 16(21), 9299.

Honey, M. (1999). *Ecotourism and Sustainable Development: Who Owns Paradise?* Washington, DC: Island Press.

Honey, M. (2008). *Ecotourism and Sustainable Development: Who Owns Paradise?* (2nd ed.). Island Press.

Jones, L., & Boyd, E. (2011). Exploring social barriers to adaptation: Insights from Western Nepal. *Global Environmental Change*, 21(4), 1262–1274. https://doi.org/10.1016/j.gloenvcha.2011.06.002

Jopp, R., DeLacy, T., & Mair, J. (2010). Developing a framework for regional destination adaptation to climate change. *Current Issues in Tourism*, 13(6), 591–605.

Kaján, E., & Saarinen, J. (2013). Tourism, climate change and adaptation: A review. *Current Issues in Tourism*, 16(2), 167–195.

López-i-Gelats, F., et al. (2016). Adapting rural livelihoods to climate variability and change: Farming systems in the Andes. *Global Environmental Change*, 39, 258–274.

Macleod, D., & Russell, R. (2010). Vulnerability of ecotourism communities to climate change in the Caribbean. *Tourism Geographies*, 12(4), 460–476.

Ministry of Natural Resources and Environment (MORNE). (2020). *Viet Nam's Updated Nationally Determined Contribution (NDC)*. Hanoi: MONRE.

Moreno, A., & Becken, S. (2009). A climate change vulnerability assessment methodology for coastal tourism. *Journal of Sustainable Tourism*, 17(4), 473–488.

Nyaupane, G. P., Morais, D. B., & Graefe, A. R. (2004). Nature-based tourism constraints: A cross-activity comparison. *Annals of Tourism Research*, 31(3), 540–555.

People's Committee of Thua Thien Hue Province. (2016). *Resolution No. 03-NQ/TU dated November 8, 2016, on the development of Thua Thien Hue into a centrally governed city based on heritage, culture, ecology, landscape, and environmental friendliness*.

Scott, D., McBoyle, G., & Minogue, A. (2008). Climate change and the future of skiing in Canada. *Tourism Management*, 29(6), 140–148.

Simpson, M. C., Gössling, S., Scott, D., Hall, C. M., & Gladin, E. (2008). *Climate Change Adaptation and Mitigation in the Tourism Sector: Frameworks, Tools and Practices*. Paris: United Nations Environment Programme (UNEP), University of Oxford, United Nations World Tourism Organization (UNWTO), and World Meteorological Organization (WMO).

The International Ecotourism Society (TIES). (2015). *What is ecotourism?* Retrieved from https://ecotourism.org/news/what-is-ecotourism/

Wearing, S., & Neil, J. (2009). *Ecotourism: Impacts, Potentials and Possibilities?* (2nd ed.). Routledge.

Zhou, D., & Ap, J. (2017). Participatory approaches to climate change adaptation in ecotourism communities. *Annals of Tourism Research*, 63, 138–151.

Zhou, D., et al. (2017). Seasonal fluctuations in tourism and economic resilience in coastal communities. *Journal of Sustainable Tourism*, 25(2), 273–289.

6 Global Research Dynamics on Climate Change and Indigenous Livelihoods

Insights from Scientometric Analysis

Md Mainul Sk and Tapas Ranjan Patra

INTRODUCTION

Climate change is among the major issues of concern in the world today, and it has effects on vulnerable people in the society, including indigenous persons whose lives depend on the natural resource base. Such groups, usually with livelihoods in agriculture, fishing, and other natural resource-related activities, have been more impacted because they depend on climate-sensitive assets and have low coping potential (Abbass et al., 2022; Weiskopf et al., 2020; Rahman et al., 2018). Norton-Smith (2016) and Rahman and Alam (2016) have underlined that these groups are now at greater risk, for instance, poor crop yield, species extinction, and frequent extreme weather events. The study on the impact of climate change in the Arctic reveals how climate change through ice melting affects traditional hunting practices disrupting food security and cultural continuity (Meakin and Kurvits, 2009). Similarly, tropical areas suffer from soil erosion, long-term drought, and erratic rainfall patterns, which further compel indigenous communities (Osman, 2014; Kronik and Verner, 2010; Suhrke and Hazarika, 1993).

Despite these adversities, indigenous people somehow have been able to be resilient through the use of their traditional ecological knowledge (TEK) in order to cope with these threats (Nelson and Shilling, 2018). TEK is a storehouse of intergenerational wisdom that discursively constructs ecological knowledge to manage resources for agriculture, sustainable food production, and conservation strategies (O'Donoghue et al., 2019). Examples from the Amazon Basin show that how rotational agroforestry lessens the effect of deforestation and maintains the health of the ecosystem (Figueiredo et al., 2020; Hecht, 1982). Similarly, works in South Asia show how tribal societies manage risks and variability in climate by storing water, practicing diversified and intercropping and shifting cultivation (Babu, 2024; Rivera-Ferre et al., 2021). Such practices not only sustain the livelihood of the communities but also support the environmental management agenda globally. However, TEK, though highly resilient and adaptive in nature, continues to confront increasing threats from climate change that are further worsening social justice and culture (Magni, 2017; Kasperson and Kasperson, 2012).

Climate variability, indigenous livelihood, and global justice have emerged as important areas of research within the discipline in the last few years. Prior research has examined indigenous peoples' vulnerability assessments (Rashidi and Lyons, 2022; Adger et al., 2009), the role of TEK in the adaptive processes (Lemi, 2019; Whyte, 2013), and the socio-political aspects of climate justice (Schlosberg et al., 2017). The global environmental justice movement focuses on the specific situation of indigenous people and supports their rights to land, resources, and sovereignty. National and global guidelines like the United Nations Declaration on the Rights of Indigenous Peoples

DOI: 10.1201/9781003620747-7

(UNDRIP) have laid down the principle that supports indigenous knowledge systems in climate change policies (Assembly, 2007). However, the analysis given by Schlosberg and Collins (2014) shows that even with such normative improvements, there is a lack of equitable climate justice with continued marginalization of indigenous peoples in decision-making practices.

Although the current literature provides an important understanding of the relationship between climate change and indigenous peoples' livelihood, it suffers from geospatial and thematic gaps. Scientists rigorously concentrate their studies on the Arctic and the Amazon while sub-Saharan Africa and South Asia are less explored (Galappaththi et al., 2019; Ford et al., 2015a). Moreover, the research tends to have short timescale perspectives that do not fully capture how climate change affects indigenous peoples' livelihoods in the long run. This means that there is a need to encourage cross-disciplinary research and knowledge co-production between scholars, policymakers, and indigenous peoples to come up with practical interventions.

While there is empirical literature analyzing climate change and indigenous peoples' livelihoods, there are few scientometric reviews to identify and synthesize global trends in research. Scientometric analysis is a powerful approach to systematically outlook the global research trends on climate change and indigenous livelihoods. This method uses quantitative tools to examine the research outputs, thematic trends, and collaboration network, which reveals research opportunities and directions to inform future work (Aria and Cuccurullo, 2017; Van Eck and Waltman, 2010). Previous scientometric analyses by Singh et al. (2022) show that the interest in this area of research is growing, with North America, South America, and Southeast Asia as the most active areas. Nevertheless, there is little research that can document the state of the art of climate change impacts on indigenous livelihoods or the ways in which the field has developed in terms of its focus areas and integration of research traditions. There is also little interaction between academic research, policy, and indigenous peoples to develop usable knowledge and support fair climate policy (Hillsdon, 2024). The lack of such analyses limits the appreciation of the trends in research on this topic over the years, the level of cross-disciplinary works, and the areas of need.

Therefore, this study seeks to fill these gaps by undertaking a scientometric review of the global literature on climate change and indigenous livelihoods. This paper aims to offer a comprehensive understanding of the field by mapping publication trends, identifying research contributors, and analyzing thematic clusters, as well as to discuss potential future directions in the field. It is envisaged that the study's findings will apprise the improvement of research relevance to the needs of indigenous communities and to the sustainable development agenda.

DATABASE AND METHODOLOGY

To map the research on climate change and indigenous livelihoods, this study uses scientometric analysis. The data used in this research were only retrieved from the Scopus database, which includes a vast number of peer-reviewed articles. The search strategy involved the use of specific keywords: (Indigenous OR Livelihood) AND (Climate change) AND PUBYEAR>2015 AND PUBYEAR<2024 AND LANGUAGE: English AND DOCTYPE: ar. Boolean operators "AND" and "OR" were used to refine the search terms. The period for this study was chosen to be from 2015 to 2024 in order to have a more recent study. All document types including journal articles, conference proceedings, and book chapters were included to ensure that the review is systematic. Only English articles were included for consistency in analysis. The final dataset was used for scientometric analysis of publication dynamics, collaboration, and thematic landscapes. The analysis was performed using two widely used sophisticated tools: VOSviewer and Biblioshiny in RStudio. VOSviewer was used to map and analyze the collaboration networks, while Biblioshiny enabled a better understanding of trends and themes, as well as citation analysis. The research design and the methodological process of the study, which includes data collection, data treatment, and the multi-level analysis, are depicted in Figure 6.1.

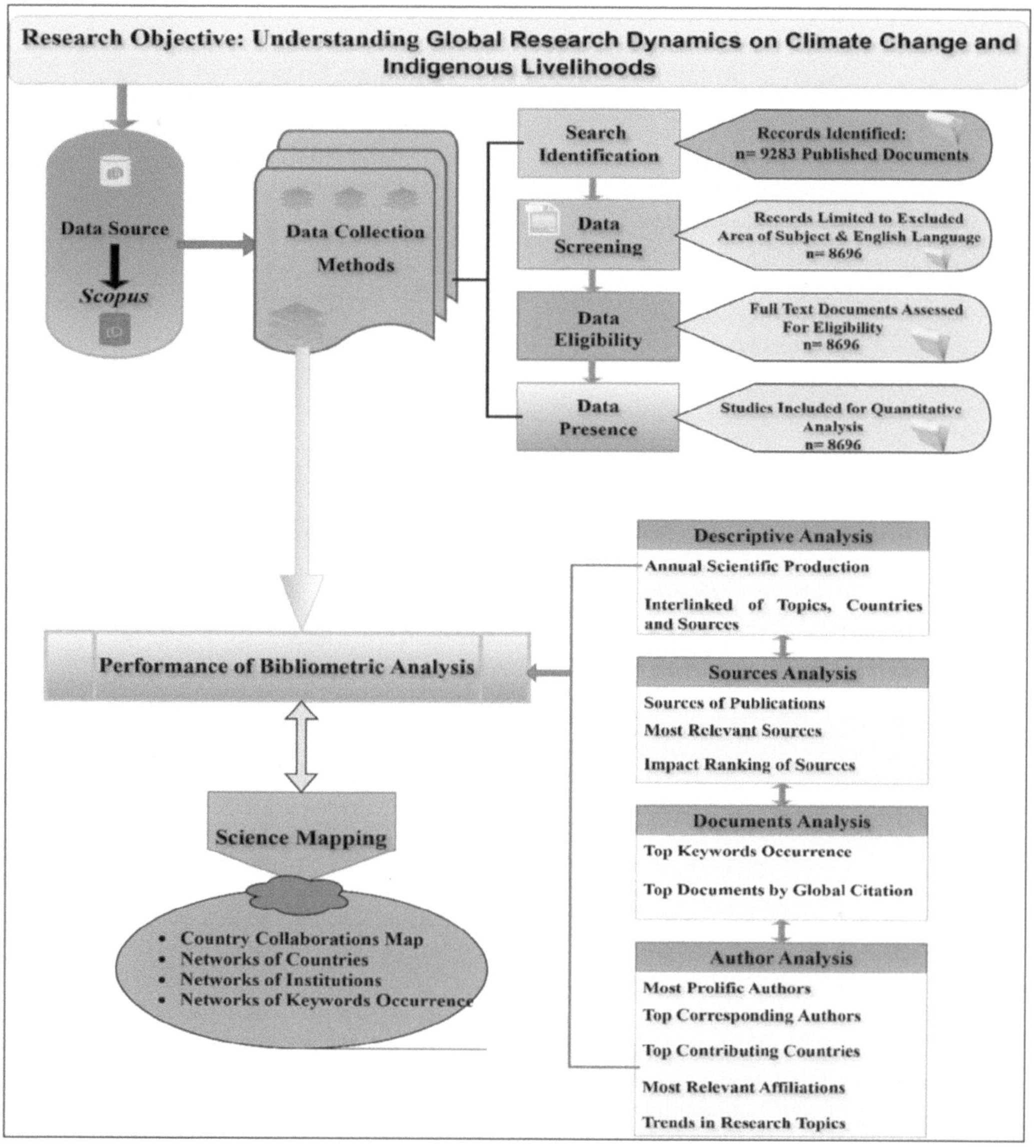

FIGURE 6.1 The research design and methodological workflow of the study.

RESULTS

Important Information

The dataset for this study comprises 2,151 sources, including journals, books, and other academic databases, highlighting the multidisciplinary nature of research in this field. A total of 8,696 documents were reviewed, and the main body of the documents was peer-reviewed journal articles as they are key documents for the dissemination of knowledge in this subject area. The growth rate per annum is 17.13%, pointing to active and sustained scholarly engagement in this area of focus on climate change and indigenous livelihoods (Table 6.1). The age of the documents is somewhat young with an average of 3.27 years, which means that most of these works have been undertaken in the recent past, and this could be in response to current world issues. Overall, an average of 17.88 citations per document indicates the level of academic recognition of research conducted in this field.

TABLE 6.1
Main Information about the Data of Climate Change and Indigenous Livelihood

Description	Results
Timespan	2015:2024
Sources (Journals, Books, etc.)	2151
Documents	8,696
Annual growth rate %	17.13
Document average age	3.27
Average citations per doc	17.88
References	511585
Document Contents	
Keywords Plus (ID)	16,850
Author's Keywords (DE)	18,426
Authors	
Authors	29,704
Authors of single-authored docs	1,096
Authors Collaboration	
Single-authored docs	1,171
Co-Authors per Doc	4.62
International co-authorships %	39.62
Document Types	
Article	8,696

Source: Prepared by the researchers based on Scopus database using Biblioshiny (in RStudio)

The analysis provided 16,850 Keywords Plus (ID) and 18,426 Author's Keywords (DE). Keywords Plus describes the overall topic of the research as indexed in the database, while Author's Keywords demonstrate the subjects that the authors are concerned with. This two-fold structure of keywords reveals the analytical framework for the topic of the study in this domain. From the analyzed dataset, there are 29,704 authors who contributed to the documents, and among them, 1,096 authored single-authored documents and 1,171 documents were single-authored in the entire dataset. These figures illustrate the essence of research that involves several teams of people working together to solve problems like climate change and indigenous peoples' welfare. In particular, each paper had 4.62 co-authors on average to reflect the interdisciplinarity of such research. In addition, 39.62% of the documents included international co-authorship, which shows that this area is an international and collaborative one.

PUBLICATION TRENDS

Number of Publications over Time

As illustrated in Figure 6.2, the overall output of publications on these topics has been rising significantly during these years, from 357 in 2015 to 1,481 in 2024. The steepest rise was seen between 2020 and 2024, which corresponds to the times of climate change and indigenous people's contributions to the sustainability and resilience narratives. Mean total citations (TC) per article

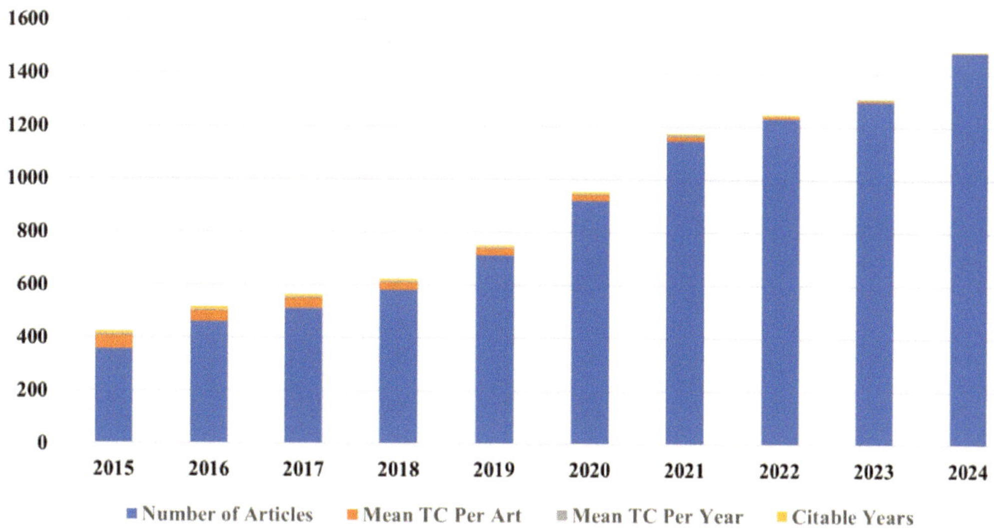

FIGURE 6.2 Publication trends over time.

TABLE 6.2
Annual Growth in Publications of Climate Change and Indigenous Livelihoods

Year	Number of Articles	Mean TC Per Art	Mean TC Per Year	Citable Years
2015	357	50.87	5.09	10
2016	461	39.95	4.44	9
2017	512	38.19	4.77	8
2018	581	30.36	4.34	7
2019	713	27.04	4.51	6
2020	918	23.89	4.78	5
2021	1147	17.71	4.43	4
2022	1230	9.47	3.16	3
2023	1296	5.29	2.64	2
2024	1481	1.14	1.14	1

also confirms the multiplier effect of the prior popular articles. Overall, articles of 2015 received the highest mean citation of 50.87, and those of 2016 and 2017 received 39.95 and 38.19 citations, respectively (Table 6.2). This is because the earlier research work laid down the theoretical and empirical background for the subsequent works. The articles published between 2023 and 2024 had a lower mean citation with 5.29 and 1.14, respectively, owing to shorter citation windows. The mean TC per Year gives a further understanding of the existence of articles in later years. The publications from the years 2017 to 2020 revealed a higher average of 4.34–4.78 citations per year than the most recent years, which were still in their initial citation period.

Citable Years, which range from 10 years for the publications of 2015 to 1 year for the publications of 2024, underline the role of time when assessing citation impact. The articles published in previous years got more time to gain citations and hence a higher mean citation rate is observed. These trends point to the increasing body of literature on climate change and indigenous peoples' livelihood systems with earlier works having a high citation impact and recent works capturing current concerns.

Altogether, this trajectory shows that more and more, the contribution of indigenous peoples' knowledge and practices is considered as valuable for the advancement of the sustainability of the world.

THREE-FIELD MAP ANALYSIS OF CLIMATE CHANGE AND INDIGENOUS RESEARCH

Figure 6.3 highlights a three-field map analysis of the countries involved, authors, and topics most frequently featured in climate change and indigenous research. China stands as the most prolific authoring country with 632 publications contributing a massive input in climate change and indigenous peoples' survival. Based on this research, the following Chinese scholars are most active in this field: Wang Y (187 publications), Zhang Y (158), and Zhang J (135). China has international research collaboration links with eight countries with outgoing links to India, United States, Germany, and Australia. Likewise, India has also published 356 articles, which also reflects India's academic contribution in the form of authors like Kumar P with 132 articles, Kumar A with 116 articles, and Pandey R with 83 articles. Indian researchers collaborate with many countries, keeping associations with seven countries. The United States is also very active, providing 182 publications and supporting nine collaborations with an outgoing focus on authors like JR (70 publications), Zhang J (30), and Ford JD (22). The analysis shows that Wang Y is one of the leading scholars in the field of climate change, co-authoring 632 papers and collaborating with China, India, and the United States. In the same way, both Zhang Y and JR have vast relations with the international market especially with China, Australia, and Canada. These authors are not only some of the most productive members of the scholarly community, but they also serve as important gatekeepers, helping to facilitate information from one country to another.

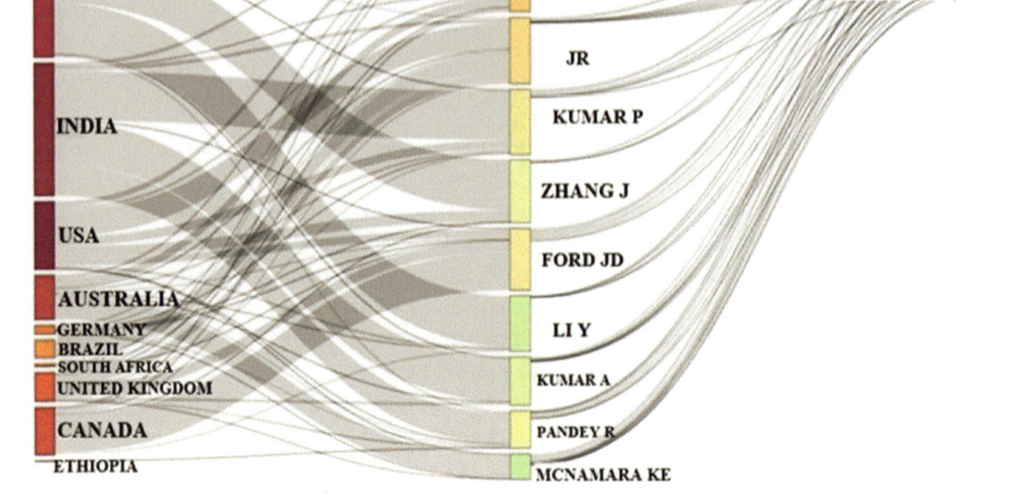

FIGURE 6.3 Three-field fold map of climate change and indigenous research.

The most frequently used research themes include "Climate Change" mentioned 84 times, "Adaptation" mentioned 27 times, "Vulnerability" 22 times, and "Resilience" 18 times. These themes are closely related to highly productive authors such as Wang Y, Zhang Y, McNamara KE, and Pandey R, thus underlining their importance for the development of the field. New categories such as "Sustainability" (6) and "Indigenous Knowledge" (4) show that more importance is being granted to incorporating local knowledge into strategies on climate change, which is a trend toward more inclusive and culturally sensitive work. The three-field map reveals a connected weave of interconnections that shows China, India, and the United States as the center of research activity. Prolific writers are instrumental in transferring information and filling the gap between geography and subject areas. The relationships between the key authors and the most cited research topics reveal a strong correlation between the research output and the scholarship maturity.

TOP CONTRIBUTING AUTHORS, INSTITUTIONS, AND COUNTRIES

Most Prolific Authors in Climate Change and Indigenous Livelihoods Research

In Table 6.3, the authors are ranked based on their total number of publications (NP), TC, and other scientometric indicators including H-Index, G-Index, and M-Index to determine the overall and cumulative contributions of authors over the years. Wang Y can be considered the most productive author with 36 articles published since 2016, 995 citations, the H-Index of 13, and the G-Index of 31. Ford JD has earned the most TC (1,018) and the greatest H-Index (21), which attest to its scientific significance, and their M-Index of 2.1 points to a stable impact over the last five years. Among all the authors, Pandey R has the highest G-Index (25) and TC (1,324), although he has only five publications. Likewise, all the authors included in the list have presented balanced contributions and have received appreciable citation scores as well as H-Index, which is evident through the works of McNamara KE, Mabhaudhi T, and Xu J. These findings highlight the contributions of these authors to the furtherance of knowledge on climate change and indigenous livelihoods.

Using the H-Index, which measures both productivity and citation impact, Ford JD's research is deemed the most influential with an H-Index of 21, while using the G-Index, which focuses on articles that are among the most cited, Pandey R research is identified as the most prolific with a G-Index of 25. The M-Index, which adjusts the H-Index to the length of a scientist's career, shows how Ford JD (2.1) and McNamara KE (1.6) have been productive scholars throughout their careers. Moreover, the charts of Publication Year Start (PY-Start) show that most active authors started contributing from 2015, which coincides with the growing focus on climate change and indigenous populations.

TABLE 6.3
Most Prolific Authors of Climate Change and Indigenous Research in the Global

Rank	Author	NP	TC	H-Index	G-Index	M-Index	PY-Start
1	Wang Y	36	995	13	31	1.444	2016
2	Zhang Y	32	562	13	23	1.444	2016
3	JR	27	467	12	21	1.5	2017
4	Ford JD	26	1018	21	26	2.1	2015
5	Pandey R	25	1324	15	25	1.5	2015
6	McNamara KE	22	744	16	22	1.6	2015
7	Kumar A	22	368	11	19	1.1	2015
8	Mabhaudhi T	21	950	12	21	1.2	2015
9	Xu J	18	737	12	18	1.2	2015
10	Singh S	18	303	10	17	1	2015

GEOGRAPHIC DISTRIBUTION OF CORRESPONDING AUTHORS

The contributions of corresponding authors in climate change and indigenous livelihoods research demonstrate the geographic spread of the authors (Table 6.4). The United States leads with 1,048 articles (12.1% of the total), showcasing a significant balance between single-country publications (SCP): The journal contains 687 SCPs and 361 multi-country publications (MCP), where 34.4% of all its publications are international collaborations. India again comes second with 657 articles (7.6%), and the emphasis is more on SCPs (537), and only 18.3% of MCP indicating a low level of international collaboration. The next two countries are Australia and China, each contributes about 5.8% of the total articles, with a significant focus on cross-national studies, at 43.8% and 44%, respectively. The United Kingdom and North America have the highest percentage of MCPs (60.6% and 52.5%, respectively), showing a rather extensive internationalization. Likewise, Germany's MCP percentage of 56.5 pointed out the country's high input in global research collaborations, even if it has lesser articles published (Figure 6.4).

TABLE 6.4
Corresponding Authors' Countries

Rank	Country	Articles	Articles %	SCP	MCP	MCP %
1	United States	1,048	12.1	687	361	34.4
2	India	657	7.6	537	120	18.3
3	Australia	505	5.8	284	221	43.8
4	China	502	5.8	281	221	44
5	United Kingdom	472	5.4	186	286	60.6
6	Canada	439	5	294	145	33
7	South Africa	345	4	255	90	26.1
8	Germany	260	3	113	147	56.5
9	Ethiopia	234	2.7	166	68	29.1
10	Ghana	164	1.9	109	55	33.5

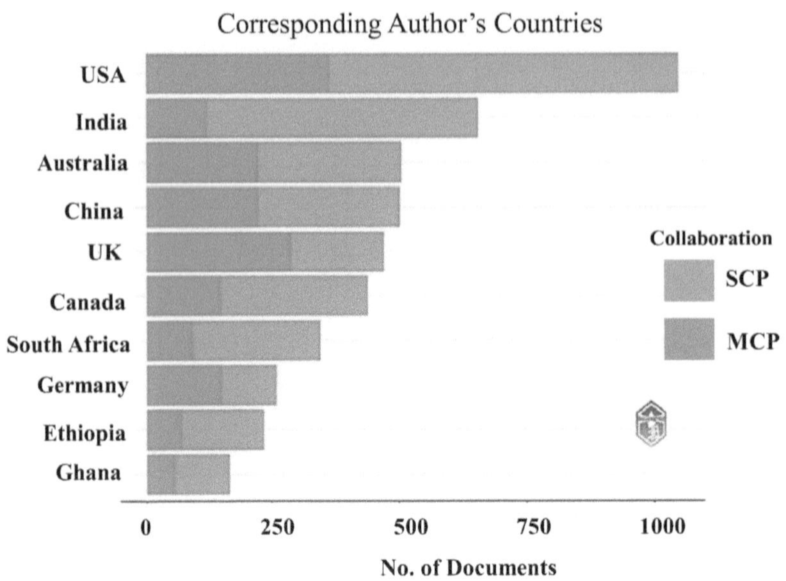

FIGURE 6.4 Country-wise research collaboration and corresponding author.

Other companies that also participated include Canada with 5%, South Africa with 4%, Ethiopia with 2.7%, and Ghana with 1.9%. Canada and South Africa are between the SCPs and MCPs, while Ethiopia and Ghana have lower MCP percentages, which shows that the research is more focused in a specific region. These findings show the importance of developed countries like the United States, the United Kingdom, and Australia, and emerging economies like India, China, and South Africa in this important research area. The levels of international collaboration also differ to show the integration and focus area of climate change and indigenous livelihoods across the globe.

MOST RELEVANT AFFILIATIONS

In Table 6.5, the authors are grouped into their affiliations and the top ten affiliations that produced the most articles are listed. The first is Addis Ababa University that has 199 publications accounting for 13% of the total, and the second is the University of California with 195 publications (Figure 6.5). Others are Kwame Nkrumah University of Science and Technology with 174 publications, the University of KwaZulu-Natal with 171 publications, and the University of British Columbia with 156 publications. Other universities of note include McGill University, with 154 publications, the University of Leeds with 135, and the University of Ghana with 134 publications.

TABLE 6.5
Most Relevant Affiliations

Rank	Affiliation	Articles
1	Addis Ababa University	199
2	University of California	195
3	Kwame Nkrumah University of Science and Technology	174
4	University of KwaZulu-Natal	171
5	University of British Columbia	156
6	McGill University	154
7	University of Leeds	135
8	University of Ghana	134
9	University of Copenhagen	125
10	The University of Queensland	122

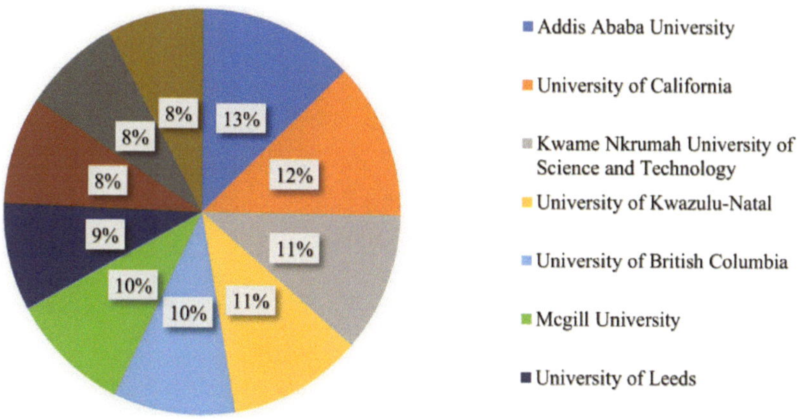

FIGURE 6.5 Global universities and share of published articles.

The University of Copenhagen and the University of Queensland take the tenth position with 125 and 122 publications, respectively. These findings are in line with the major roles played by these institutions in enhancing knowledge on climate change and indigenous peoples' livelihoods, with the majority of the institutions based in areas relevant to such studies.

MOST CITED COUNTRIES IN CLIMATE CHANGE AND INDIGENOUS LIVELIHOODS RESEARCH

Table 6.6 shows the top ten countries according to TC and average article citations (AAC), and Figure 6.6 provides a graphical comparison of their citation impact. The country that leads is the United States with a total of 20,496 citations and 19.60 average citations per article, which proves the United States' influence on the development of related research. The United Kingdom has the second-highest TC with 12,289 citations, yet it has the highest AAC standing at 26.00, which shows the great contribution and quality of the articles published. Australia ranks third with 10,820 citations and a relatively high citation with an average of 21.40 citations per article, which clearly indicates its participation in this area. Fourth and fifth are China and India, with a total of 9,312 and 8,011 citations, respectively. However, the AAC for India (12.20) are significantly lower than for China (18.50), which shows that there is scope for India to have more impact despite it producing

TABLE 6.6
Most Cited Countries

Rank	Country	TC	Average Article Citations
1	United States	20,496	19.60
2	United Kingdom	12,289	26.00
3	Australia	10,820	21.40
4	China	9,312	18.50
5	India	8,011	12.20
6	Canada	6,528	14.90
7	Germany	6,111	23.50
8	South Africa	4,253	12.30
9	Netherlands	3,637	28.60
10	Ethiopia	3,589	15.30

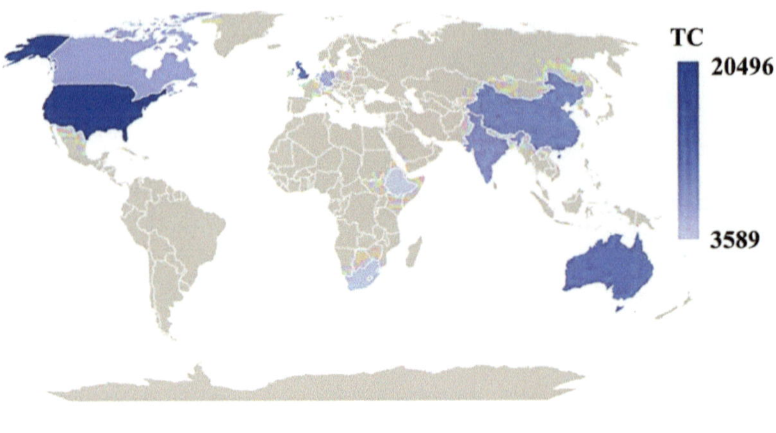

FIGURE 6.6 Countries with the highest number of citations.

a large amount of research. The same goes for Canada, which has 6,528 TC and 14.90 AAC and Germany with 6,111 TC and 23.50 AAC, as they have relatively high citation averages. Others are South Africa with 4,253 TC and 12.30 AAC, the Netherlands with 3,637 TC and 28.60 AAC, and Ethiopia with 3,589 TC and 15.30 AAC. The Netherlands has the highest AAC (28.60), even though it has the least number of citations compared to other countries. The pattern of the studies presented in this article is quite diverse in terms of geography; research was conducted in North America, Europe, Asia, and Africa. These trends illustrate the degree to which this field is international and the extent of the differential impact of scholarship across the globe.

KEY THEMES ANALYSIS

Thematic Structure of Climate Change and Indigenous Research

The climate change and indigenous research thematic map (Figure 6.7) show the conceptual framework within this field by organizing the primary themes into motor themes, basic themes, and niche or emerging themes. The most important and significant areas of the research that are indicated by the motor themes are "indigenous," "indigenous peoples," and "ecosystem services." Such terms represent the indigenous population and its connection with ecosystem services in the context of climate change issues. The basic themes, which are presented as more general areas of investigation, include concepts like "climate change," "adaptation," and "vulnerability." These are the key thematic areas that define a significant part of the research discourse in this field and capture the relationship between climate change and human adaptation. In the niche themes, which are relatively new and very specific fields of study, terms such as "adaptive capacity," exposure, and "sensitivity" are most commonly used. These concepts represent complex analyses of the ability of indigenous societies to adapt or not to the effects of climate change. Combined, the thematic map captures the relationship between these fields of research, as well as the current and emerging strands of climate change and indigenous livelihoods literature.

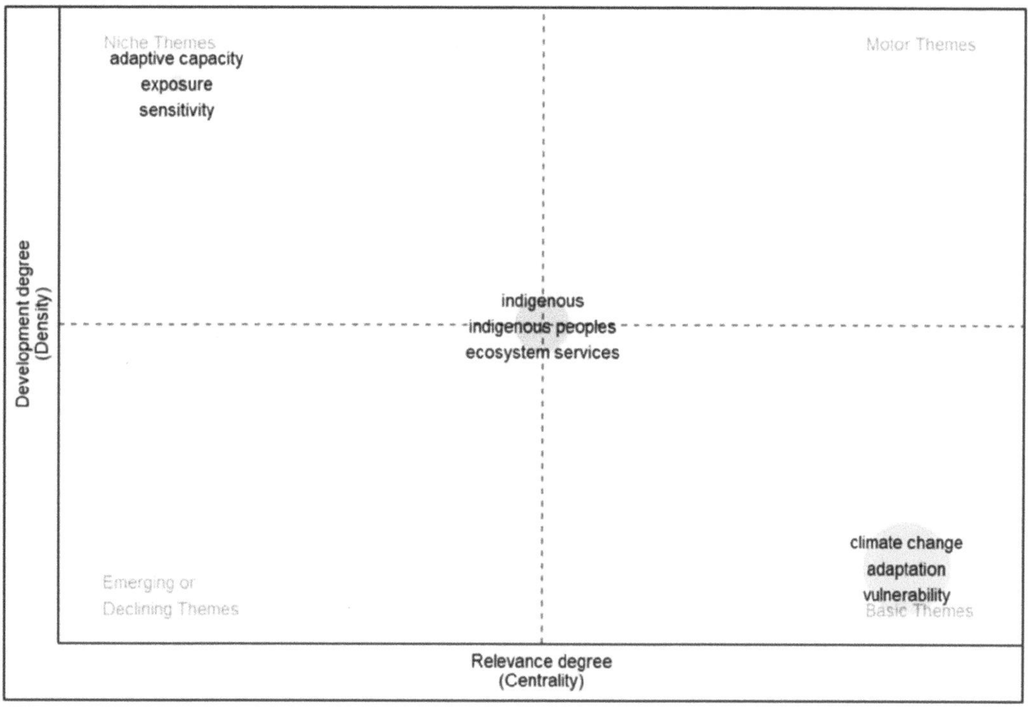

FIGURE 6.7 Thematic map of climate change and indigenous research.

MOST RELEVANT AUTHOR KEYWORDS IN CLIMATE CHANGE AND INDIGENOUS RESEARCH

Table 6.7 displays the 15 most frequent terms used in the articles analyzed, which represent the major topics and sub-fields of research in the area. The most frequent term is Climate Change with 3,071 mentions meaning that it is the main topic of the research field. Strategies related to the keyword Adaptive, including Adaptation (732), and Resilience (363), reflect the focus on how the phenomenon of climate change affects the studied systems. Furthermore, the use of terms like Vulnerability (437) and Adaptive Capacity (215) shows the interest in identifying and increasing the possibility of communities in mitigating the impacts of change in the environment. The social and economic factors are well captured in the evaluation of climate change impacts with the most used terms being Livelihoods (328), Food Security (327), and Agriculture (271) showing how climate change affects rural and indigenous people. The use of terms such as Indigenous Knowledge (used 253 times), Indigenous Peoples (used 156 times), and Indigenous (used 161 times) points to the growing understanding of indigenous people's worldviews and their contribution to the practice of sustainable adaptation. New and developing terms like Sustainability (221) and Climate Change Adaptation (221) suggest the increasing concern regarding the long-term strategies of climate change, while drought (177) is a more pointed topic showing the real problems of affected communities. In the present analysis, the identified author keywords reflect the holistic understanding of climate change issues incorporating scientific, socioeconomic, and indigenous approaches that can greatly inform strategic policy decisions and sustainable development actions.

RESEARCH COLLABORATION

Global Collaboration Networks

Table 6.8 presents the list of the ten countries that are most influential in the network based on measures including the number of clusters, betweenness centrality, closeness centrality, PageRank, and total strength, giving an idea of their impact within the research area. The United States is the most central node in the network based on all the measures: betweenness centrality, 70.671; PageRank, 0.093; and total link strength, 2,551. This is because it remains the foremost platform

TABLE 6.7
Most Relevant Authors Keywords

Rank	Words	Occurrences
1	Climate change	3,071
2	Adaptation	732
3	Vulnerability	437
4	Resilience	363
5	Livelihoods	328
6	Food security	327
7	Agriculture	271
8	Indigenous knowledge	253
9	Climate change adaptation	222
10	Sustainability	221
11	Adaptive capacity	215
12	Livelihood	209
13	Drought	177
14	Indigenous	161
15	Indigenous peoples	156

TABLE 6.8
Global Collaboration Metrics in Climate Change and Indigenous Livelihoods Research

Rank	Node	Cluster	Betweenness	Closeness	PageRank	Total Link Strength
1	United States	3	70.671	0.02	0.093	2,551
2	United Kingdom	3	44.798	0.02	0.076	2,062
3	Australia	1	19.927	0.02	0.051	1,398
4	Germany	3	15.416	0.02	0.044	1,240
5	Canada	3	7.275	0.02	0.04	1,081
6	China	1	14.372	0.02	0.038	902
7	South Africa	2	10.971	0.02	0.032	808
8	India	1	7.05	0.02	0.031	756
9	Kenya	2	3.811	0.019	0.029	742
10	Netherlands	3	3.531	0.02	0.029	697

that connects collaborations and shares information across the globe. The United Kingdom and Australia come immediately after the United States and Canada with total link strength of 2,062 and 1,398, respectively. The two countries show high collaboration density, especially in Cluster 3 for the United Kingdom and Cluster 1 for Australia.

Germany and Canada also show a strong impact. Germany has a betweenness centrality of 15.416, while Canada has a total link strength of 1,081 and is a part of Cluster 3. China, one of the countries active in Cluster 1, is also highly connected in the region and beyond, with the total link strength of 902 and the PageRank of 0.038. South Africa, India, and Kenya are of special significance in the context of the countries of the Global South. South Africa, in Cluster 2, stands seventh with a betweenness centrality of 10.971 and a total link strength of 808. India is situated in Cluster 1 and has a total link strength of 756, further indicating the country's increasing significance in the research area. Kenya that is also in Cluster 2 has a betweenness centrality of 3.811 and a total link strength of 742 showing its connection with partners within and beyond the cluster. Finishing off the top ten is the Netherlands, which has added greatly to Cluster 3 with a total link strength of 697 and a PageRank of 0.029 to further reinforce the European cooperation. The findings also underscore the multiplicity and interlinkages of the global engagement on climate change and indigenous livelihoods research with specific geographical and topical hubs leading the collaboration (Figure 6.8).

IMPACT ANALYSIS

Highly Cited Articles

Table 6.9 lists the top ten papers in terms of TC, total citations per year (TC per year), and normalized total citations (normalized TC) so as to evaluate the academic impact. The most impactful paper is Weiss et al. (2020, *Remote Sensing of Environment*) with overall 1,066 citations, 213.20 citations per year, and the highest normalized TC: 44.62. The second most cited study is Early et al. (2016, *Nature Communications*) with 896 citations, 99.56 TC per year, and a normalized TC of 22.43; this work has proven highly and continuously relevant in the area. Hamilton and Casey (2016, *Global Ecology and Biogeography*) has 774 citations and obtained the third place, as it has 86.00 TC/year and 19.38 normalized TC. Likewise, Rockström et al. (2017, *Ambio*) presents the largest number of citations with 725, a TC per year of 90.63, and a normalized TC of 18.98. Other noteworthy efforts are Biggs et al. (2015, *Environmental Science & Policy*) with 654 citations and Tellman et al. (2021, *Nature*), which, although published recently, has already received 601 citations and an annual TC of 150.25. Serdeczny et al. (2017, *Regional Environmental Change*), Pelling et al. (2015, *Climate Change*), Whyte K (2017, *English Language Notes*) and Tao et al. (2015, *Proceedings of the*

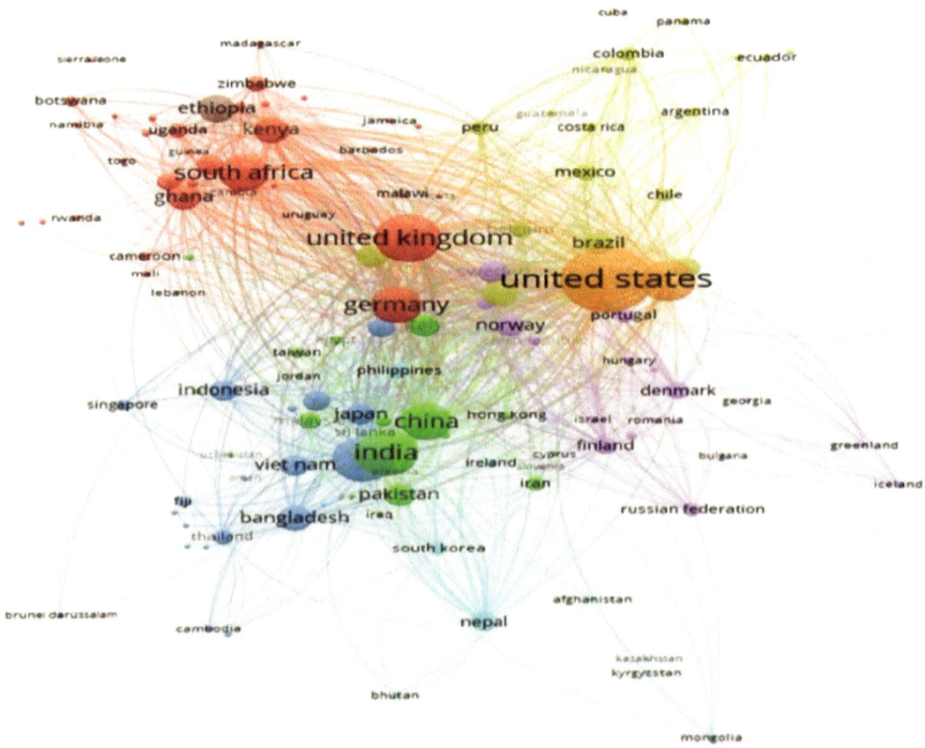

FIGURE 6.8 Visualization of global collaboration networks in climate change and indigenous livelihoods research.

TABLE 6.9
Most Cited Documents Globally

Rank	Paper	DOI	Total Citations	TC per Year	Normalized TC
1	Weiss M, Jacob F, and Duveiller G, 2020, *Remote Sens Environ*	10.1016/j.rse.2019.111402	1066	213.20	44.62
2	Early R, 2016, *Nat Commun*	10.1038/ncomms12485	896	99.56	22.43
3	Hamilton SE, 2016, *Global Ecol Biogeogr*	10.1111/geb.12449	774	86.00	19.38
4	Rockström J, 2017, *Ambio*	10.1007/s13280-016-0793-6	725	90.63	18.98
5	Biggs EM, 2015, *Environ Sci Policy*	10.1016/j.envsci.2015.08.002	654	65.40	12.86
6	Tellman B, 2021, *Nature*	10.1038/s41586-021-03695-w	601	150.25	33.93
7	Serdeczny O, 2017, *Reg Environ Change*	10.1007/s10113-015-0910-2	568	71.00	14.87
8	Pelling M, 2015, *Clim Change*	10.1007/s10584-014-1303-0	501	50.10	9.85
9	Whyte K, 2017, *Engl Lang Notes*	10.1215/00138282-55.1-2.153	474	59.25	12.41
10	Tao S, 2015, *Proc Natl Acad Sci U S A*	10.1073/pnas.1411748112	471	47.10	9.26

National Academy of Sciences) follow the last one and also have high citation impact as supported by the citation scores. These papers are most cited and show the broad and interdisciplinarity of the area, including remote sensing, ecology, policy, and humanistic. Their high TC per year and normalized TC values confirm their continued significance and their importance to the research on climate change impacts and indigenous peoples' coping strategies.

HIGHLY INFLUENTIAL JOURNALS

In Table 6.10, the ten best-ranked journals are presented according to the total articles, TC, and the key scientometric indicators: the H-Index, G-Index, and M-Index, which help to characterize the journals' productivity and scholarly impact in this field. Sustainability (Switzerland) is the most productive outlet with 277 articles published in the last five years, received 4,311 citations and an H-Index of 31 demonstrating both the quantity and quality of the output. Of the two scientific journals, *Environmental Change* and *Climatic Change* have the highest G-Index values (63), which means that many of the articles are highly cited. Interestingly, *Climatic Change* has the highest H-Index of 39 which strengthens its dominant position in the field. Other journals like *Climate and Development and Land Use Policy* have an equal distribution of articles and have the highest citation of 3,135 and 3,313 and H-Indexes of 32 and 31, respectively, proving the journals scholarly contribution. Both *PLOS One* and *Environmental Science and Policy* are also highly productive with more than 70 articles that were analyzed and G-Index higher than 39, which proves their significance in the field. The M-Index also captures the H-Index with reference to the career span, and the distribution remains stable across the scattering of the journals, with a range of 2.3 for *Ecology and Society* to a high of 3.9 for *Climatic Change*. The PY-Start shows that these journals have been active since 2015 which is consistent with the increasing focus on climate change and indigenous research. Combined, these journals provide key insights and contributions to the ongoing discussions that define this important field.

TRENDS IN RESEARCH TOPICS

An examination of the thematic focus of the key literature published in the field shows distinct trends in the areas of study within climate change and indigenous people's subsistence. Table 6.11 summarizes the frequencies of specific terms and their chronological profile, defined by the year of first quartile occurrence (Q1), median year, and third quartile (Q3). These metrics show how different topics have evolved and how critical they are to the field. Climate Change is the most discussed term in the corpus with a total frequency of 3071, which gains the highest importance from 2019 (Q1) to 2023 (Q3). Other popular themes include Adaptation (732) and Vulnerability (436), which are also evident in the literature as key considerations in understanding the effects of climate change on indigenous people with the bulk of the analysis done between 2020 and 2021. Resilience and Livelihoods, linked with adaptation strategies, also receive similar focus with median years in 2020 and 2021, respectively.

TABLE 6.10
Most Relevant Sources in Climate Change and Indigenous Research

Rank	Sources	Articles	TC	H-Index	G-Index	M-Index	PY-Start
1	*Sustainability (Switzerland)*	277	4,311	31	49	3.1	2015
2	*Climate and Development*	143	3,135	32	47	3.2	2015
3	*Regional Environmental Change*	134	4454	35	63	3.5	2015
4	*Climatic Change*	117	4,583	39	63	3.9	2015
5	*Environment, Development and Sustainability*	113	1,673	25	33	2.5	2015
6	*PLOS One*	89	1,932	26	41	2.6	2015
7	*Land Use Policy*	88	3,313	31	55	3.1	2015
8	*Science of The Total Environment*	87	3,007	31	53	3.1	2015
9	*Environmental Science and Policy*	71	2,790	26	52	2.6	2015
10	*Ecology and Society*	69	1,646	23	39	2.3	2015

TABLE 6.11
Trends Topic in Climate Change and Indigenous Research

Term	Frequency	Year (Q1)	Year (Median)	Year (Q3)
Climate Change	3,071	2019	2021	2023
Adaptation	732	2018	2021	2022
Vulnerability	436	2018	2020	2023
Resilience	363	2019	2021	2023
Livelihoods	328	2017	2020	2022
Food Security	327	2020	2022	2023
Indigenous Knowledge	253	2020	2022	2023
Adaptive Capacity	215	2018	2020	2023
Drought	177	2019	2022	2023
Redd+	71	2017	2019	2021
Nepal	64	2017	2019	2021
Tanzania	43	2016	2019	2022
Climate Resilience	35	2022	2023	2024
Mental Health	29	2020	2023	2023
Institutions	24	2017	2018	2020
Bolivia	17	2017	2018	2022
Productivity	17	2022	2023	2023
CMIP6	17	2022	2024	2024
Southern Africa	16	2017	2018	2022
Forest Governance	14	2017	2017	2020
Environmental Impacts	11	2022	2024	2024
Environmental Impact	9	2021	2024	2024
Community Forestry	8	2017	2017	2018
Social Learning	7	2016	2016	2021
Subsistence Farming	7	2016	2016	2017
APSIM	7	2016	2017	2018

Certain themes, such as Food Security (327) and Indigenous Knowledge (253), were more highlighted in 2022, which is important as the world tries to embrace traditional approaches to combat climate change. Recent topics like Climate Resilience, which has 35 mentions, and Environmental Impacts, with 11 mentions, as well as Productivity, with 17 mentions, have increased their activity and have median years in or after 2023. Geo references, such as Nepal (64), Tanzania (43), and Southern Africa (16), indicate that more work is being done in regionally based research, especially in areas that are most affected by climate change. The distribution of terms, as depicted in Figure 6.9, reveals the changing focus of research themes from basics such as adaptation and vulnerability to relatively newer areas including climate resilience and productivity. This trend can be attributed to the growth of climate change and indigenous peoples' activities and sustenance research, owing to new challenges and increased international advocacy for sustainable and fair development.

DISCUSSIONS

The findings of this scientometric analysis provide valuable insights into the global research trends in climate change and indigenous people's survival strategies. It was evident from the analysis that research productivity in the area of climate change and indigenous people's livelihoods has been increasing progressively over the years. The NP increased significantly from the period of 2015–2024 because many scholars and policymakers are now more aware of climate change and

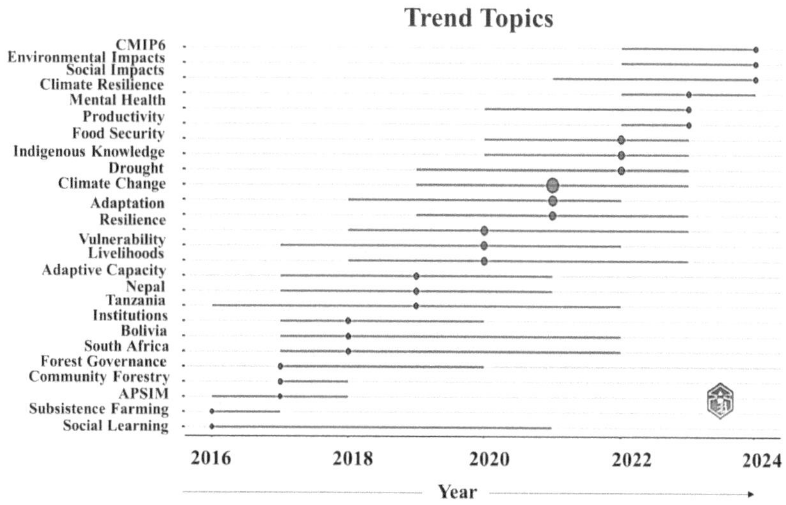

FIGURE 6.9 Changing focus of research themes on climate change and indigenous knowledge.

its effects on indigenous populations. The significant growth between 2022 and 2024 is consistent with the global focus on the Paris Agreement and the SDGs which have both increased the focus on sustainability and climate change. The yearly trends in publication show that the subject is steadily expanding. Even though the literature review shows that evidence from prior years was relatively scarce, the increase in recent years shows growing scholarship interest. The NP also shows an upward trend in the number of collaborations and the diversification of research areas. The three-field map showed how countries, authors, and topics were linked together in this literature. China, India, and the United States were identified as the leading players and were the core of the international research collaboration. These authors include Wang Y, Zhang Y, and JR, who were found to connect major themes, including Climate Change, Adaptation, Vulnerability, and Resilience, to different research clusters.

Wang Y, Ford JD, and Pandey R were found to be the authors who have contributed and published the most in the field. By using their H-Index, G-Index, and TC, they have immensely contributed to the field. Ford JD was the most prominent author with the highest citation impact, and Wang Y and Pandey R were acknowledged for their continuous contribution and focus on the topic. The analysis by country of corresponding authorship showed that the leading countries were China, India, and the United States. The countries represented here are characterized by strong academic institutions and research grants, which translates into their high productivity. New contributions from Germany, Canada, and Australia also allow for identifying the expansion of the research field. The authors from China and India were the most outstanding contributors to climate change and indigenous research, demonstrating the key position of universities and research institutions. These institutions have developed many linkages especially with their counterparts in developed countries such as the United States and Australia. The United States and Canada were the most productive countries in terms of TC, due to their continuous contributions and high-quality articles. Many of these countries have been leading the way in documenting how climate change affects indigenous people and will provide a standard for other investigations.

The thematic analysis highlighted adaptation measures, vulnerability evaluation, and resilience development of indigenous people. Sustainability and indigenous knowledge are the new fields that show a trend toward integrating indigenous knowledge into climate change policies. This transformation of thematic focus is a testimony to the rising appreciation of indigenous knowledge in sustainable development. The most frequent words in the field, like Climate Change, Indigenous Knowledge, Adaptation, and Vulnerability, are characteristic of the contemporary global agenda.

Thematic keywords also point to an increasing focus on integrated and social types of climate change resilience.

The study showed significant, normalized collaboration, where China, India, and the United States were the most active middlemen in the network. This shows cooperation between the developed and the developing countries in tackling climate issues and using the different forms of expertise. The most cited articles pointed out the basic works that have had a major impact on the development of the area. Many of these works provided solutions to problems such as climate change, indigenization of knowledge, and policy implications, which are pivotal in defining academic research. Based on Scopus, we recognized Regional Environmental Change and Climatic Change as the most significant journals that contributed to the existing knowledge on climate change and indigenous people's subsistence. These high citation impact factors show that they serve as the main outlets for contributions that can make a difference. The findings showed shifting research themes, with the prior literature concentrating on climate change impact and adaptation, and the recent literature concentrating on the sustainability and resilience. This shift is toward focusing on sustainable and gender and indigenous people's sensitive solutions.

The results of this research are consistent with and build on other scientometric and bibliometric investigations in similar disciplines (Baidya and Saha, 2024; Hsieh and Yeh, 2024). Other earlier works, including those by Ebhuoma (2024); Ford et al. (2015b), have emphasized the importance of adaptation measures and indigenous knowledge to mitigate climate change effects. This research affirms their conclusions as these themes dominate the literature of the present study. Likewise, the identification of the most productive authors in the analysis, Wang Y and Ford JD, is in line with previous studies that found this researcher's works to be most influential in the field (Borolla et al., 2024). However, this study builds on the previous research by not only tracking the global and regional collaboration networks but also the thematic trends, especially with the help of three-field map analysis. Unlike previous literature that has mainly looked at the developed world, this research shows the growing role of developing countries, especially China and India. This is a move toward more diverse research recruitment, not only in terms of inclusivity but also geographical location, as has been observed by other authors. Furthermore, the focus on such emerging areas as resilience and sustainability in this study also points to the dynamic nature of the research area that has been relatively underexplored in other bibliometric assessments. Through these findings, this study provides a more comprehensive view of the field's evolution than would be possible without them.

As more research is being conducted on climate change and indigenous livelihoods, the present scientometric analysis reveals some important research gaps. A relatively limited regional emphasis appears as one of the major limitations (Figure 6.10). This is the case despite the fact that China, India, and the United States lead the research agenda, little or no research comes from or is focused on Africa, Latin America, or small island developing states. It is in these regions that the impact of climate change is most felt, yet these areas are seldom analyzed within literature (Hussainzad and Gou, 2024; Song, 2024). Another major problem is the lack of voice given to indigenous peoples' point of view in the current literature. This paper, therefore, argues that despite the general recognition of the significance of indigenous knowledge in addressing climate change, the conceptualizations of such knowledge remain marginalized within the climate change discourse. Much of the current literature focuses on the application of scientific approaches rather than the knowledge and understanding of the communities concerned. More needs to be done to ensure that indigenous people are heard in climate change research and policy.

The analysis also shows that collaboration networks in global research are rather fragmented. However, the study shows that there are close associations between the developed economies and the emerging economics but limited ties between the developed economies and the vulnerable regions. This exclusion reduces the sharing of experiences and lessons learned, which are needed for the formulation of sound climate change adaptation measures. Expanding the existing global collaboration networks to the currently marginalized regions might create more effective and fair plans to tackle climate change. As far as the current literature is concerned, there are areas of

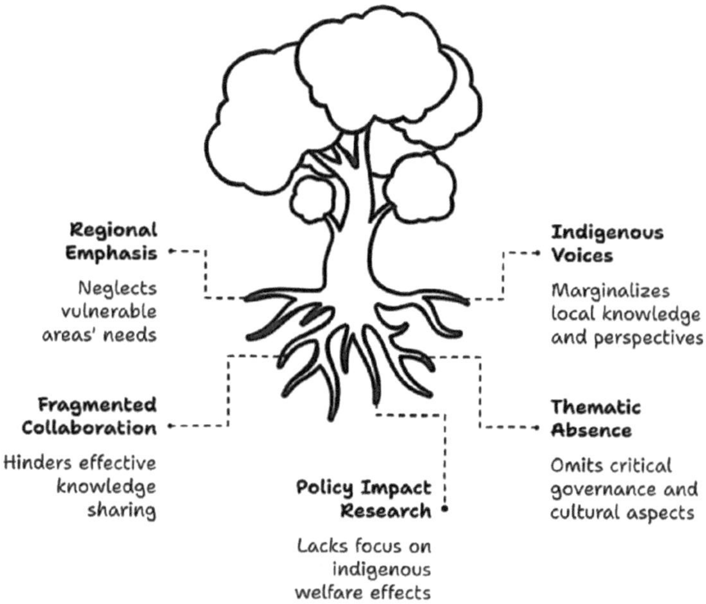

FIGURE 6.10 Limited representation in climate change research.

thematic absence. Climate change, adaptation, vulnerability, and resilience are the main themes of the research; yet, important topics such as governance, indigenous leadership, and the social–cultural aspects of climate change remain scarce. Furthermore, there is a lack of literature on the effects of climate policies on indigenous people's welfare in the future (Menzies et al., 2022).

CONCLUSION

This research work presents a detailed scientometric review of the global literature on climate change and indigenous livelihoods. Through an analysis of publication frequencies, authors, geographical locations, co-authorship, themes, and citations, the article reveals the developments, major players, and the gaps in the field. The literature exploring the impacts of climate change on indigenous people's livelihoods and the copious literature on climate change has expanded in recent years. Nevertheless, according to the analysis, there are ongoing inequalities in the geographical distribution of the analyzed publications, their thematic priorities, and the ability to encompass indigenous knowledge in mainstream scientific and political discourse. Remarkable contributions from China, India, and the United States dominate the current scholarship, while scholars from Africa, Latin America, and small island developing states, who are most affected by climate change, are relatively scarce. However, there is limited analysis of critical issues including governance, socio-cultural issues, and the effects of climate actions on the lives of indigenous people in the adapted and resilient societies. Factors such as the limited incorporation of indigenous knowledge systems and the still weak and fragmented global cooperation networks call for a more comprehensive and integrative research agenda.

These conclusions for this study provide important implications to further research and policy on climate change and indigenous peoples' livelihoods. Closing the geographic divide means focusing on neglected areas particularly the most disadvantaged ones including Africa, Latin America, and small island developing states. Integration of the indigenous knowledge systems and support for community-led initiatives must be encouraged in order to increase the appropriateness of adaptation measures. Broadening the existing global collaboration networks to engage vulnerable regions and intensifying interdisciplinary work on the governance, socio-cultural aspects, and future climate

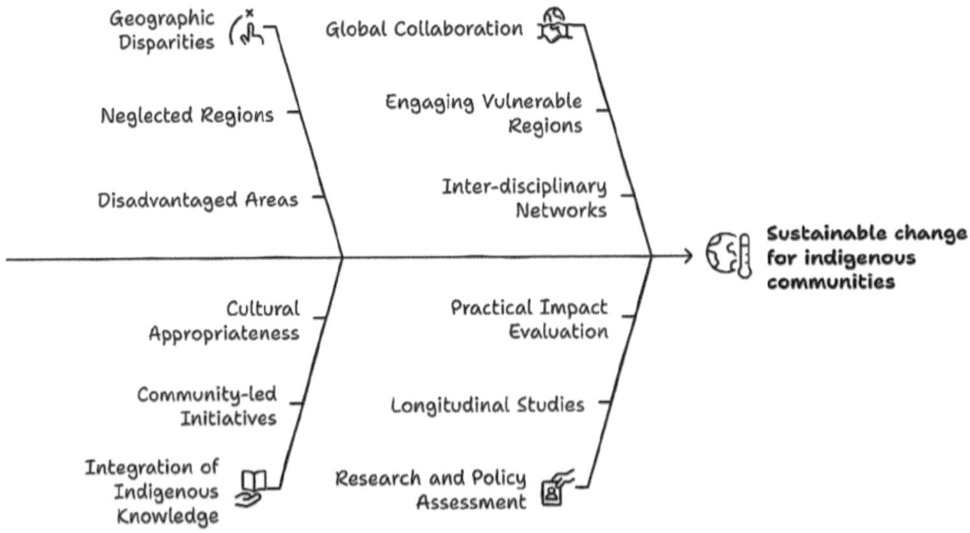

FIGURE 6.11 Bridging gaps in climate policy for indigenous peoples.

change consequences is critical. Moreover, increasing the availability of longitudinal research and assessing the practical impact of climate change policies can contribute to closing the gap between theory and implementation and support positive and sustainable change for indigenous communities (Figure 6.11).

REFERENCES

Abbass, K., Qasim, M. Z., Song, H., Murshed, M., Mahmood, H., & Younis, I. (2022). A review of the global climate change impacts, adaptation, and sustainable mitigation measures. *Environmental Science and Pollution Research*, *29*(28), 42539–42559. https://doi.org/10.1007/s11356-022-19718-6

Adger, W. N., Dessai, S., Goulden, M., Hulme, M., Lorenzoni, I., Nelson, D. R., & Wreford, A. (2009). Are there social limits to adaptation to climate change? *Climatic Change*, *93*, 335–354. https://doi.org/10.1007/s10584-008-9520-z

Aria, M., & Cuccurullo, C. (2017). Bibliometrix: An R-tool for comprehensive science mapping analysis. *Journal of Informetrics*, *11*(4), 959–975. https://doi.org/10.1016/j.joi.2017.08.007

Assembly, U. G. (2007). United Nations declaration on the rights of indigenous peoples. *UN Wash*, *12*, 1–18. https://www.converge.org.nz/pma/DRIPGA.pdf

Babu, P. K. (2024). Traditional agri/horticultural crops, farming practices, and conservation in the Himalayan region. In A. Borthakur & P. Singh (Eds.), *The Resilience of Traditional Knowledge Systems for a Sustainable Future* (pp. 59–90). Cham: Springer. https://doi.org/10.1007/978-3-031-56858-9_3

Baidya, A., & Saha, A. K. (2024). Exploring the research trends in climate change and sustainable development: A bibliometric study. *Cleaner Engineering and Technology*, 100720. https://doi.org/10.1016/j.clet.2023.100720

Biggs, E. M., Bruce, E., Boruff, B., Duncan, J. M., Horsley, J., Pauli, N., ... & Imanari, Y. (2015). Sustainable development and the water–energy–food nexus: A perspective on livelihoods. *Environmental Science & Policy*, *54*, 389–397.

Borolla, F. V., Rahmayanti, H., Hidayat, O. S., & Abdullah, K. H. (2024). Indigenous knowledge and sustainable resource management: A bibliometric review. *Fronteira: Journal of Social, Technological and Environmental Science*, *13*(4), 88–103. https://orcid.org/0000-0002-5035-9105

Early, R., Bradley, B. A., Dukes, J. S., Lawler, J. J., Olden, J. D., Blumenthal, D. M., ... & Tatem, A. J. (2016). Global threats from invasive alien species in the twenty-first century and national response capacities. *Nature Communications*, *7*(1), 12485.

Ebhuoma, E. E. (2024). Indigenous knowledge and natural infrastructure resilience to climate change in developing countries: a bibliometric analysis. *Frontiers in Environmental Economics*, *3*, 1295690. https://doi.org/10.3389/frevc.2024.1295690

Figueiredo, R. D. O., Cak, A., & Markewitz, D. (2020). Agricultural impacts on hydrobiogeochemical cycling in the Amazon: Is there any solution? *Water, 12*(3), 763. https://doi.org/10.3390/w12030763

Ford, J. D., Berrang-Ford, L., Biesbroek, R., Araos, M., Austin, S. E., & Lesnikowski, A. (2015a). Adaptation tracking for a post-2015 climate agreement. *Nature Climate Change*, *5*(11), 967–969. https://doi.org/10.1038/nclimate2744

Ford, J. D., McDowell, G., & Pearce, T. (2015b). The adaptation challenge in the Arctic. *Nature Climate Change*, *5*(12), 1046–1053. https://doi.org/10.1038/nclimate2723

Galappaththi, E. K., Ford, J. D., Bennett, E. M., & Berkes, F. (2019). Climate change and community fisheries in the arctic: A case study from Pangnirtung, Canada. *Journal of Environmental Management, 250*, 109534. https://doi.org/10.1016/j.jenvman.2019.109534

Hamilton, S. E., & Casey, D. (2016). Creation of a high spatio-temporal resolution global database of continuous mangrove forest cover for the 21st century (CGMFC-21). *Global Ecology and Biogeography*, *25*(6), 729–738.

Hecht, S. B. (1982). Agroforestry in the Amazon Basin: Practice, theory and limits of a promising land use. *Amazonia: Agriculture and Land Use Research*, 331–371. https://books.google.co.in/books?hl=en&lr=&id=dJOaBQAAQBAJ&oi=fnd&pg=PA331&dq

Hillsdon, M. (2024, November 5). Society watch: Why respecting Indigenous peoples' rights is key to the energy transition. *Reuters*. https://www.reuters.com/sustainability/society-equity/society-watch-why-respecting-indigenous-peoples-rights-is-key-energy-transition-2024-11-05

Hsieh, Y. L., & Yeh, S. C. (2024). The trends of major issues connecting climate change and the sustainable development goals. *Discover Sustainability*, *5*(1), 31. https://doi.org/10.1007/s43621-024-00183-9

Hussainzad, E. A., & Gou, Z. (2024). Climate risk and vulnerability assessment in informal settlements of the global south: A critical review. *Land*, *13*(9), 1357. https://doi.org/10.3390/land13091357

Kasperson, R. E., & Kasperson, J. X. (2012). Climate change, vulnerability and social justice. In *Social Contours of Risk* (pp. 301–321). Routledge. https://www.taylorfrancis.com/chapters/edit/10.4324/9781849772549-20/

Kronik, J., & Verner, D. (2010). *Indigenous Peoples and Climate Change in Latin America and the Caribbean*. World Bank Publications. https://books.google.co.in/books?hl=en&lr=&id=MdAAGCpSXrUC&oi=fnd&pg=PR5&dq

Lemi, T. (2019). The role of traditional ecological knowledge (TEK) for climate change adaptation. *International Journal of Environmental Sciences & Natural Resources*, *18*(1), 28–31. https://www.researchgate.net/profile/Tamiru-Lemi/publication/339505270

Magni, G. (2017). Indigenous knowledge and implications for the sustainable development agenda. *European Journal of Education*, *52*(4), 437–447. https://doi.org/10.1111/ejed.12238

Meakin, S., & Kurvits, T. (2009). Assessing the impacts of climate change on food security in the Canadian Arctic. *Affairs, Ia N (Ed.). GRID-Arendal,* Ottawa, Canada, *46*. https://www.cakex.org/sites/default/files/documents/foodsec_updt_LA_lo.pdf

Menzies, A. K., Bowles, E., Gallant, M., Patterson, H., Kozmik, C., Chiblow, S., & Popp, J. N. (2022). "I see my culture starting to disappear": Anishinaabe perspectives on the socioecological impacts of climate change and future research needs. *Facets*, *7*(1), 509–527. https://doi.org/10.1139/facets-2021-0066

Nelson, M. K., & Shilling, D. (Eds.). (2018). *Traditional Ecological Knowledge: Learning from Indigenous Practices for Environmental Sustainability*. Cambridge University Press. https://books.google.co.in/books?hl=en&lr=&id=QjdqDwAAQBAJ&oi=fnd&pg=PA141&dq

Norton-Smith, K. (2016). *Climate Change and Indigenous Peoples: A Synthesis of Current Impacts and Experiences* (Vol. 944). United States Department of Agriculture, Forest Service, Pacific Northwest Research Station.

O'Donoghue, R., Sandoval-Rivera, J. C. A., & Payyappallimana, U. (2019). Landscape, memory and learning to change in changing worlds: Contemplating intergenerational learning and traditional knowledge practices within social-ecological landscapes of change. *Southern African Journal of Environmental Education*, *35*. https://doi.org/10.4314/sajee.v35i1.10

Osman, K. T. (2014). *Soil Degradation, Conservation and Remediation* (Vol. 820). Dordrecht: Springer Netherlands. https://doi.org/10.1007/978-94-007-7590-9

Pelling, M., O'Brien, K., & Matyas, D. (2015). Adaptation and transformation. *Climatic Change*, *133*(1), 113–127.

Rahman, H. T., Robinson, B. E., Ford, J. D., & Hickey, G. M. (2018). How do capital asset interactions affect livelihood sensitivity to climatic stresses? Insights from the northeastern floodplains of Bangladesh. *Ecological Economics*, *150*, 165–176. https://doi.org/10.1016/j.ecolecon.2018.04.006

Rahman, M. H., & Alam, K. (2016). Forest dependent indigenous communities' perception and adaptation to climate change through local knowledge in the protected area—A Bangladesh case study. *Climate*, *4*(1), 12. https://doi.org/10.3390/cli4010012

Rashidi, P., & Lyons, K. (2022). Democratizing global climate governance? The case of indigenous representation in the Intergovernmental Panel on Climate Change (IPCC). In *Economics and Climate Emergency* (pp. 284–299). Routledge. https://www.taylorfrancis.com/chapters/edit/10.4324/9781003174707-21

Rivera-Ferre, M. G., Di Masso, M., Vara, I., Cuellar, M., López-i-Gelats, F., Bhatta, G. D., & Gallar, D. (2021). Traditional agricultural knowledge in land management: the potential contributions of ethnographic research to climate change adaptation in India, Bangladesh, Nepal, and Pakistan. *Climate and Development, 13*(7), 644–661. https://doi.org/10.1080/17565529.2020.1848780

Rockström, J., Williams, J., Daily, G., Noble, A., Matthews, N., Gordon, L., ... & Smith, J. (2017). Sustainable intensification of agriculture for human prosperity and global sustainability. *Ambio, 46*(1), 4–17.

Schlosberg, D., & Collins, L. B. (2014). From environmental to climate justice: climate change and the discourse of environmental justice. *Wiley Interdisciplinary Reviews: Climate Change, 5*(3), 359–374. https://doi.org/10.1002/wcc.275

Schlosberg, D., Collins, L. B., & Niemeyer, S. (2017). Adaptation policy and community discourse: risk, vulnerability, and just transformation. *Environmental Politics, 26*(3), 413–437. https://doi.org/10.1080/09644016.2017.1287628

Serdeczny, O., Adams, S., Baarsch, F., Coumou, D., Robinson, A., Hare, W., ... & Reinhardt, J. (2017). Climate change impacts in Sub-Saharan Africa: From physical changes to their social repercussions. *Regional Environmental Change, 17*(6), 1585–1600.

Singh, V. K., Singh, P., Uddin, A., Arora, P., & Bhattacharya, S. (2022). Exploring the relationship between journals indexed from a country and its research output: an empirical investigation. *Scientometrics, 127*(6), 2933–2966. https://doi.org/10.1007/s11192-022-04366-x

Song, K. (2024). Knowledge of climate change in developing countries and its effects. *Journal of Student Research, 13*(2). https://doi.org/10.47611/jsrhs.v13i2.6806

Suhrke, A., & Hazarika, S. (1993, June). *Pressure Points: Environmental Degradation, Migration and Conflict*. Cambridge, MA: American Academy of Arts and Sciences. https://www.cmi.no/publications/file/1374%20pressure-points-environmental-degradation.pdf

Tao, S., Fang, J., Zhao, X., Zhao, S., Shen, H., Hu, H., ... & Guo, Q. (2015). Rapid loss of lakes on the Mongolian Plateau. *Proceedings of the National Academy of Sciences, 112*(7), 2281–2286.

Tellman, B., Sullivan, J. A., Kuhn, C., Kettner, A. J., Doyle, C. S., Brakenridge, G. R., ... & Slayback, D. A. (2021). Satellite imaging reveals increased proportion of population exposed to floods. *Nature, 596*(7870), 80–86.

Van Eck, N., & Waltman, L. (2010). Software survey: VOSviewer, a computer program for bibliometric mapping. *Scientometrics, 84*(2), 523–538. https://doi.org/10.1007/s11192-009-0146-3

Weiskopf, S. R., Rubenstein, M. A., Crozier, L. G., Gaichas, S., Griffis, R., Halofsky, J. E.,... & Whyte, K. P. (2020). Climate change effects on biodiversity, ecosystems, ecosystem services, and natural resource management in the United States. *Science of the Total Environment, 733*, 137782. https://doi.org/10.1016/j.scitotenv.2020.137782

Weiss, M., Jacob, F., & Duveiller, G. (2020). Remote sensing for agricultural applications: A meta-review. *Remote Sensing of Environment, 236*, 111402.

Whyte, K. P. (2013). On the role of traditional ecological knowledge as a collaborative concept: A philosophical study. *Ecological Processes, 2*, 1–12. https://doi.org/10.1186/2192-1709-2-7.

Whyte, K. (2017). Indigenous climate change studies: Indigenizing futures, decolonizing the Anthropocene. *English Language Notes, 55*(1), 153–162.

7 Assessing the Challenges of Coastal Areas of Mousuni Island in the Indian Sundarbans under Climate Change and Sustainable Tourism-Based Human Adaptation

Abhay Sankar Sahu, Milan Ghosh, Selim Molla, Jaysikha Bhattacharyya, and Mouparna Ghosh

INTRODUCTION

Sundarbans, the world's largest mangrove forest, is a vulnerable delta region of the Indian sub-continent, which consists of several islands. It is outspread between India and Bangladesh, while latter comprises the largest proportion of it. Situated over the coast of the Bay of Bengal, this region has to cope with severe problems of coastal erosion, inundation, loss of land, etc. due to numerous hydro-geomorphic and climatic factors, and human interference also. Sundarbans is renowned for its unique flora and fauna as well as its extraordinary culture, and many tourists come here to experience this. Thus, tourism-related activities have flourished in Sundarbans both in India and in Bangladesh. It has been observed that many tourists are now interested in visiting the rich estuarine-based, less developed, less connected and less-known Sundarbans region for its natural beauty, where commercial or mass tourism cannot be developed. Thus, various forms of local small-scale tourism activities through the direct involvement of some local people, such as camp tourism, rural tourism, etc. have been developed here, which are broadly considered as eco-friendly tourism. Nascent tourist spots are now emerging here and there over the vulnerable coastal region of the Bay of Bengal, which signifies mainly different forms of rural tourism. Mousuni Island, situated in the Indian Sundarban delta, is a part of this vulnerable coastal region where camp tourism has emerged as a significant unplanned tourism activity along the seaside, exhibiting an exclusive local rural culture. Many of the local rural residents of this island are now dependent upon tourism-based activities for their livelihood. However, this tourism is being affected by the vulnerable nature of the island, mainly caused by climate change. Sea-level rise, severe coastal erosion and loss of land are the major issues of Mousuni Island, which are some of the consequences of climate change, and it could result in the entire island sinking within 80–100 years (Karmakar et al., 2020). Thus, in this context, the scenario of the coastal areas of Mousuni Island and the sustainability of its present camp tourism in this adverse environment has been assessed here for a sustainable human adaptation model.

DOI: 10.1201/9781003620747-8

At present, eco-friendly types of tourism or ecotourism have been given utmost priority for the sustainable development of different regions, especially in vulnerable and least developed regions like Sundarbans. It is viewed as ecologically sustainable nature-based tourism associated with teaching and understanding the natural environment, conservation of natural resources and sustainable development of the local community (Australia Department of Tourism, 1994, as cited in Diamantis, 1999). Camp tourism system, a kind of nature-based tourism, has mostly emerged in areas with low connectivity due to hard relief or vast water bodies. Traditionally, it is mainly involved with tenting or temporary accommodation systems and outdoor recreation in a rural environment (Brooker & Joppe, 2014). Rural tourism in rural areas with low connectivity has gained a lot of attention. Villages with attractive scenic beauty are nowadays becoming a destination for recreation and tourism. Tourism in rural areas creates job opportunities for local people and thus revitalises the local economy with the enhancement of the rural lifestyle in underprivileged areas (Liu, 2006).

This chapter presents the extension and impact of a very pronounced issue, that is, coastal erosion due to several hydro-geomorphic, climatic as well as human-induced factors, which cause severe environmental damage on Mousuni Island. Besides coastal erosion, present tourism-based activities on this island are also responsible for environmental deterioration. In this context, two principal models or approaches have been used in this study to evaluate such deterioration, which are the Digital Shoreline Analysis System (DSAS) and the Scenic Quality Model. In order to assess the coastal erosion on Mousuni Island, DSAS, an augmentation to the ArcGIS platform, has been used to analyse the temporal changes of the coastline of this island from 2006 to 2021. Samanta et al. (2017) have estimated that Mousuni Island has lost about 5.24 km^2 of its land between 1968 and 2012, and its total area has been reduced by 15.63%. The impact of present tourism-based activities on the scenic quality of this island, especially in the coastal areas, has been reflected through using the Scenic Quality Model of Integrated Valuation of Ecosystem Services and Trade-offs (InVEST) software. In this vulnerable background under the present scenario of climate change, the sustainability of the tourism-based human adaptation model on Mousuni Island has been assessed for sustainable development of this island.

STUDY AREA

Mousuni Island, a *gram panchayat* in Namkhana block of Kakdwip sub-division in South 24 Parganas district, West Bengal, India (Figure 7.1), is a small, vulnerable island located in the Indian Sundarban delta region, approximately 120 km south of Kolkata. Muriganga River borders this island to the west and north-west, Pitt's Creek or Chinai River to the east and the Bay of Bengal to the south. The present 24.11 km^2 area of this island is divided into four revenue *mouzas*, namely, Mousani or Mousuni, Bagdanga, Kusumtala and Baliara, where the total population residing on this island is 22,073 (Census of India, 2011). Here, the focus of this study is paid to the southernmost *mouza* that is, Baliara, as it has been highly vulnerable for many years, confronting coastal erosion and cyclonic hazards during the monsoon period especially and also maximum tourism-based adaptation is observed here due to its highest proximity to the sea. The population of this 12.10 km^2 *mouza* is 8,672, with a total number of 1,746 households and a population density of almost 717 persons per km^2 (Census of India, 2011).

Mousuni Island faces direct impact of climate change through frequent occurrences of natural hazards and storm surges, which directly affect the poor, local rural people living near the bottomland or coastal areas of this island, forcing them to relocate or migrate and become ecological refugees (Mukherjee & Siddique, 2018). Climate change-induced natural hazards lead to severe coastal erosion, huge destruction of properties and natural resources, flood and salt water intrusion, inundation and loss of land on Mousuni Island, which has a direct impact on the lives of rural people of this island, mainly engaged in farming and fishing for their livelihoods. Fishing and agricultural activities are directly affected by the above phenomenon, which has made the lives

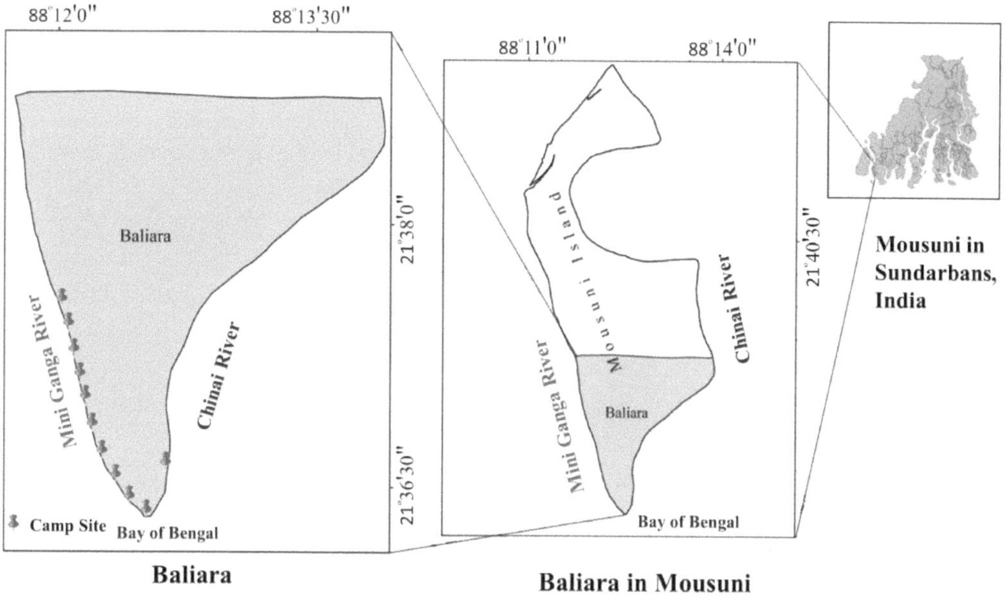

FIGURE 7.1 Study area, prepared based on the survey of India, 2011. (Google Maps, 2023; Google Earth Imagery, 2024.)

of poor people hard on Mousuni Island and made them both ecologically and socio-economically vulnerable (Mukherjee & Siddique, 2018).

In this vulnerable environment of Mousuni Island, the local rural people have been forced to choose tourism as alternative for their livelihood in collaboration with external agencies. In this context, a question arises regarding the sustainability or eco-friendliness of this new and unplanned tourism, and attention should also be given to the analysis of the vulnerability of this island for its sustainable development. In this background, this study is relevant to focus on the above issues.

DATABASE AND METHODOLOGY

This research has been conducted utilising both secondary and primary data sources. Mousuni Island's shoreline has been changed due to continuous erosion and deposition through time. A meticulous field survey has been done among camp-owners, tourists and local people with organised questionnaires. Different secondary data sources have been followed to understand the history of coastal issues here, and models like DSAS and Scenic Quality Model using InVEST have also been considered. Finally, for understanding the present situation of the study area, especially the socio-economic condition of local people and to review the reasons behind opting for tourism activity as a new dimension of human adaptation on Mousuni Island, which is not very old on a temporal scale, an extensive field survey was carried out through door-to-door visits.

DIGITAL SHORELINE ANALYSIS SYSTEM

DSAS is a software extension developed by ESRI Geographic Information System (GIS) as an augmentation to the ArcGIS platform, facilitating statistical analysis of shoreline dynamics. It can examine changes in coastlines over time by utilising historical shoreline positions represented in the form of a vector layer. Multispectral satellite images have been used to extract shorelines of Mousuni Island, West Bengal (Table 7.1).

TABLE 7.1
Description of Satellite Data

Satellite and Sensor	Date	Path & Row	Band Used	Spatial Resolution (m)
Landsat-5 TM	2006/04/23	Path: 138 Row: 044	VNIR	30
Landsat-7 ETM+	2011/04/26	Path: 138 Row: 044	VNIR	30
Landsat-8 OLI level-1	2016/04/06	Path: 138 Row: 044	VNIR	30
Landsat-8 OLI level-1	2021/04/25	Path: 138 Row: 044	VNIR	30

Note: All data are sourced from USGS EarthExplorer (2023).

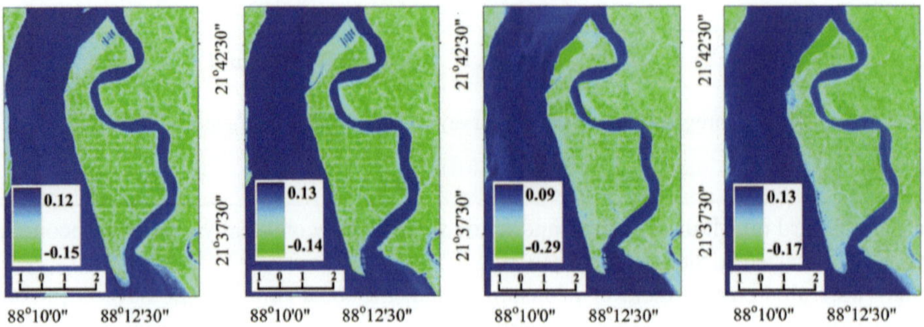

FIGURE 7.2 (a) NDWI of 2006, (b) NDWI of 2011, (c) NDWI of 2016, (d) NDWI of 2021. (Based on satellite images mentioned in Table 7.1.)

It is very difficult to extract the shoreline correctly because it is the transitional zone between land and seawater (Maiti & Bhattacharya, 2009). In this chapter, the Normalised Difference Water Index (NDWI) (Figure 7.2a–d) has been adopted. NDWI is an index-based measurement of the moisture content on the Earth's surface. The following equation has been used to calculate NDWI:

$$\text{NDWI} = \frac{\text{Green} - \text{NIR}}{\text{Green} + \text{NIR}} \tag{7.1}$$

Here, green and near infrared bands are represented by green and NIR, respectively. NDWI ranges between –1 and +1 where a value close to +1 signifies water bodies, while a value close to –1 ensures non-water bodies (McFeeters, 1996). However, the most difficult aspect is determining a suitable threshold point within the range of –1 to +1, which can effectively distinguish water bodies from land areas (Sahu, 2014a; Sahu, 2014b; Sahu, 2016).

After extracting shorelines, a baseline has been created on the landward side at a buffer distance of 500 m. Around 625 transect lines have been drawn. The end point rate (EPR) technique is used to assess the extent of shoreline change over a specific time. EPR is able to measure the rate of change in shoreline position. Typically, it is calculated by measuring the change in shoreline position between the starting and ending years of the study and then dividing that value by the total number of years in the study period, as outlined below:

$$\text{EPR} = \frac{D_1 - D_2}{t_1 - t_2} \tag{7.2}$$

Here, $D_1 - D_2$ represents the change in coastline position between two consecutive years, namely t_1 and t_2.

EPR is generally more reliable for short-term analysis. Net Shoreline Movement (NSM) is also used as a crucial statistical tool employed for monitoring and forecasting shoreline. It represents the distance between shoreline positions measured in two consecutive years. The following equation has been used to determine NSM:

$$\text{NSM} = (D_y - D_0)m \tag{7.3}$$

Here, D_y and D_o refer to the shoreline positions of the most recent and the earliest years, respectively.

The DSAS tool can both calculate historical shoreline change rates and forecast future changes—such as projections for the next 10 or 20 years—by analysing past shoreline data. This forecasting is performed using the Kalman filter (Kalman, 1960), which combines observed shoreline positions with predicted values. To ensure accurate predictions of future shoreline changes, the Kalman filter employs the linear regression method.

SCENIC QUALITY MODEL USING INVEST

The Scenic Quality Model using InVEST has been applied to identify the scenic beauty spots most impacted within the tourist destination. InVEST is an open-source software model for spatial assessment of ecosystem services relying on GIS and developed by the Natural Capital Project of Stanford University (available at – https://naturalcapitalproject.stanford.edu/software/invest).

Three main requirements for scenic quality assessment using InVEST are the vector layer of the study area, features impacting scenic quality as a vector layer and a digital elevation model. Here, scenic quality is determined based on three parameters, which are illegal soil mining, construction of artificial structures and embankments. In this model, these parameters are used as point features, which negatively affect the scenic beauty of the area. These point features have been identified and incorporated from Google Earth Imagery, 2023. The DEM map of Mousuni Island has been collected from USGS SRTM, 2014. Every point feature has been incorporated based on the Wang et al. (2000) reference-plane viewed algorithm. The weightage and aggregate of raster value of each point feature have been calculated and are divided into quartiles to create a raster, which represents visual quality. The weighted sum of all point features has been computed by incorporating the weight and the sum of all visible raster points.

RESULTS

Shoreline Analysis

The shorelines have been extracted from NDWI maps of the years 2006, 2011, 2016 and 2021 (Figure 7.3). From 2006 to 2021, the total length of the shoreline was decreased by 2.24 km. The maximum decrease in shoreline length (1.4 km) is observed between 2006 and 2011. The analysis reveals a consistent decline in total land area from 2006 to 2021, primarily due to erosion outweighing accretion during this period. Maximum erosion observed in the south-western portion of the study area, and consequently shoreline shifted landward direction. The entire coast of Mousuni Island has been segmented into four sections, that is, zones A, B, C and D (Figure 7.4).

Figure 7.5 illustrates the spatial distribution of NSM between 2006 and 2021. To assess NSM, around 625 transects were generated at 50-m spacing. In zone A, between transects 2 and 61, NSM values range from –500 me to –350 m, with an average shoreline retreat of approximately –424.94 m. Transects numbered 62–74, 134–142, and 165–204 also exhibit negative NSM values, falling within the range of –350 to –200 m, which are mainly located near zone B (Figure 7.5). A mixing of positive

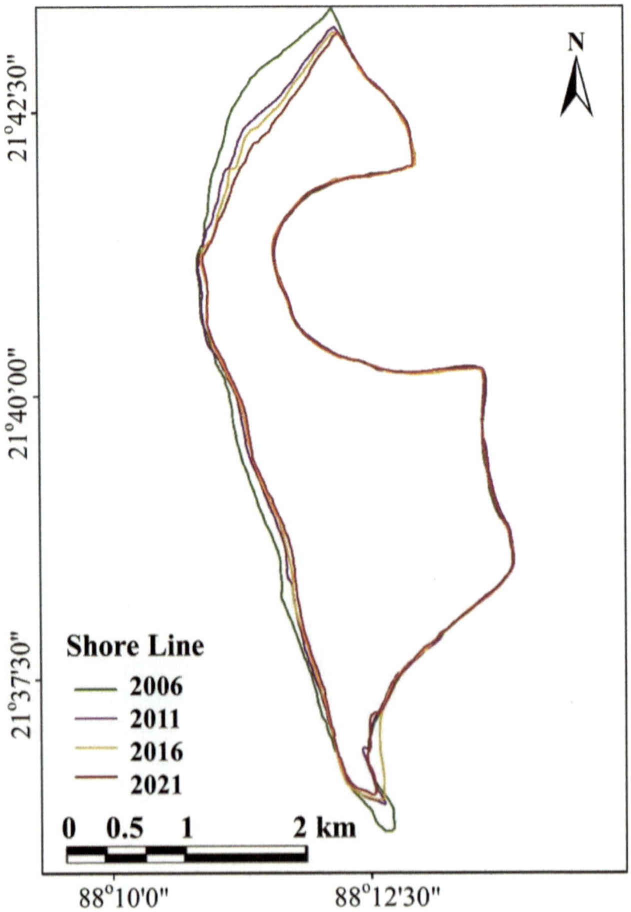

FIGURE 7.3 Temporal position of shorelines between 2006 and 2021.

and negative values of NSM is found from 286 to 625 transect numbers. The highest positive value of NSM is found in the D zone while negative values of NSM are observed in A, B and C zones. Spatial variation of EPR along Mousuni Island is shown in Figures 7.6 and 7.7. The EPR shows notably high negative values between transects 12 and 46 that are mainly located near zone A in the northern part and zone C in the southernmost part of the island. The EPR varies between –51.89 m and –34.66 m, with an average value of –42.31 m. Moderate negative EPR values range from –25 m to –5 m, which have been observed in zone B in the south-western part of the area. A mixing EPR value has been observed along the zone D from transect number 286–625 in the eastern part of the island.

The study calculated the total amount of eroded land between 2006 and 2021 is 3.77 km^2 (Figure 7.8). Maximum erosion took place during 2006–2011, when cyclone 'Aila' struck on May 25, 2009, along the coast of Mousuni Island. Consequently, a massive land loss (2.94 km^2) happened with cyclonic storm along the sea-facing coast of the Mousuni Island. Between 2016 and 2021, Mousuni Island has lost about 1 km^2 of the land area where cyclone 'Amphan' struck on May 20, 2020, which is mainly responsible for coastal erosion along the western part of the island. Between 2011 and 2016, no major cyclone struck over the coast of Mousuni Island, therefore, minor land loss was found during this period. Erosion is predominantly observed in the western and southern regions of the area under study. The rate of erosion between 2006 and 2021 is high in zones A and -C as these coastal belts are directly exposed to the sea. The rate of erosion is significantly lower in

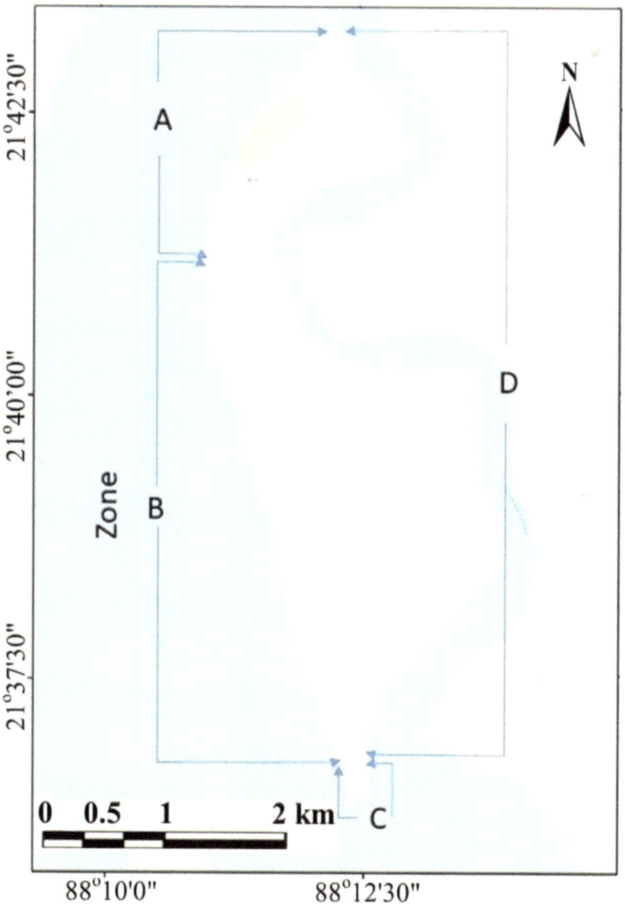

FIGURE 7.4 Segmentation of the coast of Mousuni Island.

the eastern part of the area because wave actions are less effective as the area is not fully exposed to the sea. Sediments carried by rivers are deposited along the eastern portion of the area, while wave action and storm surges have consistently contributed to erosion in the western part. As erosion is dominant over accretion, the total area of the island is decreasing at a remarkable rate.

Using the EPR model with the Beta Shoreline Forecasting tool in ArcGIS (Figure 7.9), future shoreline positions for 10-year and 20-year projections have been estimated. The model demonstrated that the total area will be decreased to 22.99 km^2 and 20.92 km^2 by the year 2031 and 2041, respectively (Figure 7.10). The model also predicted that the total length of the shoreline of the year of 2031 and 2041 will be 32.99 and 34.72 km, respectively, with a net change of –1.34 between 2021 and 2031, while 0.397 between 2031 and 2041.

SCENIC QUALITY ASSESSMENT

The Scenic Quality Model has been applied here to find out the impacts of newly proposed development projects based on their visual quality in nearshore, offshore, as well as backshore zones. The main development project is the tourist camps mainly along the south-western portion of Mousuni Island, specifically in Baliara *mouza*. The tourist camps have emerged due to the charming sea beauty, newly grown mangrove forests and an open and long sea beach. The scenic quality is excellent in the southern part of the island, especially in Baliara. The gradual

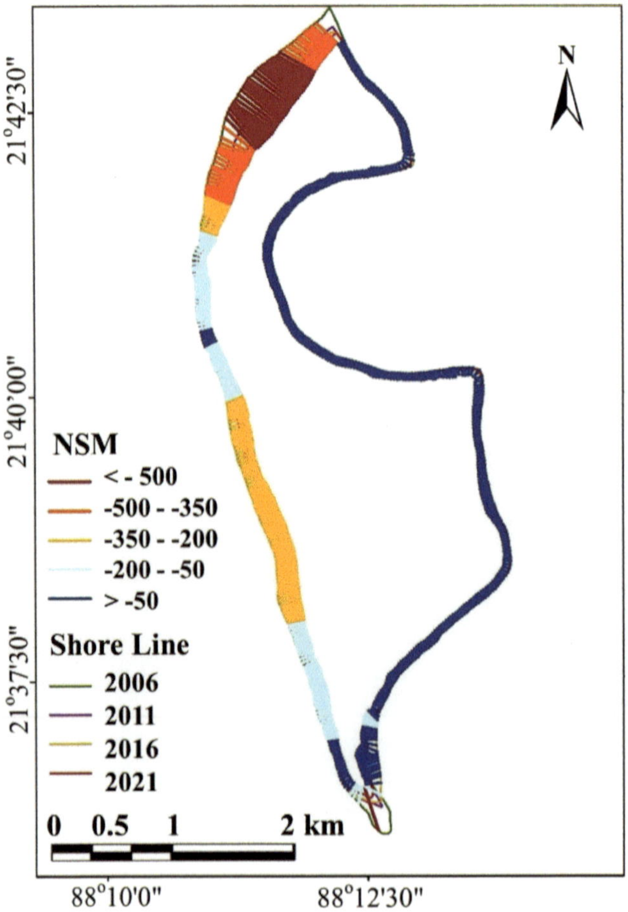

FIGURE 7.5 Spatial distribution of NSM.

development of tourism can partially support local people for their livelihood, but it also modifies the environment, which negatively affects the natural scenic quality of the area. Here, such kind of features are illegal soil mining, construction of embankments and development of new concrete structures. On the basis of these three parameters, here scenic quality is assessed through maps (Figure 7.11).

By imposing a Scenic Quality Model, it is found that scenic quality is mostly deteriorated in the southern, south-eastern and south-western parts of the island due to heavy soil mining for the temporary protection from storm surges and land filling for tourist camps. As the south-western part of the island faces continuous erosion, illegal soil mining is very high in this part. Another reason for scenic quality deterioration is the construction of concrete structures for tourist camps. Most of these structures are developed along the south-western portion of this island. Natural scenic beauty has been enhanced in recent times over the newly emerged land in the eastern part of the island due to the plantation of mangroves, but embankment along the eastern bank severely marred natural scenic beauty. The value of scenic quality is highest in the south-western and eastern portion (14–30) and therefore these zones are categorised under the highly negative impact class, that is, class 4 and the rest of the portion belongs to low to moderate class zones. Moreover, natural scenic beauty is relatively good in the western and northern portions of the island with scenic quality value ranges from 2.94 to 14. The main reason for this condition is the almost complete absence of tourist camps in that portion of the island.

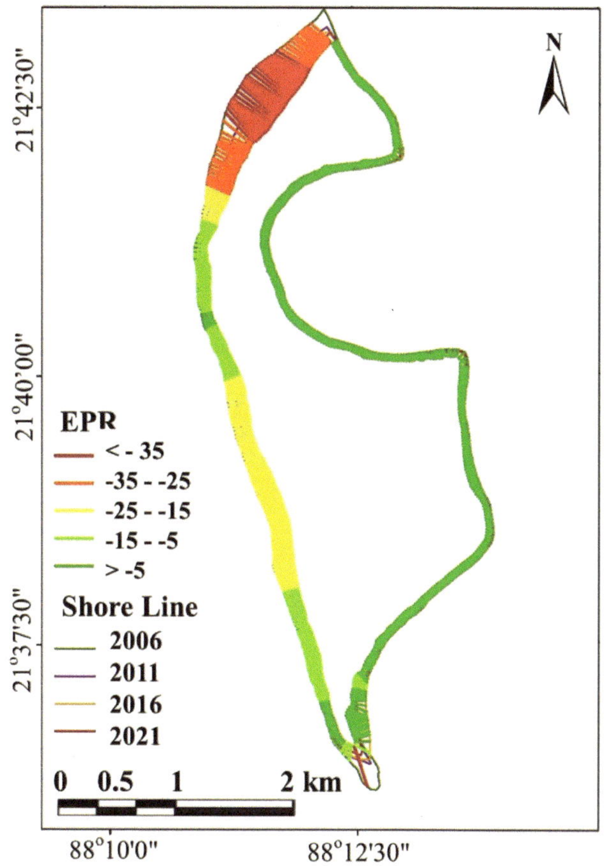

FIGURE 7.6 Spatial variation of EPR.

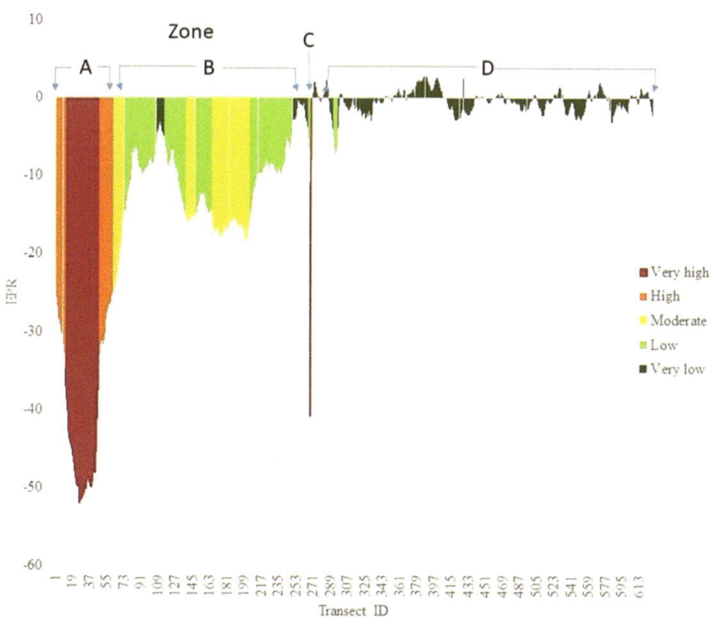

FIGURE 7.7 Transect-wise variation of EPR.

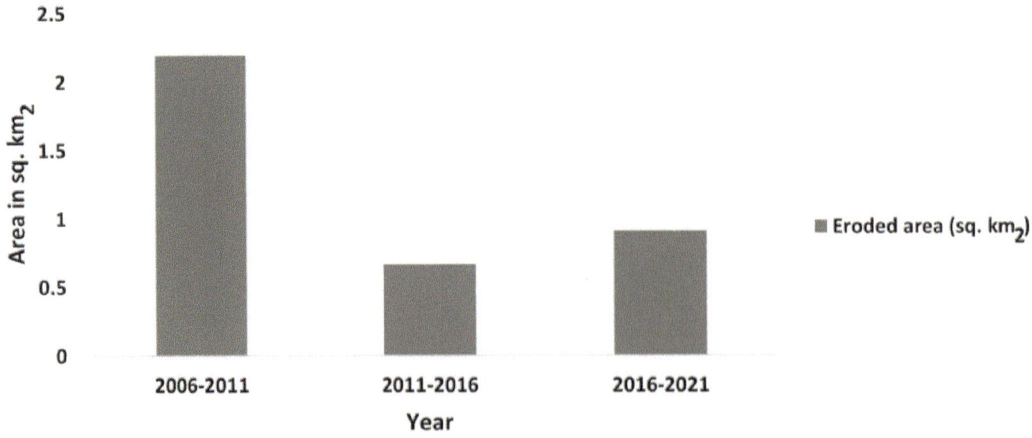

FIGURE 7.8 Total amount of erosion from 2006 to 2021.

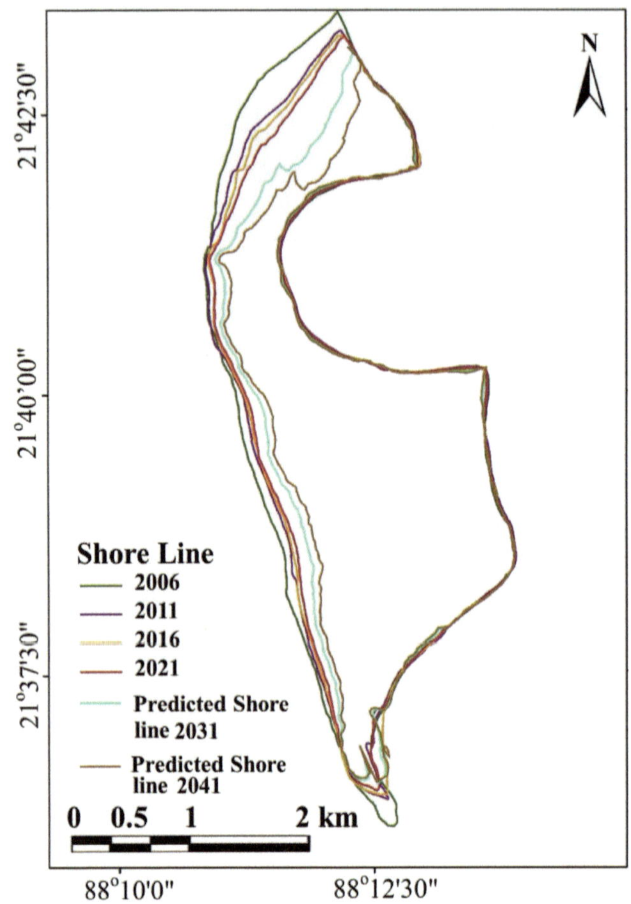

FIGURE 7.9 Predicted shoreline of year 2031 and 2041.

It is observed that tourist camps at Baliara are developed based on the scenic beauty, but now they are majorly responsible for the diminution of the daintiness of the island. Scenic quality class (Figure 7.11) in the extreme south-eastern and eastern portion has shown very high visual impact (>75 percentile) and the northern portion shows low visual impact (<25 percentile). Except for scenic

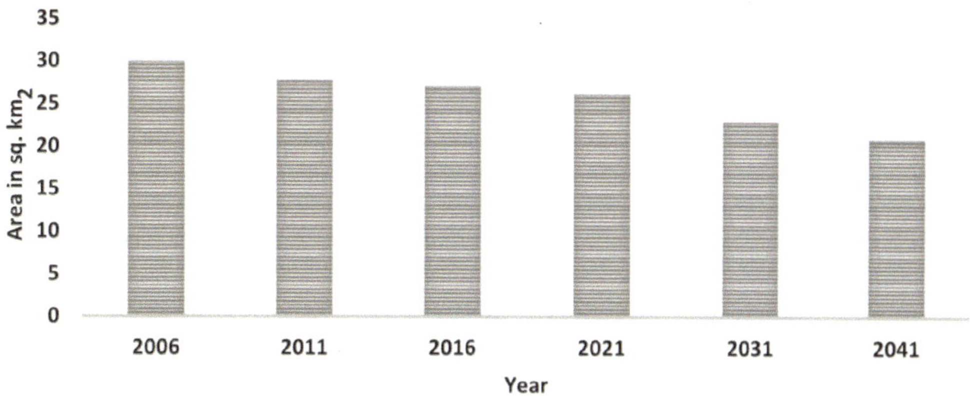

FIGURE 7.10 Total area from 2006 to 2041.

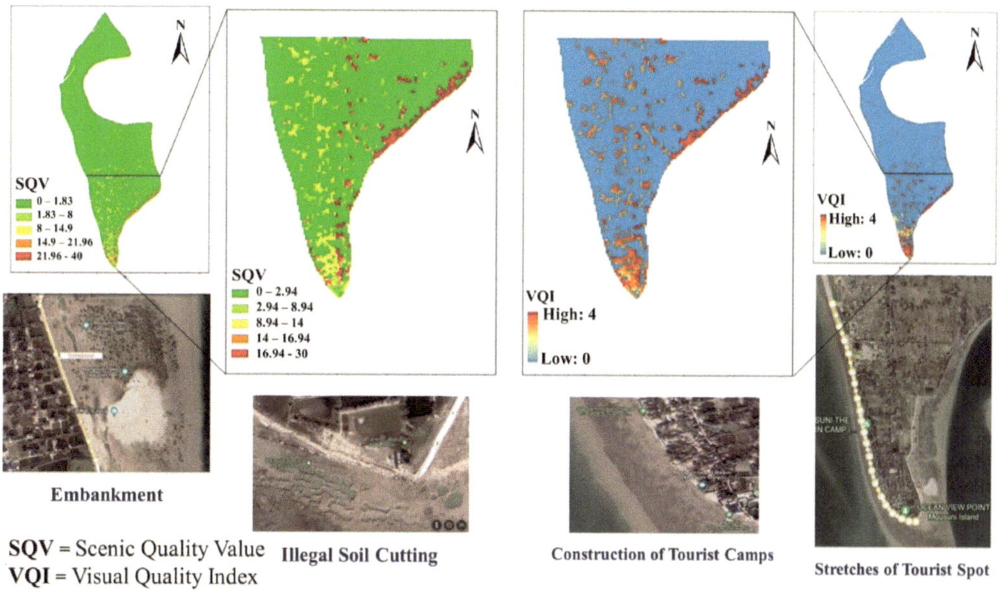

FIGURE 7.11 Scenic quality of Mousuni Island. (Based on Survey of India, 2011; Google Maps, 2023; Google Earth Imagery, 2024.)

quality, the beach ecological-ecosystem condition is also deteriorated due to illegal soil mining and the construction of concrete structures and accumulation of waste like plastic, glass, nets, etc. on the beach. The development of tourist camps and storm surge protections through embankments is mainly responsible for the deterioration of natural scenic beauty in Baliara, and thus the natural environment is gradually changed to artificial structures. It makes the area economically vulnerable as well as susceptible to natural hazards.

DISCUSSIONS

Challenges of Coastal Areas of Mousuni Island under Climate Change

From the above results, it is clear that Mousuni Island is facing severe coastal erosion, shrinking of its total area, and environmental deterioration, which accelerates the vulnerability of this island and it may continue in future if proper measures are not taken. The DSAS in the previous section of this study shows contemporary and future prediction of coastal erosion, which is highly remarkable for management and future planning of this island. Among the four revenue mouzas of Mousuni Island, Baliara is the southernmost and most vulnerable mouza as it experiences the maximum impacts of climatic hazards and disasters, and the highest amount of coastal erosion due to its highest proximity to sea, which leads to loss of land, a huge number of destruction of properties and displacement of the inhabitants, especially in the coastal areas. Baliara mouza has lost about 2.40 km^2 of its land since 1942 due to coastal erosion, which has caused every villager to lose some of their land to sea, as 51% of the families of this mouza stay in highly vulnerable areas (WWF-India, 2010). Such consequences are effects of changing patterns of temperature and rainfall, increased frequency of natural hazards and disasters (such as severe cyclones, storm surges, floods, etc.), loss of coastal land, etc., due to climate change (Mukherjee & Siddique, 2018). Repeated occurrences of cyclones and storm surges lead to frequent breaches in the poor embankment, especially in Baliara mouza, where only 1.8 km out of a total 8.49 km of embankment is brick-paved and the rest of the part is just susceptible earthwork, while the total length of the coastline of Baliara is 11.06 km (WWF-India, 2010). Frequent breaching of embankments leads to floods, saltwater intrusion and inundation of agricultural land and harming rural economic activities and the livelihoods of poor rural people and forcing them to relocate from coastal stretches.

TOURISM-BASED HUMAN ADAPTATION ON MOUSUNI ISLAND

Climate change and unplanned human activities have made Mousuni Island ecologically and socio-economically vulnerable. Due to the highest proximity to sea and presence of most of the tourism-related activities, the southern part of this island, especially Baliara mouza, experiences the highest vulnerable situations. Here, poor condition or breaching of embankments or absence of embankments results in heavy flood and inundation, accelerates the erosional activities, destroys a huge number of properties and land resources, which forces the poor rural residents of this island to relocate from their original residences, especially from coastal areas. It directly affects the livelihoods of the residents as most of the people are poor and largely dependent upon farming and fishing-related activities for their livelihood, which are largely affected by the vulnerable nature of this island. It has been seen that above 68% families in Baliara mouza are below the poverty line, and about 99% families reside in mud houses (WWF-India, 2010). The literacy rate among the people of Baliara is about 68.69%, while the male literacy rate is slightly higher than the female literacy rate (Census of India, 2011).

The present scenario of workers or the working population in Baliara mouza is also not progressive by nature. From Table 7.2, it can be inferred that a large number of the population of Baliara is unemployed or underemployed as the total number of non-workers is much higher than the total number of workers, which indicates a large dependency rate among the population of this mouza. Here, the female dependent population is higher than the male dependents. This scenario stipulates a poor socio-economic condition of the residents of Baliara, which is highly needed to be improved. Here, agricultural activity is the main economic activity as most people are engaged in it, however, it is highly affected by frequent floods and inundation or seawater intrusion, which leads to the destruction of crops, salinisation of soil, and makes it infertile or useless for agriculture for a long duration. A high level of coastal erosion also results in the loss of agricultural land. Cyclones, storms and other natural hazards also affect fishing-related activities, which is another important economic activity for the residents of this island. Each year, a considerable number of young individuals migrate from this island to other regions for educational or professional purposes. Therefore, in this adverse situation, alternatives are needed for the improvement of the socio-economic condition for the people of Baliara, and that's why people here are now being engaged to tourism-related activities as an alternative economic activity for their livelihood since the previous decade.

A small-scale immature tourism, camp tourism, has been developed on Mousuni Island, mainly along the coastal stretches, due to the existence of a natural sea beach, a planted mangrove forest,

TABLE 7.2
Scenario of Working Population in Baliara Mouza, Mousuni Island

Workers	Total No. of Persons	Total No. of Males	Total No. of Females
Main Workers	**1,909**	**1,733**	**176**
Cultivators	676	624	52
Agricultural labourers	558	529	29
Household industry workers	58	26	32
Other workers	617	554	63
Marginal Workers	**808**	**478**	**330**
Cultivators	217	122	95
Agricultural labourers	345	248	97
Household industry workers	109	10	99
Other workers	137	98	39
Total Workers	**2,717**	**2,211**	**506**
Non-workers	**5,955**	**2,147**	**3,808**

Source: Last Census of India, 2011.

the attraction of various migratory birds, its natural scenic beauty and resources. The Scenic Quality Model using InVEST in the previous section shows the impact of human interference on the natural environment, involving camp tourism and artificial engineering protection works. Here, camp tourism has been established along the sea beach in an unplanned manner. Accommodations are provided in cottages (both permanent and temporary structures), mud houses and tents, where tourists are served locally cooked food, giving this tourism on Mousuni Island a unique social identity. Due to the proximity of the sea beach, most of these camps are situated in Baliara, allowing tourists to enjoy sea views and the local rural life with fresh homemade local foods in a natural setting. Consequently, Mousuni Island has now become a favourite weekend tourist destination for people from Kolkata and nearby areas. Although various external agencies operate this camp tourism of this island, a significant number of local people are now also involved in this camp business in several ways such as camp owner or its manager, cooking and serving food, cleaning and washing, supply of necessary materials to the camps, overall maintenance, arrangement of some recreational activities for the tourists, etc. But, it may be pointed out that, this relatively new, unplanned camp tourism of Mousuni Island is facing numerous challenges and difficulties since its inception due to the island's vulnerability and hazard-prone nature, lack of funding and various administrative issues. It has not yet received full authorisation or recognition from the government due to different law and order issues. Additionally, this tourism is not fully sustainable, and the lack of facilities and communication gaps are the reasons it has not attracted the full attention of tourists yet. The present form of tourism has been largely responsible for the deterioration of the island's natural scenic beauty, as demonstrated by the Scenic Quality Model. Therefore, this unauthorised, unplanned and unsustainable camp tourism should be transformed into authorised, planned, and sustainable eco-friendly tourism. This would benefit the local population and conserve nature through effective utilisation of local resources, while guaranteeing maximum involvement of local residents in tourism-related activities. The government should take immediate action to develop a mature eco-friendly tourism on Mousuni Island through the maximum involvement of local people in order to protect both the island and its inhabitants, as it is one of the best alternatives for sustainable development of this island. This approach may also encourage youth of this island to choose tourism as a profession, potentially reducing the rate of outmigration from Mousuni Island in future.

CONCLUSION

Climate change and human activities are primarily accountable for the ecological and socio-economic vulnerable situation of Mousuni Island, such as severe coastal erosion and loss of land, frequent occurrences of natural hazards, breaching of the poor coastal embankments, floods, huge destruction of properties and habitats, salt water intrusion and inundation of agricultural land, displacement of the poor rural inhabitants and disruption in their daily lives and livelihoods. In this vulnerable environment, tourism-based human adaptation is seen mainly in the coastal areas of this island as an alternative for the livelihood of local people. But, the main problem is such unplanned and unauthorised tourism is not so eco-friendly till now. Present tourism-related activities are responsible for the deterioration of natural scenic beauty as well as environmental quality, which is also an emerging issue on Mousuni Island and it needs proper focus and attention. Despite several negative impacts on natural environment due to physical and anthropogenic reasons, in order to save human livelihoods in such a vulnerable island like Mousuni Island in the Sundarban delta region, development of eco-friendly as well as sustainable tourism infrastructures as an alternative may help to reduce the vulnerability of this island and to combat climate change, which will induce sustainable development of this area. Tourism-based human adaptation on Mousuni Island is an emerging issue in different fields of physical and social sciences, and proper focus should be given to it by the authorities and researchers for the sake of saving the environment alongside the welfare of the local community of this island.

ACKNOWLEDGEMENTS

The first author would like to acknowledge the RUSA Component-10 (Entrepreneurship and Career Hub), University of Kalyani, for the financial support as Principal Investigator [Project sanction order: IP/RUSA(C-10)/09/2021, dated: 26-11-2021] to carry out the research work presented in this chapter. The authors are also grateful to the editor(s) for accepting our chapter. We are likewise acknowledging the tourists, camp-owners and local people of Mousuni Island for their assistance in conducting the field survey.

REFERENCES

Brooker, E., & Joppe, M. (2014). A critical review of camping research and direction for future studies. *Journal of Vacation Marketing*, 1–17. https://doi.org/10.1177/1356766714532464

Census of India. (2011). *District Census Handbook, South Twenty Four Parganas, Village and Town Wise Primary Census Abstract* (Directorate of Census Operations), 20(XII-B). West Bengal.

Diamantis, D. (1999). The concept of ecotourism: Evolution and trends. *Current Issues in Tourism*, *2*(2 & 3), 93–122. https://doi.org/10.1080/13683509908667847

Kalman, R. E. (1960). A new approach to linear filtering and prediction problems. *Journal of Basic Engineering*, *82*(1): 35–45. https://doi.org/10.1115/1.3662552

Karmakar, M., Dey, P., & Roy, M. (2020). Rise of sea level and the sinking islands of sundarban region: A study of Mousuni Island in India. *Journal of Global Resources*, *6*(1), 233–237.

Liu, A. (2006). Tourism in rural areas: Kedah, Malaysia. *Tourism Management*, *27*, 878–889. https://doi.org/10.1016/j.tourman.2005.05.007

Maiti, S., & Bhattacharya, A. K. (2009). Shoreline change analysis and its application to prediction: A remote sensing and statistics based approach. *Marine Geology*, *257*, 11–23. https://doi.org/10.1016/j.margeo.2008.10.006

McFeeters, S. K. (1996). The use of the Normalized Difference Water Index (NDWI) in the delineation of open water features. *International Journal of Remote Sensing*, *17*(7), 1425–1432. https://doi.org/10.1080/01431169608948714

Mukherjee, N., & Siddique, G. (2018). Climate change and vulnerability assessment in Mousuni Island: South 24 Parganas District. *Spatial Information Research*, *26*, 163–174. https://doi.org/10.1007/s41324-018-0168-0

Sahu, A. S. (2014a). A study on Moyna Basin water-logged areas (India) using remote sensing and GIS methods and their contemporary economic significance. *Geography Journal, 2014*, 401324, 1–9. https://dx.doi.org/10.1155/2014/401324

Sahu, A. S. (2014b). Identification and mapping of the water-logged areas in Purba Medinipur part of Keleghai river basin, India: RS and GIS methods. *International Journal of Advanced Geosciences*, *2*(2), 59–65. https://doi.org/10.14419/ijag.v2i2.2452

Sahu, A. S. (2016). Water inundation hazard and micro level basin hydromorphological relationships: A geographical assessment. *Modeling Earth Systems and Environment*, *2*, 1–11. https://doi.org/10.1007/s40808-016-0081-2

Samanta, B., Das, S., & Hazra, S. (2017). Micro level vulnerability assessment of a community living in Mousuni Island in the Indian Sundarban: An Integrated Study Employing Geoinformatics. In S. Hazra, A. Mukhopadhyay, A. R. Ghosh, D. Mitra, & V. K. Dadhwal (Eds.), *Environment and Earth Observation: Case Studies in India*, 195–213. Springer Nature, Cham, Switzerland. https://doi.org/10.1007/978-3-319-46010-9_13

Survey of India. (2011). Village-Mouza database of West Bengal. Accessed from https://onlinemaps.surveyofindia.gov.in/Digital_Product_Show.aspx.

USGS EarthExplorer. (2023). Access link: https://earthexplorer.usgs.gov

Wang, J., Robertson, G.J., & White, K. (2000). Generating viewsheds without using sightlines. *Photogrammetric Engineering & Remote Sensing*, *66*(1), 87–90.

WWF-India. (2010). *Sundarbans: Future Imperfect (Climate Adaptation Report)*. https://library.wmo.int/records/item/42662-sundarbans-future-imperfect-climate-adaptation-report.

Part 2

Urbanization, Public Health, and Microclimate Challenges in a Changing Climate

8 Climate Change and Its Multidimensional Factors

Khemlata Soni, Narayan Lal Panwar, and Pranay Rajendra Lanjekar

INTRODUCTION

Climate change (CC) is one of the most critical world challenges of the 21st century, significantly impacting environmental, social, and economic systems. It states that long-term shifts in temperature, rainfall patterns, and extreme weather events are mainly driven by anthropological activities such as the excessive emission of greenhouse gases (GHGs), deforestation, industrialization, and unsustainable agricultural practices (Abbass et al., 2022). The rapid increase in global temperatures, exacerbated by the Industrial Revolution and the ever-growing energy demand, has intensified environmental stress, threatening ecosystems and human livelihoods worldwide. The significance of CC is far-reaching, with rising sea levels endangering coastal cities, extreme heat waves increasing health risks, and unpredictable weather patterns disrupting food and water security. Additionally, biodiversity loss and economic instability are becoming more prevalent as ecosystems struggle to adapt to these changes. Addressing these issues requires a comprehensive approach, incorporating early warning systems, policy interventions, and sustainable technological advancements to mitigate the crisis effectively.

To better understand the scale and complexity of CC, it is essential to examine the multidimensional nature of its drivers and impacts, which include environmental, socioeconomic, political, technological, and cultural dimensions (IPCC, 2021). Scholars such as O'Neill et al. (2017) and Filho et al. (2022) have emphasized that CC cannot be solely attributed to environmental degradation but is also deeply influenced by inequality, consumption patterns, urbanization, governance failures, and globalization. For instance, studies from different regions of the world have shown that socioeconomic vulnerability, population density, infrastructure limitations, and political inaction exacerbate the impacts of climate events (Shah et al., 2021). Therefore, a global literature review reveals that CC is not only a scientific concern but also a complex sociopolitical issue requiring transdisciplinary solutions. Figure 8.1 shows the average number of deaths in a decade worldwide due to natural disasters.

The awareness and scientific study of CC have evolved over decades, with landmark initiatives such as the Intergovernmental Panel on Climate Change (IPCC), the Paris Agreement, and the Lancet Commission playing crucial roles in monitoring and mitigating its impacts. These efforts emphasize the importance of reducing fossil fuel dependency, encouraging renewable energy, and implementing adaptive strategies to safeguard communities against climate-related risks. A thorough understanding of the multidimensional causes and impacts of CC is essential for developing effective mitigation and adaptation strategies. Since CC intersects with environmental, socioeconomic, political, technological, and cultural factors, a holistic perspective is necessary to identify intervention points and craft-targeted solutions. This chapter explores the complex drivers of CC, its diverse consequences, and the integrated approaches required to combat its challenges. By analyzing the interconnections among different sectors, it aims to provide a comprehensive overview of CC as a global issue and highlight potential strategies for building a sustainable future.

DOI: 10.1201/9781003620747-10

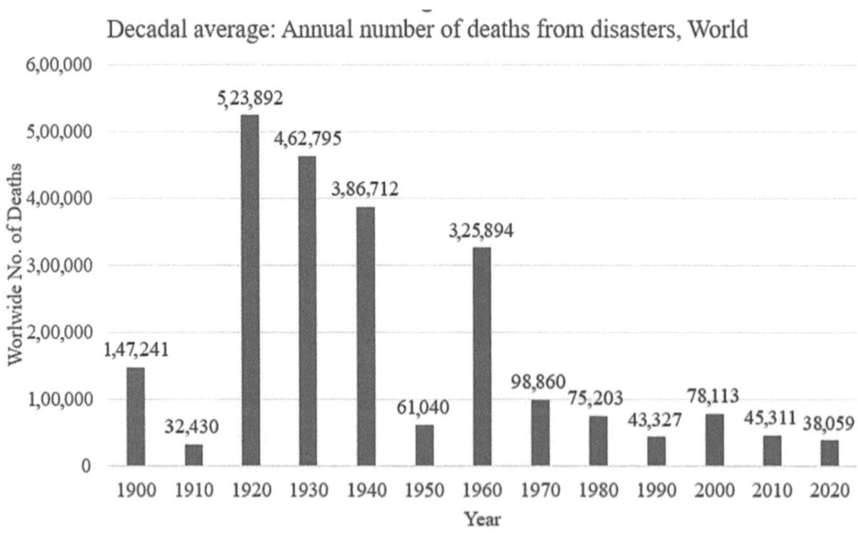

FIGURE 8.1 Decadal average: Annual number of deaths from disasters, World. (Data Source: Our World in Data based on EM-DAT, CRED.)

STATUS OF CLIMATE CHANGE

CC is primarily driven by anthropogenic actions such as deforestation, burning of fossil fuels, and industrial processes, which have led to significant and lasting disruptions in global climate patterns. Its extensive effects encompass increasing sea levels, escalating temperatures, changing precipitation patterns, more frequent and severe storms, and disruptions in ocean currents. These transformations significantly impact multiple sectors, such as public health, agriculture, water resources, biodiversity, and energy production. For example, variations in temperature and irregular rainfall influence crop yields, while ocean acidification endangers marine ecosystems and fisheries. CC is intensifying natural disasters such as floods, droughts, and storms, severely impacting millions of lives worldwide. Vulnerable communities, including marginalized groups, indigenous populations, and low-income households, face disproportionate challenges, further deepening existing social and economic inequalities. As one of the most pressing global crises, CC demands compelling action to reduce GHG emissions, implement adaptive strategies, and protect the planet for future generations.

CC is primarily driven by GHG, which originates from both natural systems and anthropogenic activities. Natural sources include forest fires, oceans, earthquakes, mud volcanoes, permafrost, wetlands, and volcanic eruptions (Yue & Gao, 2018). In contrast, human-induced emissions are largely associated with industrial activities, energy production, and changes in forestry, land use, and land-use patterns (Gries et al., 2019). In 2023, GHG emissions reached a record high of 53.0 Gt CO_2eq (excluding emissions from land use, forestry, and land-use change). This marks a 1.9% increase (994 Mt CO2eq) compared to 2022 levels. The world's largest GHG emitters in 2023 were China, the United States, India, the EU27, Russia, and Brazil. Together, these nations accounted for 49.8% of the global population, 63.2% of global GDP, 64.2% of fossil fuel consumption, and 62.7% of total GHG emissions. India witnessed the highest relative growth (+6.1%), while China experienced the largest absolute increase, adding 784 Mt CO_2eq. Fossil CO_2 remained the dominant contributor to global emissions, making up 73.7% of the total, followed by CH_4 (18.9%), N_2O (4.7%), and F-gases (2.7%). Since 1990, global fossil CO_2 emissions have surged by 72.1% (Crippa, 2024). Table 8.1 shows GHG emission trends in various countries from 1990 to 2022.

On the other hand, human-induced emissions mainly stem from energy production, industrial operations, and activities linked to forestry, land use, and land-use changes (Edenhofer et al., 2014). Figure 8.2 shows the main causes of CC.

TABLE 8.1
Historical Greenhouse Gas (GHG) Emission (Mt CO_2eq) Trends (1990–2022)

Country/ Region	1990	2000	2005	2015	2020	2021	2022	Change (1990–2022)
China	4,073.56	5,425.51	8,431.92	13,479.88	14,879.56	15,632.89	15,684.63	+285%
United States	6,233.03	7,033.02	7,314.45	6,870.34	6,558.10	6,340.50	6,412.30	+2.9%
India	1,008.23	1,301.10	1,619.45	2,306.78	2,412.34	2,654.89	2,816.45	+179%
EU27	4,915.14	4,513.34	4,597.10	3,922.02	3,427.44	3,617.74	3,587.80	–27%
Russia	3,315.45	2,100.34	2,200.45	2,300.78	2,400.89	2,450.67	2,500.34	–24.6%
Brazil	1,200.34	1,300.45	1,400.56	1,500.67	1,600.78	1,650.89	1,700.12	+41.7%
Global total	33,268.12	36,991.71	42,318.43	50,134.38	50,632.31	53,056.61	53,786.04	+61.8%

Source: EDGAR 2024 report.

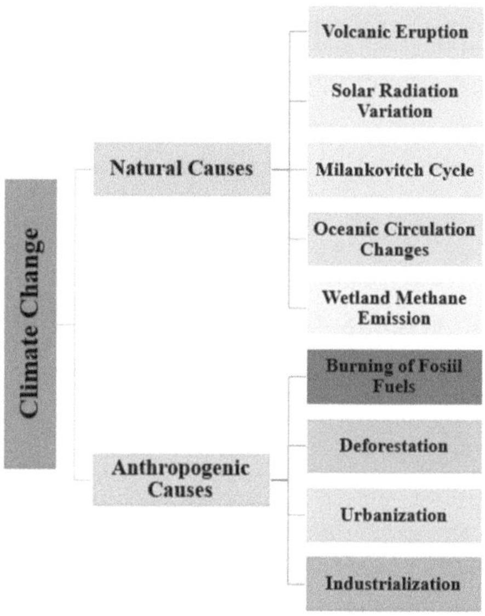

FIGURE 8.2 Causes of climate change.

NATURAL FACTORS OF CLIMATE CHANGE

Human activities are a major driver of CC, and natural forces have long influenced Earth's climate, shaping global temperatures, atmospheric composition, and weather patterns over millions of years. These processes operate across vast timescales, contributing to both warming and cooling trends. Among the key natural factors, volcanic eruptions release GHGs and aerosols, temporarily cooling the planet, while solar radiation variations alter Earth's temperature by changing the Sun's energy output. Milankovitch cycles, driven by shifts in Earth's orbit and axial tilt, trigger long-term climate fluctuations. Additionally, oceanic circulation patterns regulate heat distribution worldwide, affecting regional and global climate conditions, while wetlands naturally emit methane, a potent GHG that influences atmospheric warming. Recognizing these natural contributors is crucial for differentiating between human-induced and naturally occurring CCs.

VOLCANIC ERUPTIONS AS A NATURAL CAUSE OF CLIMATE CHANGE

Volcanic eruptions are one of the most significant natural contributors to CC, primarily through the release of sulfur dioxide (SO_2) and other aerosols into the atmosphere (Gao et al., 2021). When injected into the stratosphere, SO_2 undergoes chemical reactions to form sulfate aerosols, which have a profound impact on global climate. These aerosols scatter incoming solar radiation, leading to a cooling effect on the Earth's surface (Huhtamaa et al., 2022). This phenomenon, known as volcanic forcing, can persist for several years, depending on the magnitude of the eruption and the altitude at which aerosols are dispersed. Figure 8.3 shows volcanic gases in the atmosphere.

Major historical eruptions, such as the 1991 Mount Pinatubo eruption, have demonstrated the climate-altering potential of volcanic activity. The Pinatubo event released approximately 20 million tons of SO_2, leading to a temporary global temperature reduction of about 0.5°C over the following two years (Chylek et al., 2020). In addition to cooling effects, volcanic eruptions can also disrupt atmospheric circulation patterns, influencing monsoon activity, precipitation levels, and even ozone depletion in the stratosphere. Beyond the short-term cooling effects, volcanic activity can also contribute to long-term climate variability. Large-scale eruptions have been linked to historical periods of significant climatic shifts, like the Little Ice Age, which was partially attributed to a series of eruptions during the 13th–19th centuries. These events highlight the complex interplay between volcanic activity and climate systems, where eruptions can trigger cascading effects, including changes in ocean heat content, shifts in the Intertropical Convergence Zone (ITCZ), and disruptions in atmospheric circulation.

While volcanic eruptions act as a cooling agent in the short term, their overall contribution to CC remains dependent on the frequency and magnitude of events. Understanding these impacts is crucial for refining climate models and predicting potential future disruptions caused by large-scale volcanic activity.

SOLAR RADIATION VARIATIONS AND THEIR IMPACT ON CLIMATE CHANGE

Solar radiation is the primary driver of Earth's climate, influencing atmospheric circulation, ocean currents, and temperature distribution. Variations in solar activity, such as sunspot cycles, changes in solar irradiance, and cosmic ray flux, contribute to natural climate variability on timescales

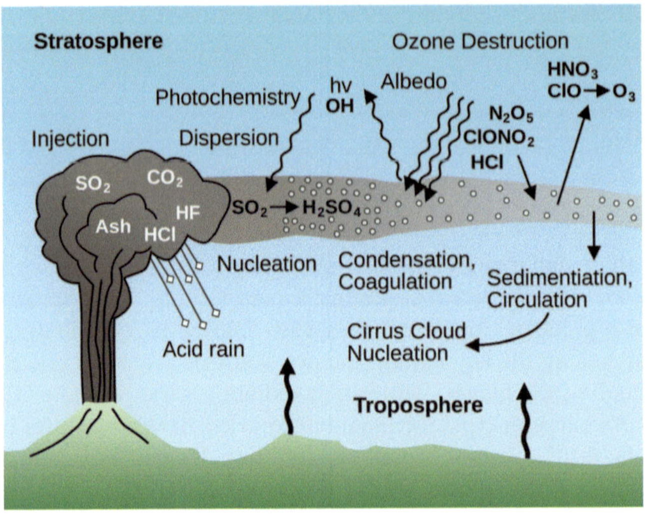

FIGURE 8.3 Volcanic gases reaction in the atmosphere. (Source: Modified from USGS, 2021.)

ranging from decades to millennia. While the Sun's energy output remains relatively stable compared to other climate forcings, even small fluctuations can trigger significant climate responses, particularly when combined with feedback mechanisms in the Earth's atmosphere and oceans.

The Sun follows an approximately 11-year cycle of activity, marked by fluctuations in sunspot numbers. In high sunspot activity, the Sun emits slightly more radiation, increasing Earth's energy input. Conversely, periods of low sunspot activity correlate with reduced solar irradiance. Observations from space-based radiometers, such as the Solar Radiation and Climate Experiment (SORCE), confirm that total solar irradiance varies by about 0.1% over the solar cycle (Lean, 2010). This variation, though small, can influence atmospheric stratospheric ozone levels, temperatures, and oceanic heat uptake. Phases of reduced solar activity, like the Maunder Minimum (1645–1715), have been linked to lower global temperatures, as evidenced during the Little Ice Age. Similarly, high solar activity periods, such as the Medieval Warm Period, have coincided with slightly warmer global temperatures. These historical correlations suggest that solar variability plays a role in modulating natural climate fluctuations, though it is not the sole driver of long-term climate trends.

Solar forcing refers to changes in the amount of solar energy received by Earth. While direct changes in solar irradiance are small, they can interact with atmospheric circulation patterns and feedback mechanisms. Key processes include:

- **Stratospheric Ozone and UV Radiation:** Solar ultraviolet (UV) radiation varies more significantly than total solar irradiance (up to 10% over the solar cycle). Changes in UV radiation affect stratospheric ozone concentrations, leading to alterations in atmospheric temperature gradients and jet stream behavior (Lean, 2010).
- **Cloud Formation and Cosmic Ray Flux:** Some studies suggest that solar activity influences cloud cover through its modulation of cosmic rays. When solar activity is high, the solar wind shields Earth from cosmic rays, reducing cloud formation and potentially leading to warming. Conversely, during low solar activity, increased cosmic ray flux enhances cloud formation, leading to cooling effects (Svensmark et al., 2009).
- **Ocean–Atmosphere Interactions:** Solar variability can influence ocean circulation patterns, such as the El Niño-Southern Oscillation (ENSO). Studies indicate that solar maxima may enhance El Niño conditions, while solar minima can contribute to La Niña-like cooling patterns (Meehl et al., 2009).

Proxy data from ice cores, tree rings, and sediment records provide insights into long-term solar variability. Reconstructions suggest that solar activity has been relatively high over the past century compared to earlier periods, but modern observations indicate no significant upward trend in solar irradiance over the past several decades (Lean & Rind, 2008). Although solar variability has influenced the climate in the past, its contribution to recent global warming is minimal compared to anthropogenic GHG emissions. The IPCC (2023) estimates that solar forcing accounts for less than 10% of observed 20th- and 21st-century warming, whereas GHG emissions are the dominant driver. Recent observations suggest that the Sun may be entering a period of reduced activity similar to the Dalton Minimum (1790–1830). Some scientists hypothesize that if a prolonged solar minimum occurs, it could slightly offset anthropogenic warming, but not enough to counteract the projected increase in global temperatures due to GHG (Lean, 2010).

Solar radiation variations, driven by sunspot cycles, UV fluctuations, and cosmic ray interactions, play a role in natural climate variability. While historical data indicate that changes in solar activity have influenced past climate events, contemporary research confirms that anthropogenic factors, particularly GHG emissions, are the primary drivers of current global warming. Understanding the interplay between solar- and human-induced climate forcings is crucial for improving climate models and forecasting future CCs.

MILANKOVITCH CYCLES AND THEIR EFFECT ON CLIMATE

Earth's climate undergoes long-term variations because of changes in its orbit around the Sun, known as Milankovitch cycles (Buis, 2020). Originally proposed by Serbian mathematician Milutin Milankovitch in the early 20th century, these cycles explain how changes in Earth's eccentricity, axial tilt, and precession influence the distribution of solar radiation over time. These gradual orbital shifts, occurring over tens to hundreds of thousands of years, have been crucial in shaping the natural transitions between ice ages and interglacial periods in Earth's history (NASA, 2020). One of the key components of Milankovitch cycles is eccentricity, which refers to the shape of Earth's orbit around the Sun. Over 100,000 years, Earth's orbit shifts between a more circular and a more elliptical shape. When the orbit is nearly circular, the amount of solar energy received throughout the year remains relatively stable, leading to milder seasonal variations. However, when the orbit becomes more elliptical, the difference between Earth's closest approach to the Sun (perihelion) and its farthest distance (aphelion) increases, amplifying seasonal temperature contrasts. This variation influences the intensity of summers and winters, particularly in the higher latitudes, and plays a role in the growth and retreat of ice sheets.

Another crucial factor is axial tilt, or obliquity, which refers to the angle at which Earth's axis is inclined relative to its orbit. Currently, Earth's axial tilt is 23.5°, but it fluctuates between 22.1° and 24.5° over a cycle of 41,000 years. When the tilt angle is higher, seasonal differences become more extreme, resulting in warmer summers and colder winters. In contrast, when the tilt angle decreases, summers become cooler, allowing ice sheets to persist and expand, leading to glaciation. The role of axial tilt is particularly significant in determining whether ice ages occur, as cooler summers in high-latitude regions prevent ice from melting, allowing glacial accumulation over thousands of years. The third major component, precession, is often described as the wobble of Earth's axis. This slow movement, occurring over 26,000 years, changes the timing of when seasons occur about Earth's position in its orbit. Currently, the Northern Hemisphere experiences summer when Earth is farthest from the Sun, leading to relatively mild summers. However, in the opposite phase of precession, when summer coincides with perihelion, solar radiation is stronger, leading to warmer summers that can contribute to ice sheet melting. Precession interacts with eccentricity and axial tilt, influencing the strength of seasons and the overall climate balance.

Milankovitch cycles have been linked to past climate transitions, particularly the alternation between glacial (ice age) and interglacial (warm) periods. During glacial periods, colder summers due to changes in axial tilt and precession prevent ice from melting, leading to the accumulation of ice sheets over millennia. In contrast, interglacial periods, like the one we are currently experiencing (the Holocene, which began around 11,700 years ago), occur when orbital conditions allow for warmer summers and widespread ice sheet melting. The Last Glacial Maximum (LGM), which occurred 20,000 years ago, was a period when ice sheets covered vast regions of North America and Eurasia. As the climate warmed due to changes in Earth's orbit, ice sheets began to retreat, marking the transition to the present interglacial phase.

Scientific evidence supporting Milankovitch cycles comes from ice core samples, deep-sea sediments, tree rings, and lake sediments, which provide records of past climate variations. Oxygen isotope analysis from ocean sediments has revealed patterns of glacial and interglacial periods that align closely with Milankovitch cycle predictions. Additionally, Antarctic ice cores show a correlation between past temperature changes and differences in solar radiation caused by Earth's orbital changes. These records confirm that Earth's climate history has been strongly influenced by orbital shifts over geological timescales. Although Milankovitch cycles play a crucial role in shaping Earth's long-term climate, they operate over extremely slow timescales, spanning tens of thousands to hundreds of thousands of years. In contrast, the current trend of global warming is occurring at an unprecedented rate, far exceeding the natural temperature variations driven by orbital changes.

According to the IPCC (2021), GHG emissions from human activities are the dominant force behind recent CC, overwhelming the expected cooling trends based on Milankovitch cycles. If natural orbital influences were the primary driver of today's climate, Earth would be experiencing a slow cooling trend rather than rapid warming. In conclusion, Milankovitch cycles are fundamental to understanding Earth's long-term climate variations, influencing ice ages and interglacial periods through changes in eccentricity, axial tilt, and precession. These slow orbital shifts alter the distribution of solar energy, triggering climate feedback mechanisms that amplify their effects. While these cycles have dictated natural CC over millions of years, they cannot explain the rapid CCs observed in recent decades, highlighting the dominant role of human-induced GHG emissions in today's global warming.

OCEANIC CIRCULATION CHANGES AND THEIR IMPACT ON CLIMATE

The ocean serves as a vast reservoir of heat, moisture, and carbon, playing a fundamental role in shaping Earth's climate. Oceanic circulation patterns, driven by wind, temperature gradients, and salinity differences, act as a global conveyor belt, redistributing heat between the poles and the equator. These circulation changes influence temperature, monsoons, precipitation patterns, and even extreme weather events. Among the most significant and well-studied phenomena are the Atlantic Meridional Overturning Circulation (AMOC), the Pacific Decadal Oscillation (PDO), the ENSO, and the Indian Ocean Dipole (IOD). Each of these patterns contributes to natural climate variability over different timescales, but their interactions with human-induced CC pose emerging challenges for weather predictability and global climate stability.

ENSO is a key climate phenomenon influencing global weather patterns through its three phases: El Niño (warm phase), La Niña (cool phase), and neutral conditions. This disrupts weather patterns, leading to droughts in South Asia, heavy rainfall in the Americas, and increased Pacific hurricanes. In contrast, La Niña strengthens trade winds, pushing warm water westward and bringing cold, nutrient-rich water to the South American coast. This results in stronger monsoons in South Asia, droughts in the US, and more Atlantic hurricanes. ENSO fluctuations impact food security, water availability, and ecosystems, making it a crucial focus of climate monitoring. Beyond ENSO, other oceanic circulation patterns shape climate variability. The AMOC, including the Gulf Stream, transports warm tropical water northward, but Arctic ice melt is slowing this system, potentially leading to harsher winters in Europe, rising the US sea levels, and altered rainfall in Africa and South America. The PDO, a long-term cycle of North Pacific temperature shifts, can either intensify or moderate global warming effects. Similarly, the IOD influences monsoons, with a positive phase causing drought in Australia and heavy rains in East Africa, while a negative phase brings the opposite effects. These oceanic shifts play a crucial role in long-term climate patterns and extreme weather events.

CC is altering these oceanic circulation patterns in unpredictable ways. Rising sea surface temperatures could intensify ENSO events, leading to more frequent extreme weather conditions like droughts, floods, and hurricanes. The weakening of AMOC, caused by Arctic ice melt, could result in regional cooling in Europe but accelerated warming in the tropics, altering weather patterns that have remained stable for centuries. Additionally, shifts in ocean salinity and density, influenced by glacial meltwater, threaten to disrupt thermohaline circulation, a key driver of long-term climate stability. Another emerging concern is the increase in marine heatwaves, which disrupt marine ecosystems, cause coral bleaching, and affect fisheries. Understanding oceanic circulation changes is essential for predicting and mitigating CC impacts. These ocean-driven climate patterns shape regional weather extremes, influence water availability, and affect global food security. As CC accelerates, disruptions in oceanic circulation could intensify extreme weather events, alter precipitation patterns, and threaten biodiversity and human livelihoods. Improved climate models and better monitoring of ocean–atmosphere interactions will be critical in preparing for future climatic shifts driven by oceanic circulation changes.

WETLAND METHANE EMISSIONS AND THEIR EFFECT ON CLIMATE CHANGE

Wetlands are among the largest natural sources of methane (CH_4), GHGs with a global warming potential (GWP) approximately 28 times higher than CO_2 over 100 years (IPCC, 2023). While wetlands act as important carbon sinks, storing vast amounts of organic carbon, they also release CH_4 due to anaerobic microbial decomposition of organic matter in waterlogged conditions. This dual role makes wetlands significant in the global climate system, contributing both to carbon sequestration and GHG emissions. Understanding wetland methane emissions is crucial for assessing their net climate impact and developing effective mitigation strategies.

Methane production in wetlands occurs primarily through anaerobic microbial activity in oxygen-deprived environments. The process begins with the decomposition of organic matter, followed by methanogenesis, where specialized archaea convert acetate, hydrogen, and carbon dioxide into CH_4 (Conrad, 2020). The methane produced is then released into the atmosphere through three major pathways: diffusion, where methane slowly escapes through water; ebullition, where methane bubbles up from sediments; and plant-mediated transport, where vascular plants such as sedges facilitate CH_4 release through their roots and stems. The rate of CH_4 emission rate depends on factors such as temperature, water level, and availability of organic matter. Wetlands contribute approximately 20%–30% of methane emissions globally, releasing an estimated 150–200 Tg CH_4 per year (Saunois et al., 2020). The emission rates vary across different types of wetlands, with tropical wetlands being the largest contributors due to their high temperatures and organic productivity. Boreal and temperate wetlands, though smaller contributors are susceptible to climate-induced changes such as permafrost thawing, which can accelerate methane release (Schuur et al., 2015). As global temperatures rise, wetland methane emissions are expected to increase, further intensifying the greenhouse effect and contributing to CC. Figure 8.4, shows methane emission from wetlands.

One of the most concerning aspects of wetland methane emissions is their role in climate feedback loops. Rising temperatures enhance microbial activity in wetlands, leading to greater CH_4 production. This, in turn, increases atmospheric methane concentrations, intensifying the greenhouse effect and further warming the planet. In Arctic regions, permafrost thawing releases previously

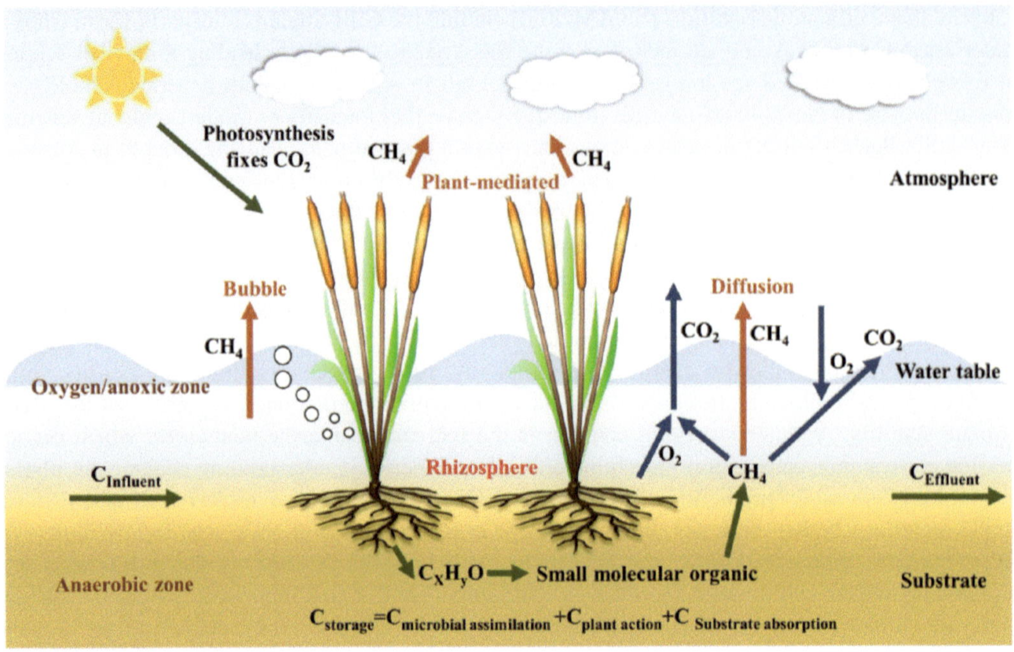

FIGURE 8.4 Methane emission from wetland. (Source: Modified from Yu et al., 2023.)

trapped carbon in the form of methane and CO_2, exacerbating global warming. This self-reinforcing cycle makes wetlands a crucial area of study in CC research. In addition to natural processes, human activities significantly influence wetland methane emissions. Wetland drainage and land conversion for agriculture and urban development reduce CH_4 emissions but release stored CO_2, leading to long-term climate impacts. Changes in precipitation patterns and rising global temperatures may also expand wetland areas in some regions, leading to increased methane emissions. Furthermore, nutrient enrichment from agricultural runoff can alter microbial activity, potentially enhancing methane production. These interactions highlight the complex relationship between wetlands and CC, emphasizing the need for balanced conservation and mitigation efforts.

To mitigate wetland methane emissions while preserving their ecological benefits, several strategies can be employed. Restoration and protection of natural wetlands can help maintain their carbon sequestration function while minimizing GHG fluxes. Water level management in human-made wetlands, such as rice paddies and reservoirs, can be optimized to reduce CH_4 emissions. Additionally, integrating wetland methane emissions into climate models will improve predictions and inform better mitigation strategies. In conclusion, wetlands play a dual role in CC as both carbon sinks and methane sources. While they store large amounts of carbon, their CH_4 emissions contribute significantly to global warming. The interaction between wetlands and CC is complex, involving natural processes, anthropogenic influences, and feedback mechanisms. Managing wetland methane emissions requires a combination of conservation, policy interventions, and scientific research to ensure that these valuable ecosystems contribute to CC mitigation rather than exacerbating global warming. Future research should focus on refining emission models, implementing adaptive management strategies, and incorporating wetland dynamics into global climate policies.

ANTHROPOGENIC FACTORS OF CLIMATE CHANGE

Anthropogenic activities have become the leading force driving CC, significantly altering the Earth's natural balance. Over the past two centuries, rapid industrialization, urban expansion, and reliance on fossil fuels have led to an unprecedented rise in GHG emissions, intensifying global warming and its associated impacts. The burning of fossil fuels for energy and transportation releases vast amounts of CH_4, CO_2, and N_2O into the atmosphere, trapping heat and disrupting climate stability. Deforestation further exacerbates this crisis by reducing the planet's ability to absorb CO_2, accelerating atmospheric carbon buildup. Meanwhile, urbanization replaces natural landscapes with heat-retaining infrastructure, contributing to the urban heat island effect, air pollution, and altered precipitation patterns. Additionally, industrialization has driven large-scale emissions from manufacturing, agriculture, and waste management, further intensifying environmental degradation. These human-induced activities are the main catalysts of modern CC, requiring urgent global action to mitigate their long-term consequences.

IMPACT OF BURNING FOSSIL FUELS ON CLIMATE CHANGE

The burning of fossil fuels is the main driver of global CC, releasing vast amounts of CO_2 and other GHG into the atmosphere. These emissions intensify the greenhouse effect, trapping heat and leading to a significant rise in global temperatures. Since the Industrial Revolution, the Earth's average temperature has risen by approximately 1°C. If global warming surpasses 1.5°C, it could lead to severe impacts, including rising sea levels, loss of biodiversity, extreme weather events, and food shortages. According to the IPCC, fossil fuel emissions are the primary driver of CC, contributing 89% of global CO_2 emissions. Despite increasing awareness, fossil fuels remain the dominant energy source, meeting around 80% of the world's energy demand. Figure 8.5 shows the fossil fuel consumption rate per capita in leading countries.

Fossil fuels are formed over millions of years from the decomposition of organic matter. While they have fueled industrial growth and modern civilization, their widespread use has come at a

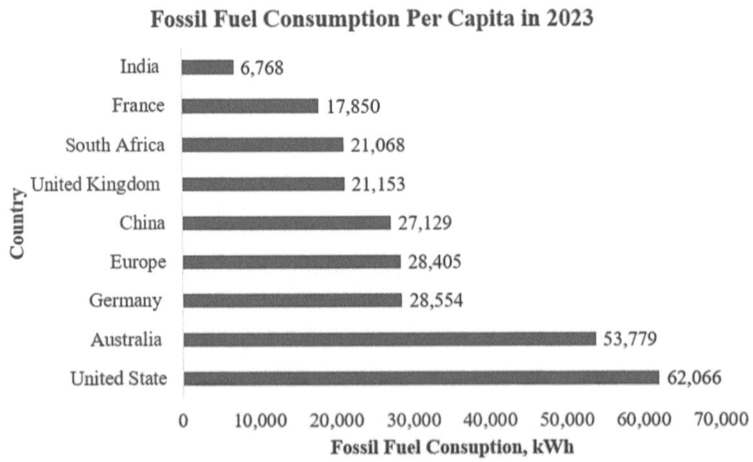

FIGURE 8.5 Fossil fuel consumption per capita, 2023. (Source: https://ourworldindata.org/fossil-fuels.)

heavy environmental cost. Coal, the dirtiest of all fossil fuels, is responsible for over 0.3°C of the 1°C increase in global temperatures, making it the single largest contributor to global warming. It is primarily used for electricity generation, emitting high levels of carbon dioxide and other pollutants. Oil, another major carbon emitter, contributes approximately one-third of the world's total carbon dioxide emissions. In addition to its role in CC, oil spills have severely damaged marine ecosystems, causing long-term harm to biodiversity. Natural gas, often promoted as a cleaner alternative, still accounts for one-fifth of total carbon emissions. Furthermore, methane leaks from natural gas infrastructure significantly exacerbate global warming, as CH_4 is a potent GHG with a much higher short-term warming potential than CO_2.

To prevent catastrophic CC, prompt action is needed to reduce fossil fuel dependency. The Paris Agreement (2015) marked a global commitment to limit warming by reducing carbon emissions. However, recent reports indicate that global fossil fuel production remains on track to exceed safe limits by 2030. Despite these warnings, governments and industries continue to expand fossil fuel extraction, undermining climate goals. A transition toward renewable energy sources, stricter environmental policies, and sustainable practices is critical to mitigating the long-term impacts of CC and ensuring a safer future for the planet.

DEFORESTATION AND ITS IMPACT ON CLIMATE CHANGE

Deforestation, the large-scale removal of forests, is a major contributor to global CC. Forests serve as natural carbon sinks, capturing CO_2 from the atmosphere and playing a crucial role in climate regulation. However, deforestation and forest fires release this stored carbon back into the air, contributing significantly to GHG emissions. Deforestation not only reduces the Earth's ability to absorb CO_2 but also disturbs ecosystems, accelerates land degradation, and alters local and global weather patterns. Forests play a vital role in the carbon cycle, absorbing and storing large amounts of CO_2 through photosynthesis. When trees are cleared for urban expansion, agriculture, or logging, this stored carbon is released, contributing to rising atmospheric CO_2 levels. According to scientific estimates, deforestation accounts for nearly 10% of global carbon emissions, making it a significant driver of CC (LSE, 2023). In tropical regions, where deforestation rates are highest, the loss of forests has drastically reduced the planet's ability to sequester carbon, intensifying the greenhouse effect and leading to increased global temperatures.

Deforestation affects not only carbon emissions but also local and global climate systems. Forests influence temperature regulation, humidity levels, and rainfall patterns by releasing water vapor through transpiration. Large-scale deforestation reduces this moisture release, leading to drier climates and increased vulnerability to droughts. Furthermore, deforestation can alter monsoon cycles and disrupt weather patterns, leading to risky climate events such as heatwaves, storms, and prolonged dry seasons. In tropical rainforests like the Amazon, deforestation has been linked to a decline in rainfall, which in turn affects agriculture, water availability, and biodiversity. The destruction of forests leads to severe biodiversity loss, as forests are home to over 80% of terrestrial species (WWF, 2020). The loss of vegetation and habitat results in species extinction and disrupts ecological interactions essential for climate regulation. Many species that help in maintaining soil health, pollination, and pest control face habitat destruction, further worsening environmental degradation.

Forests act as natural barriers against soil erosion by stabilizing the land with their root systems. Deforestation exposes the soil to direct wind and water erosion, reducing its fertility and increasing susceptibility to floods and landslides. Without tree cover, the land loses its ability to retain moisture, leading to desertification and declining agricultural productivity. This has significant socioeconomic consequences, particularly in regions where communities depend on forests for their livelihoods. Recognizing the severe impact of deforestation, global initiatives such as REDD+ (Reducing Emissions from Deforestation and Forest Degradation) and the Paris Agreement emphasize the importance of preserving forests to combat CC. Many countries have committed to reforestation and afforestation projects to restore lost forest cover and enhance carbon sequestration. However, despite these efforts, deforestation rates remain high, particularly in tropical regions. Urgent policy measures, sustainable land-use practices, and community-based forest conservation strategies are essential to mitigate the adverse effects of deforestation on CC.

Deforestation remains a critical environmental challenge with far-reaching climate implications. From increasing CO_2 emissions to disrupting global weather patterns, the large-scale destruction of forests accelerates CC and threatens ecological balance. Sustainable forest management, afforestation initiatives, and stricter environmental policies are vital to reducing deforestation and preserving forests as key climate regulators. Protecting and restoring forests is not only crucial for biodiversity and soil conservation but also an essential strategy in the fight against global warming.

URBANIZATION AND ITS IMPACT ON CLIMATE CHANGE

Urbanization, characterized by the rapid expansion of cities and the replacement of natural landscapes with built environments, significantly influences local and global climate systems. As urban areas grow, they modify atmospheric conditions, alter regional weather patterns, and contribute to CC through increased GHG emissions, heat retention, and air pollution. The impact of urbanization extends beyond city limits, affecting temperature regulation, precipitation patterns, and the frequency of extreme weather events.

One of the most well-documented climate effects of urbanization is the Urban Heat Island (UHI) effect, where cities experience higher temperatures than surrounding rural areas. This occurs because urban surfaces, such as concrete and asphalt, absorb and retain heat, reducing nighttime cooling. Additionally, the high-energy consumption in urban areas releases anthropogenic heat, further intensifying warming. Studies indicate that UHI contributes to increased heat waves, putting urban populations at greater risk of heat-related illnesses and energy demand surges for cooling. Urbanization affects atmospheric moisture levels, leading to the Urban Moisture Island (UMI) or Urban Dry Island (UDI) effect, depending on the level of evapotranspiration and water retention in city landscapes. The modification of natural land cover, reduced vegetation, and extensive impervious surfaces alter the hydrological cycle, impacting humidity and local rainfall. Urban areas also disrupt natural wind patterns, influencing cloud formation, storm intensity, and precipitation distribution. Some studies suggest that urbanization can enhance localized convective storms, increasing the risk of urban flooding due to inadequate drainage infrastructure.

Urban centers are major contributors to air pollution, releasing significant amounts of GHGs, aerosols, and particulate matter. High concentrations of carbon dioxide (CO_2), nitrogen oxides (NO_x), sulfur dioxide (SO_2), and black carbon from transportation, industry, and residential heating systems exacerbate global warming. Additionally, urban air pollution interacts with climate systems by affecting cloud formation and radiative forcing, leading to changes in regional climate conditions. The expansion of urban areas has been linked to an increased frequency and intensity of extreme weather events, including heavy rainfall, heat waves, and storms. The modification of land surfaces influences atmospheric dynamics, sometimes intensifying weather extremes. Urban environments with poor green cover and inadequate climate adaptation strategies face higher vulnerability to extreme hydrometeorological events, such as hurricanes, typhoons, and intense monsoons. Addressing the climate challenges associated with urbanization requires a shift toward sustainable urban planning. Strategies such as green infrastructure, urban reforestation, reflective surfaces, and energy-efficient building designs can help mitigate UHI effects. Additionally, improved public transportation, stricter emissions regulations, and investments in renewable energy can reduce the carbon footprint of urban areas. Policies such as climate-resilient infrastructure planning, flood management systems, and urban climate monitoring are crucial for minimizing climate-related risks.

Urbanization is a key driver of CC, influencing temperatures, precipitation, air quality, and extreme weather patterns. While cities contribute significantly to environmental challenges, they also present opportunities for climate adaptation and mitigation. Implementing sustainable urban development strategies is important to reducing urban climate impacts and ensuring a resilient future for growing populations.

INDUSTRIALIZATION AND ITS IMPACT ON CLIMATE CHANGE

Industrialization has played a crucial role in transforming societies by driving economic growth, technological advancements, and improved living standards. The shift from agrarian economies to industrialized societies led to significant developments in manufacturing, transportation, and urbanization. However, these advancements came at a considerable environmental cost. Industrialization, particularly since the Industrial Revolution, marked the beginning of large-scale fossil fuel consumption, which continues to be the driving force behind CC. The consequences of this transformation are still evident today, as industries remain one of the largest contributors to global GHG emissions, resource depletion, and environmental degradation.

The Industrial Revolution, which began in the late 18th century, laid the foundation for modern industrialization by introducing mechanized production powered by coal, oil, and gas. Before this period, atmospheric CO_2 levels remained relatively stable. However, the widespread adoption of coal-powered factories, steam engines, and mechanized industries led to a substantial increase in CO_2 emissions. This marked the beginning of a long-term reliance on fossil fuels, which continues to fuel global warming today. Over the past two centuries, human activities have released approximately 2.3 trillion tonnes of CO_2 into the atmosphere (IPPC, 2023). While natural systems like oceans and forests have absorbed some of these emissions, a significant portion remains in the air, trapping heat and accelerating CC. The Earth's average surface temperature has already increased by 1.1°C since the start of industrialization, contributing to extreme weather events, rising sea levels, and widespread ecological disruptions.

The impact of industrialization on CC is largely driven by the rise in GHGs. Apart from CO_2, other pollutants such as methane (CH_4) and nitrous oxide (N_2O) have seen significant increases due to industrial activities. CH_4 concentrations have more than doubled since pre-industrial times, while N_2O levels have increased by around 20%. Since the 18th century, atmospheric CO_2 levels have increased from 275 parts per million (ppm) to 415 ppm, with the annual rate of increase accelerating over time. Currently, human activities release more than 30 billion tonnes of CO_2 annually, pushing global climate conditions toward irreversible tipping points (Anderson, 2024). The early industrial era focused primarily on economic progress and technological advancements, with little awareness of the environmental impact. The long-term consequences of industrialization became apparent much later, as scientific research revealed links between human activities and CC, ozone layer depletion, and biodiversity loss. The burning of fossil fuels for energy, transportation, and industrial processes has significantly altered the planet's natural balance. The continued reliance on oil, coal, and natural gas has resulted in deforestation, habitat destruction, and air pollution, further exacerbating CC. Industrial waste and chemical byproducts have also contaminated water sources, reduced air quality, and contributed to the acidification of oceans.

The Industrial Revolution was led by developed nations, such as the United States, the United Kingdom, and Germany, which have historically contributed nearly 79% of global CO_2 emissions (Little, 2023). However, the impacts of CC disproportionately affect developing countries, where communities are more vulnerable to rising temperatures, extreme weather events, and food insecurity. While wealthier nations have benefited from industrial progress, developing countries now face pressure to limit their reliance on fossil fuels while still striving for economic growth. Transitioning to cleaner energy sources requires large-scale investments in renewable technologies, which many low-income nations cannot afford. This creates an ongoing debate about climate justice and the responsibility of industrialized countries to support sustainable development in less-developed regions.

Although the environmental damage caused by industrialization cannot be entirely reversed, strategic mitigation efforts can significantly reduce future climate risks. Some of the most effective approaches include:

- **Adaptation of Renewable Energy:** Shifting from fossil fuels to solar, hydro, wind, and geothermal energy can decrease emissions and reliance on non-renewable resources.
- **Implementing Green Technology:** Industries can adopt energy-efficient manufacturing processes, carbon capture technologies, and sustainable materials to lower their carbon footprint.
- **Enhancing Circular Economy Practices:** Encouraging recycling, reusing materials, and minimizing industrial waste can help conserve resources and reduce pollution.
- **Strengthening Climate Policies:** Governments must enforce stricter environmental regulations, carbon pricing mechanisms, and emission reduction targets to hold industries accountable for their environmental impact.

Industrialization has played a transformative role in shaping modern economies and improving human living standards. However, its environmental cost remains one of the greatest challenges of the 21st century. The continued reliance on fossil fuels and industrial emissions has accelerated CC, causing lasting damage to ecosystems, weather patterns, and global temperatures. Addressing these issues requires a global commitment to sustainable development, technological innovation, and climate policies that balance economic growth with environmental preservation. By investing in renewable energy, cleaner technologies, and climate adaptation strategies, societies can mitigate the long-term effects of industrialization and work toward a more sustainable future.

CONCLUSIONS

CC is a multifaceted global crisis driven by both natural and anthropogenic factors, with far-reaching consequences for economies, ecosystems, and societies. Natural processes such as volcanic eruptions, solar radiation variations, and Milankovitch cycles have historically influenced Earth's climate, but the rapid acceleration of global warming in recent decades is predominantly due to anthropogenic activities. The burning of fossil fuels, urbanization, deforestation, and industrialization has significantly increased GHG emissions, leading to unprecedented environmental stress. The impacts of CC are already evident in rising global temperatures, extreme weather events, biodiversity loss, and socioeconomic disruptions, particularly affecting vulnerable communities. Addressing CC requires a comprehensive and integrated approach that combines mitigation and adaptation strategies. Reducing GHG emissions through the transition to renewable energy, sustainable land-use practices, and green technologies is essential. Additionally, enhancing climate resilience through policy interventions, early warning systems, and community-based conservation efforts is crucial for safeguarding ecosystems and human livelihoods. The findings of this chapter underscore the importance of global cooperation, scientific innovation, and sustainable development in combating CC. By adopting a holistic perspective and implementing targeted solutions, we can mitigate the adverse effects of CC and work toward a sustainable and resilient future for generations to come.

ACKNOWLEDGMENTS

This work was supported by the Department of Renewable Energy Engineering, College of Technology and Engineering, Maharana Pratap University of Agriculture and Technology, Udaipur, India.

REFERENCES

Abbass, K., Qasim, M. Z., Song, H., Murshed, M., Mahmood, H., & Younis, I. (2022). A review of the global climate change impacts, adaptation, and sustainable mitigation measures. *Environmental Science and Pollution Research*, *29*(28), 42539–42559. https://doi.org/10.1007/s11356-022-19718-6

Anderson, K. (2024). *What Was the Industrial Revolution's Environmental Impact?* Greenly Institute. https://greenly.earth/en-us/blog/ecology-news/what-was-the-industrial-revolutions-environmental-impact

Buis, A. (2020). Milankovitch (orbital) cycles and their role in earth's climate. *Dataho.In*. https://dataho.in/wp-content/uploads/2023/01/Milankovitch_Orbital_Cycles.pdf

Chylek, P., Folland, C., Klett, J. D., & Dubey, M. K. (2020). CMIP5 climate models overestimate cooling by volcanic aerosols. *Geophysical Research Letters*, *47*(3). https://doi.org/10.1029/2020GL087047

Conrad, R. (2020). Importance of hydrogenotrophic, aceticlastic, and methylotrophic methanogenesis for CH_4 production in terrestrial aquatic environments. *Pedosphere*, *30*(1), 25–39. https://doi.org/10.1016/S1002-0160(18)60052-9.

Crippa, M. G. (2024). GHG emissions of all world countries. In *Publications Office of the European Union*. https://joint-research-centre.ec.europa.eu

Filho, W. L., Setti, A. F. F., Azeiteiro, U. M., Lokupitiya, E., Donkor, F. K., Etim, N. A. N. A., Matandirotya, N., Olooto, F. M., Sharifi, A., Nagy, G. J., & Djekic, I. (2022). An overview of the interactions between food production and climate change. *Science of the Total Environment*, *838*. https://doi.org/10.1016/j.scitotenv.2022.156438

Gao, C., Ludlow, F., Matthews, J. A., Stine, A. R., Robock, A., Pan, Y., Breen, R., Nolan, B., & Sigl, M. (2021). Volcanic climate impacts can act as ultimate and proximate causes of Chinese dynastic collapse. *Communications Earth and Environment*, *2*(1). https://doi.org/10.1038/s43247-021-00284-7

Gries, T., Redlin, M., & Ugarte, J. E. (2019). Human-induced climate change: The impact of land-use change. *Theoretical and Applied Climatology*, *135*(3), 1031–1044.

Huhtamaa, H., Stoffel, M., & Corona, C. (2022). Recession or resilience? Long-range socioeconomic consequences of the 17th century volcanic eruptions in northern Fennoscandia. *Climate of the Past*, *18*(9), 2077–2092. https://doi.org/10.5194/cp-18-2077-2022

Intergovernmental Panel on Climate Change (IPCC). (2023). Climate change 2021 — The physical science basis. In *Climate Change 2021 — The Physical Science Basis*. https://doi.org/10.1017/9781009157896

IPCC. (2021). Climate change widespread, rapid, and intensifying – IPCC. Accessed from https://www.ipcc.ch/2021/08/09/ar6-wg1-20210809-pr/.

Lean, J. L. (2010). Cycles and trends in solar irradiance and climate. *Wiley Interdisciplinary Reviews: Climate Change*, *1*(1), 111–122. https://doi.org/10.1002/wcc.18

Lean, J. L., & Rind, D. H. (2008). How natural and anthropogenic influences alter global and regional surface temperatures: 1889 to 2006. *Geophysical Research Letters*, *35*(18). https://doi.org/10.1029/2008GL034864

Little, B. (2023). How early signs of climate change date back to the industrial revolution. *History*, 1–3. https://www.history.com/news/industrial-revolution-climate-change

LSE. (2023). What is the role of deforestation in climate change and how can "Reducing Emissions from Deforestation and Degradation" (REDD+) help? *The London School of Economics*. https://www.lse.ac.uk/granthaminstitute/explainers/whats-redd-and-will-it-help-tackle-climate-change/

Meehl, G. A., Arblaster, J. M., Matthes, K., Sassi, F., & Van Loon, H. (2009). Amplifying the pacific climate system response to a small 11-year solar cycle forcing. *Science*, *325*(5944), 1114–1118. https://doi.org/10.1126/science.1172872

NASA. (2020). *Why Milankovitch (Orbital) Cycles Can't Explain Earth's Current Warming*. Blog. https://climate.nasa.gov/blog/2949/why-milankovitch-orbital-cycles-cant-explain-earths-current-war

O'Neill, B. C., Oppenheimer, M., Warren, R., Hallegatte, S., Kopp, R. E., Pörtner, H. O., Scholes, R., Birkmann, J., Foden, W., Licker, R., Mac H, K. J., Marbaix, P., Mastrandrea, M. D., Price, J., Takahashi, K., Van Ypersele, J. P., & Yohe, G. (2017). IPCC reasons for concern regarding climate change risks. *Nature Climate Change*, *7*(1), 28–37. https://doi.org/10.1038/nclimate3179

Saunois, M., Stavert, A. R., Poulter, B., Bousquet, P., Canadell, J. G., Jackson, R. B., Raymond, P. A., Dlugokencky, E. J., Houweling, S., Patra, P. K., Ciais, P., Arora, V. K., Bastviken, D., Bergamaschi, P., Blake, D. R., Brailsford, G., Bruhwiler, L., Carlson, K. M., Carrol, M., … Zhuang, Q. (2020). The global methane budget 2000—2017. *Earth System Science Data*, *12*(3), 1561–1623. https://doi.org/10.5194/essd-12-1561-2020

Schuur, E. A., et al. (2015). Climate change and the permafrost carbon feedback. *Nature*, *520*(7546), 171–179. https://www.embase.com/search/results?subaction=viewrecord&from=export&id=L603666723%0A. https://dx.doi.org/10.1038/nature14338

Shah, A., Nazari, M., Antar, M., Msimbira, L. A., Naamala, J., Lyu, D., Rabileh, M., Zajonc, J., & Smith, D. L. (2021). PGPR in agriculture: A sustainable approach to increasing climate change resilience. *Frontiers in Sustainable Food Systems*, *5*. https://doi.org/10.3389/fsufs.2021.667546

Svensmark, H., Bondo, T., & Svensmark, J. (2009). Cosmic ray decreases affect atmospheric aerosols and clouds. *Geophysical Research Letters*, *36*(15). https://doi.org/10.1029/2009GL038429

USGS. (2021). Volcanoes can affect climate. *U.S. Department of the Interior*. https://www.usgs.gov/natural-hazards/volcano-hazards/volcanoes-can-affect-climate

WWF. (2020). *Why Forests are so Important*. World Wide Fund for Nature. https://wwf.panda.org/discover/our_focus/forests_practice/importance_forests/

Yu, G., Chen, J., Wang, G., Chen, H., Huang, J., Li, Y., Wang, W., Song, F., Ma, Y., Wang, Q., Wang, M., Ling, T., Shu, Z., Sun, J., & Yu, Z. (2023). Recent advances in constructed wetlands methane reduction: Mechanisms and methods. *Frontiers in Microbiology*, *14*. https://doi.org/10.3389/fmicb.2023.1106332

Yue, X. L., & Gao, Q. X. (2018). Contributions of natural systems and human activity to greenhouse gas emissions. *Advances in Climate Change Research*, *9*(4), 243–252. https://doi.org/10.1016/j.accre.2018.12.003

9 Heat Index in a Warming India

Computation, Characterization, Trend Analysis, and Spatial Disparities

Usman Mohseni and Vinnarasi Rajendran

INTRODUCTION

Extreme heat events are becoming more common and intense worldwide, with direct and indirect consequences for human health and ecosystems. Extreme heat instances are hazardous events that create substantial health problems (Kent et al., 2014). These occurrences contribute to the warming of the planet. The primary driver of global warming includes the growing emissions of greenhouse gases through diverse human-induced sources, which leads to greater amounts of pollution (Fischer et al., 2007). According to the IPCC, the global mean temperature is projected to go above 1.5°C by the end of this century (Krinner et al., 2013). As the global mean temperature rises, HW, which is one of the most fatal events will also increase (Mcgregor et al., 2015). This trend of rising temperatures will continue aggressively unless necessary measures are taken to control pollution levels (Pattnayak et al., 2017). On the other hand, modernization and industrialization, while enhancing human life, have severe environmental impacts, consuming substantial energy and causing air pollution and atmospheric heat, which trigger respiratory diseases (Chen et al., 2015; Luan et al., 2019). Consequently, the average temperature is increasing daily, heightening the likelihood of severe heat wave (HW) events, which subsequently raise the mortality rate (Pattnayak et al., 2019). Climate change is considered the primary driver of these extreme events occurring with increasing frequency worldwide (Frich et al., 2002).

Heat stress (HS) can have a severe impact on humans, livestock, crops, and infrastructure. HS can cause heat cramps, exhaustion, and dehydration, as well as heatstroke, which can be fatal if not properly managed (Fischer & Schär, 2010; Wang et al., 2016). As a result, analyzing HS is critical for comprehending its implications. The Occupational Safety and Health Administration (OSHA) defines HS as a condition that happens when the human body is not able to sufficiently cool itself by sweating, causing an increase in core body temperature that ultimately leads to heat-related illnesses. Heat index (HI), often known as "feels-like temperature," is used to measure HS by taking temperature and relative humidity into consideration. The HI helps inform HS prevention strategies, such as adjusting outdoor activities, staying hydrated, seeking shade or air conditioning, and implementing heat emergency plans. The other consequences of heat events extend beyond human health (Kotharkar et al., 2021; Pascal et al., 2021; Zhang et al., 2019), increasing the risk of other disasters. Heat can exacerbate drought, creating conditions conducive to wildfires, and significantly affecting the food and agricultural sector, environment, and ecosystems (Luan et al., 2019).

HS typically occurs in summer and is less severe in early monsoon. During the summer, the HI is significantly higher than the observed temperature. HI, which accounts for both humidity and temperature is expressed by the numerical value of "HI" (Montero et al., 2013). HI is a method that is used to determine the possible risk of HS. In addition, the Steadman Scheme is used to calculate HI. This scheme is applied to the meteorological data and is subjected to numerous regressions,

 DOI: 10.1201/9781003620747-11

including the Poisson Regression analysis technique (Steadman, 1979). HI must be carefully studied to investigate and manage the health concerns associated with heat. If the HI for a certain place is known, the severity of the negative impacts of excessive heat can be thoroughly investigated. The formula used to calculate HI has been utilized by many scientists and investigators across the world for evaluating the impact of HS on humans.

In recent years, few studies have focused specifically on understanding the HS, its impact on human life, and its predictability in India. According to Mishra et al. (2020), intensive irrigation cools the land surface by 1°C and the air by 0.5°C, but it increases low-level moist enthalpy, raising specific and relative humidity and thus moist HS metrics. This results in increased moist HS in India due to irrigation. Another study by Kumar et al. (2022) on HS in four different cities of India found varying risks: Delhi, despite higher temperatures, had lower HS due to lower relative humidity (RH), while Chennai had maximum risks due to high temperatures and RH. Mumbai and Kolkata had lower extreme HS risk due to relatively lower temperatures. Coastal areas were identified as more impacted by HS, experiencing frequent extreme conditions. Climatologically, HW are prevalent from March to June, particularly in the north, northwest, central, and eastern coastal regions. Studies indicate increasing frequency, duration, and impacts of HW in central and northwestern India (Goyal et al., 2023). Interest in HW studies surged post-2015 after HW was declared a national disaster by the National Disaster Management Authority (NDMA), following over 2,000 deaths that year. The number of affected states rose from 9 in 2015 to 23 in 2020, prompting more focused studies on HW trends and regional characterization (NDMA, 2019).

In this respect, this chapter aims to identify and analyze monthly trends, characterization, and spatial variations of HI for the pre-monsoon (March–May) and early summer monsoon (June–August) seasons during 1981–2023 across India. This study will assist the researchers and policy-makers in comprehending the value of HI as a critical tool by assisting individuals and authorities to assess the risk of heat-related illnesses and take appropriate measures to protect public health during hot and humid conditions, as well as to mitigate the HS impact in India.

STUDY AREA AND DATA COLLECTION

The study area encompasses the entire nation of India over 43 years from 1981 to 2023, focusing on the summer and early monsoon seasons (March–August). This research was conducted at the state level across India, a country with a vast geographical expanse of 3.27 million square kilometers, comprising 28 states and 8 union territories (UTs) (Figure 9.1). India's diverse climate includes a wide temperature range, with eastern states generally experiencing higher temperatures than western, yet lower than those in the northern region. Kerala and Tamil Nadu maintain humid temperatures throughout the year. In this study, HI was computed over India, which has a high population density and is highly dependent on agriculture. HI was evaluated using Steadman's approach and categorized using the criteria provided by the India Meteorological Department (IMD). This study employed daily gridded maximum temperature (T_{max}) and RH data with a resolution of 0.5°×0.625° from the Modern-Era Retrospective Analysis for Research and Applications, version 2 (MERRA-2). Figure 9.2 illustrates the spatial distribution of the mean daily T_{max} and RH across India from March to August during the years 1981–2023. T_{max} ranges from 19.79°C to 50.96°C, while RH spans from 3.5% to 100%. The months of April, May, and June were identified as the hottest, with July and August being the most humid. Higher temperatures were typically observed in the coastal and central regions, whereas cooler temperatures prevail in the northernmost part of the country. The East Coast and southern regions exhibit the highest humidity levels.

MERRA-2, which succeeded the original MERRA dataset, benefits from advancements in the assimilation system that facilitate the incorporation of various datasets. These datasets provide daily and monthly time series data for different climatic parameters, including precipitation,

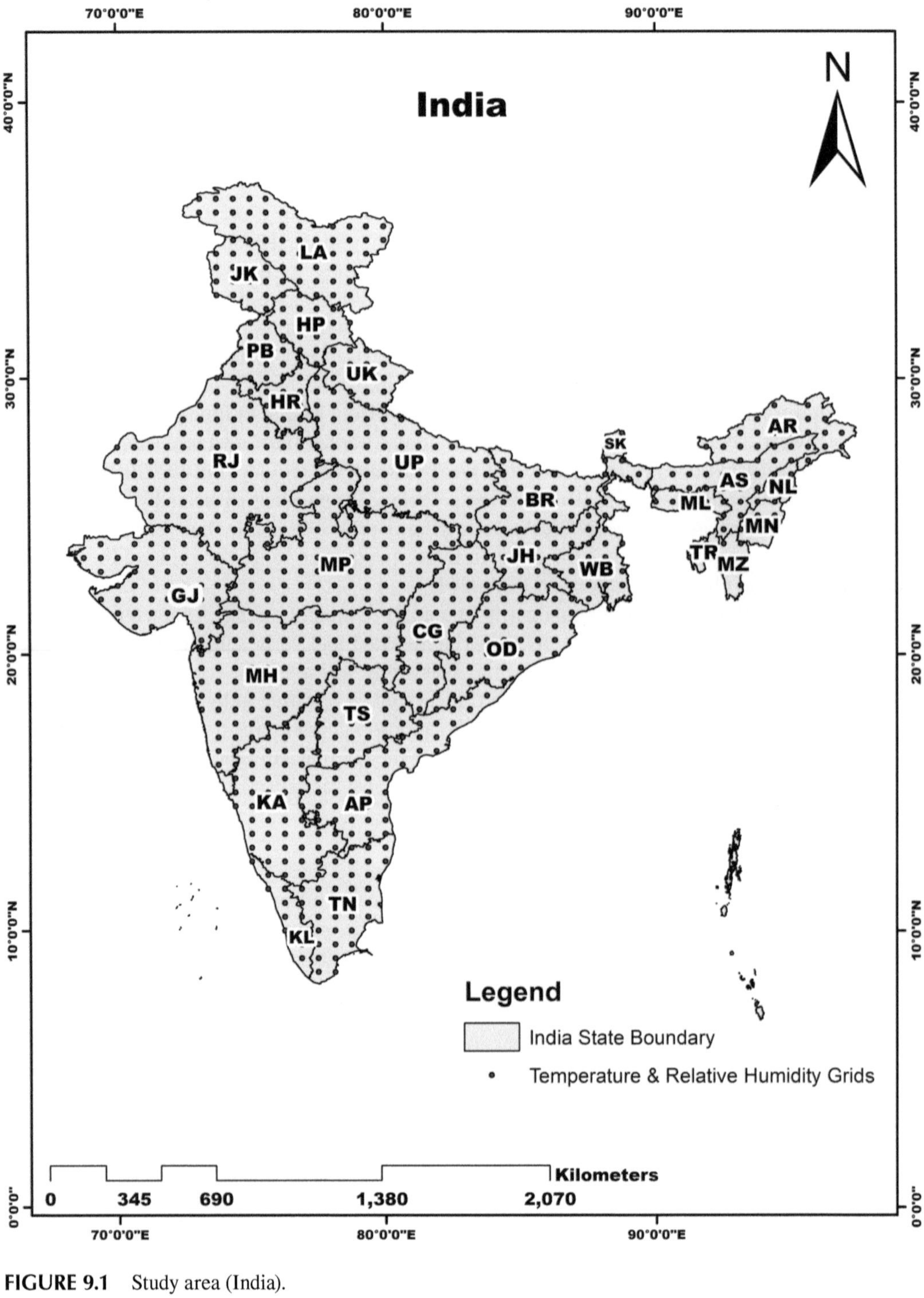

FIGURE 9.1 Study area (India).

temperature, RH, soil moisture, wind speed, and ice sheet cover globally. The daily T_{max} and RH data at the specified grid size were downloaded from the Goddard Earth Sciences (GES) Data and Information Services Center (DISC) (http://disc.sci.gsfc.nasa.gov/mdisc/) from 1981 to 2023. This data was available in NetCDF format (.nc). The data was spatially subsetted to focus

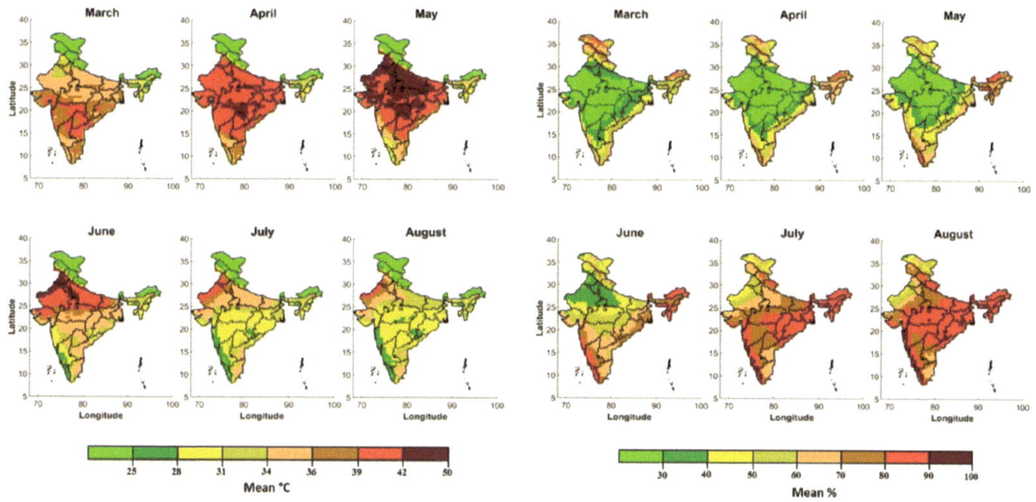

FIGURE 9.2 The mean of the maximum temperature (T_{max} in °C) and relative humidity (RH in %) for the months from March to August from 1981 to 2023 over India.

on the Indian subcontinent, covering the area between latitudes 8°4′28″N to 37°17′53″N and longitudes 68°7′53″E to 97°24′47″E.

METHODOLOGY

As mentioned in the Introduction, this study aims to identify and analyze monthly trends, characterization, and spatial variations of HI over India. To achieve these, the methodology is divided into the following steps: (i) computation of HI using Steadman's approach, (ii) characterization of HI, (iii) trend analysis, and (iv) spatial variation of HI over India.

COMPUTATION OF HI USING STEADMAN'S APPROACH

In this study, Steadman's approach was used to evaluate the HI, the algorithm of which is explained in detail (Steadman, 1979), where Steadman constructed a table that uses RH and temperature to produce the "apparent temperature" or the temperature the body "feels." This approach has been used by the National Weather Service (NWS), National Oceanic and Atmospheric Administration (NOAA), USA, and now also adopted by the IMD, for calculating HI. The formula to determine HI is applicable over a particular threshold value of temperature based on the T and RH, as stated in the equation below.

$$\begin{aligned} HI = {} & -42.379 + 2.04901523*T + 10.14333127*RH - 0.22475541*T*RH \\ & - 0.00683783*T^2 - 0.05481717*RH^2 + 0.00122874*T^2*RH \\ & + 0.00085282*T*RH^2 - 0.00000199*T^2*RH^2 \end{aligned} \tag{9.1}$$

where HI = Heat Index in °F; RH = Relative Humidity in %, T = Ambient Dry Bulb Temperature in °F.

Based on HI, the heat index chart provides different categories like "Category I (green color)," "Category II (yellow color)," "Category III (orange color)," and "Category IV (red color)". By understanding these categories and their respective warnings, essential measures can be taken to mitigate the effects of HS. IMD has suggested color codes for different categories of HI that can be seen in Table 9.1.

TABLE 9.1
Different Categories of HI

Category	Color Code	HI Value	Warning
I	Green	<35	Safe
II	Yellow	35≥HI<45	Be alert
III	Orange	45≥HI<55	Be prepared
IV	Red	HI≥55	Take action

CHARACTERIZATION OF HI

In this study, the HI is analyzed in terms of frequency, where HI are counted and classified into different categories as outlined in Table 9.1, for the period from 1981 to 2023, covering March–August. HS events are identified as daily events when the HI values fall within, below, or above the specified ranges provided in Table 9.1. Consequently, the frequency of HI is directly proportional to the duration due to the criteria used for their identification. The monthly variations in the frequency of HI from March to August across India have been thoroughly examined and presented. This analysis can be utilized to identify trends in HI over the study period.

TREND ANALYSIS

The Mann–Kendall test is employed to detect trends in time series data, evaluating whether there is a monotonic upward or downward trend over time without assuming any specific distribution for the data. It depends on the ranks of the samples rather than actual data. This makes it resistant to extremes and non-normally distributed data. This feature makes the Mann–Kendall test ideal for evaluating extreme values like HS, which frequently lack an evenly distributed normal distribution. Thus, to assess long-term monthly trends in different categories of HI, a non-parametric Mann–Kendall trend test with a 95% confidence level was utilized. Because of its robustness and dependability, the Mann–Kendall test has been widely employed in climate studies for trend detection (Srinivas et al., 2020; Subash & Sikka, 2014).

RESULTS AND DISCUSSIONS

Computation of HI

The HI was calculated using daily T_{max} and RH for March–August from 1981 to 2023 across India, using Steadman's technique. Based on the approach provided by the NWS in 2011, a MATLAB function called "HeatIndex" was developed to compute and categorize the HI, keeping in view the guidelines set forth by the IMD. Detailed information about this algorithm is available at NWS (NOAA, 2021). The daily HI for India was calculated and categorized using the developed "HeatIndex" function. According to the findings, the western regions and eastern coastlines of India had the greatest HI values, while the northeastern and northern areas had the lowest HI values. The states along the eastern coastlines, which include Odisha and Andhra Pradesh, and the western region of Gujarat possess the highest HI values. On the other hand, the northeastern regions like Sikkim and Arunachal Pradesh as well as the northern states like Ladakh and Jammu & Kashmir exhibit the lowest HI values.

To help in the classification of HI, a HI chart specifically for India has been produced. It can aid in comprehending how hot it feels under various circumstances. This chart represents categories and their corresponding warnings: "Category I (Safe)," "Category II (Be Alert)," "Category III (Be Prepared)," and "Category IV (Take Action)" as illustrated in Figure 9.3. This chart aids in taking appropriate

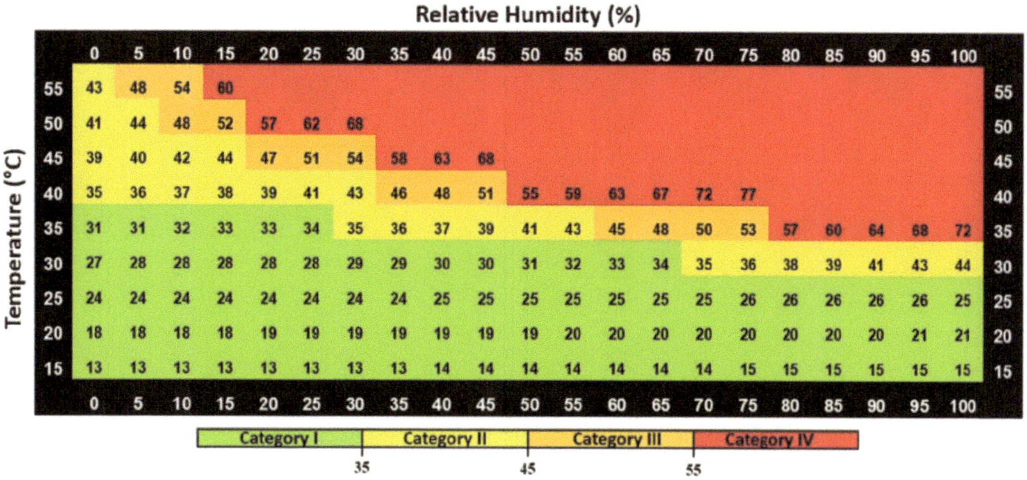

FIGURE 9.3 Heat index chart.

measures to mitigate the impact of HS. The recommended warnings for each category of HI can be effectively utilized in the development of heat action plans.

FREQUENCY OF HI

Monthly Occurrences of Category I HI

The monthly climatology of HI belonging to Category I across India indicates that HS are less frequent in April–June, while they are more frequent in March, July, and August (Figure 9.4). Notably, Category I HI consistently remained at or above 30 events per year in the northern and northeastern regions during the duration of the research period. In March, the average number of HI in this category varies geographically: from 20 to 30 events per year in the northwestern states, 10–20 events per year in central states, and 1–10 events per year in the coastal states of Gujarat, Maharashtra, and Odisha. The southern states, however, experienced very few occurrences of this category of HI. During April, May, and June, there was a notable decline in the frequency of HI. The analysis reveals that western, central, and east coastal states showed a reduction in the frequency. As the southwest monsoon typically begins around early June, an increase in HI was observed in the southern state of Kerala and the west coast states of Karnataka and Maharashtra. During the monsoon season, the HI tends to retreat, resulting in an increased frequency in July and August. The regions most frequently affected by Category I HI include the northern regions (Jammu and Kashmir, Himachal Pradesh, Uttarakhand, and Ladakh) and the northeastern states (Sikkim and Arunachal Pradesh) of India.

MONTHLY OCCURRENCES OF CATEGORY II HI

HI categorized as Category II ($35 \geq HI < 45$) exhibits distinct spatial variations across India. These are less prevalent during the peak heat months of May and June and are more frequent in March, April, July, and August (Figure 9.4). Throughout the study period, the northern and northeastern regions consistently experienced fewer than one Category II HI event per year. In March, the frequency of HI in this category varies significantly by region: the western states of Maharashtra and Karnataka, along with parts of Telangana and Chhattisgarh, record 20–30 events per year. The central states observe 10–20 events per year, while the east coastal state of Andhra Pradesh and the southern state of Tamil Nadu report 1–10 events per year. April sees a two-fold increase in HI in central and northwestern states, compared to March. Despite May and June being the hottest

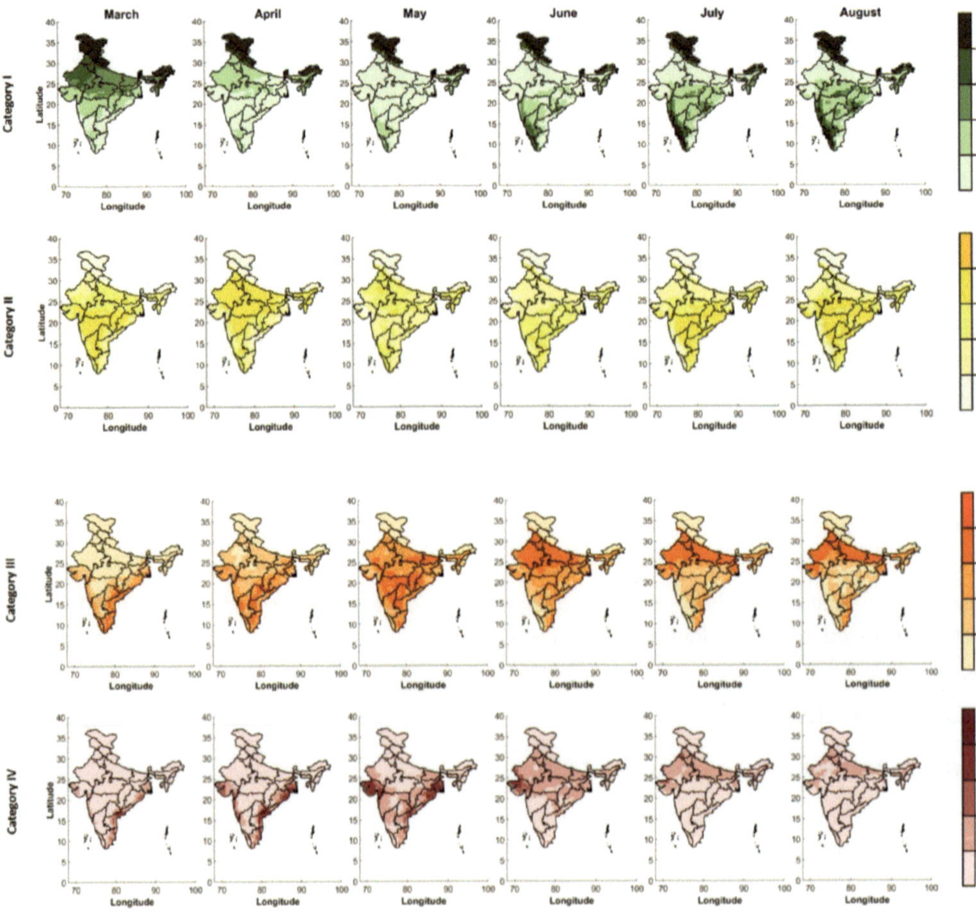

FIGURE 9.4 Monthly average occurrences of different categories of HI over India during 1981–2023 (March–August).

months, the frequency of HI notably declines during this period. Conversely, during the monsoon months of July and August, the retreat of HI leads to an increased frequency. The analysis highlights that during the summer, the regions most frequently affected by Category II HI are the western states (Maharashtra and Karnataka) and the central states (Madhya Pradesh and Uttar Pradesh). In contrast, during the monsoon season, the most affected regions shift to include Maharashtra, Telangana, Chhattisgarh, and the states along the eastern coastlines, which include Odisha and Andhra Pradesh.

MONTHLY OCCURRENCES OF CATEGORY III HI

The monthly climatology of Category III HI in India reveals distinct spatial patterns. These are most frequent during the peak summer months of May and June, and less prevalent in March, April, July, and August (Figure 9.4). Throughout the study period, HI in Category III ($45 \geq HI < 55$) consistently remained below one event per year in the northern and northeastern regions. In March, the geographical distribution of Category III HI shows considerable variability. Regions along the western coastal states such as Gujarat, Maharashtra, Karnataka, and Kerala experience between 1 and 10 events per year. In contrast, the coastal states of Odisha and Andhra Pradesh, along with the southern state of Tamil Nadu, observe 10–20 events per year. In April, the frequency of HI doubles across most parts of the country, except in the northern and northeastern regions. The hottest months, May and June, show a significant increase in the frequency. Notably, June and July recorded the highest occurrence, with 20–30 events per year, particularly in the northwestern states (Rajasthan, Punjab, and Haryana), as well as Uttar Pradesh and Bihar, marking them as the most affected regions. During the monsoon months of July and August, the frequency of HI decreases as HI retreats.

MONTHLY OCCURRENCES OF CATEGORY IV HI

The analysis of Category IV HI events in India indicates that these are less frequent in March, July, and August, while their occurrence is more pronounced in April, May, and June (Figure 9.4). In March, the frequency remained below 1 per year across most of the study area, except in a few parts of the eastern coastal states where the frequency ranged from 1 to 20 events per year. In April, the spatial distribution of Category IV HI increased along the regions of Gujarat and the western and eastern coastal states. The hottest months, May and June, exhibited a significant rise in the frequency. However, May recorded the highest occurrences, with 20–30 events per year, particularly in the states of Gujarat, Odisha, West Bengal, and Andhra Pradesh. During the monsoon months, the frequency decreased, maintaining a range of 1–10 events per year in the northwestern and central states. The analysis reveals that the states of Gujarat, Odisha, West Bengal, and Andhra Pradesh were the most affected by Category IV HI.

TREND ANALYSIS OF HI

The Mann–Kendall test was applied at a 95% confidence level to each of the 916 grid points to examine the trends in monthly occurrences of different categories of HI from March to August for the period 1981–2023 (Figure 9.5). The results of the monthly trend analysis highlight the dynamic nature of the HI across different regions in India. The detailed results for Category I HI are explained here. For monthly occurrences of Category I HI, no trend was observed in the northern states, including Ladakh, Jammu and Kashmir, Himachal Pradesh, and Uttarakhand, as well as in the northeastern state of Arunachal Pradesh throughout the period. In March, the trend analysis reveals a significant decreasing trend (red regions) in the northern (including states like Punjab, Haryana, and Uttar Pradesh) and central parts of India (including Madhya Pradesh and Chhattisgarh). This suggests that these regions experience a drop in Category I HI. Conversely, several pockets in the south (including Kerala, Tamil Nadu, and Telangana) display an increasing trend (blue regions), indicating a rise. The month of April shows a pronounced decreasing trend spreading across the central (including Madhya Pradesh and Maharashtra), northern (including Delhi, Haryana, and Uttar Pradesh), and parts of western India (including Gujarat and Rajasthan). This pattern suggests a continued drop in the HI in these areas. In contrast, the southern states (inclusive of Kerala, Karnataka, and Tamil Nadu) and eastern coastal states (inclusive of Odisha and Andhra Pradesh) exhibit an increasing trend, indicating a rise in HI. The trend of declining HI continues throughout May, mostly affecting the northwest states (Rajasthan, Punjab, and Haryana). The southern states (Karnataka, Kerala, and Tamil Nadu) and certain regions of other states (Maharashtra and Bihar) continue to show a rising trend, suggesting an increase in events. A notable shift occurs in June, with a substantial portion of the central and northern regions, including Madhya Pradesh and Maharashtra as well as Uttarakhand and Himachal Pradesh, exhibiting an upward trend. This might indicate that the HI is rising in certain regions. Karnataka, Tamil Nadu, and Andhra Pradesh are among the southern regions that still have red areas, which show a downward trend. In July, the increasing trend in the HI is more widespread, covering most of the central (including Maharashtra and Madhya Pradesh), western part of Gujarat, and northern parts of India (including Punjab and Haryana). The southern regions (including Karnataka, Kerala, Tamil Nadu, Telangana, and Andhra Pradesh) show a decreasing trend. August continues to show an increasing trend in the northwestern

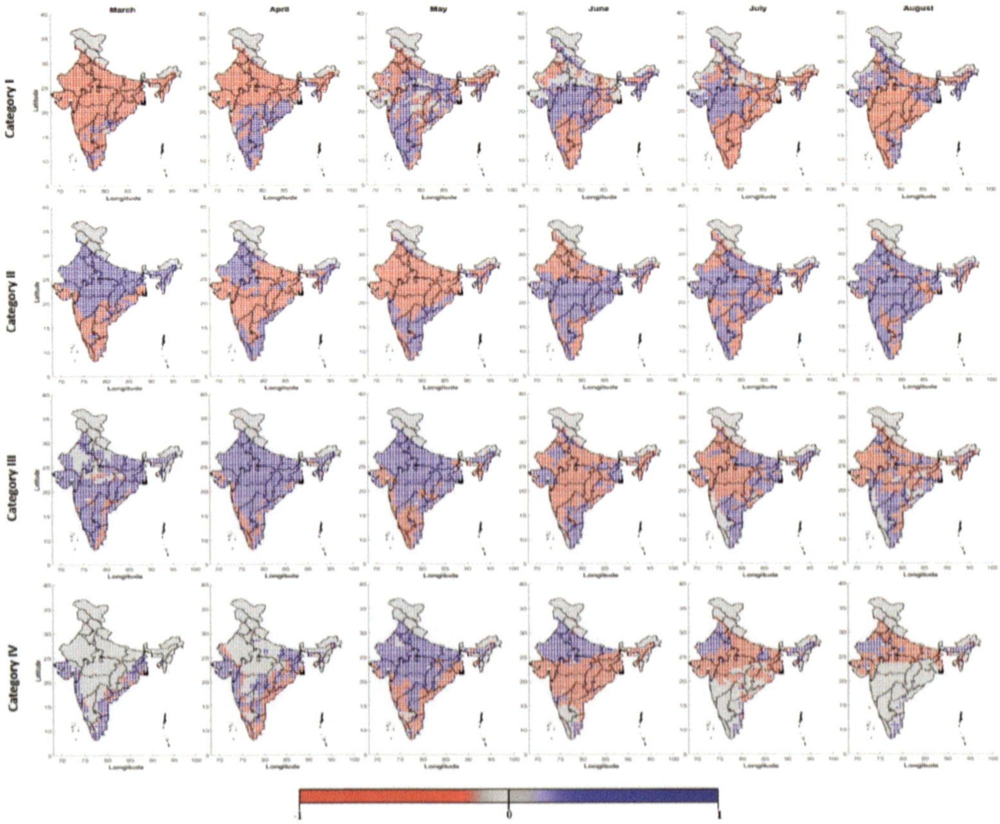

FIGURE 9.5 Monthly trend for different categories of HI during 1981–2023 (March–August).

states of Rajasthan, the western state of Gujarat, and southern parts of northern states (including Jammu and Kashmir, Himachal Pradesh, and Uttarakhand), indicating a rise in the HI. The southern parts (including Karnataka, Kerala, and Telangana), and central states (including Madhya Pradesh, Maharashtra, and Chhattisgarh) show a decreasing trend. Whereas, the eastern coastal parts (including Tamil Nadu, Andhra Pradesh, Odisha, and West Bengal) have mixed trends with both red (decreasing) and blue (increasing) regions, suggesting localized variations. Areas with no significant trend (gray regions) are observed, particularly in the western and northwestern parts (including Gujarat and Rajasthan).

Overall, the trend analysis for Category I HI over the months from March to August shows distinct seasonal variations. The pre-monsoon months (March–May) are characterized by a decreasing trend in the HI in the northern (including states like Punjab, and Haryana) and central regions of India (inclusive of Uttar Pradesh, Madhya Pradesh, and Chhattisgarh), while the monsoon months (June–August) show an increasing trend in these regions. The southern regions (including Kerala, Tamil Nadu, and Karnataka) exhibit mixed trends throughout the period, indicating localized variations. Spatial trend patterns for other categories of HI can be seen in Figure 9.5. These findings highlight the dynamic nature of the HI across different regions in India.

SPATIAL VARIATION OF HI

The spatial variations in HI provide valuable insights into the trends in HI occurrences. This study conducted a spatial analysis for all decadal years from 1981 to 2023, revealing significant changes in the distribution of HI events (Figure 9.6). Over the northern, northwest, northeast, and central

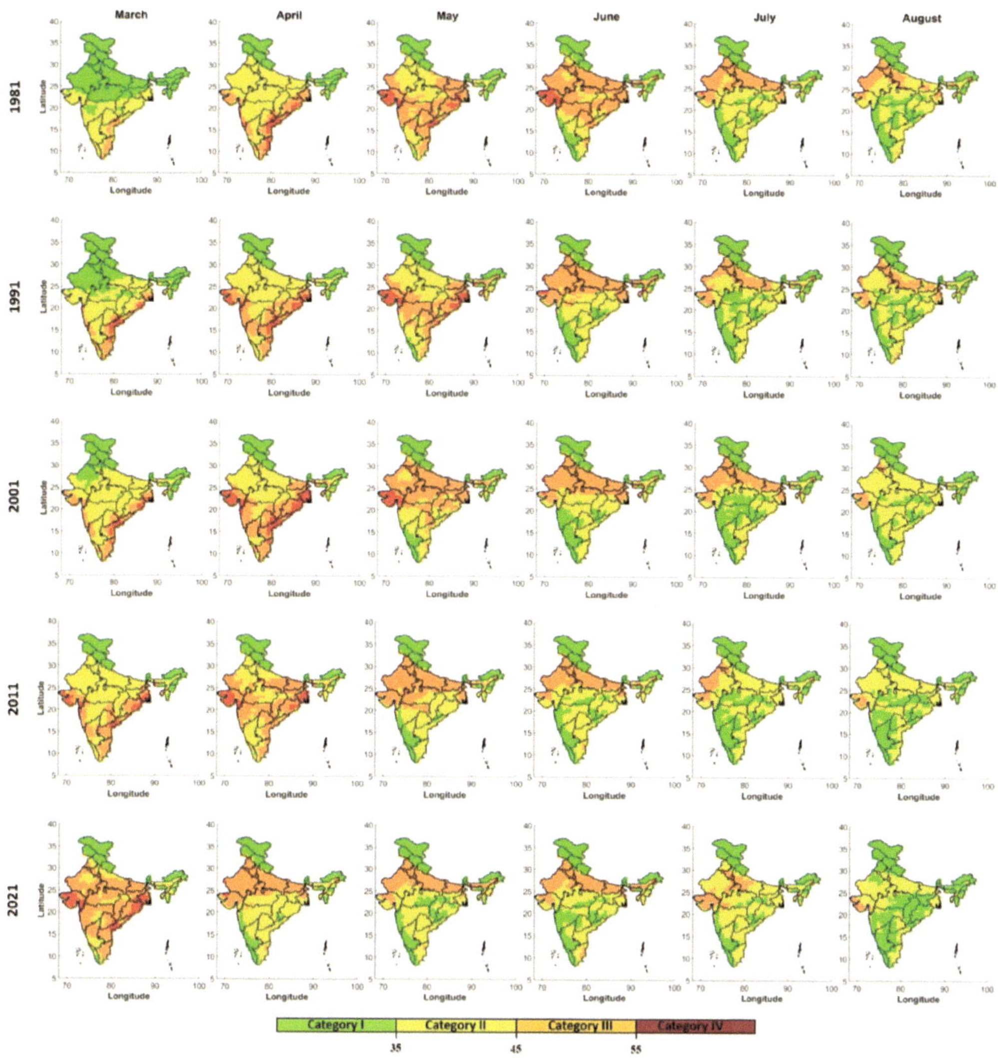

FIGURE 9.6 Monthly spatial variation of HI for the decadal years of 1981, 1991, 2001, 2011, and 2021 for March–August.

states in March, there was a noticeable drop in Categories I and II HI. Conversely, Category III HI increased over coastal, central, and northwestern states, while Category IV HI rose along the western state of Gujarat and the coastal states of Andhra Pradesh, Odisha, and West Bengal. This shift aligns with a warming trend toward northern India and the south-central area post-1980s, as highlighted in the trend analysis of T_{max} over India (Ross et al., 2018). In April, Category I remained spatially constant over the northern states throughout the period. However, Category II decreased spatially in the northwest and central parts of India. Category III increased along the coastal states, which are most humid in April. Category IV initially rose until 2010 but declined thereafter due to decreasing RH, as indicated by a change point analysis of humidity trends in April 2011 (Chithra et al., 2018).

May and June, the hottest months, exhibited significant spatial variations over the period. Category I showed no change in the northern states, while Category II shifted from the northwestern and central states to the southern and eastern coastal states. In contrast, Category III exhibited an opposite shift to Category II. Category IV declined spatially, with the eastern coastal states no longer being the most affected regions during these months. This shift can be attributed to synoptic factors like the current high temperatures, anomalies in the direction of the winds, and the eastward recurrence of cyclonic storms across the Bay of Bengal (Kumar Jenamani, 2012). During the monsoon months of July and August, there was an increase in the spatial variation of Category I in the western, southern, and central states. Category III declined spatially over the central and northwestern states as the HI tends to retreat during the monsoon season. The observed decrease in HI can be linked to the declining trend in T_{max} between 1950 and 2016 (Ross et al., 2018). Mukherjee and Mishra (2018) noticed a similar drop, attributing it to the combination of increasing air aerosol deposition in the Indo-Gangetic area and cooling by evaporation from enhanced latent heat flow as a result of intense irrigation. This cooling effect influences the surface energy budget, thus reducing surface temperatures. Consequently, the eastern region does not frequently experience HW and is not considered a protruding HW area, as also noted by Ratnam et al. (2016) and Rohini et al. (2016).

CONCLUSIONS

The present study focuses on identifying and analyzing trends, characterizing, and observing spatial variations of the HI during the summer (March–June) and early monsoon (July–August) seasons from 1981 to 2023 across India. The analysis reveals a significant increase in HI, indicating potential future escalation. The following conclusions are drawn from this study:

1. The highest HI values were recorded in Gujarat, Odisha, and Andhra Pradesh, while the lowest were in Ladakh, Jammu & Kashmir, Sikkim, and Arunachal Pradesh.
2. The characteristics of HI have significantly evolved, becoming more recurrent and spreading into areas formerly unaffected or minimally affected by HI.
3. A warming trend is evident in the northwestern and eastern coastal plains of India, whereas a gradual cooling phase is observed in the northeastern and southwestern regions.
4. The northwest, central, and states along the eastern coastlines were identified as the most affected regions, posing a direct health risk to the inhabitants.
5. The rise of Category IV HI in the eastern coastal states is concerning because these states are more vulnerable to HS because of excessive humidity, which might have a serious negative impact on public health.
6. The early start of summer has led to an increase in the spatial variation of Categories III and IV HI in March in recent years, which has prompted a warning to the population to be prepared and take action against HS. Additionally, the spread of Category III HI increased over the northwestern and central regions in May, following the warming trend, and is expected to continue rising without a mitigation strategy.

In India, where a significant portion of the population is involved in activities that are outdoors, the growing exposure to extreme heat events underscores the necessity to acknowledge HS as a serious health risk. To mitigate the impact of HS, the recommended warnings for each category of HI can be effective in the development of heat action plans. These should include efficient early warning systems, focused mapping of susceptible populations, and public education initiatives to ensure proper heat-related health care. By addressing these issues urgently, the adverse effects of HS can be reduced, thereby protecting public health and improving resilience against future HS events.

ACKNOWLEDGMENTS

The authors would like to acknowledge Ms. Shivani Verma, Ms. Shreya Mittal, and Ms. Parul Kairo, undergraduate students from the Department of Civil Engineering at the Indian Institute of Technology Roorkee, who have contributed to this work. The first author would also like to acknowledge the support of IIT Roorkee and the Ministry of Human Resource Development, Government of India, through the Prime Minister's Research Fellowship (PMRF) for carrying out this research for Grant No. PM-31-22-785.

CONFLICT OF INTEREST

The authors declare no conflict of interest.

DATA AVAILABILITY STATEMENT

The study utilizes freely available dataset from the Modern-Era Retrospective Analysis for Research and Applications, version 2 (MERRA-2) (https://disc.gsfc.nasa.gov/information/glossary?title=MERRA-2). These data are available from the corresponding author upon reasonable request.

REFERENCES

Chen, K., Bi, J., Chen, J., Chen, X., Huang, L., & Zhou, L. (2015). Influence of heat wave definitions to the added effect of heat waves on daily mortality in Nanjing, China. *Science of the Total Environment, 506–507*, 18–25. https://doi.org/10.1016/j.scitotenv.2014.10.092

Chithra, N. R., Shahul, D., Muralidhar, S., Unnikrishnan, U., & Akhil Rajendran, K. (2018). Trend analysis of climatic variables in the Indian subcontinent. *Mausam, 69*(3), 375–386. https://doi.org/10.54302/mausam.v69i3.327

Fischer, E. M., & Schär, C. (2010). Consistent geographical patterns of changes in high-impact European heatwaves. *Nature Geoscience, 3*(6), 398–403. https://doi.org/10.1038/ngeo866

Fischer, E. M., Seneviratne, S. I., Vidale, P. L., Lüthi, D., & Schär, C. (2007). Soil moisture-atmosphere interactions during the 2003 European summer heat wave. *Journal of Climate, 20*(20), 5081–5099. https://doi.org/10.1175/JCLI4288.1

Frich, P., Alexander, L. V., Della-Marta, P., Gleason, B., Haylock, M., Tank Klein, A. M. G., & Peterson, T. (2002). Observed coherent changes in climatic extremes during the second half of the twentieth century. *Climate Research, 19*(3), 193–212. https://doi.org/10.3354/cr019193

Goyal, M. K., Singh, S., & Jain, V. (2023). Heat waves characteristics intensification across Indian smart cities. *Scientific Reports, 13*(1), 1–16. https://doi.org/10.1038/s41598-023-41968-8

Kent, S. T., McClure, L. A., Zaitchik, B. F., Tiffany T. Smith, & Julia M. Gohlke. (2014). Heat waves and health outcomes in Alabama (USA): The importance of heat wave definition. *Environmental Health Perspectives, 122*(2), 151–159.

Kotharkar, R., Ghosh, A., & Kotharkar, V. (2021). Estimating summertime heat stress in a tropical Indian city using Local Climate Zone (LCZ) framework. *Urban Climate, 36*(June 2020), 100784. https://doi.org/10.1016/j.uclim.2021.100784

Krinner, G., Germany, F., Shongwe, M., Africa, S., France, S. B., Uk, B. B. B. B., Germany, V. B., Uk, O. B., France, C. B., Uk, R. C., Canada, M. E., Erich, M., Uk, R. W. L., Uk, S. L., & Lucas, C. (2013). Long-term climate change: Projections, commitments and irreversibility. In *Climate Change 2013 the Physical Science Basis: Working Group I Contribution to the Fifth Assessment Report of the Intergovernmental Panel on Climate Change, 9781107057*, 1029–1136. https://doi.org/10.1017/CBO9781107415324.024

Kumar Jenamani, R. (2012). Analysis of ocean-atmospheric features associated with extreme temperature variation over east coast of India-A special emphasis to Orissa heat waves of 1998 and 2005. *Mausam, 63*(3), 401–422.

Kumar, P., Rai, A., Upadhyaya, A., & Chakraborty, A. (2022). Analysis of heat stress and heat wave in the four metropolitan cities of India in recent period. *Science of the Total Environment, 818*. https://doi.org/10.1016/j.scitotenv.2021.151788

Luan, G., Yin, P., Wang, L., & Zhou, M. (2019). Associations between ambient high temperatures and suicide mortality: A multi-city time-series study in China. *Environmental Science and Pollution Research, 26*(20), 20377–20385. https://doi.org/10.1007/s11356-019-05252-5

McGregor, G. R., Bessemoulin, P., Ebi, K. L., & Menne, B. (Eds.). (2015*). Heatwaves and health: guidance on warning-system development* (No. 1142). Geneva: World Meteorological Organization.

Mishra, V., Ambika, A. K., Asoka, A., Aadhar, S., Buzan, J., Kumar, R., & Huber, M. (2020). Moist heat stress extremes in India enhanced by irrigation. *Nature Geoscience, 13*(11), 722–728. https://doi.org/10.1038/s41561-020-00650-8

Montero, J. C., Miron, I. J., Criado, J. J., Linares, C., & Díaz, J. (2013). Difficulties of defining the term, heat wave, in public health. *International Journal of Environmental Health Research*, *23*(5), 377–379. https://doi.org/10.1080/09603123.2012.733941

Mukherjee, S., & Mishra, V. (2018). A sixfold rise in concurrent day and night-time heatwaves in India under 2°C warming. *Scientific Reports*, *8*(1), 1–9. https://doi.org/10.1038/s41598-018-35348-w

NDMA. (2019). *National Guidelines for Preparation of Action Plan-Prevention and Management of Heat Waves* (p. 40). https://ndma.gov.in/sites/default/files/PDF/Guidelines/heatwaveguidelines.pdf

NOAA. (2021). *Heat Index Calculator*. National Weather Service, U.S. Department of Commerce, 1–2. https://www.calculator.net/heat-index-calculator.html

Pascal, M., Wagner, V., Alari, A., Corso, M., & Le Tertre, A. (2021). Extreme heat and acute air pollution episodes: A need for joint public health warnings? *Atmospheric Environment*, *249*(October 2020). https://doi.org/10.1016/j.atmosenv.2021.118249

Pattnayak, K. C., Abdel-Lathif, A. Y., Rathakrishnan, K. V., Singh, M., Dash, R., & Maharana, P. (2019). Changing climate over chad: Is the rainfall over the major cities recovering? *Earth and Space Science*, *6*(7), 1149–1160. https://doi.org/10.1029/2019EA000619

Pattnayak, K. C., Kar, S. C., Dalal, M., & Pattnayak, R. K. (2017). Projections of annual rainfall and surface temperature from CMIP5 models over the BIMSTEC countries. *Global and Planetary Change*, *152*, 152–166. https://doi.org/10.1016/j.gloplacha.2017.03.005

Ratnam, J. V., Behera, S. K., Annamalai, H., Ratna, S. B., Rajeevan, M., & Yamagata, T. (2016). ENSO's far reaching connection to Indian cold waves. *Scientific Reports*, *6*, 1–12. https://doi.org/10.1038/srep37657

Rohini, P., Rajeevan, M., & Srivastava, A. K. (2016). On the variability and increasing trends of heat waves over India. *Scientific Reports*, *6*, 1–9. https://doi.org/10.1038/srep26153

Ross, R. S., Krishnamurti, T. N., Pattnaik, S., & Pai, D. S. (2018). Decadal surface temperature trends in India based on a new high-resolution data set. *Scientific Reports*, *8*(1), 2–11. https://doi.org/10.1038/s41598-018-25347-2

Srinivas, R., Singh, A. P., Dhadse, K., & Magner, J. (2020). Hydroclimatic river discharge and seasonal trends assessment model using an advanced spatio-temporal model. *Stochastic Environmental Research and Risk Assessment*, *34*(2), 381–396. https://doi.org/10.1007/s00477-020-01780-6

Steadman, R. G. (1979). The assessment of sultriness. Part I: A temperature-humidity index based on human physiology and clothing science. *Journal of Applied Meteorology*, *18*, 1–7.

Subash, N., & Sikka, A. K. (2014). Trend analysis of rainfall and temperature and its relationship over India. *Theoretical and Applied Climatology*, *117*(3–4), 449–462. https://doi.org/10.1007/s00704-013-1015-9

Wang, Y., Berardi, U., & Akbari, H. (2016). Comparing the effects of urban heat island mitigation strategies for Toronto, Canada. *Energy and Buildings*, *114*, 2–19. https://doi.org/10.1016/j.enbuild.2015.06.046

Zhang, J., Chen, Q., Wang, Q., Ding, Z., Sun, H., & Xu, Y. (2019). The acute health effects of ozone and PM 2.5 on daily cardiovascular disease mortality: A multi-center time series study in China. *Ecotoxicology and Environmental Safety*, *174*(January), 218–223. https://doi.org/10.1016/j.ecoenv.2019.02.085

10 Cities, Urbanization, and Environmental Vulnerable Community

A Case Study of Bulandshahr City

Zeba Khanam

INTRODUCTION

Vulnerability of urban areas can be traced by the process of urbanization, which consequently affects the urban environment adversely (Lankao and Qin, 2011). It continues to haunt developing countries further despite several initiatives adopted to reduce vulnerability (Singh et al., 2012). Urbanization in South Asia is a global concern. However, the South Asian countries have a low share of the urban population; the absolute number is significantly high. For example, only 35.4% of the Indian population is urban, but the size of the urban population is 493.2 million, which is 1.5 times more than the total population of the United States, the third most populated country in the world (World Bank, 2022). In India, the urbanization process has been characterized as lopsided, where big cities like Delhi and Mumbai have grown significantly over the past decades. However, studies observed that small and medium cities are also growing at an unprecedented rate (Dutta et al., 2013; Rumbach, 2016). Therefore, the assessment of urban environmental vulnerability in smaller urban areas is relevant. The results will benefit stakeholders in urban governance in planning, monitoring, and evaluating changes over time at local and regional scales (Bulkele and Tuts, 2013).

Urban areas are complex eco-environments factors that involve various anthropogenic activities (Zhang et al., 2006), which directly and indirectly impact the environment. Environmental deterioration in the form of increasing air pollution, water contamination, inadequate green space, and the concentration of hazardous wastes has become a major concern for urban sustainability (Nyambod, 2010). The widespread change in LULC, including urban sprawl (Johnson, 2001), loss of vegetation (Richards and Belcher, 2019), prime agricultural lands (Doygun, 2009), and natural wetlands (Mao et al., 2018), in urban areas, as evident from significant change in thermal signature, often represents changes in urban eco-environmental conditions at micro-scales (Adger, 2000; Miller and Small, 2003; Mallick et al., 2013). The increase in urban population in these cities is compensated by the spatial growth of urban built-up areas both in vertical and horizontal spheres (Huang et al., 2021; Zambon et al., 2019). The horizontal spilling of the urban build-up and increase in population density put considerable pressure on urban land and environment (Yu et al., 2021). The rapid and unplanned urbanization process has also adversely affected urban green cover in most cities.

In developing and low-income countries, the lack of infrastructural services is further exacerbated by unplanned development in urban areas (Roy, 2021). Unplanned urbanization burdens the cities' resources, such as water, land, and agriculture, through construction activities, industrialization, and commercialization (Alam et al., 2006). As a result of urban expansion and urbanization (including the urban sprawl), the transportation network has also spread out in urban areas, consequently, infrastructural networks have expanded to form broad economic and community areas (Jenelius, 2010). Water logging (WL) and unmanaged solid waste (SW)

DOI: 10.1201/9781003620747-12

generation are the consequence of haphazard development (Roy, 2021). This urbanization overburdens not only the drainage system due to encroachment, but it also impacts the quality of runoff. Therefore, the quality of surface water impacts groundwater through discharge, eventually (Birawat et al., 2021).

Identifying the vulnerable localities can be helpful for the sustainable development of an urban area (Turner et al., 2003). A comprehensive analysis of the least developed localities of Bulandshahr city is required to help the local authorities involve poor and marginalized communities in the process of urban planning and policy implementation. As urban areas are complex systems, this chapter has adopted a multi-criteria decision-making (MCD) method, which is widely followed for vulnerability assessment (Dammalage and Jayasinghe, 2019). The MCD approach considers multiple factors in a complex problem, and categorizes them into a hierarchical structure of importance, then finally calculates the weight of all criteria to define the synthetic index of vulnerability (Benmoussa et al., 2019; Tercan and Dereli, 2020). Further, integrating the MCD results into a geographic information system (GIS) allows the researcher to produce a vulnerability map. The MCD method and GIS together provide a powerful decision support system (Kazemi et al., 2016). As a result, there is an increasing application of MCD in various disciplines, including resource management and spatial planning. This chapter incorporates eight criteria to identify vulnerable areas in Bulandshahr city; the eight criteria are LULC, Land Surface Temperature LST, NDVI, NDBI, road density (RD), water quality, WL, and SW. We also used analytical hierarchy process (AHP) for determining the weightage of each criterion in the decision-making, and integrated it with the GIS for spatial visualization of the vulnerable areas and the vulnerable community of the city.

The vulnerable population in Bulandshahr, particularly those residing in urban slum areas, faces a complex set of environmental and socio-economic challenges. These informal settlements are characterized by high population density, inadequate housing, limited access to clean water and sanitation, and poor waste disposal systems. Often located on marginal lands such as floodplains, near open drains, or alongside industrial zones, these slums are highly exposed to environmental risks like waterlogging, contamination, and extreme heat. Residents, many of whom belong to low-income or marginalized social groups, are disproportionately affected by environmental degradation and lack the resources to adapt or relocate. Studies including those focused on the broader NCR and Bulandshahr specifically (Khanam et al., 2026; Khanam et al., 2022) highlight that slum dwellers are among the most environmentally vulnerable communities due to their dependence on degraded natural resources, lack of infrastructure, and exclusion from formal planning processes. Addressing the needs of these communities is essential for equitable urban development and for building resilience against climate-related and anthropogenic hazards.

STUDY AREA

Bulandshahr city (Figure 10.1), the administrative center of Bulandshahr district in western Uttar Pradesh, is situated at 280 4'N and 770 52'E and 225.5 m above mean sea level (Bulandshahr Municipality, 2018). Delhi, India's National Capital, is situated 68 km^2 away from the Bulandshahr city. Therefore, the urban landscape of Bulandshahr city has undergone severe transformation as a result of its proximity to the National Capital Region. The population of the city has increased from 1,27,201 people in 1991 to more than 2,30,024 in the 2011 census. The quality and comfort of urban living in Bulandshahr city have suffered due to rapid urbanization. Bulandshahr was chosen for this study because, in the past few decades, population growth was rapid and it has resulted in unchecked spatial development and cluttered urbanization. Second, the proximity to Delhi NCR and the development of industry raises important concerns regarding the city's future urban planning. Urban planners, decision-makers, and stakeholders might benefit from the analysis of urban patterns and identification of environmentally susceptible areas by using cutting-edge geospatial approaches (Khanam et al., 2022).

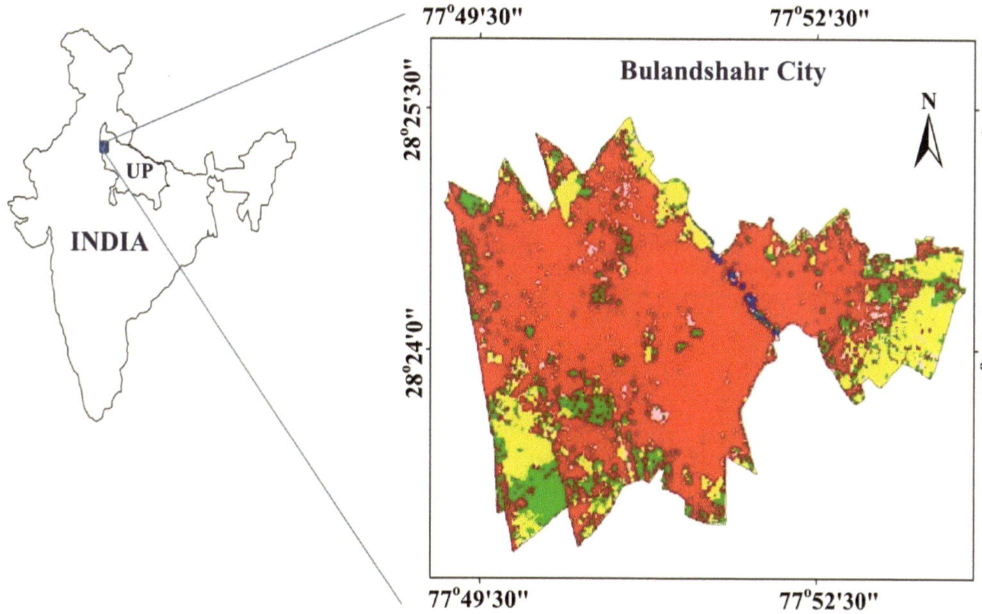

FIGURE 10.1 Study area (Bulandshahr city).

METHODS AND MATERIALS

Overview

The map of urban vulnerability can be used in all stages of disaster management, including preparedness, prevention, operations, mitigation, relief, and recovery (Zhang et al., 2014). Integrating urban planning and hazard mitigation efforts is essential for building sustainable communities in urban areas (Campbell, 1996). Thus, vulnerability maps can be employed as part of a scenario modeling strategy for urban planning. Figure 10.2 depicts the comprehensive methodologies adopted for environmental vulnerability assessment in the present study. It has used seven parameters for environmental vulnerability assessment, that is, LULC, LST, NDBI, NDVI, water quality, RD, WL, and SW. Cloud-free and haze-free satellite images (Landsat 8) were selected for the study. The satellite images were downloaded from the online source, USGS Earth Explorer website (https://earthexplorer.usgs.gov); (The details of satellite imagery used in the work are given in Table 10.1). The land cover, NDVI and NDBI, and LST maps were generated from Landsat-8 imagery. They were verified by Google Earth Image. RD map was generated using OpenStreetMap and line density in ArcGIS. A water quality map was generated by analyzing different water quality parameters (collected from government offices, Bulandshahr city) along with IDW and WQI to assess water quality. WL and SW generation maps were prepared by field survey data taken during the period of 2019 in Bulandshahr city (Pic-1,2,3). A questionnaire survey was employed in 37 wards of the Bulandshahr city, which includes the socio-economic conditions and WL and SW-related information. All the criteria components were integrated using AHP and Weighted Overlay techniques to assess the environmentally vulnerable areas. We employed three different software systems to process the data, that is, Microsoft Excel, ERDAS, and ArcGIS.

SELECTING PARAMETERS FOR URBAN VULNERABILITY ASSESSMENT

Land Use Land Cover

LULC changes are not only key factors among wide-ranging determinants of land surface parameters (such as LST, surface runoff, and evapotranspiration), but play a significant role in affecting determinants of land surface functioning (Lambin et al., 2003). Many studies worldwide have considered

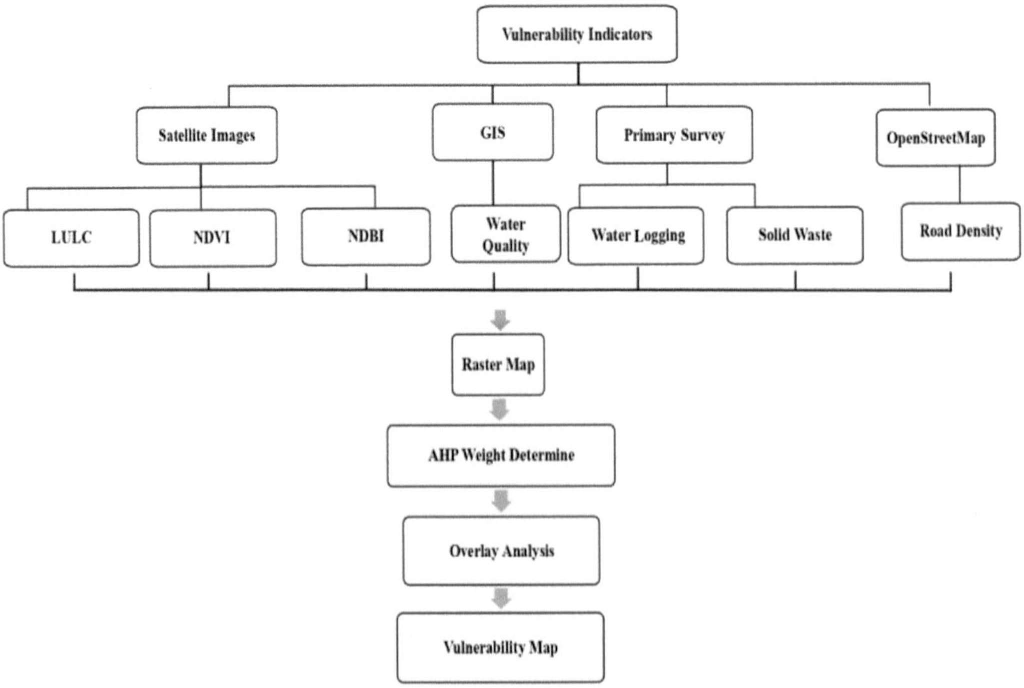

FIGURE 10.2 Study flow chart.

TABLE 10.1
Details of the Landsat Imageries Used in the Present Study

Satellite	Sensors	Date	No. of Spectral Band	Spatial Resolution	Source
Landsat 8	Operational Land Imager (OLI) Thermal Infrared Sensor (TRS)	March 8, 2003	11	15 M	USGS

LULC as one of the parameters of vulnerability assessment (Khosravi et al., 2017; Choudhary et al., 2018; Sarif and Gupta, 2022). Liou et al. (2017) in their study concluded that land modification due to increased anthropogenic activities can intensify the rate and extent of eco-environmental vulnerability.

Land Surface Temperature Estimation

Urban environments exhibit increased land surface temperature due to the presence of concrete, impervious surfaces, and building materials that retain heat and increase the sensitivity (Inostroza et al., 2016). The additive effects of UHI exacerbate extreme heat events, like heat waves, in urban areas, and expose a large population to different health complications (Wong and Yu, 2005; Yuan and Bauer, 2007). Heat wave (extreme heat) is a critical and deteriorating threat to human health; especially in urban areas, LST ranged from 35°C to 55°C indicated a potential heat hazard (Chakraborty and Li, 2022). The following sections will describe the LST retrieval procedure from Landsat 8.

MWA (Mono Window Algorithm) for LST

Two thermal infrared bands on the Landsat 8 TIR sensor were used with the split window technique to retrieve LST. Du et al. (2015) in their study used the split window algorithm with two thermal bands (bands 10 and 11) to extract LST from Landsat 8 imagery. Other than this, for Landsat 8 data, several studies, including Rozenstein et al. (2014), used the two-factor split window technique. However, the USGS has lately advised consumers to proceed with TIRS band 10 data and published a warning about the unpredictability of Landsat thermal band 11 readings on their website (Wang et al., 2015). Thermal band 10 of Landsat 8 have been employed for this purpose to retrieve the LST after the MWA (Qin et al., 2001; Liu and Zhang, 2011; Singh et al., 2017).

LST Retrieval from Landsat 8

Using band 10 of the Landsat 8 TIRS data, the Radiative Transfer equation was applied to retrieve LST. This approach was chosen because, in contrast to the Single Channel method and Qin's split window algorithm, which have low and moderate accuracy, respectively (Liu and Zhang, 2011), it has a better level of accuracy.

Conversion of Digital Number to Radiance

The following procedure was used to translate the Digital Number of band 10 (Table 10.2) into spectral radiance measurements:

$$\mathrm{TOA}(L) = M_L * Q_{\mathrm{cal}} + sA_L \tag{10.1}$$

The Retrieval of Brightness Temperature

According to Liu and Zhang (2011), the TIR band (Band 6) from Landsat 7 measures radiation coming from the Earth's surface between the spectral range of 10.4 and 12.5 m. The TIR band from Landsat 8 (Band 10) captures radiation coming from the Earth's surface between the spectral range of 10.6 and 11.9 m. The inverse of Plank's function was used to calculate the temperature values and determine the radiance values of infrared bands (Kumari et al., 2018).

$$\mathrm{BT} = \left(K_2/(\ln(K_1/L)+1)\right) - 273.15 \tag{10.2}$$

Estimation of NDVI

According to Weng (2004), the amount of vegetation in an area has a significant impact on its temperature. Since the normalized difference vegetation index (NDVI) is one of the most widely used indices for measuring vegetation, it is calculated using the following equation in the current study.

Calculation of the P_v (Proportion of Vegetation)

To measure P_V, the Landsat 8 dataset was utilized. From NDVI, it is possible to determine the satellite's pixel-scale vegetation cover fraction (Sobrino et al., 2004).

$$P_V = \mathrm{Square}\left(\frac{(\mathrm{NDVI} - \mathrm{NDVI}_{\min})}{(\mathrm{NDVI}_{\max} - \mathrm{NDVI}_{\min})}\right) \tag{10.3}$$

The Calculation of Land Surface Emissivity

Calculating (land surface emissivity) LSE (ε) is crucial for determining the LST of a land surface in an area. An object's emissivity is determined by one of its thermophysical properties. The NDVI

TABLE 10.2
Radiance Constant Value of Satellite Data (Landsat 7 & 8)

Satellite Sensor	K_1	K_2
Landsat 8 OLI/TIRS	774.89	1321.08

threshold technique can be used to determine the emissivity of various land surfaces throughout the spectral range of 10–12 m because emissivity varies with wavelength. The following equation can be used to calculate emissivity.

$$\varepsilon = 0.004 * P_v + 0.986 \tag{10.4}$$

Retrieval of LST

Finally, subsequent equation is utilized to convert LST from brightness temperature

$$\text{LST} = \text{BT}/(1 + \lambda * (\text{BT}/\rho) * \ln(e) \tag{10.5}$$

where

BT = brightness (at satellite temperature)
λ = wavelength of emitted radiance (0.00115)
ρ = 14,380 (Constant)
e = emissivity

Normalized Difference Built-up Index (NDBI)

Increase in urban population is compensated by the expansion of the built-up land area (Wang et al., 2011). The uncontrolled and unplanned expansion of urban built-up areas results in the emergence of various environmental problems (Sun et al., 2012; Saha and Pal, 2019). According to Arif et al. (2022), the higher level of vulnerabilities is triggered by uncontrolled population increase and unclear regulations. For this reason, a built-up distribution parameter is needed for environmental vulnerability assessment. There are numerous indices available to highlight the developed aspects of a location. NDBI is utilized in this analysis for the same. To calculate NDBI, Zha et al.'s (2003) approach is used in conjunction with shortwave infrared and near-infrared bands. SWIR reflections from built-up and arid areas are higher than NIR reflections. The infrared spectrum is not reflected by aquatic bodies, while green surfaces reflect more NIR than SWIR. NDBI's values range from 1 (plus) to 1 (negative). Values > 0 display aquatic bodies and vegetation, while all +1 values depict built-up areas and desolate lands.

$$\text{NDBI} = \frac{(\text{SWIR} - \text{NIR})}{\text{SWIR} + \text{NIR}} \tag{10.6}$$

Normalized Difference Vegetation Index

Vegetation indices in RS data provide indicators of vegetation stability on a global scale. NDVI is a standard index that showcases information on the biomass and greenness of the vegetation (Pettorelli et al., 2005). It is regarded as an effective tool for assessing the influence of human activities and natural seasonal changes on the ecological environment (Liu et al., 2021). Ma et al. (2020), and Zhang et al. (2022) have identified ecological vulnerability using NDVI. The most popular index (Purevdorj et al., 1998) used to determine how much of an area is covered by vegetation is the NDVI. Based on the Gao et al. (2002) equation, NDVI values were examined utilizing the R and IR bands of Landsat satellites. Landsat 8 multi-temporal images (from 2003 and 2018) were used to determine NDVI using an algorithm. Higher near-infrared wavelengths are reflected by greenery and flora (Liu and Weng, 2008; Zheng et al., 2011; Mallick, 2014).

$$\text{NDVI} = \frac{(\text{NIR} - \text{R})}{\text{NIR} + \text{R}} \tag{10.7}$$

Solid Waste Generation

Urban areas are hubs of industrial and household SW generation, which pose significant risk to environment (Tsheleza et al., 2019). The cost of environment-friendly SW management to protect human health has become costly for urban government worldwide (Kubanza and Simatele, 2016; Gutberlet, 2018). According to Hoorenweg et al. (2013), the generation of SW has increased tenfold worldwide over the last few decades, and it is projected to double by 2025. The impacts of SW mismanagement on the environment and human health are equally visible in less developed and advanced nations (Gutberlet, 2018). Urban settlements are the major source of SW generation in urban built-up areas. The socio-economic conditions of a household plays significant role as a risk factor as well as key predictors of vulnerabilities to SW-induced environmental hazards (Van Beukering et al. 1999; Miller and Spoolman, 2011). Many researches on various factors contributing to waste generation in urban areas indicate that the socio-economic conditions of a household and their link with SW generation have a positive relation. The management behaviors are also important for understanding urban vulnerability to SW mismanagement and resultant environmental problems (Gutberlet, 2018).

Water Quality of Bulandshahr City

Water quality assessment is important to reduce the level of water pollution so that sustainable water management strategies can be developed for groundwater management (Heiß et al., 2020; Haied et al., 2021). Regular monitoring of water quality is essential to explore the new technologies and advancements in water quality assessment (Udeshani et al., 2020). WQI (water quality index) is one of the techniques to determine the quality of water in various European countries, not only Europe, but several studies in India have used this technique for drinking water quality assessment (Ponsadailakshmi et al., 2018; Yadav et al., 2018; Khangembam and Kshetrimayum et al., 2019). These studies employed several water quality parameters, that is, PH, EC, total hardness, calcium, magnesium, chloride, iron, manganese, zinc, and lead as parameters (Tirkey et al., 2017; Ponsadailakshmi et al., 2018).

Water Quality Index

This technique employed several steps to attain WQI

$$Wi = \frac{wi}{\sum_{i=1}^{n} wi}$$

$$qi = \frac{Ci}{Si} \times 100$$

$$\text{WQI} = \sum_{i=1}^{n} Si \tag{10.8}$$

Water Logging Condition in Bulandshahr City

People are migrating in large number toward industrial areas for better employment opportunities and improved standard of living (Ahmed, 2015). As a result, it is affecting the hydrology of an area because of high-level consumption of water, runoff generation, which is challenging the current drainage network functioning of cities (Browne et al., 2021). Due to extra burden on the drainage system, many cities worldwide, and also in India, are facing urban flooding or WL conditions (Gupta and Nair, 2011; Seidl et al., 2020). Consequently, WL has been used as a vulnerability assessment parameter to measure efficient urban functioning (Yin et al., 2011; Lin et al., 2018; Shi, 2012).

Road Density in Bulandshahr City

To date, very few landscapes remain devoid of roads, and as the human population continues to increase, it is likely that RD, will increase as well, whether it be by new developments, road improvements, or road expansion schemes. The development and presence of roads can reduce landscape permeability, lead to habitat loss, and increase habitat fragmentation (Bennett, 2017). Several studies (Tiburan et al., 2009; Sahoo et al., 2016; Iqra et al., 2020) regarding the assessment of environmental vulnerability have taken RD into consideration. RD map in this particular study is prepared from road length per square kilometer of the study area. The map of RD was prepared with the help of a tool of GIS, that is, line density (spatial analyst tool, GIS). Density is calculated in units of length per unit of area. For better accuracy, the roads of Bulandshahr city have been extracted with the help of the OpenStreetMap website.

METHODS FOR VULNERABILITY ASSESSMENT

Analytical Hierarchy Process

AHP is a MCD methodology developed by Thomas L. Saaty in the 1970s. AHP provides a systematic approach for evaluating and prioritizing alternatives based on a set of criteria or factors.

The key idea behind AHP is that complex decision problems can be broken down into a hierarchy of criteria and alternatives. The criteria represent the factors or attributes that are important for the decision, while the alternatives are the options or choices being considered. By hierarchically structuring the decision problem, AHP helps to organize and analyze the various elements involved. AHP involves a process of pairwise comparisons, where decision-makers systematically evaluate the relative importance or preference between pairs of criteria or alternatives. The comparisons are typically done using a scale that captures the strength of preference or importance, such as Saaty's 1–9 scale (Saaty, 1991).

After pairwise comparisons, mathematical calculations are applied to derive priority scores for the criteria and alternatives. These priority scores represent the relative importance or preference of each element about the decision problem. AHP also incorporates a consistency check, which involves examining the consistency of the pairwise comparisons. Consistency ratios are calculated to ensure that the judgments made by the decision-maker are coherent and reliable. If inconsistencies are detected, adjustments or revisions may be necessary (Table 10.3).

The final output of the AHP process is an overall ranking or evaluation of the alternatives based on the defined criteria. This ranking helps decision-makers make informed choices by considering the relative importance of each criterion and alternative. AHP has been widely applied in various fields, including business, engineering, project management, resource allocation, and strategic planning (Chanu and Bakimchandra, 2022). It provides a structured and quantitative approach to decision-making, allowing decision-makers to consider multiple criteria and complex trade-offs systematically. AHP can be effectively applied in vulnerability assessment to prioritize vulnerabilities based on their significance and criticality. The pairwise ranking in the AHP has to be consistent

TABLE 10.3
Pairwise Comparison Scale

Verbal Judgment	Numeric Value
Extremely important	9
	8
Very strongly more important	7
	6
Strongly more important	5
	4
Moderately more important	3
	2
Equally important	1

Source: Adopted from Saaty and Wind (1980).

TABLE 10.4
Vulnerability Criteria

Main Criteria	Subcriteria	Ranges	Area (ha)	Percentage
LULC	Built-up		13.03	79.30
	Vegetation		2.04	11.44
	Agriculture		1.79	9.94
	Vacant		1.05	5.83
	River/water		0.09	0.39
LST	Very low	26.61–28.63	98.00	5.87
	Low	28.64–30.13	137.00	8.19
	Medium	30.14–31.12	219.00	13.14
	High	31.13–32.11	617.00	37.00
	Very high	32.12–37.07	597.00	35.80
NDVI	Very low	<0.1	692.00	41.49
	Low	0.1–0.2	656.00	39.36
	Medium	0.2–0.3	169.00	10.16
	High	0.3–0.4	120.00	7.17
	Very high	>0.4	31.00	1.89
NDBI	Very low	0.0–0.1	125.00	7.49
	Low	0.1–0.2	121.00	7.25
	Medium	0.2–0.3	251.00	15.05
	High	0.3–0.4	623.00	37.35
	Very high	>0.4	548.00	32.85
Water quality	Excellent	0–25	55.92	3.35
	Good	26–50	786.01	47.12
	Acceptable	51–75	672.06	40.29
	Bad	76–100	131.46	7.88
Water logging	Very low	0–15		8.00
	Low	15–30		7.00
	Medium	31–45		15.00
	High	46–60		37.00
	Very high	>60		33.00
Solid waste	Very low	0–15		9.87
	Low	15–30		8.19
	Medium	31–45		15.14
	High	46–60		36.00
	Very high	>60		30.80

with all pairwise comparisons. Several previous studies have mentioned the parameters are LULC, NDVI, NDBI, LST, RD, SW, and Urban Drainage (Tables 10.4 and 10.5). We adopted the highly influential parameter after extensive literature reviews and additional expert evaluation (Åström et al., 2011) (Table 10.6).

RESULTS AND DISCUSSION

LULC

The evaluation of LULC changes, which is based on maximum likelihood classification, summarizes the LULC classes in Bulandshahr city for 2019. LULC classes were divided into five categories, and overall accuracy was 89 percent, whereas the kappa coefficient is 0.85. In Bulandshahr, the built-up

TABLE 10.5
Details of Concentration of Different Physico-Chemical Parameters of Water Quality at Different Stations in Bulandshahr City

		Physical Test				Chemical Test						
		PH	TDS	Turbidity		Chloride	Total Hardness	Sulfate	Iron	Nitrate	Alkalinity	Arsenic
S. No.	Place Name	6.5–8.6	2000	5	Color	1,000	600	400	1	45	600	0
1	2	3	4	5	6	7	8	9	10	11	12	13
1	Firoz Market	6.8	854	15.89	Colorless	245	780	134.3	2.49	Nil	550	0
2	Government Hospital	7.74	263	3.1	Colorless	40	250	73.92	Nil	7.618	280	0
3	Mandi Fateh Ganj	7.21	740	3.56	Colorless	165	490	194.1	1.11	21.93	420	0
4	Collectorate	7.35	415	2.21	Colorless	110	320	67.9	Nil	44.06	360	0
5	Mandi Fateh Ganj	7.26	503	2.21	Colorless	75	330	88.86	1	19.62	360	0
6	Teachers Colony (Judges Compound)	7.74	257	2.46	Colorless	30	260	49.7	0.48	4.41	260	0
7	Adarsh Nagar	7.39	274	2.85	Colorless	35	240	50.13	Nil	1.503	250	0
8	Ambedkar Park	7.24	339	1.54	Colorless	55	300	65.88	Nil	10.96	310	0
9	Government Hospital	7.43	464	1.64	Colorless	75	310	86.47	1.8	29.06	370	0
10	Khand Gang Colony	6.97	445	1.56	Colorless	70	250	68.7	0.99	11.98	280	0
11	Yamunapuram	7.66	245	2.54	Colorless	35	240	53.99	1.15	4.88	260	0
12	Yamunapuram	7.65	159	2.52	Colorless	30	190	33.55	Nil	2.20	180	0
13	Government Hospital	7.16	520	6.36	Colorless	90	380	92.14	0.82	32.35	380	0
14	Bhoor Road	7.78	265	1.53	Colorless	20	210	6.48	2.01	1.11	40	0
15	Female Hospital	7.51	223	3.46	Colorless	40	170	81.9	0.49	2.83	270	0
16	Female Hospital	7.58	226	3.95	Colorless	55	170	76.17	0.37	1.744	260	0
17	Vikas Bhawan	6.76	531	2.52	Colorless	85	360	88.65	1.41	34.72	700	0
18	Government Hospital	7.26	456	3.33	Colorless	80	280	330	0.28	18.90	520	0

TABLE 10.6
Pairwise Comparison Matrix

	C1	C2	C3	C4	C5	C6	C7	C8	W_i
C1	**1**	**3**	**4**	**3**	**4**	**3**	**3**	**3**	**0.26**
C2	0.33	1	3	0.33	2	0.33	2	0.33	0.12
C3	0.25	0.33	1	0.5	2	0.5	2	0.5	0.09
C4	0.33	3	2	1	2	0.33	0.33	0.33	0.07
C5	0.25	0.5	0.5	0.5	1	0.33	0.5	0.33	0.08
C6	0.33	3	2	3	3	1	2	2	0.17
C7	0.33	0.5	0.5	3	2	0.5	1	0.33	0.08
C8	0.33	3	2	3	3	0.5	3	1	0.13

Source: Calculated by the author.
Note: (i) C1=LULC; C2=LST; C3=Solid waste; C4 Water quality; C5=Road density; C6=NDBI; C7=Drainage; C8=NDVI. (ii) CI=0.02; CR=0.01. (iii) *Wi*=Calculated weight.

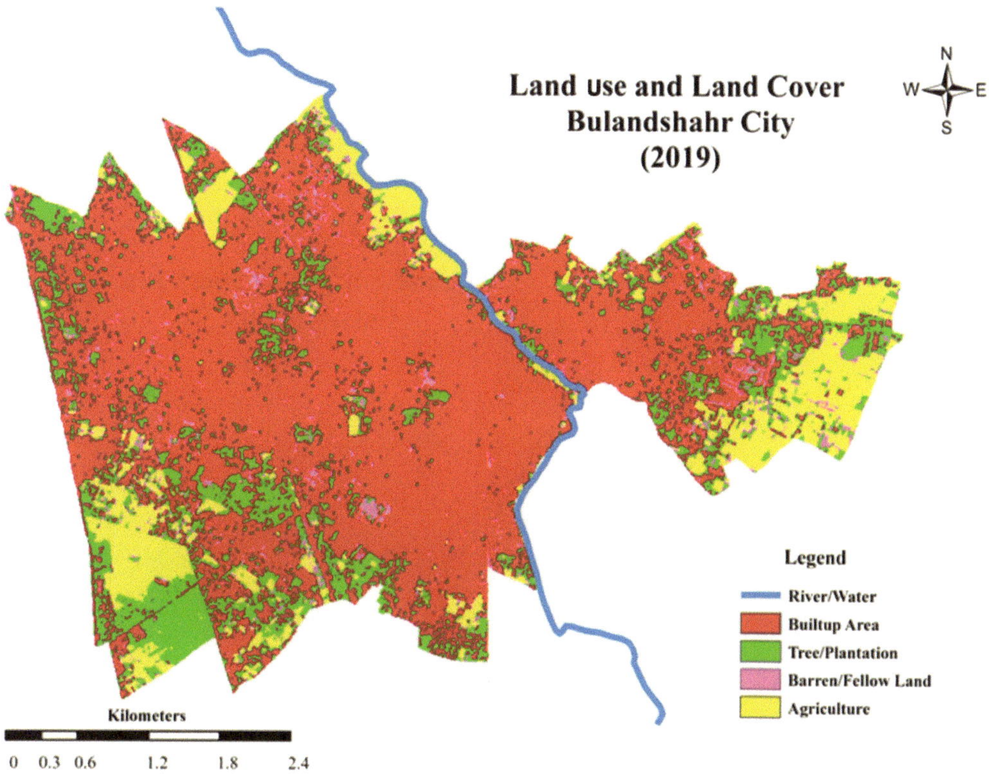

FIGURE 10.3 LULC pattern of Bulandshahr city.

wwarea makes up roughly 72% of the case study's overall area. New development in the study has led to growth in the built-up area due to new construction sites are visible across the city (Table 10.3 and Figure 10.3); 11 percent of the entire land is currently covered by natural vegetation. Contrary to built-up land, vacant land, agricultural land, and natural vegetation have a low percentage of the total study area, that is, 1600 ha. The built-up region of the case study was limited to the old or inner section of the city, as can be observed in the map of the city. The city initially developed close to Kali Nadi's (River's) water source, but later it began to expand along its main thoroughfares. Nearly all of the land inside the municipal boundaries is covered with built-up area.

Land Surface Temperature

To evaluate the LST of the study area, the LST values for Bulandshahr city for the year 2019 were quantitatively extracted from ETM/OLI (TIRS) images. In Bulandshahr city, the LST values vary from 26.44 to 37.07 for the year 2019 (Figure 10.4). In 2019, it was discovered that the city's small central area had a higher land surface value and had significantly taken over the entire city. Due to urbanization in Bulandshahr city, the city's center is vulnerable to temperature increases. It was found that the differences in surface radiance temperature between built-up areas and vegetation-covered areas can reach up to 7°C, as illustrated in the cross-section profile of the city in Figure 10.5.

Vegetation Pattern of Bulandshahr City

Around the globe, there is an increase in urbanization, particularly in developing countries where agricultural land is dominant. As a result of new development, urban areas have almost completely lost their natural vegetation. For the study area's NDVI analysis, Landsat 8 imageries from the year 2019 were used (Figure 10.6). The NDVI images were then reclassified into five categories, each

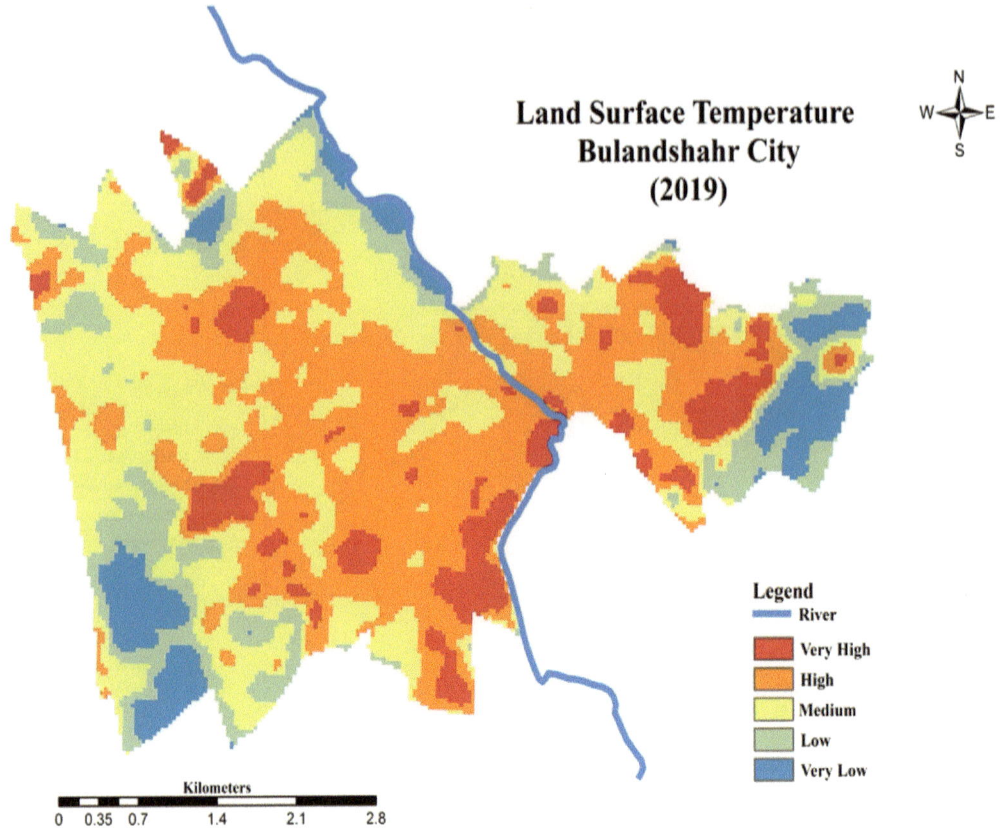

FIGURE 10.4 LST distribution of Bulandshahr city.

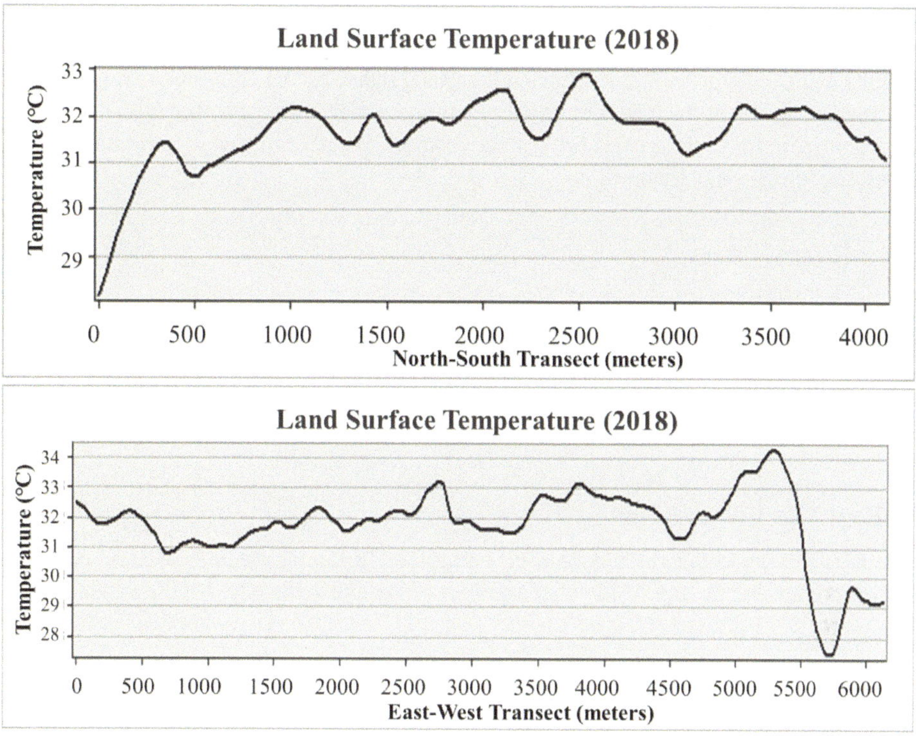

FIGURE 10.5 LST profile of Bulandshahr city (east and west).

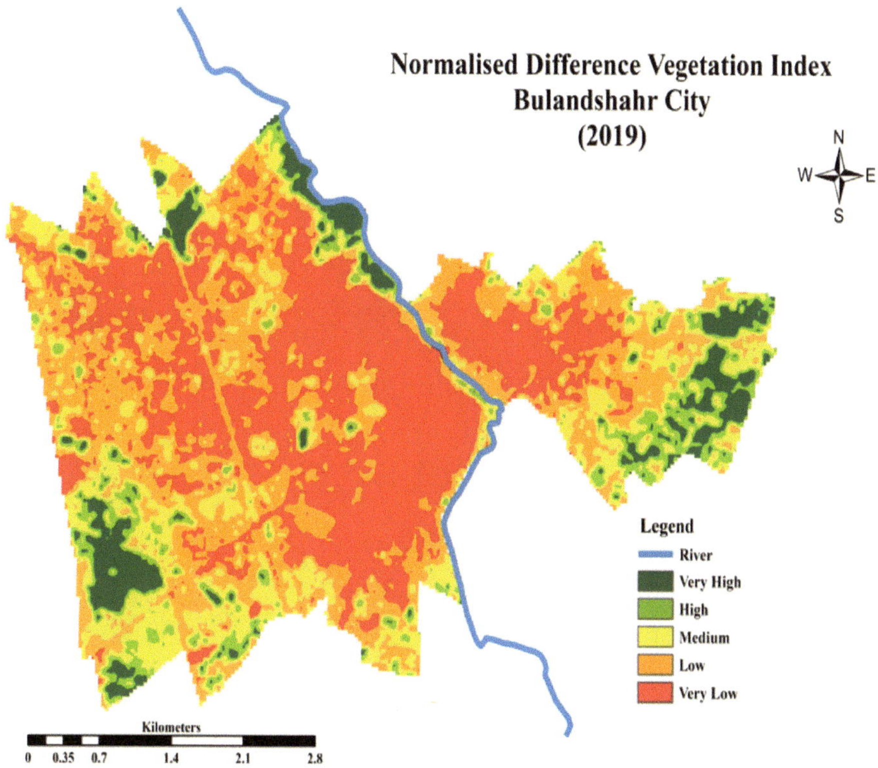

FIGURE 10.6 Vegetation distribution of Bulandshahr city.

referring to a different level of vegetation cover in the city, from very low (0.1) to low (0.1–0.2), medium (0.2–0.3), high (0.3–0.4), and very high (>0.4). Lowest NDVI values were found in the city's central, built-up, predominantly residential area. Higher NDVI values were reported in the N, NE, SW, and W regions of the city. The city's vegetation pattern indicates that high ranges of NDVI values are dominant for agricultural and vegetative areas. Highly built-up area predominates in the region with the lowest NDVI values.

Pattern of Built-Up Area

The expansion of built-up surfaces in the city is one of the most significant factors influencing the land surface temperature of Bulandshahr; hence, NDBI has been computed for the year 2018 to track any changes in built-up area in the city (Figure 10.7). The values of the NDBI range from –0.57 to 0.36 for the year 2019 shows that the built-up area will predominate. In comparison to the north-west and south-east parts of the city, the central and historic areas of the city have greater NDBI values.

Water Quality of Bulandshahr City

To assess the water quality of Bulandshahr city, a total of 18 samples were taken into consideration. To get a comprehensive picture of the water quality of Bulandshahr city, WQI has been calculated.

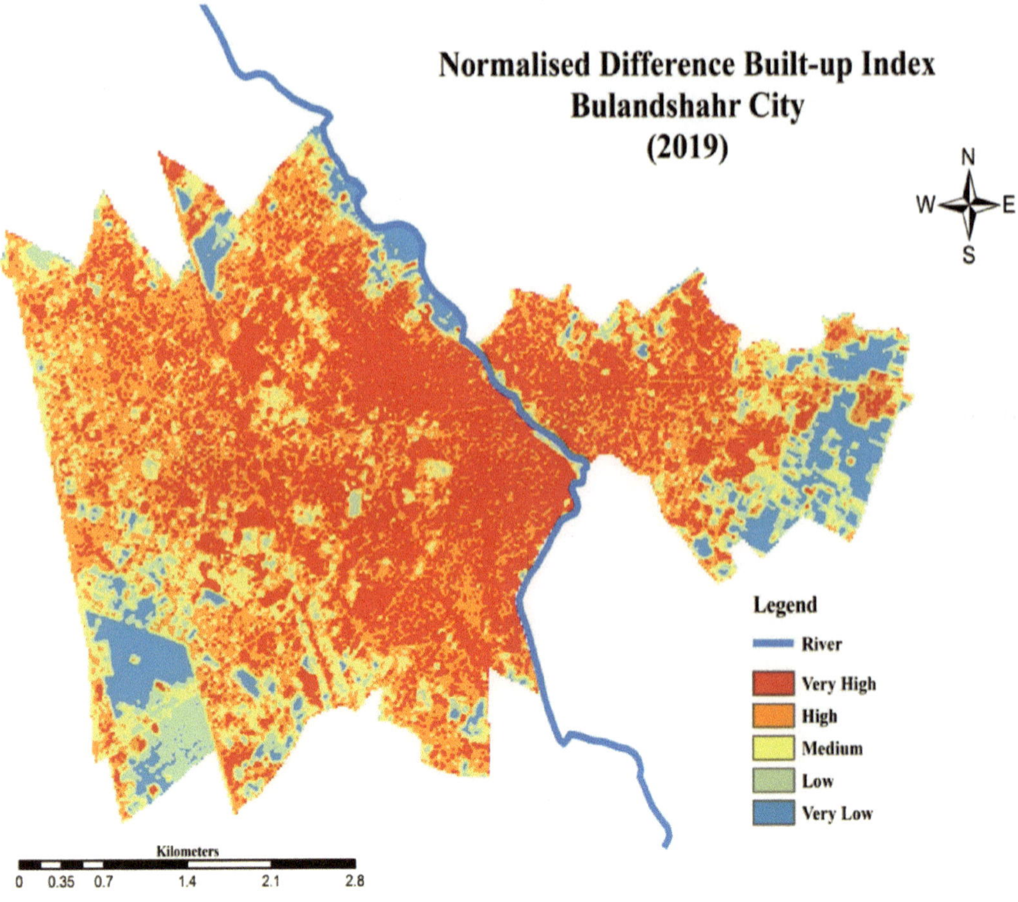

FIGURE 10.7 Built-up/settlement distribution of Bulandshahr city.

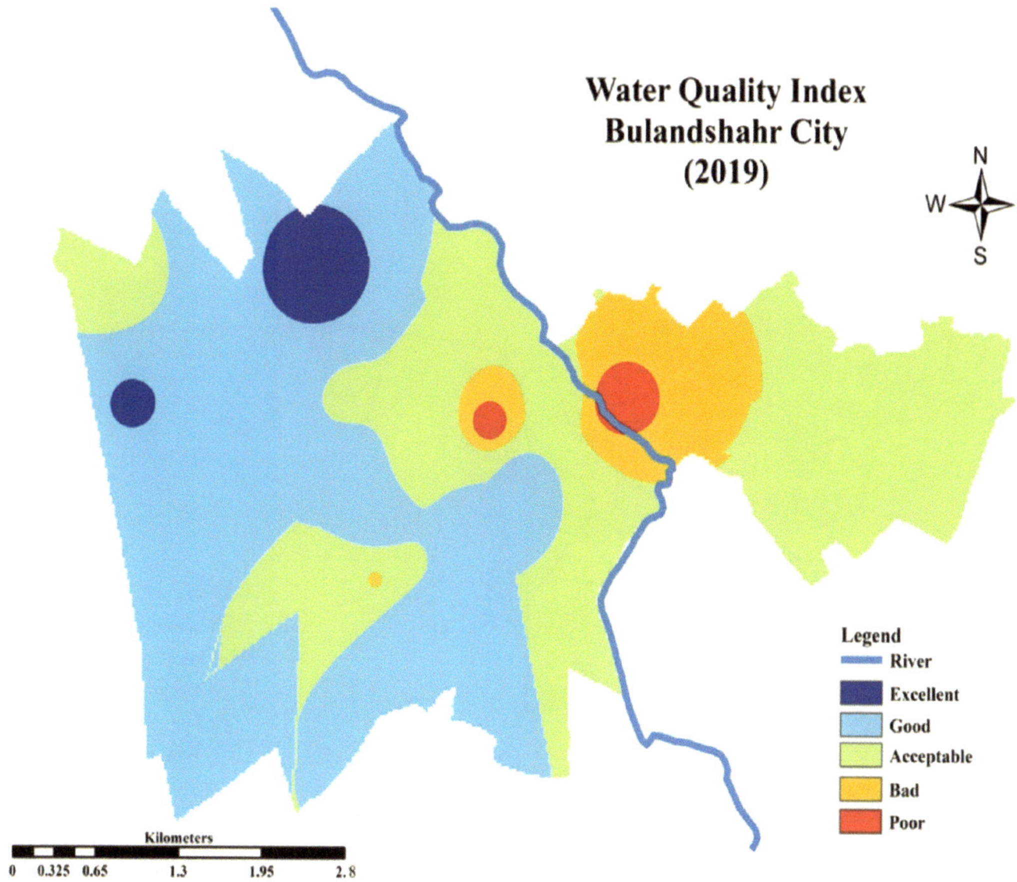

FIGURE 10.8 Water quality of Bulandshahr city.

From the analysis, it was observed that eight samples fall into poor to very poor water quality, while the remaining 10 samples secure their positions into the "good" category (Table 10.6). The spatial pattern of water quality of Bulandshahr city reveals that the eastern peripheral and newly developed areas have poor water quality due to improper water supply (Figure 10.8). The other reason for the contamination of water quality in these areas is the location of Kali Nadi, which flows through the middle of the city.

Road Density of Bulandshahr City

The RD of Bulandshahr city is 251 km^2, which is calculated by dividing the total length of the roads by the total area of a particular spatial unit. The high density of roads can be observed in the old/middle part of the city; however, roads are not accessible throughout the city. The major chauras of the city are highly congested in the city at the peak time (Figure 10.9). Results indicate that the percentage of built-up area was higher in the city, which is not considered suitable for city health.

Water Logging in Bulandshahr City

In this chapter, three characteristics of drains were analyzed using Rank Score: (i) attended or unattended, (ii) spillover or contained, and (iii) stagnant or flowing. At first, numbers were assigned to

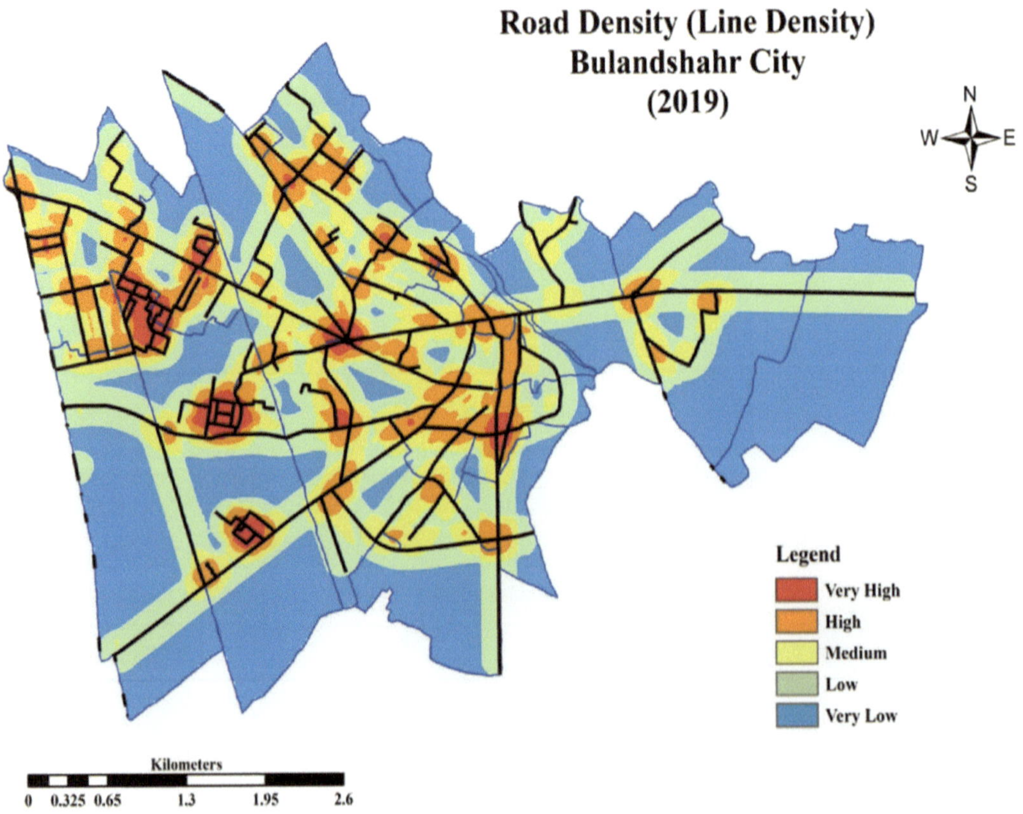

FIGURE 10.9 Road density of Bulandshahr city.

each category of each characteristic of the drain based on a ranking system, and then a total score of each category was calculated. Based on these scores, five qualitative categories were formed from very good to very bad. WL occurs due to an improper drainage system and choking of drains, which causes stagnation of the water. Due to the lack of drains, excess siltation, choking, and improper cleaning of drains turn every rainfall in the city into a disaster. Figure 10.10 reveals that the older city has a better slope that facilitates drainage; the middle and mostly, the peripheral region of the city, without any plan, is a major factor behind WL.

Solid Waste Management

In Bulandshahr city, the total SW generation per day is 80 metric tonnes, in which 60 metric tonnes is wet waste and the remaining 20 metric tonnes is dry waste (Bulandshahr Municipality, 2018). Major sources of SW generation in the city were households, commercial and industrial establishments, and hospitals. During the survey, it was observed that sweepers (safaiwalas) were not regular as well and the provision and maintenance of the dustbin are not up to the mark. Residents complained that there is no proper function of the facilities, as no dustbins were provided by the municipality (Figure 10.11). Heaps of garbage were observed all around the city, which further substantiated the claims of the municipality of regular disposal of SW. Results reveal that eastern peripheral areas were most deprived in terms of SW management.

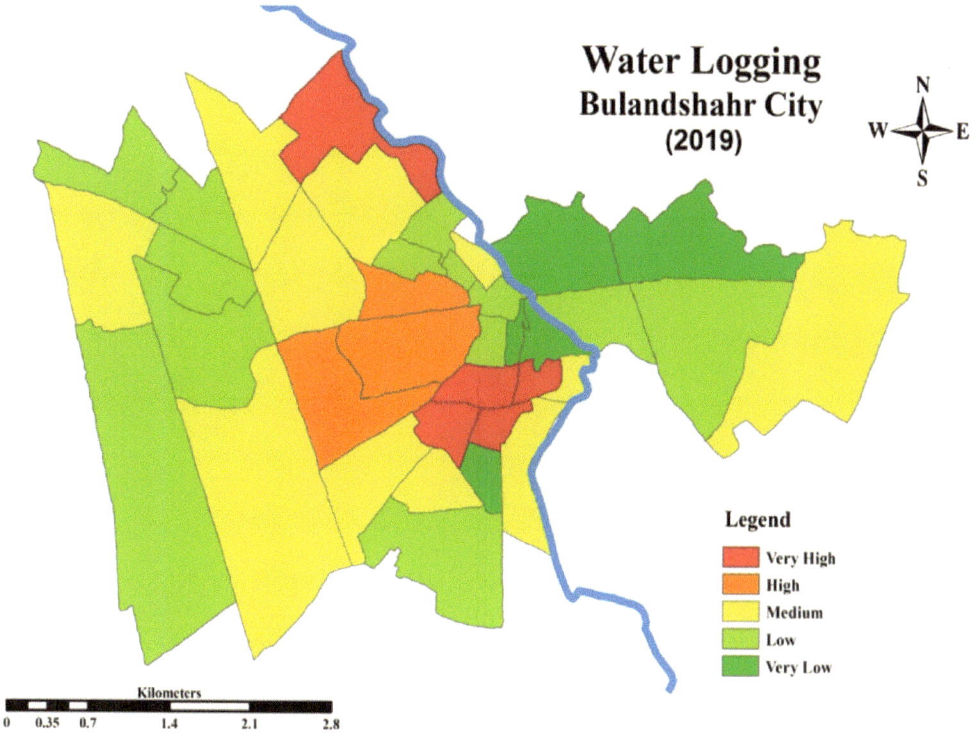

FIGURE 10.10 Water logging in Bulandshahr city (ward wise).

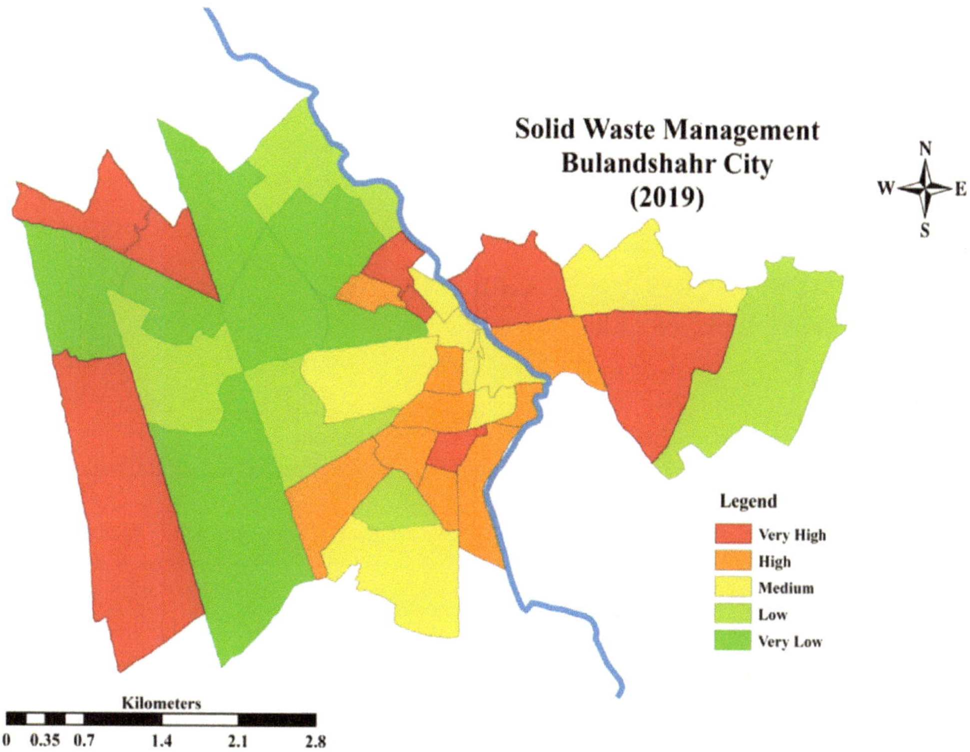

FIGURE 10.11 Solid waste generation in Bulandshahr city (ward wise).

FIGURE 10.12 Overlay maps of seven criteria.

Vulnerability Analysis of Bulandshahr City

Areas representing high vulnerability face lots of environmental problems, municipal negligence, congestion and haphazard growth, and unplanned urbanization (Figures 10.12 and 10.13). Problems such as high density of built-up areas, narrow streets, and high temperatures are the environmental risks these areas are prone to, in comparison to their counterparts. Some wards that have developed in the last few years also come under this category because these areas, being away from the city center, were almost neglected by municipality services. Urban poor are the most vulnerable population to environmental degradation because access to safe drinking water, water supply, and other facilities is luxury for these people. Wards such as Yamunapuram, Nazimpura, Hari Enclave, Shanti Niketan, Civil Line I, and Civil Line II are least vulnerable areas.

THE URBAN SLUMS OF BULANDSHAHR: MAPPING ENVIRONMENTAL VULNERABILITY

Location: Bulandshahr, Uttar Pradesh, Northern India

Issue: Heat stress, water logging, solid waste accumulation, and low vegetative cover (Figure 10.14)

Community Involved: Slum dwellers and informal urban settlements

The vulnerability assessment conducted in Bulandshahr city revealed a high degree of environmental risk concentrated within its urban slums and informal settlements. Using remote sensing indicators such as NDVI, NDBI, LST, and field-based data on waterlogging and SW accumulation, the study identified areas with extreme exposure to urban environmental hazards. Results showed that regions with low vegetation cover (NDVI$<$0.2), high built-up density (NDBI$>$0.4), and elevated land surface temperatures were directly correlated with areas inhabited by low-income groups. These communities also experienced frequent water stagnation and waste dumping, contributing to deteriorating health and hygiene conditions. Pairwise comparisons using the AHP confirmed that

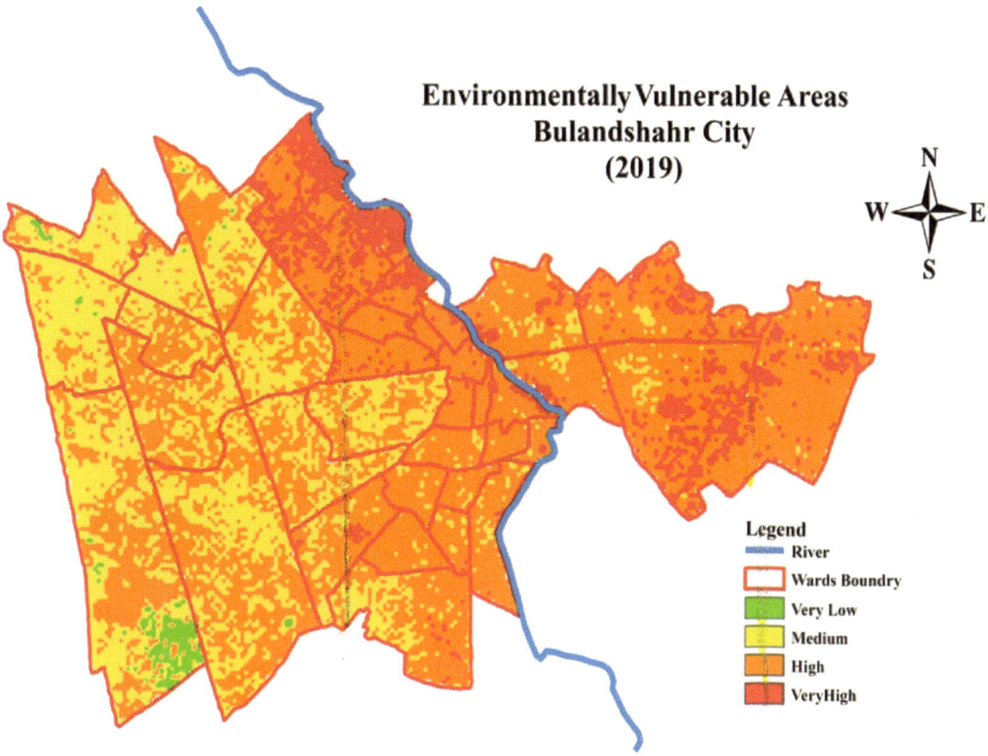

FIGURE 10.13 Identified environmental vulnerable zones of Bulandshahr city.

built-up density and waste accumulation were the most critical factors in determining vulnerability, with a consistency ratio of 0.01 ensuring reliable weighting. The findings emphasize that urban slums in Bulandshahr are not only socially and economically marginalized but also face spatially embedded environmental risks that require urgent, localized mitigation strategies. Community feedback from field surveys indicated a willingness to participate in SW management and drainage maintenance if supported by municipal authorities, highlighting the potential for bottom-up resilience planning.

CONCLUSION

This chapter aims to identify the Bulandshahr city's environmental susceptible zones. A total of seven criteria were used to identify the city's environmentally vulnerable zones. This chapter finds out that the city's oldest and peripheral neighborhoods, where individuals from low-income groups reside, were the most vulnerable areas. It's because the city's low-income neighborhoods have been the most neglected and are at the greatest environmental risk. These environmental concerns can be addressed through the improvement of the low-income neighborhoods of the city. Therefore, sustainable approaches, along with proper planning and diligent implementation of developmental programs, can help achieve sustainability in small cities like Bulandshahr. In addition, significant public engagement along with government efforts could result in a significant development of the urban ecosystem.

FIGURE 10.14 Dilapidated houses, outskirts low vegetation cover, and improper waste dumping.

REFERENCES

Adger, W. N. (2000). Institutional adaptation to environmental risk under the transition in Vietnam. *Annals of the Association of American Geographers*, *90*(4), 738–758.

Ahmed, B. (2015). Landslide susceptibility modelling applying user-defined weighting and data-driven statistical techniques in Cox's Bazar Municipality, Bangladesh. *Natural Hazards*, *79*(3), 1707–1737.

Alam, M. J., Alam, M. J. B., Rahman, M. H., Khan, S. K., & Munna, G. M. (2006). Unplanned urbanization: Assessment through calculation of environmental degradation index. *International Journal of Environmental Science & Technology*, *3*, 119–130.

Arif, N., Wardhana, A., & Martiana, A. (2022, February). Spatial analysis of the urban physical vulnerability using remote sensing and geographic information systems (case study: Yogyakarta City). In *IOP Conference Series: Earth and Environmental Science* (Vol. 986, No. 1, p. 012067). IOP Publishing.

Åström, D. O., Bertil, F., & Joacim, R. (2011). Heat wave impact on morbidity and mortality in the elderly population: A review of recent studies. *Maturitas*, *69*(2), 99–105.

Bardhan, R., Debnath, R., & Bandopadhyay, S. (2016). A conceptual model for identifying the risk susceptibility of urban green spaces using geo-spatial techniques. *Modeling Earth Systems and Environment*, *2*(3), 1–12.

Benmoussa, N., Elyamami, A., Mansouri, K., Qbadou, M., & Illoussamen, E. (2019). A multi-criteria decision making approach for enhancing university accreditation process. *Engineering, Technology & Applied Science Research*, *9*(1), 3726–3733.

Bennett, V. J. (2017). Effects of road density and pattern on the conservation of species and biodiversity. *Current Landscape Ecology Reports*, *2*(1), 1–11.

Birawat, K. K., Nachiyar, M. C., Mayaja, N. A., & Srinivasa, C. V. (2021). Impact of urbanisation on lakes—a study of Bengaluru lakes through water quality index (WQI) and overall index of pollution (OIP). *Environmental Monitoring and Assessment*, *193*(7), 1–20.

Browne, S., Lintern, A., Jamali, B., Leitão, J. P., & Bach, P. M. (2021). Stormwater management impacts of small urbanising towns: The necessity of investigating the 'devil in the detail'. *Science of the Total Environment*, *757*, 143835.

Bulkeley, H., & Tuts, R. (2013). Understanding urban vulnerability, adaptation and resilience in the context of climate change. *Local Environment*, *18*(6), 646–662.

Campbell, S. (1996). Green cities, growing cities, just cities? Urban planning and the contradictions of sustainable development. *Journal of the American Planning Association*, *62*(3), 296–312.

Chakraborty, A., & Li, X. (2022). Exploring the heterogeneity in relationship between heat exposure and land development in Mumbai, India: A framework to address urban vulnerability in developing megacities. *Journal of Environmental Planning and Management*, *65*(2), 337–355.

Chanu, M. L., & Bakimchandra, O. (2022). Landslide susceptibility assessment using AHP model and multi resolution DEMs along a highway in Manipur, India. *Environmental Earth Sciences*, *81*(5), 156.

Choudhary, K., Boori, M. S., & Kupriyanov, A. (2018). Spatial modelling for natural and environmental vulnerability through remote sensing and GIS in Astrakhan, Russia. *The Egyptian Journal of Remote Sensing and Space Science*, *21*(2), 139–147.

Dammalage, T. L., & Jayasinghe, N. T. (2019). Land-use change and its impact on urban flooding: A case study on Colombo district flood on May 2016. *Engineering, Technology & Applied Science Research*, *9*(2), 3887–3891.

Doygun, H. (2009). Effects of urban sprawl on agricultural land: a case study of Kahramanmaraş, Turkey. *Environmental Monitoring and Assessment*, *158*, 471–478.

Du, C., Ren, H., Qin, Q., Meng, J., & Zhao, S. (2015). A practical split-window algorithm for estimating land surface temperature from Landsat 8 data. *Remote Sensing*, *7*(1), 647–665.

Dutta, S., Bardhan, S., & Bhaduri, S. (2013). Patterns of urbanization and environmental quality in the context of Indian cities. *Environment and Urbanization Asia*, *4*(2), 287–299.

Gao, F., Jin, Y., Schaaf, C. B., & Strahler, A. H. (2002). Bidirectional NDVI and atmospherically resistant BRDF inversion for vegetation canopy. *IEEE Transactions on Geoscience and Remote Sensing*, *40*(6), 1269–1278.

Gupta, A. K., & Nair, S. S. (2011). Urban floods in Bangalore and Chennai: Risk management challenges and lessons for sustainable urban ecology. *Current Science*, *100* (11), 1638–1645.

Gutberlet, J. (2018). Waste in the city: Challenges and opportunities for urban agglomerations. *Urban Agglomeration*, *21*, 191.

Haied, N., Khadri, S., Foufou, A., Azlaoui, M., Chaab, S., & Bougherira, N. (2021). Assessment of groundwater pollution vulnerability, hazard and risk in a semi-arid region. *Journal of Ecological Engineering*, *22*(10), 1–13.

Heiß, L., Bouchaou, L., Tadoumant, S., & Reichert, B. (2020). Index-based groundwater vulnerability and water quality assessment in the arid region of Tata city (Morocco). *Groundwater for Sustainable Development*, *10*, 100344.

Hoornweg, D., Bhada-Tata, P., & Kennedy, C. (2013) Environment: Waste production must peak this century. *Nature*, *502*(7473), 615–617.

Huang, X., Li, J., Yang, J., Zhang, Z., Li, D., & Liu, X. (2021). 30 m global impervious surface area dynamicss and urban expansion pattern observed by Landsat satellites: From 1972 to 2019. *Science China Earth Sciences*, *64*(11), 1922–1933.

Inostroza, L., Palme, M., & De La Barrera, F. (2016). A heat vulnerability index: Spatial patterns of exposure, sensitivity and adaptive capacity for Santiago de Chile. *PLOS One*, *11*(9), e0162464.

Iqra, N. A., Zzaman, R. U., Rodrigues, E. A., Haque, A., Khan, M. S. A., & Hussain, M. A. (2020). Environmental vulnerability assessment in the south-west coastal region of Bangladesh using principal component analysis. *Coastal Engineering Proceedings*, (36v), 24–24.

Jenelius, E. (2010). Large-scale road network vulnerability analysis. KTH, Royal Institute of Technology. Doctoral Thesis in Infrastructure with specialisation in transport and location analysis.

Johnson, M. P. (2001). Environmental impacts of urban sprawl: A survey of the literature and proposed research agenda. *Environment and Planning A*, *33*(4), 717–735.

Kazemi, H., Sadeghi, S., & Akinci, H. (2016). Developing a land evaluation model for faba bean cultivation using geographic information system and multi-criteria analysis (A case study: Gonbad-Kavous region, Iran). *Ecological Indicators*, *63*, 37–47.

Khanam, Z., Siddiqui, M. A., Ajmal, U., Jamal, S., & Ali, S. K. A. (2026). Spatiotemporal monitoring of land surface temperature for assessing heat-related health risks. In Q. M. Ali, R. Ahmad, I. Ali, & P. Chatterjee (Eds.), *Sustainable Development: A Geospatial and Statistical Perspective* (pp. 161–176). Apple Academic Press.

Khanam, Z., Siddiqui, M. A., Faridi, R. A., Ali, S. A., & Khan, Z. (2022). Analysing land use land cover changes and urban built-up growth using NDBI and Shannon's entropy model: A spatio-temporal analysis of Bulandshahr city, *Asian Profile*, *50*(4), India.

Khangembam, S., & Kshetrimayum, K. S. (2019). Evaluation of hydrogeochemical controlling factors and water quality index of water resources of the Barak valley of Assam, Northeast India. *Groundwater for Sustainable Development*, *8*, 541–553.

Khosravi, H., Haydari, E., Shekoohizadegan, S., & Zareie, S. (2017). Assessment the effect of drought on vegetation in desert area using landsat data. *The Egyptian Journal of Remote Sensing and Space Science*, *20*, S3–S12.

Kubanza, N. S., & Simatele, D. (2016). Social and environmental injustices in solid waste management in sub-Saharan Africa: A study of Kinshasa, the Democratic Republic of Congo. *Local Environment*, *21*(7), 866–882.

Kumari, B., Tayyab, M., Shahfahad, Salman, Mallick, J., Khan, M. F., & Rahman, A. (2018). Satellite-driven land surface temperature (LST) using Landsat 5, 7 (TM/ETM+ SLC) and Landsat 8 (OLI/TIRS) data and its association with built-up and green cover over urban Delhi, India. *Remote Sensing in Earth Systems Sciences*, *1*(3), 63–78.

Lambin, E., Geist, H., & Lepers, E., 2003. Dynamics of land-use and land-cover change in tropical regions. *Annual Review of Environment and Resources*, *28*, 205–241.

Lankao, P. R., & Qin, H. (2011). Conceptualizing urban vulnerability to global climate and environmental change. *Current Opinion in Environmental Sustainability*, *3*(3), 142–149.

Lin, T., Liu, X., Song, J., Zhang, G., Jia, Y., Tu, Z.,... & Liu, C. (2018). Urban waterlogging risk assessment based on internet open data: A case study in China. *Habitat International*, *71*, 88–96.

Liou, Y. A., Nguyen, A. K., & Li, M. H. (2017). Assessing spatiotemporal eco-environmental vulnerability by Landsat data. *Ecological Indicators*, *80*, 52–65.

Liu, K., Li, X., & Wang, S. (2021). Characterizing the spatiotemporal response of runoff to impervious surface dynamics across three highly urbanized cities in southern China from 2000 to 2017. *International Journal of Applied Earth Observation and Geoinformation*, *100*, 102331.

Liu, L., & Zhang, Y. (2011). Urban heat island analysis using the Landsat TM data and ASTER data: A case study in Hong Kong. *Remote Sensing*, *3*(7), 1535–1552.

Ma, J., Zhang, C., Guo, H., Chen, W., Yun, W., Gao, L., & Wang, H. (2020). Analyzing ecological vulnerability and vegetation phenology response using NDVI time series data and the BFAST algorithm. *Remote Sensing*, *12*(20), 3371.

Mallick, J., Rahman, A., & Singh, C. K. (2013). Modeling urban heat islands in heterogeneous land surface and its correlation with impervious surface area by using night-time ASTER satellite data in highly urbanizing city, Delhi-India. *Advances in Space Research*, *52*(4), 639–655.

Mao, D., Wang, Z., Wu, J., Wu, B., Zeng, Y., Song, K.,... & Luo, L. (2018). China's wetlands loss to urban expansion. *Land Degradation & Development*, *29*(8), 2644–2657.

Miller, G. T., & Spoolman, S. (2011). *Living in the Environment: Principles, Connections, and Solutions*. 17th International edition, Cengage Learning, Brooks/Cole, Pacific Gove, CA.

Miller, R. B., & Small, C. (2003). Cities from space: Potential applications of remote sensing in urban environmental research and policy. *Environmental Science & Policy*, *6*(2), 129–137.

Nyambod, E. M. (2010). Environmental consequences of rapid urbanisation: Bamenda City, Cameroon. *Journal of Environmental Protection*, *1*(01), 15.

Pettorelli, N., Vik, J. O., Mysterud, A., Gaillard, J. M., Tucker, C. J., & Stenseth, N. C. (2005). Using the satellite-derived NDVI to assess ecological responses to environmental change. *Trends in Ecology & Evolution*, *20*(9), 503–510.

Purevdorj, T. S., Tateishi, R., Ishiyama, T., & Honda, Y. (1998). Relationships between percent vegetation cover and vegetation indices. International journal of remote sensing, 19(18), 3519-3535.

Ponsadailakshmi, S., Sankari, S. G., Prasanna, S. M., & Madhurambal, G. (2018). Evaluation of water quality suitability for drinking using drinking water quality index in Nagapattinam district, Tamil Nadu in Southern India. *Groundwater for Sustainable Development*, *6*, 43–49.

Purevdorj, T. S., Tateishi, R., Ishiyama, T., & Honda, Y. (1998). Relationships between percent vegetation cover and vegetation indices. *International Journal of Remote Sensing*, *19*(18), 3519–3535.

Qin, Z., Karnieli, A., & Berliner, P. (2001). A mono-window algorithm for retrieving land surface temperature from Landsat TM data and its application to the Israel-Egypt border region. *International Journal of Remote Sensing*, *22*(18), 3719–3746.

Richards, D. R., & Belcher, R. N. (2019). Global changes in urban vegetation cover. *Remote Sensing*, *12*(1), 23.

Roy, S. (2021). Sustainable urbanisation in Bangladesh and Dhaka: challenges and way forward [Background Paper]. 23rd ASEF Summer University: Liveable Cities for a Sustainable Future. Asia-Europe Foundation. https://asef.org/wp-content/uploads/2021/11/ASEFSU23-Background-Paper_Sustainable-Urbanisation-in-Bangladesh-Dhaka.pdf.

Rozenstein, O., Qin, Z., Derimian, Y., & Karnieli, A. (2014). Derivation of land surface temperature for Landsat-8 TIRS using a split window algorithm. *Sensors*, *14*(4), 5768–5780.

Rumbach, A. (2016). Decentralization and small cities: Towards more effective urban disaster governance? *Habitat International*, *52*, 35–42.

Saha, T. K., & Pal, S. (2019). Exploring physical wetland vulnerability of Atreyee river basin in India and Bangladesh using logistic regression and fuzzy logic approaches. *Ecological Indicators*, *98*, 251–265.

Sahoo, S., Dhar, A., & Kar, A. (2016). Environmental vulnerability assessment using Grey Analytic Hierarchy Process based model. *Environmental Impact Assessment Review*, *56*, 145–154.

Sarif, M., & Gupta, R. D. (2022). Evaluation of seasonal ecological vulnerability using LULC and thermal state dynamics using Landsat and MODIS data: A case study of Prayagraj City, India (1987–2018). *Environmental Science and Pollution Research*, *29*(51), 77502–77535.

Saaty, T. L. (1991). Some mathematical concepts of the analytic hierarchy process. *Behaviormetrika*, *18*(29), 1–9.

Saaty, T. L., & Wind, Y. (1980). Marketing applications of the analytic hierarchy process. *Management Science*, *26*(7), 641–658.

Shi, Y. (2012). Risk analysis of rainstorm waterlogging on residences in Shanghai based on scenario simulation. *Natural Hazards*, *62*(2), 677–689.

Singh, P., Kikon, N., & Verma, P. (2017). Impact of land use change and urbanization on urban heat island in Lucknow city, Central India. A remote sensing based estimate. *Sustainable Cities and Society*, *32*, 100–114.

Singh, K. K., Vogler, J. B., Shoemaker, D. A., & Meentemeyer, R. K. (2012). LiDAR-Landsat data fusion for large-area assessment of urban land cover: Balancing spatial resolution, data volume and mapping accuracy. *ISPRS Journal of Photogrammetry and Remote Sensing*, *74*, 110–121.

Sobrino, J. A., Jiménez-Muñoz, J. C., & Paolini, L. (2004). Land surface temperature retrieval from LANDSAT TM 5. *Remote Sensing of Environment*, *90*(4), 434–440.

Sun, F., Sun, W., Chen, J., & Gong, P., 2012. Comparison and improvement of methods for identifying water bodies in remotely sensed imagery. *International Journal of Remote Sensing*, *33*(21), 6854–6875. https://doi.org/10.1080/01431161.2012.692829.

Tercan, E., & Dereli, M. A. (2020). Development of a land suitability model for citrus cultivation using GIS and multi-criteria assessment techniques in Antalya province of Turkey. *Ecological Indicators*, *117*, 106549.

Tiburan, C. L., Kobayashi, S., Mizuno, K., & Saizen, I. (2009). Developing a spatial-based approach for vulnerability assessment of Philippine watersheds and its potential in disaster management. *Disaster Management and Human Health Risk: Reducing Risk, Improving Outcomes*, *110*, 21.

Tirkey, P., Bhattacharya, T., Chakraborty, S., & Baraik, S. (2017). Assessment of groundwater quality and associated health risks: A case study of Ranchi city, Jharkhand, India. *Groundwater for Sustainable Development*, *5*, 85–100.

Tsheleza, V., Nakin, M. D., Ndhleve, S., Kabiti, H. M., & Musampa, C. M. (2019). Vulnerability of growing cities to solid waste-related environmental hazards: The case of Mthatha, South Africa. *Jamba: Journal of Disaster Risk Studies*, *11*(1), 1–10.

Turner, B. L., Kasperson, R. E., Matson, P. A., McCarthy, J. J., Corell, R. W., Christensen, L.,... & Schiller, A. (2003). A framework for vulnerability analysis in sustainability science. *Proceedings of the National Academy of Sciences*, *100*(14), 8074–8079.

Udeshani, W. A. C., Dissanayake, H. M. K. P., Gunatilake, S. K., & Chandrajith, R. (2020). Assessment of groundwater quality using water quality index (WQI): A case study of a hard rock terrain in Sri Lanka. *Groundwater for Sustainable Development*, *11*, 100421.

Van Beukering, P., Sehker, M., Gerlagh, R., & Kumar, V. (1999). *Analysing Urban Solid Waste in Developing Countries: A Perspective on Bangalore, India*. Collaborative Research in the Economics of Environment and Development.

Wang, P., Li, Z., Gao, W., 2011. Rapid shrinking of glaciers in the Middle Qilian Mountain region of Northwest China during the last50 years. *Journal of Earth Science*, *22*(4), 539–548. https://doi.org/10.1007/s12583-011-0195-4.

Wang, F., Qin, Z., Song, C., Tu, L., Karnieli, A., & Zhao, S. (2015). An improved mono-window algorithm for land surface temperature retrieval from Landsat 8 thermal infrared sensor data. *Remote Sensing*, *7*(4), 4268–4289.

Wong, N. H., & Yu, C. (2005). Study of green areas and urban heat island in a tropical city. *Habitat International*, *29*(3), 547–558.

World Bank. (2022). India Urban Population. Accessed from https://www.macrotrends.net/global-metrics/countries/ind/india/urban-population.

Yadav, K. K., Gupta, N., Kumar, V., Choudhary, P., & Khan, S. A. (2018). GIS-based evaluation of groundwater geochemistry and statistical determination of the fate of contaminants in shallow aquifers from different functional areas of Agra city, India: Levels and spatial distributions. *RSC Advances*, *8*(29), 15876–15889.

Yin, Z. E., Yin, J., Xu, S., & Wen, J. (2011). Community-based scenario modelling and disaster risk assessment of urban rainstorm waterlogging. *Journal of Geographical Sciences*, *21*(2), 274–284.

Yu, L., Lyu, Y., Chen, C., & Choguill, C. L. (2021). Environmental deterioration in rapid urbanisation: Evidence from assessment of ecosystem service value in Wujiang, Suzhou. *Environment, Development and Sustainability*, *23*(1), 331–349.

Yuan, F., & Bauer, M. E. (2007). Comparison of impervious surface area and normalized difference vegetation index as indicators of surface urban heat island effects in Landsat imagery. *Remote Sensing of environment*, *106*(3), 375–386.

Zambon, I., Colantoni, A., & Salvati, L. (2019). Horizontal vs vertical growth: Understanding latent patterns of urban expansion in large metropolitan regions. *Science of the Total Environment*, *654*, 778–785.

Zha, Y., Gao, J., & Ni, S. (2003). Use of normalized difference built-up index in automatically mapping urban areas from TM imagery. *International Journal of Remote Sensing*, *24*(3), 583–594.

Zhang, X., Liu, K., Li, X., Wang, S., & Wang, J. (2022). Vulnerability assessment and its driving forces in terms of NDVI and GPP over the Loess Plateau, China. *Physics and Chemistry of the Earth, Parts A/B/C*, *125*, 103106.

Zhang, Y., Yang, Z., & Yu, X. (2006). Measurement and evaluation of interactions in complex urban ecosystem. *Ecological Modelling*, *196*(1–2), 77–89.

Zhang, Z., Demšar, U., Rantala, J., & Virrantaus, K. (2014). A fuzzy multiple-attribute decision-making modelling for vulnerability analysis on the basis of population information for management. *International Journal of Geographical Information Science*, *28*(9), 1922–1939.

11 Living with Heat
Exploring Remote Sensing Techniques for Urban Microclimate Study in Ghaziabad City, India

Sadaf

INTRODUCTION

Since local microclimates have a cumulative effect on global climate change, microclimates are the most crucial component of contemporary global climate research. The average temperature of the Earth has significantly increased due to climate change (IPCC, 2014). Anthropogenic causes (such as emission of greenhouse gases and urbanization) and natural factors (such as radiation from the sun and climate variability) combine to cause a rise in land surface temperature (LST). The increase in greenhouse gases resulting from human activities is the primary driver of the average increase in LST.

Changes in land use/land cover that are having a significant impact on the climate, are among the most evident consequences of human intervention in the natural environment (Teferi & Abraha, 2017). Throughout human history, one of the main drivers of land use and land cover (LULC) that has significantly influenced climate change is urbanization. Environmental degradation is mostly caused by changes in land cover, and the situation is made worse by rapid urbanization (Ho et al., 2016; Singh et al., 2017; Kikon et al., 2016; Zhou et al., 2016). Local, regional, and global LSTs have been rising due to LULC changes caused by the growth of impermeable surfaces such as parking lots, buildings, asphalt, and concrete paths, and other surfaces (Degefu et al., 2023). The loss of water and agricultural land are linked to higher solar radiation absorption, which raises LST and has detrimental impacts on the climate and human well-being (Fadhil et al., 2023; Owen et al., 1998). Surface and air temperatures can rise to a few degrees compared to the undeveloped surroundings due to natural or man-made deviations in the LULC pattern (Sherafati et al., 2018; Sultana & Satyanarayana, 2020).

One of the primary drivers of climate change is the development of urban heat islands (UHI), an environmental phenomenon associated with urbanization. When the surface and atmosphere of a city are significantly warmer compared to nearby natural areas, this phenomenon is denoted by the term UHI (Rotach et al., 2005; Voogt & Oke, 2003). The LST increase is causing an increase in the UHI phenomenon relative to the surrounding rural area. Inadvertent climate alteration occurs when the surfaces and atmosphere of urbanized areas show warmer characteristics compared with those exhibited in nonurbanized counterparts (Aboelata & Sodoudi, 2020). The degree of UHI is correlated with the condition of water, vegetation, and built-up regions, as well as how these factors vary gradually (Chen et al., 2006).

The UHI has a major detrimental impact on ecosystems, economic growth, climate, and human health and well-being in general (Zhou et al., 2014). The effects of shifting land cover on cities' surface temperatures and their implications for human health are growing more significant issues as metropolitan areas continue to expand quickly. In order to develop climate-resilient communities and suitable urban settings for inhabitants, it is essential to mitigate the root cause of rising temperatures and implement problem-specific plans.

DOI: 10.1201/9781003620747-13

Ghaziabad is a dynamic city that is urbanizing significantly. It is an important industrial center and a developing NCR suburb, which contributes to the growth of built-up regions and higher population density. Over the past few decades, urban spatial spaces have grown at an accelerated rate. Industrialization, economic expansion, and rural-to-urban migration are the main causes of urbanization. The city expanded along key road corridors and highways toward the northwest. The hydrological services, vegetation cover, and local climate scales are all drastically altered by urbanization (Kumar et al., 2025). Unplanned urbanization, however, can occasionally have unintended consequences that worsen the economy and accelerate environmental deterioration. A rise in LST in urban areas compared to nearby rural areas is the most evident meteorological effect of urbanization (Njoku & Tenenbaum, 2022). Alteration in LULC brought about by urbanization poses a threat to urban dwellers' outdoor thermal comfort and LSTs (Patel et al., 2024). Significant effects on the LST are caused by alterations in land cover and land use; therefore, understanding the dynamics of built-up areas and green spaces in urban areas is essential to understanding how they affect UHIs and how they relate to the LST. Thus, the goal of this study was to evaluate how urban LULC alteration affected the microclimate changes in Ghaziabad city by taking into account the indices of LULC, UHI, LST, NDBI, and NDVI. This study was conducted in order to provide planning and decision-making information about the study area's LULC status, LST, and UHI.

STUDY AREA

The Indian state of Uttar Pradesh comprises the city of Ghaziabad, which is a part of the Delhi National Capital Region. It is Western Uttar Pradesh's largest city and the administrative center of the Ghaziabad district. It is situated between latitudes 28°61′ and 28°79′ North and longitudes 77°24′ and 78°55′ East (Figure 11.1). The municipal corporation encompasses 100 wards. The city has excellent road and rail connections and is a significant rail junction in North India. It is a part of Uttar Pradesh's Meerut Division. The city, which is located in the Upper Gangetic Plains, is divided into two main sections, Cis-Hindon on the east and Trans-Hindon on the western side, by the Hindon River. The climate in Ghaziabad is subtropical, with three seasons that are distinct: the monsoon season, cold winters, and hot summers. Summer temperatures typically range from 35°C to 45°C, and they are hot and humid. The winter months are comparatively chilly, with temperatures of about 10°C. Much-needed rainfall is brought during the monsoon season, which normally lasts from June to early July (City Disaster Management Report, Ghaziabad, 2012). The data showed that there were 885 girls for every 1,000 males in the 1,648,643-person Ghaziabad Municipal Corporation. The literacy rate in Ghaziabad was 84.78%. Children under the age of six comprised 12.67% of the population. Scheduled Tribes and Scheduled Castes comprised 0.18% and 13.68% of the population, respectively (District Census Handbook, 2011). One of the most populous towns in Uttar Pradesh is Ghaziabad. With 67.55% of the people residing in urban areas, Ghaziabad, Uttar Pradesh, is a substantially urbanized city. Globalization, industrialization, and modernization are the main drivers of this quick growth, which has resulted in a higher population density and an expansion of built-up regions at the expense of rural land (Kumar and Siddiqui, 2022). Ghaziabad is a significant industrial area and an expanding NCR satellite town, which adds to the area's urbanization. Temperature changes and greenhouse gas emissions in cities are found to be substantially affected by urbanization. Large dwellings and more impermeable regions, together with continuous energy use in residences, businesses, and industries, are the land use patterns that contribute to the rise in heat. Rapid urbanization causes substantial changes in LULC, which have an adverse effect on the city's LST.

METHODS AND MATERIALS

DATA SOURCES

This analysis was conducted using Landsat image TMs from 1994 and Landsat 9 OLI-2/TIRS-2 images from 2024 (Table 11.1). For all yearly images, the satellites maintained a spatial resolution of 30 m and continuously followed a predetermined path and row, 140 and 46. The Landsat images

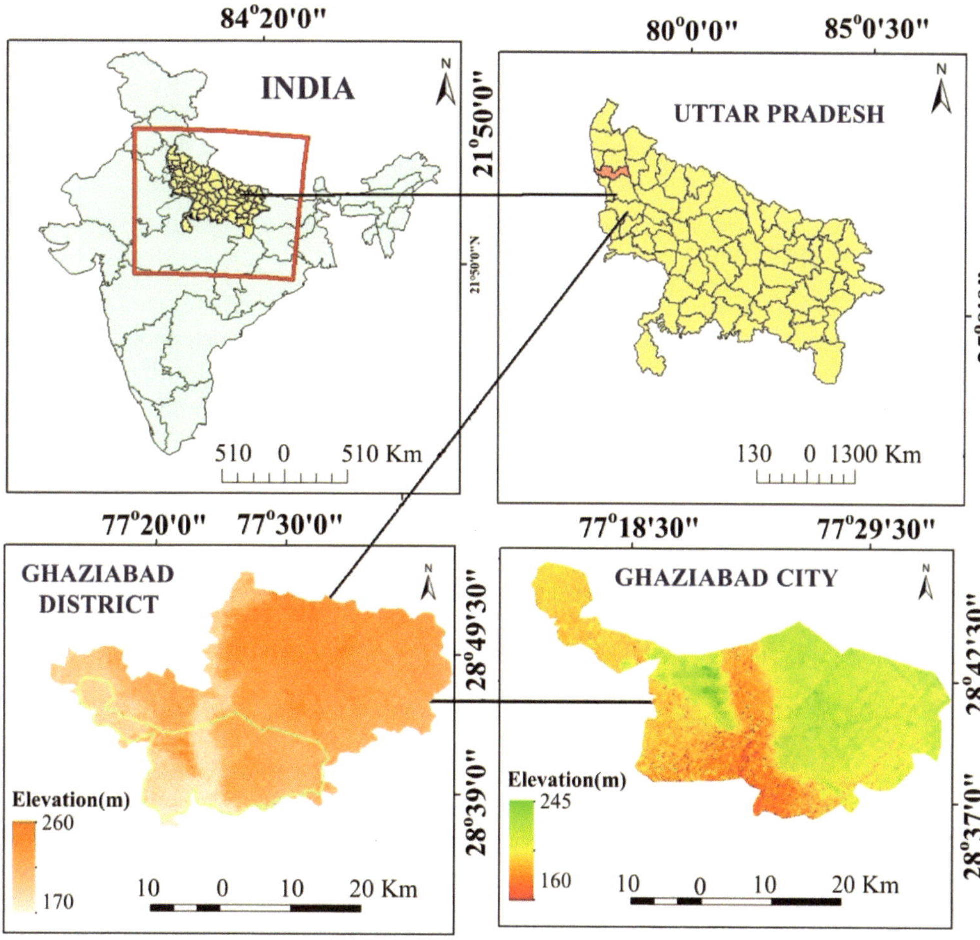

FIGURE 11.1 Study area: Ghaziabad city.

TABLE 11.1
Characteristics of Used Landsat Satellite Image

Satellites	Sensors of the Satellites	Date of Data Acquirement	Path and Row of the Satellites	Spatial Resolution (m)	Cloud Cover
Landsat 9	TIRS-2/OLI-2	29/02/2024	146/040	30	Nil
Landsat 5	TM	18/02/1994	146/040	30	Nil

were obtained from the freely accessible USGS Earth Explorer website (https://earthexplorer.usgs.gov). All of the Landsat images were downloaded in February, during the dry season, and without any cloud cover (0%) in the sky. Satellite images were projected using the UTM Zone 43 N projection and the WGS 84 datum. The Landsat images were utilized to calculate UHI, LST, and NDVI as well as to analyze LULC changes.

LAND USE MAPS: SUPERVISED CLASSIFICATION METHOD

The Landsat images that corresponded to 1994 and 2024 were classified to create the LULC maps for those years. A supervised maximum likelihood classifier was used to classify each image separately using the ERDAS Imagine 15 simulation software. The MLC determines the likelihood that a cell belongs to each LULC class based on its attribute values. A candidate cell for classification is then assigned to the class that has the greatest likelihood or maximum likelihood to be assigned to that particular subclass (Duda et al., 2000). In order to ensure accuracy, 100–200 samples had been collected from each class, depending on the class area. Five main LULC classes—vegetation, fallow land, built-up areas, water bodies, and agricultural land—were recognized and categorized.

ACCURACY ASSESSMENT

One of the most important procedures for assessing the reliability as well as the quality of the results is the accuracy evaluation of maps of LULC and classifications. In order to determine the accuracy of the outcomes of classification, the classified LULC samples are compared with reference data. The goal is to examine the overall accuracy, producer accuracy, and user accuracy to determine the level of accuracy of the result of the classification error matrix. The Kappa coefficient concept is frequently used to assess the precision of image-based classification data. The value of the Kappa coefficient ranges from –1 to 1, with 0 denoting no distinction between trained and random classifications (Moharir et al., 2025). Using Google Earth Pro and topographical sheets for 1994 and 2024, 330 training sample points were collected in order to evaluate the precision of LULC classifications for both satellite images, which were divided into five classes: built-up, agricultural land, water bodies, vegetated, and fallow land. Eq. (11.1) provides the following equation for the Kappa coefficient:

$$\text{Kappa coefficient}(k) = \frac{n\sum_{i=1}^{r} x_{ij} - \sum_{i=1}^{r}(x_i - x_{+i})}{n^2 - \sum_{i=1}^{r}(x_i - x_{+i})} \tag{11.1}$$

In the equation above, "r" stands for the quantity of rows, and "x_{ii}" for the total number of values in the "ith" row as well as column. "x_i" and "x_{+i}" stand for the totals of row "i" and column "i," respectively. The integer "n" represents all of the observations (samples/pixels).

ESTIMATION OF LST

The precision of evaluating the impact of LST and LULC has significantly increased with the use of remotely sensed, or RS, sensors. Thermal infrared bands, which are utilized by Landsat 9 and Landsat 5 for imaging the Earth's surface, were used to compute the LST of Ghaziabad city over a period of two years (1994 and 2024). Band 10 and Band 6 are the thermal bands for Landsat 9 and Landsat 5, respectively. These bands quantify the amount of heat radiation emitted by the Earth's surface. The basic steps listed below were used in the current effort to transform these DNs into LST.

FOR LANDSAT 5 THEMATIC MAPPER (1994)

A few formulas were used to construct LST maps, which were based on a Landsat 5 image. These calculations are solved using ArcGIS 10.8's raster calculator.

For Landsat 5, DNs are converted to spectral radiance employing Eq. (11.2)

$$L_\lambda = \left(\frac{\text{LMAX}_\lambda - \text{LMIN}_\lambda}{\text{QCALMAX} - \text{QCALMIN}}\right) * (\text{QCAL} - \text{QCALMIN}) + \text{LMIN}_\lambda \tag{11.2}$$

In this case, L_λ stands for the spectral radiation at the aperture of the sensor; LMAX_λ signifies the spectral radiance in ($W/(m^2 sr\mu m)$) scaled to QCALMAXIn QCALMIN, the spectral radiance is represented by LMIN_λ; In DN, QCAL stands for the quantified calibrated pixel value; QCALMAX stands for the greatest quantified calibrated pixel value in DN, which corresponds to $LMAX_\lambda$; in DN, QCALMIN stands for the Lowest Quantized Calibrated Pixel Value, which corresponds to $LMIN_\lambda$.

Using Eq. (11.3) to convert the brightness temperatures in Kelvin from radiation

$$BT = \frac{K2}{l_n\left(\frac{K1}{L\lambda}+1\right)} \tag{11.3}$$

where

In Kelvin, *BT* stands for the effective at-satellite temperature.

The calibrated constant 2 is denoted by *K*2.

The calibrated constant 1 is denoted by *K*1.

The spectral radiance is expressed in (W/(m^2 sr μm)) with L_λ.

Using Eq. (11.4) to convert Kelvin into degrees Celsius

$$\text{Temperature}(°\text{C}) = \text{“Temperature in Kelvin”} - 273.15 \tag{11.4}$$

REGARDING THE LANDSAT 9 OPERATIONAL LAND IMAGER-2

A few algorithms were used to create LST maps, and LST maps were produced using Band 10 of Landsat 9 imagery. The equations are as follows:

Using Eq. (11.5), calculate the TOA (Top of Atmosphere) spectral radiance

$$\text{TOA}(L_\lambda) = \text{ML} \times \text{Qcal+AL} \tag{11.5}$$

where

Radiance_MULT_BAND_x, denoted by ML, is the band-specific multiplication rescaling factor of the metadata, where x is the band's number. Band 10 is related to Qcal. With x being the band number, the band-specific additional rescaling factor (AL) of the metadata is RADIANCE_ADD_BAND_x.

Transformation from TOA to BT using Eq. (11.6)

$$BT = \left(\frac{K2}{l_n\left(\frac{K1}{L\lambda}+1\right)}\right) - 273.15 \tag{11.6}$$

where

when x is the thermo band number as well, the band-specific thermal transition constant retrieved from metadata is K1_CONSTANT_BAND_x. K2 represents the band-specific temperature conversion constant. (K2_CONSTANT_BAND_x) in the metadata, where x is the thermal band numbers. L_λ represents the spectral radiance of TOA.

NDVI is calculated using the following Eq. (11.7)

$$\text{NDVI} = (\text{Band5} - \text{Band4}) / (\text{Band5} + \text{Band4}) \tag{11.7}$$

Calculating the vegetation P_V percentage using Eq. (11.8)

$$P_V = \left(\frac{(\text{NDVI} - \text{NDVImin})}{(\text{NDVImax} - \text{NDVImin})}\right)^2 \tag{11.8}$$

The emissivity ε is computed using Eq. (11.9)

$$\varepsilon = (0.004 * P_V) + 0.986 \tag{11.9}$$

LST computation using Eq. (11.10)

$$\text{LST} = \left(\frac{BT}{1 + \left(\frac{\lambda * BT}{\alpha} \right) ln\ \varepsilon} \right) \tag{11.10}$$

where

the wavelength of the radiation that is released is denoted by λ. Let α stand for the *hc/k* (1.438×10^{-2} mK) and *h* stand for the Planck's constant value (6.626×10^{-34} J/s). The constant known as Boltzmann (1.38×10^{-23}J/K) is represented by *k*, and the velocity of light (2.998×108 m/s) is represented by *c*. The map of LST was finally generated.

NDVI ESTIMATION

One of the intrinsic properties of nature is the NDVI, which is a critical surface measure based on the amount of light emitted at a given point in time (Weng et al., 2004). In order to determine how well vegetation interacts with the electromagnetic spectrum, the vegetation health index, or NDVI, compares the amount of red light that the vegetation absorbs to the amount of near-infrared light that the vegetation reflects. The NDVI scale goes from –1 to +1, with a number near zero denoting a built-up region and a value near +1 denoting a healthy and green vegetative cover. The NDVI's formula is provided below; it makes use of the red regions and near-infrared of the electromagnetic spectrum.

$$\text{NDVI} = \frac{(\text{NIR} - \text{RED})}{(\text{NIR} + \text{RED})} \tag{11.11}$$

NDBI ESTIMATION

The NDBI is spectral index that measures the amount of built-up or urbanized land in a landscape and is obtained from satellite imagery. Because built-up surfaces absorb and retain heat, dense urban growth is indicated by high NDBI values, and vice versa. This tends to result in greater LST. The NDBI can be computed using Eq. (11.12) below (Zha et al., 2003):

$$\text{NDBI} = \frac{(\text{SWIR} - \text{NIR})}{(\text{SWIR} + \text{NIR})} \tag{11.12}$$

URBAN HEAT ISLAND INDEX

It is difficult to evaluate the total UHI impact in a given region utilizing LST values over an extended period since the intensity of the UHI anomaly might change dramatically over time within a certain area. The UHI index was suggested as a method for improving forecasts in order to solve this problem (Aneesh et al., 2016). This equation can be used to calculate it:

$$\text{UHI} = \frac{T_s - T_m}{SD} \tag{11.13}$$

where *SD* stands for standard deviation, T_s for land surface temperature, and T_m for the study area's mean land surface temperature.

REGRESSION ANALYSIS

This study uses a scatter plot with a linear regression line to illustrate the relationships between several variables, such as the relationship between NDVI and LST and the association between NDBI and LST. Using Microsoft Excel 2023, the scatter plot was created, incorporating the linear regression equation.

$$Y_I = B_O + B_1 X_I \tag{11.14}$$

Y represents the value of the dependent variable (Y), which is what is anticipated or explained for observation i; B_O is the interception; B_1 is the regression slope estimate; and X_i is the value of X for observation i.

RESULTS AND DISCUSSIONS

Land Use and Land Cover Maps for the Years 1994 and 2024

Between 1994 and 2024, Ghaziabad's LULC underwent significant change due to the city's fast industrialization, population expansion, migration from rural areas, economic growth, and shifting agricultural patterns. In Ghaziabad city, urbanization has altered the dynamics of LULC, as seen in Figures 11.2–11.4. The land transformation process, which took place in the city of Ghaziabad over a two-decade period, from 1994 to 2024, has been examined in this study using five main land classes. Table 11.2 shows that the built-up area expanded from 141.79 to 197.03 km^2, indicating the substantial urbanization of the area. Additionally, from 1994 to 2024, the percentages of fallow land fell by −3.14 km² (0.46%) and −13.54 km² (2.02%), respectively. However, the amount of land used for farming has also declined substantially, going from 117.17 to 73.47 km^2, correspondingly demonstrating the quickening rate of urbanization in Ghaziabad.

Furthermore, the research area's vegetation cover and water bodies have slightly decreased from 4.33 to 4.07 km^{2}and 18 to 17.13 km^2, respectively. According to the pattern, during the 1990–2021 research years, the area of other land classifications dropped, while only built-up land increased. The main causes of the area's changing LULC, as well as the situation of diminishing agricultural land and vegetation, are urbanization and industrialization.

Accuracy Assessment of the LULC Classification

To provide a credible LULC classification result, classification accuracy assessment has become essential. Tables 11.2 and 11.3 display the results of the matrix of confusion that was utilized to

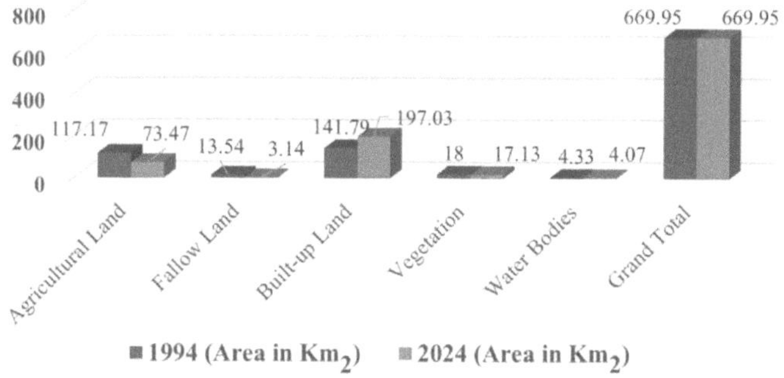

FIGURE 11.2 LULC change detection of Ghaziabad city (1994–2024).

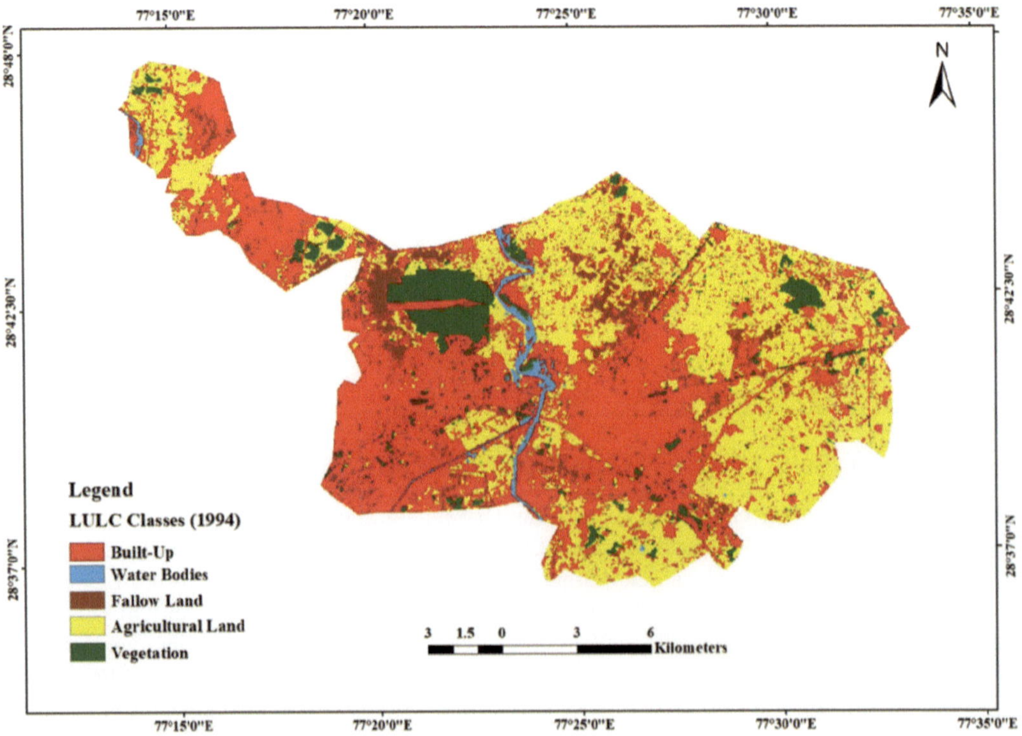

FIGURE 11.3 LULC map of Ghaziabad city (1994).

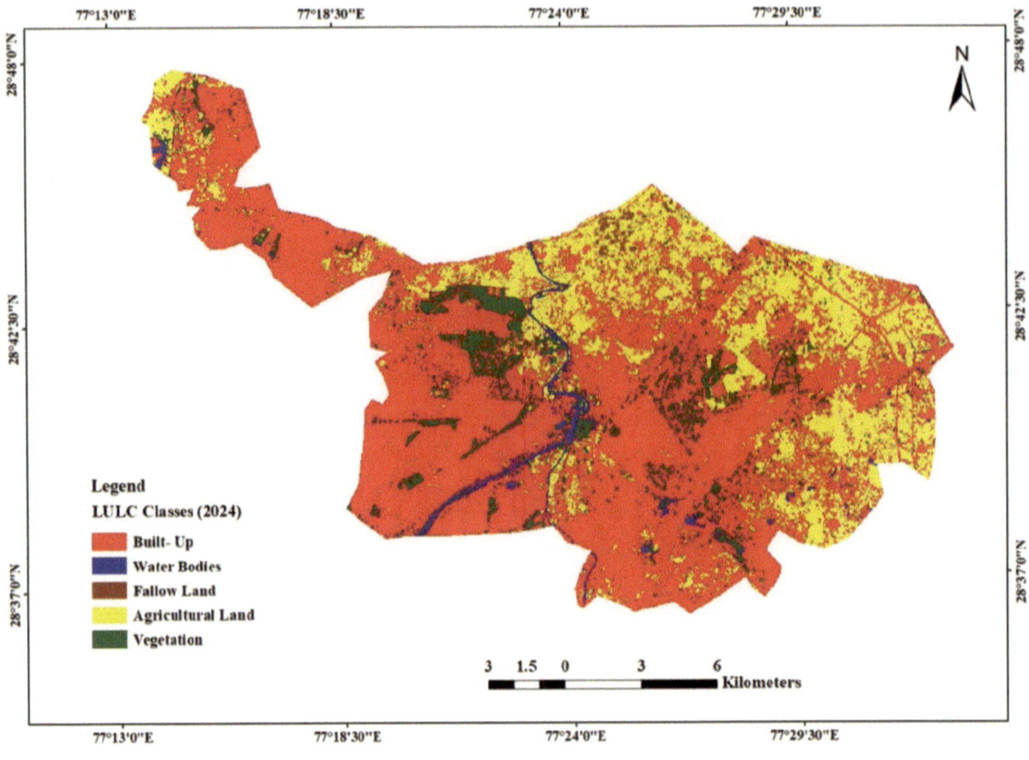

FIGURE 11.4 LULC map of Ghaziabad city (2024).

TABLE 11.2
Accuracy Assessment and Landsat 5-TM Classification (1994)

Classes	Fellow Land	Built-up	Vegetation	Agricultural Land	Water Bodies	Kappa Coefficient (*k*)
Fellow land	29	0	0	3	0	86%
Built-up	0	27	1	0	0	
Vegetation	0	0	27	0	0	
Agricultural land	3	6	1	30	2	
Water bodies	0	0	4	0	31	
Total	33	33	33	33	33	
User's accuracy	90.62%	96.42%	100%	71.42%	88.57%	
Producer's accuracy	87.87%	81.81%	81.81%	90.90%	93.93%	
Overall accuracy	87.27%					

TABLE 11.3
Classification of Accuracy Assessment of Landsat 9-OLI-2/TIRS-2 (2024)

Class Name	Fellow Land	Built-up	Vegetation	Agricultural Land	Water Bodies	Kappa Coefficient (*k*)
Fellow land	22	1	0	0	0	86.09%
Built-up	0	28	0	0	0	
Vegetation	0	0	33	1	0	
Agricultural land	11	4	0	32	2	
Water bodies	0	0	0	0	31	
Total	33	33	33	33	33	
User's accuracy	96%	100%	97%	65%	100%	
Producer's accuracy	67%	85%	100%	97%	94%	
Overall accuracy	88.48%					

evaluate the LULC classification's accuracy in the present study. With Kappa values of 86% and 86.09%, respectively, the overall accuracy of the categorized LULC classes for the years 1994 and 2024 was 87.27% and 88.48%, respectively. These results show a high level of accuracy and lie within an acceptable range.

Land Surface Temperature

The LST pattern of Ghaziabad city between 1994 and 2024 is shown in Figures 11.5 and 11.6. In 1994, the estimated LST values were 13.30°C–24.97°C, and in 2024, they were 19.77°C–33.01°C. According to the findings, the lowest temperatures in 1994 and 2024 were 13.30°C and 19.77°C, respectively. In 1994 and 2024, the highest recorded temperatures were 24.97°C and 33.01°C, respectively.

The increase in urban development in the research region between 1994 and 2024, which was marked by the construction of roads, built-up areas, highways, buildings, concrete constructions, and other impermeable substrates, subsequently caused the rise in LST. Compared to neighboring rural areas, these urban characteristics' higher thermal conductivity causing them to collect greater amounts of sunlight during the daytime and release it progressively as emissions at night, raising

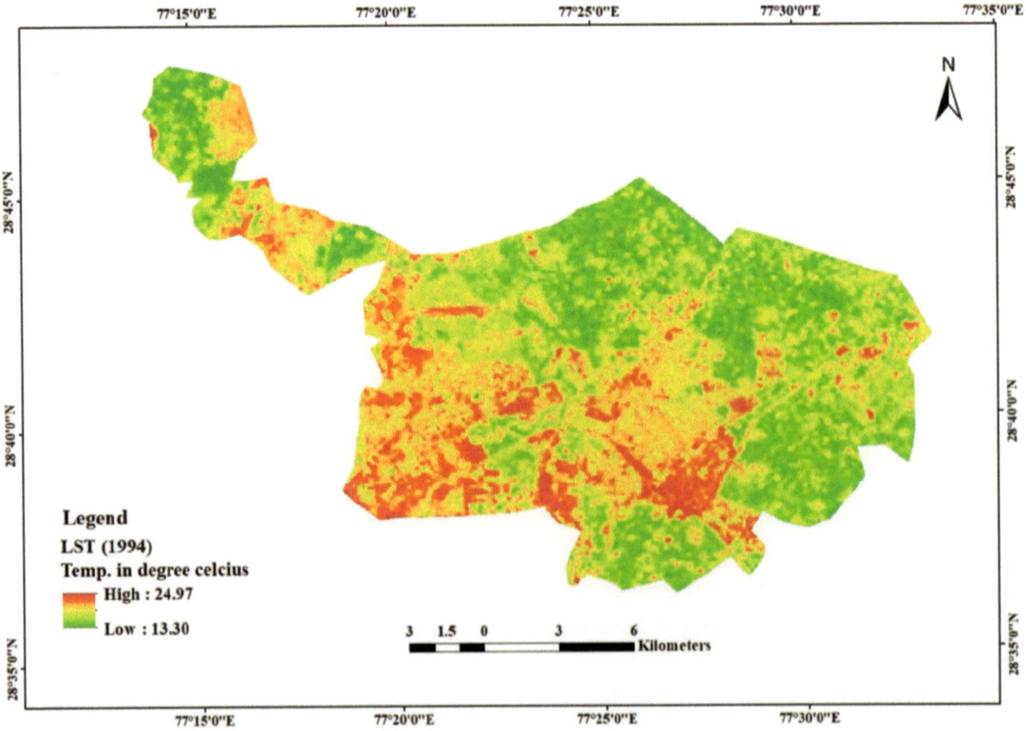

FIGURE 11.5 LST map of Ghaziabad city (1994).

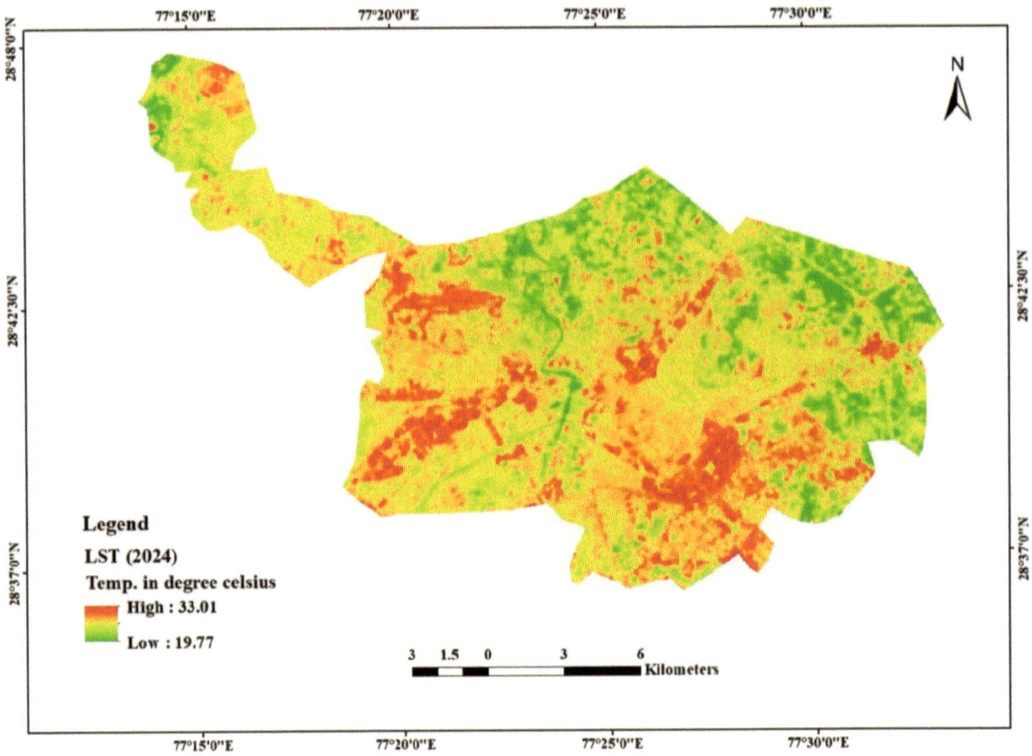

FIGURE 11.6 LST map of Ghaziabad city (2024).

the temperature. The eastern portion of the research area, which has a higher level of vegetation and agriculture, is where this temperature was primarily recorded. The western and southern regions had the fastest rates of urban expansion. The suitability of the land and the accessibility of transit, along with various demographic variables, all contributed to a rise in this direction. Real estate predominates in certain locations, along with other restrictions like factories, roads, and industries, which resulted in significant LST rises over the course of the study. The maps in Figures 11.4 and 11.5 made it easy to see the spatiotemporal pattern of the land surface temperature's variance. Similarly, the findings of this study demonstrated that, as a result of the city's increasing urbanization in all directions, the greatest LST expansion was seen in the majority of Ghaziabad.

Urban Heat Island

The eastern portion of the city, which had vegetation and agricultural land cover, showed the least amount of UHI fluctuations between 1994 and 2024. The study's UHI profile was taken from the city's western to eastern regions (Figures 11.7 and 11.8). According to the findings, the UHI impact increased more frequently as it moved from the city's eastern region into its center. The development of commercial centers, infrastructure, built-up areas, and industries in the west and northwest regions of Ghaziabad city resulted in very significant spatiotemporal variability in the UHIs. LULC change and urban growth were linked to this spatiotemporal variance. Thus, the study found that industrial zones, adjacent roadways, and densely populated areas all enhance the impacts of high UHI.

Relationship between NDVI and LST

The most generally used and well-liked technique for determining the density and distribution of vegetation in any area is the NDVI (Shahfahad et al., 2020). Since the quantity of vegetation present

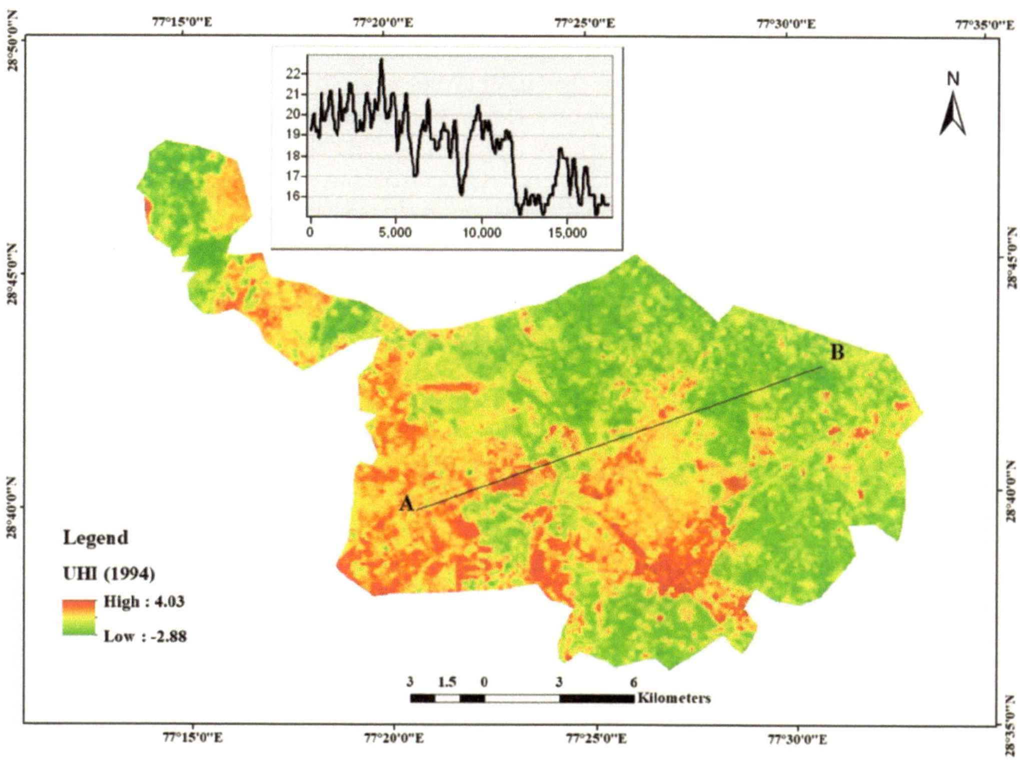

FIGURE 11.7 UHI map of Ghaziabad city (1994).

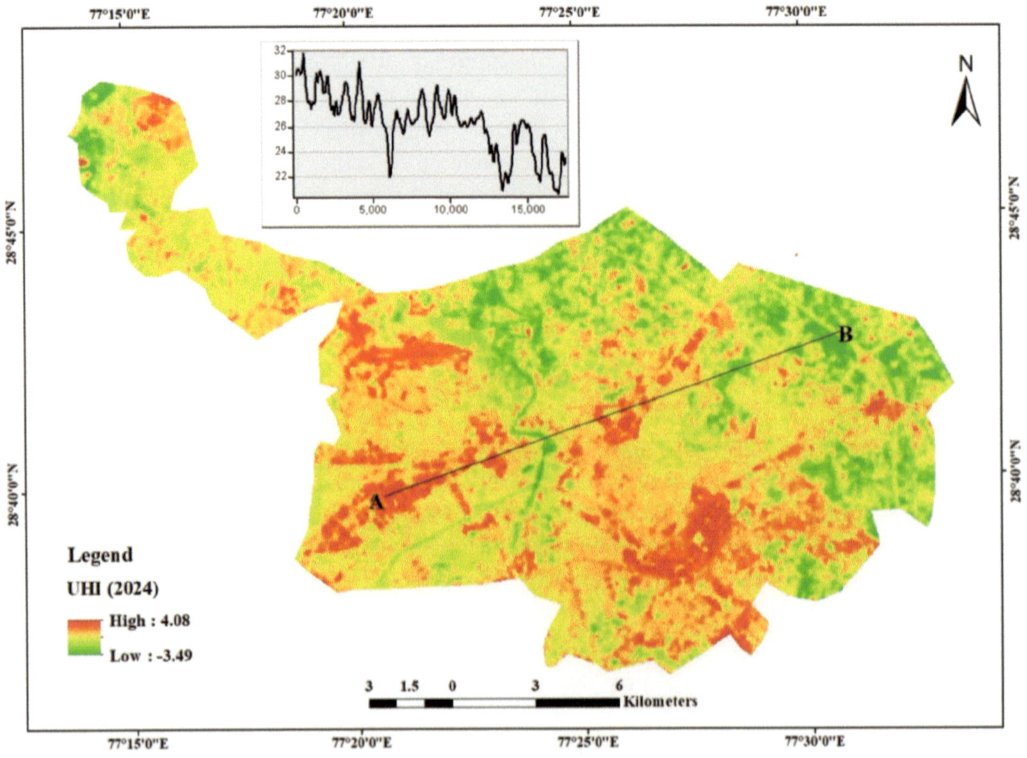

FIGURE 11.8 UHI map of Ghaziabad city (2024).

is crucial to understanding long-term urban expansion, predicting LST, and eventually determining the UHI effect, it is necessary to estimate the NDVI (Weng et al., 2004). NDVI, which is obtained from Landsat imagery, is used for this. Figures 11.8 and 11.9 illustrate the NDVI models that were derived from Landsat satellite imagery for 1994 and 2024. In 1994, the reported NDVI values were between 0.70 and –0.27, and in 2024, they were between 0.55 and –0.14 (Figures 11.9 and 11.10). The NDVI map shows both bare ground and vegetation cover; exposed rock material is represented by the lowest NDVI value, while dense vegetation is represented by the maximum NDVI value.

Regression analysis has been utilized to get the NDVI coefficient. For the years 1994 and 2024, linear regression analysis shows a contrary relationship between LST and NDVI, with respective R^2 values of 0.63 and 0.41 (Figures 11.11 and 11.12). Moreover, the inverse relationship between LST and NDVI implies that lower LST values are linked to greater vegetation cover. It is important to note that NDVI has an immense influence on how the land use/land cover (LULC) evolves since the urbanization of regions with lower NDVI values causing the vegetation density to drop, which raises LST values. Increased urbanization and the consequent decrease in vegetation cover are the causes of the decline in values for the NDVI between 1994 and 2024. The locations with the greatest LST values also have the least NDVI values, highlighting the contrast among regions with higher LST compared to those with lower temperatures according to vegetation density.

Relationship between NDBI and LST

Figures 11.14 and 11.15 show the association between the LST and the normalized difference built-up index (NDBI) for the periods 1994 and 2024.

The city's west and southwest have seen a sharp rise in built-up areas as a result of significant urbanization. As a result of this change, numerous urban green areas are being turned into

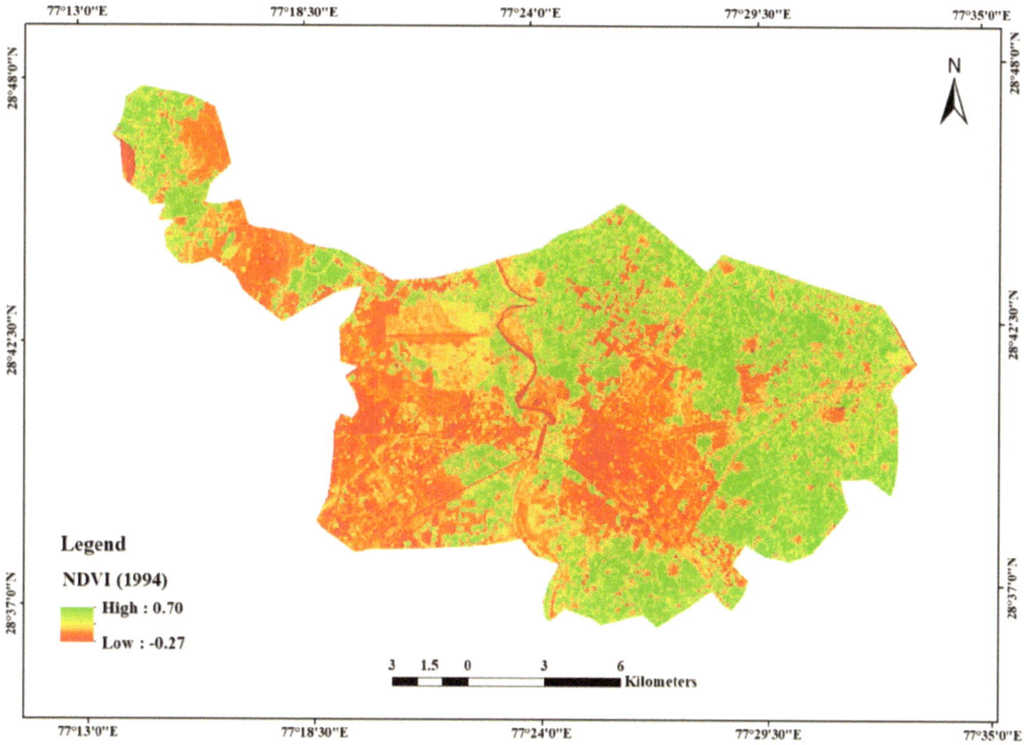

FIGURE 11.9 NDVI map of Ghaziabad city (1994).

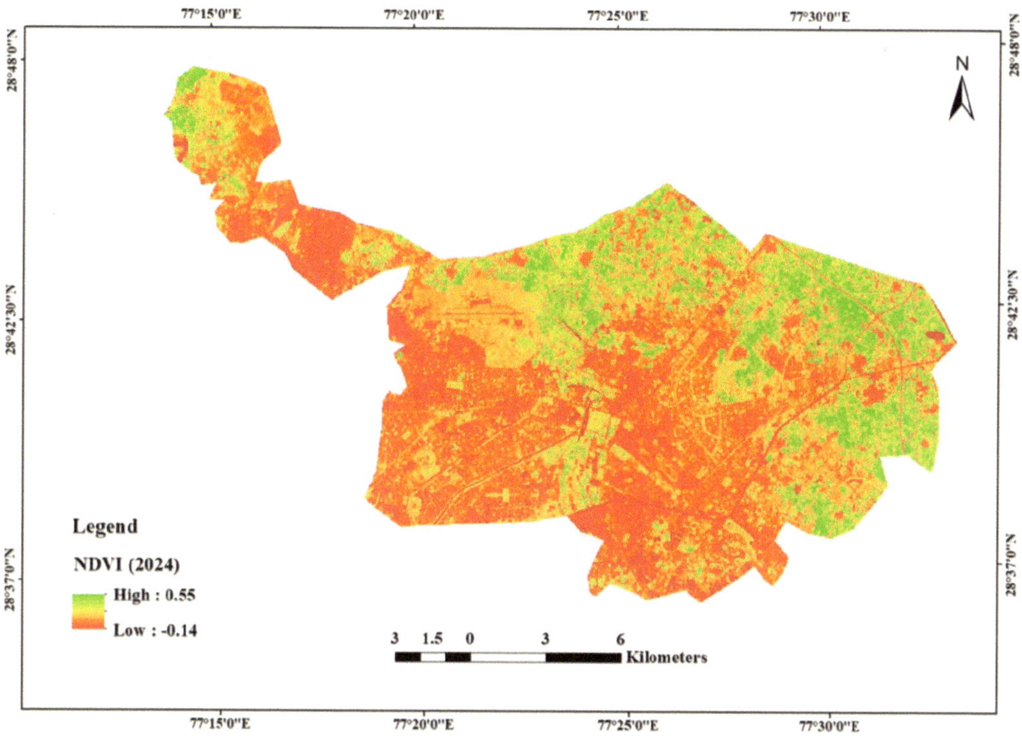

FIGURE 11.10 NDVI map of Ghaziabad city (2024).

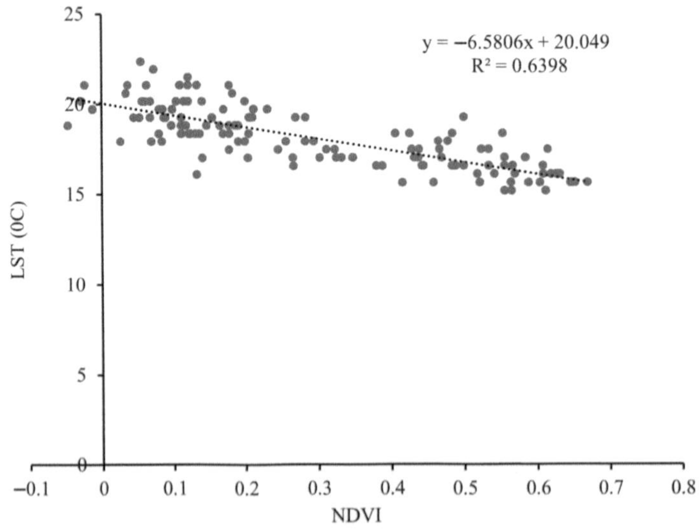

FIGURE 11.11 Association of LST and NDVI (1994).

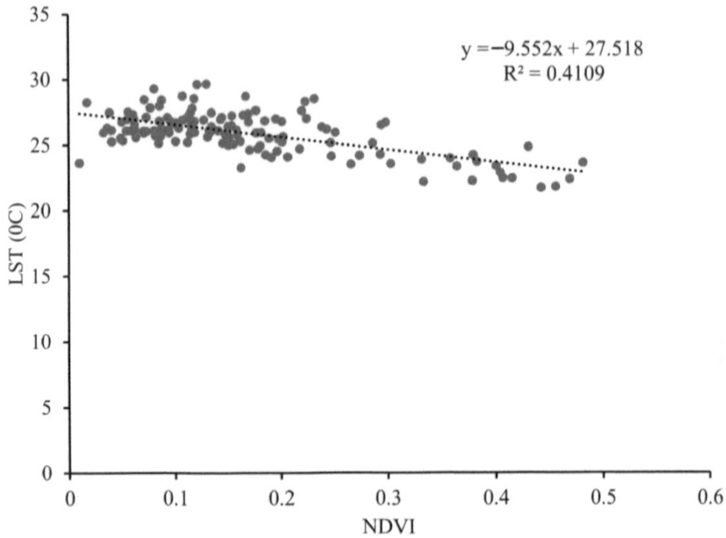

FIGURE 11.12 Association of LST and NDVI (2024).

highways, roads, household communities, and business hubs. Since it has been one of the primary causes of the rise in LST, the study of the spatiotemporal NDBI patterns is very crucial in this context. In 1994, the NDBI models showed recorded values between −0.28 and +0.56, and in 2024, they showed values between −0.28 and +0.49 (Figures 11.13 and 11.14).

While lower NDBI values show the presence of various land use categories, higher values indicate dense and sparse communities. In 1994 and 2024, respectively, LST and NDBI showed a positive correlation with R^2 values of 0.6398 and 0.5224 (Figures 11.15 and 11.16). This indicates that the LST of the affected area rises as built-up increases. It is interesting, therefore, that NDBI has a positive relationship with LST and contributes to the formation of a heat island in the city.

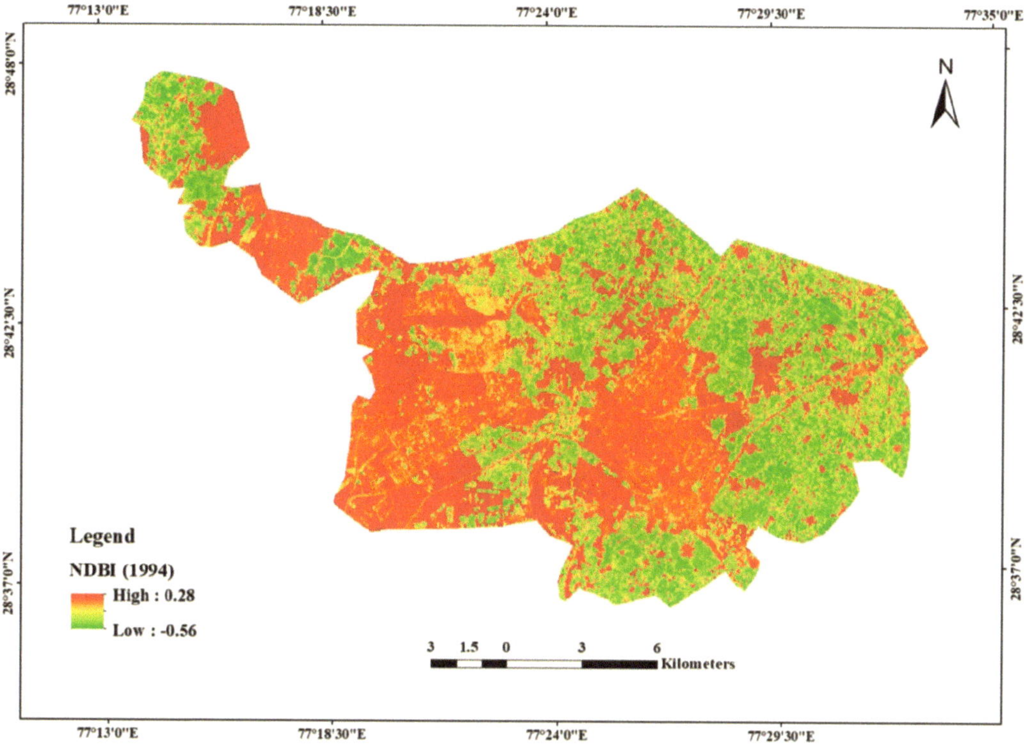

FIGURE 11.13 NDBI map of Ghaziabad city (1994).

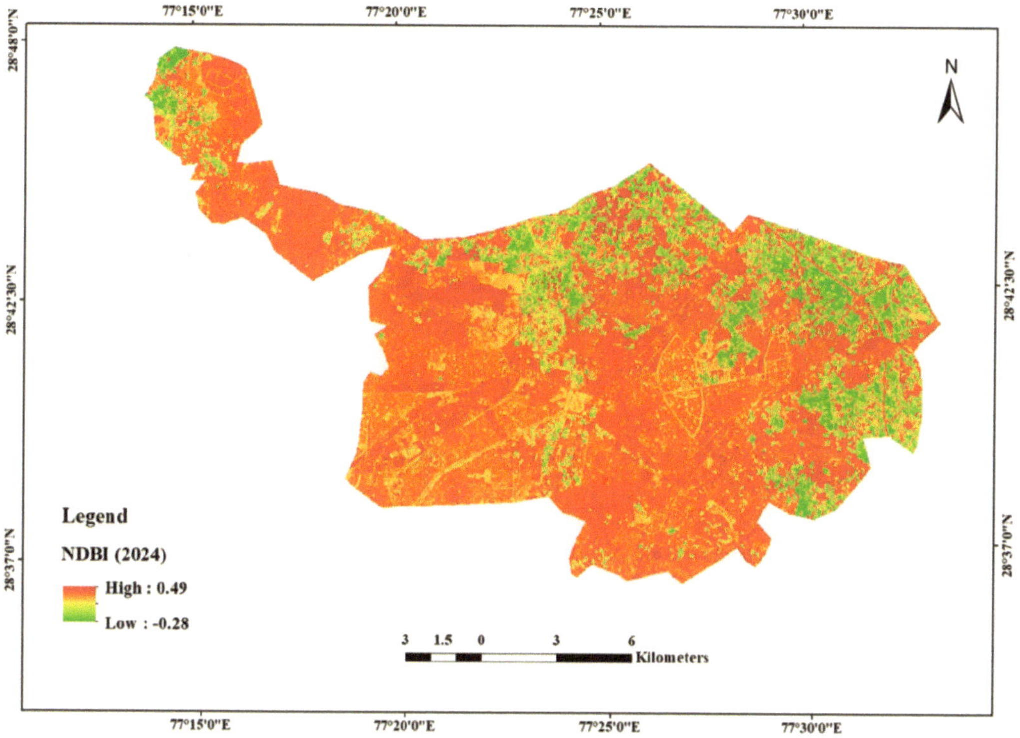

FIGURE 11.14 NDBI map of Ghaziabad city (2024).

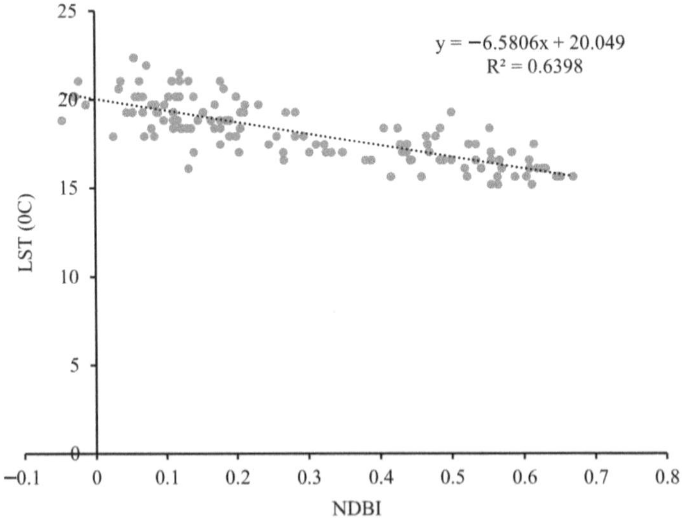

FIGURE 11.15 Association of LST and NDBI (1994).

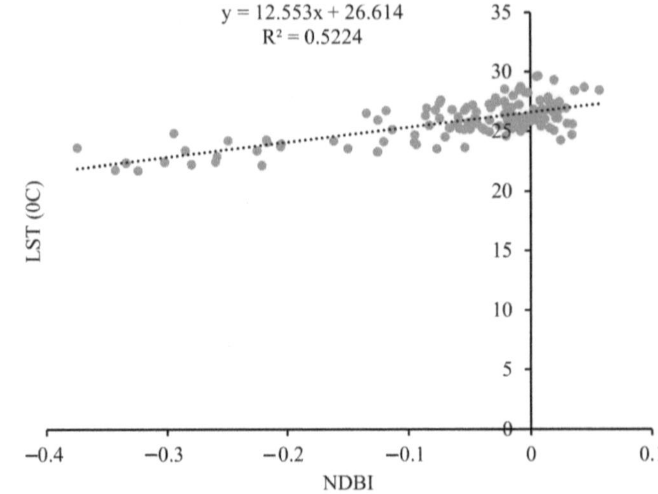

FIGURE 11.16 Association of LST and NDBI (2024).

CONCLUSION

When examining how changes in land use/land cover influenced the microclimate dynamics in Ghaziabad city, Uttar Pradesh, the advancements in remote sensing and geographic information systems (GIS) generated compelling findings. In order to ascertain and evaluate the consequences of land use/land cover shift on the microclimate patterns in Ghaziabad city, data from Landsat for the years 1994 (TM) and 2024 (OLI-2) were utilized to generate significant details on LULC and to understand the associations of LST with NDBI and NDVI. The study found that built-up areas were growing quickly and that the most significant deviations that might be the main sources of the impact of UHI in Ghaziabad city were the degradation of agricultural land and vegetation cover. This demonstrates unequivocally how significant changes in the land cover and LST have been brought about by the growth of the built-up land. In this respect, the distribution of LST in Ghaziabad was highly correlated with the distribution of indices of urbanization like (NDVI) vegetation cover and (NDBI) built-up areas, with $R^2=0.5398$ and 0.41108 for 1994 and 2024, respectively. The study also found that the increasing built environment, declining vegetation cover, and scant distribution of urban green spaces, which make LST and UHI issues more sensitive, are all getting worse these days. In order to ensure the sustainable growth of Ghaziabad city, these issues must be resolved.

REFERENCES

Aboelata, A., Sodoudi, S. (2020). Evaluating the effect of trees on UHI mitigation and reduction of energy usage in different built-up areas in Cairo. *Building and Environment*, 168, 106490.

Aneesh, M., Khandelwal, S., Kaul, N. (2016). Spatial and temporal variations of urban heat island effect and the effect of percentage impervious surface area and elevation on land surface temperature: Study of Chandigarh City India. *Sustainable Cities and Society*, 26, 264–277.

Chen, X. L., Zhao, H. M., Li, P. X., Yin, Z. Y. (2006). Remote sensing image-based analysis of the relationship between urban heat island and land use/cover changes. *Remote Sensing of Environment*, 104(2),133–146. https://doi.org/10.1016/j.rse.2005.11.016.

City Disaster Management Report, Ghaziabad (2012). The disaster Management plan of Ghaziabad city 2012–13. www.ghaziabad.nic.in

Degefu, M. A., Argaw, M., Feyisa, G. L., & Degefa, S. (2023). Dynamics of green spaces- land surface temperature intensity nexus in cities of Ethiopia. *Heliyon*, 9(2), e13274.

District Census Handbook: Ghaziabad (PDF) (2011). Censusindia.gov.in. Registrar General and Census Commissioner of India.

Duda, R. O., Hart, P. E., & Stork, D. G. (2000). *Pattern Classification*. 2nd ed., Wiley-Interscience, New York, 153.

Fadhil, M., Hamoodi, M. N., Ziboon, A. R. T. (2023). Mitigating urban heat island effects in urban environments: Strategies and tools. In *IOP Conference Series: Earth Environmental Science*, 1129:12025, IOP Publishing.

IPCC (2014). *Climate Change 2014: Impacts, Adaptation and Vulnerability. Part B: Regional Aspects*. Cambridge University Press, Cambridge, and New York.

Kikon, N. et al. (2016). Assessment of urban heat islands (UHI) of Noida City, India using multi-temporal satellite data. *Sustainable Cities and Society*, 22, 19–28. https://doi.org/10.1016/j.scs.2016.01.005.

Kumar, T. L., Bharath, J., Filho, W. L., Barbosa, H., Rao, K. K. (2025). Impact of climate change induced heat stress on the people working in the coastal cities of India. *Natural Hazards*, 121(2), 1803–1817.

Kumar, A., Siddiqui, A. R. (2022). Urban growth dynamics of ghaziabad district using remote sensing. *International Journal of Economic Perspectives*,16(10),96–103.

Moharir, K. N. et al. (2025). Estimation of land surface temperature and LULC changes impact on groundwater resources in the semi-arid region of Madhya Pradesh, India. *Advances in Space Research*, 75(1), 233–247. https://doi.org/10.1016/j.asr.2024.09.025.

Njoku, E. A., Tenenbaum, D. E. (2022). Quantitative assessment of the relationship between land use/land cover (LULC), topographic elevation and land surface temperature (LST) in Ilorin, Nigeria. *Remote Sensing Applications: Society and Environment*, 27, 100780. https://doi.org/10.1016/j.rsase.2022.100780.

Owen, T. W., Carlson, T. N., Gillies, R. R. (1998). An assessment of satellite remotely-sensed land cover parameters in quantitatively describing the climatic effect of urbanization. *International Journal of Remote Sensing*, 19(9),1663–1681.

Patel, S., Indraganti, M., Jawarneh, R. N. (2024). A comprehensive systematic review: Impact of Land Use/Land Cover (LULC) on Land Surface Temperatures (LST) and outdoor thermal comfort. *Building and Environment*, 249, 111130. https://doi.org/10.1016/j.buildenv.2023.111130.

Rotach, M. W. et al. (2005). BUBBLE — An urban boundary layer meteorology project. *Theoritical and Applied Climatology*, 81(3), 231–261. https://doi.org/10.1007/s00704-004-0117-9.

Shahfahad, Kumari, B., Tayyab, M., Ahmed, I. A., Baig, M. R. I., Khan, M. F., Rahman, A. (2020). Longitudinal study of land surface temperature (LST) using mono- and split-window algorithms and its relationship with NDVI and NDBI over selected metro cities of India. *Arabian Journal of Geosciences*, 13, 1040.

Sherafati, S., Saradjian, M. R., & Rabbani, A. (2018). Assessment of surface urban heat island in three cities surrounded by different types of land-cover using satellite images. *Journal of the Indian Society of Remote Sensing*, 46(7), 1013–1022.

Singh, P., Kikon, N., Verma, P. (2017). Impact of land use change and urbanization on urban heat island in Lucknow city, Central India. A remote sensing-based estimate. *Sustainable Cities and Society*, 32,100–114. https://doi.org/10.1016/j.scs.2017.02.018.

Sultana, S., Satyanarayana, A. V. N. (2020). Assessment of urbanisation and urban heat island intensities using landsat imageries during 2000 —2018 over a sub-tropical Indian city. *Sustainable Cities and Society*, 52, 101846.

Teferi E., Abraha H. (2017). Urban heat island effect of Addis Ababa city: Implications of urban green spaces for climate change adaptation. In *Climate Change Adaptation in Africa: Fostering Resilience and Capacity to Adapt* (Filho, W. L., Belay, S., Kalangu, J., Menas, W., Munishi, P. & Musiyiwa, K., eds), Springer, Cham, pp. 539–552.

Voogt, J. A., Oke, T.R. (2003). Thermal remote sensing of urban climates. *Remote Sensing of Environment*, 86(3), 370–384. https://doi.org/10.1016/S0034-4257(03)00079-8.

Weng, Q., Lu, D., Schubring, J. (2004). Estimation of land surface temperature–vegetation abundance relationship for urban heat island studies. *Remote Sensing of Environment*, 89(4):467–83.

Zha, Y., Gao, J., Ni, S. (2003). Use of normalized difference built-up index in automatically mapping urban areas from TM imagery. International Journal of Remote Sensing, 24(3), 583–594.

Zhou, D., Zhang, L., Hao, L., Sun, G., Liu, Y., Zhu, C. (2016). Spatiotemporal trends of urban heat island effect along the urban development intensity gradient in China. *Science of the Total Environment*, 544, 617–626.

Zhou, D., Zhao, S., Liu, S., Zhang, L., Zhu, C. (2014). Surface urban heat island in China's 32 major cities: Spatial patterns and drivers. *Remote Sensing of Environment*, 152, 51–61. https://doi.org/10.1016/j.rse.2014.05.017.

12 Navigating the Climate Crisis
Public Health Challenges and the Role of Community Empowerment

Mariya Hasnat, Mohd Kashif Khan, and Sania Kulsum

INTRODUCTION

Climate change is one of the most significant global challenges of the 21st century, with far-reaching consequences for ecosystems, economies, and societies. However, its effects on community well-being are among the most pressing concerns. As global temperatures rise, extreme weather events such as hurricanes, droughts, and floods have become more frequent and intense, resulting in widespread disruptions to human health and well-being. In addition to these extreme events, long-term shifts in climate patterns, such as rising sea levels, changing precipitation patterns, and the warming of oceans, are contributing to an evolving landscape of health risks. The primary driver of global climate shifts is the increase in greenhouse gases (GHGs), particularly carbon dioxide (CO_2), methane (CH_2), and nitrous oxide (N_2O), which trap heat in the Earth's atmosphere (Watts et al., 2018). These gases primarily result from human activities, including the burning of fossil fuels for energy, deforestation, industrial processes, and agricultural practices. As the concentration of these gases in the atmosphere rises, the Earth's average temperature continues to climb, leading to a cascade of environmental and health effects. Alongside GHG emissions, air pollution is another major environmental factor linked to global climate shifts (Costello et al., 2009). The burning of fossil fuels not only releases carbon emissions but also contributes to the formation of particulate matter (PM) and ozone, which are harmful to human health. Long-term exposure to poor air quality can lead to respiratory and cardiovascular diseases, exacerbate asthma, and even increase mortality rates. Heatwaves, which are becoming more frequent due to global climate shifts, also exacerbate the effects of air pollution, leading to additional health complications (Smith & Bradley, 2016).

The intensification of extreme weather events due to global climate shifts has profound implications for community well-being. More frequent and intense heatwaves increase the risk of heat-related illnesses, such as heat exhaustion, heat stroke, and dehydration. Additionally, severe storms and flooding can directly injure or kill people and destroy infrastructure, including health facilities, disrupting medical care. Flooding also increases the risk of waterborne diseases, as contaminants spread through water systems (Lloyd et al., 2011). Climate change affects the habitats of disease vectors such as mosquitoes, ticks, and other insects, leading to shifts in the geographic distribution of diseases like malaria, dengue, Lyme disease, and Zika virus (Haines & Ebi, 2019). Warmer temperatures and altered rainfall patterns create favorable conditions for these vectors to thrive in areas previously not affected by such diseases, putting new populations at risk. Changes in climate patterns disrupt agricultural productivity, leading to food insecurity and malnutrition, particularly in developing countries. Droughts, floods, and shifts in growing seasons reduce crop yields and threaten food supplies. Additionally, water sources are increasingly affected by global climate shifts, with rising temperatures and changing rainfall patterns influencing water availability. Contaminated water supplies, exacerbated by extreme weather events, increase the risk of waterborne diseases. The emotional and psychological toll of global climate shifts is often overlooked (Luber & McGeehin, 2008). The stress and anxiety caused by extreme weather events, displacement due to natural disasters, and the gradual realization of environmental degradation can have

DOI: 10.1201/9781003620747-11

significant mental health effects. People affected by these events may experience post-traumatic stress disorder (PTSD), depression, and anxiety, especially among at-risk groups such as children and the elderly. As global climate shifts intensifies, more people are forced to migrate due to rising sea levels, desertification, and extreme weather. This movement of people—often from rural to urban areas—puts strain on already at-risk populations and community well-being systems. Migrants are at higher risk of exposure to infectious diseases, poor sanitation, and inadequate health care services, leading to health inequities (Rao & Berman, 2019).

ADDRESSING CLIMATE CHANGE AND PUBLIC HEALTH

While global climate shifts present significant health challenges, they also present an opportunity for innovation in community well-being policy, environmental justice, and community adaptive strength. To effectively address these challenges, it is essential to integrate climate adjustment and reduction approaches into community well-being planning. In particular, community capacity building plays a crucial role in developing solutions that are not only sustainable but also equitable. Communities that are actively engaged in decision-making processes, that advocate for healthier environments, and that support adjustment approaches, such as disaster preparedness, can help reduce the impact of global climate shifts on health (Patel & Kleinman, 2018). Community-based approaches to climate adaptive strength focus on local knowledge, addressing social and economic disparities, and building capacity at the grassroots level to mitigate the health effects of global climate shifts (Horton & Lo, 2015). By understanding the multifaceted causes and consequences of global climate shifts, we can take action to reduce health risks and ensure that the most at-risk populations are protected. Empowering communities to take control of their health and environment is key to mitigating the adverse effects of global climate shifts, ensuring that we move toward a healthier, more sustainable future for all (Srinivasan, 2017).

CLIMATE CHANGE AND COMMUNITY EMPOWERMENT

As global climate shifts accelerate, their effects are disproportionately felt by marginalized and at-risk communities around the world. From extreme weather events to changes in food and water security (Berkes et al., 2003), these populations are often the hardest hit, despite contributing the least to the problem. In light of this, community capacity building becomes a vital tool in mitigating the health and social effects of global climate shifts. Rather than relying solely on top-down, government-driven solutions, empowering local communities to take control of their own climate adaptive strength is both an effective and equitable strategy (Moseley & Kanter, 2020).

COMMUNITY EMPOWERMENT

Community capacity building refers to the process of enabling individuals and groups to gain control over their lives, environments, and decisions, particularly those that affect their well-being. This capacity building is crucial in global climate shift adjustment, as communities that are aware of the risks and have the resources to address them are better positioned to reduce vulnerability and enhance adaptive strength. Community capacity building involves several key components: For capacity building to be effective, it must be comprehensive. This means involving all members of a community, particularly those who are marginalized, such as women, children, the elderly, and low-income households in decision-making processes (World Health Organization (WHO), 2015). By ensuring that diverse voices are heard, solutions can be tailored to meet the specific needs of all community members. One of the first steps in community capacity building is education. Communities must be informed about the science behind global climate shifts, its potential effects, and the measures they can take to mitigate risks. This knowledge helps build adaptive strength and encourages proactive responses to climate challenges. Local capacity-building programs can equip individuals with practical skills such as disaster preparedness, sustainable farming techniques, and water conservation practices that strengthen the community's overall ability to adapt to environmental changes (Cannon & Müller-Mahn, 2010). Empowerment also involves ensuring that communities have access to the resources they need to adapt. This includes access to financial resources, technology, infrastructure, and support systems. For instance, providing communities with the tools to monitor local weather patterns, access clean water, or improve agricultural practices can drastically improve their ability to cope with climate disruptions (Adger et al., 2005). Empowered communities are often at the forefront of innovation, but collaboration with governmental organizations, NGOs, and the private sector is essential. These partnerships can help leverage external resources, expertise, and funding to strengthen local efforts. Effective collaboration can lead to the development of climate adaptive strength plans that incorporate both local knowledge and technical expertise, creating more sustainable solutions (Patz & Balbus, 2014).

THE ROLE OF COMMUNITY EMPOWERMENT IN CLIMATE CHANGE ADAPTATION

Climate change adjustment is most effective when it is locally driven. Communities often have valuable local knowledge and expertise about their environment that can be harnessed to develop appropriate responses. Whether it's adopting new agricultural techniques, creating early warning systems for extreme weather, or protecting natural resources like forests and wetlands, empowered communities are better able to develop and implement climate adjustment approaches that are suited to their unique circumstances.

Empowerment also fosters social networks within communities. These networks can be instrumental in responding to climate-related challenges, such as coordinating evacuation plans during a flood or organizing mutual aid groups during a heatwave. Strong community bonds enable people to rely on each other in times of crisis, providing the social capital necessary for collective action (Bina, 1992). Empowered communities are more likely to engage in climate advocacy, calling for policies and actions that address the root causes of global climate shifts and their disproportionate effects on at-risk groups. Through community-driven advocacy, local populations can push for more equitable climate policies, including better access to health care, social protection, and disaster risk reduction programs. Advocacy can also help raise awareness at local, national, and international levels, driving policy change that supports at-risk communities. Empowerment helps communities transition to sustainable, climate-resilient economies. By fostering green jobs, promoting eco-friendly agricultural practices, and supporting renewable energy initiatives, empowered communities can create economic opportunities that are both environmentally sustainable and financially beneficial. These initiatives not only help mitigate the effects of global climate shifts but also promote local economic development, reducing poverty and increasing self-sufficiency (Shah & Desai, 2014).

CASE STUDIES IN COMMUNITY EMPOWERMENT (GLOBAL)

1. **Community-Led Disaster Preparedness in the Philippines:** In the Philippines, a country highly at risk to typhoons, local communities have taken charge of disaster preparedness programs. The "Community-Based Disaster Risk Management" approach has empowered local residents to identify hazards, develop early warning systems, and implement evacuation plans (Figure 12.1). These community-driven efforts have significantly reduced casualties and losses during storms and have improved overall community adaptive strength.
2. **The Role of Indigenous Knowledge in Brazil's Amazon:** Indigenous communities in Brazil have long used sustainable farming and land-management practices that are deeply connected to the natural environment. As global climate shifts threaten their land and livelihood, many of these communities are working to protect their territories and share their knowledge with broader society (Figure 12.2). By embracing indigenous knowledge,

FIGURE 12.1 Community-led disaster preparedness in the Philippines. (Source: resilience library.)

FIGURE 12.2 Role of indigenous knowledge in Brazil's Amazon. (Source: EOS.)

FIGURE 12.3 Green energy project in rural Africa. (Source: World Bank.)

they are helping to shape global climate shift reduction approaches, such as deforestation reduction and biodiversity conservation, that benefit both the environment and local communities (UNDP, 2020).

3. **Green Energy Projects in Rural Africa**: In many parts of Africa, local communities have adopted renewable energy solutions, such as solar power, to address energy scarcity (Figure 12.3). These community-led energy projects reduce reliance on fossil fuels and help mitigate the effects of global climate shifts while providing economic opportunities. In some cases, local women have taken the lead in installing and managing solar systems, empowering them economically and socially (World Bank, 2020; World Bank, 2022).

CLIMATE CHANGE AND WOMEN EMPOWERMENT IN INDIA

Climate change is a significant challenge facing India, and its effects are disproportionately felt by women, particularly in rural and marginalized communities. These effects, such as erratic weather patterns, floods, droughts, and rising temperatures, directly affect women's livelihoods, health, and well-being (India Climate Dialogue, 2020). Empowering women to actively participate in global climate shift reduction and adjustment efforts is essential for building a more resilient and sustainable future for both women and their communities. This intersection between global climate shifts and women's capacity building offers an opportunity to foster gender equality, strengthen adaptive strength, and drive sustainable development in India (Reddy, 2016).

CASE STUDIES IN INDIA

India, a country highly at risk to the effects of global climate shifts, has witnessed numerous local and community-driven initiatives aimed at enhancing climate adaptive strength. These initiatives demonstrate the power of community capacity building in addressing the effects of global climate shifts and their related health risks (Swaminathan, 2019). Below are some notable case studies that showcase how community engagement has played a critical role in mitigating climate effects and building adaptive strength.

The Coastal Communities of Gujarat: Mangrove Restoration for Disaster Resilience

Location: Gujarat, Western India

Issue: Coastal Erosion, Cyclones, and Sea Level Rise

Community Involved: Local Fisherfolk and Coastal Villages

Gujarat's coastal communities have long faced the dual threats of coastal erosion and frequent cyclones, particularly in the aftermath of Cyclone Okhi in 2017. The mangrove ecosystems, which once acted as natural buffers against these storms, were degraded due to urbanization, industrialization, and deforestation. In response, local communities, with support from non-governmental organizations (NGOs), have initiated large-scale mangrove restoration projects. Local fisherfolk, women, and youth have been trained to restore and protect these vital ecosystems, which act as a natural shield, reduce salinity intrusion, and prevent coastal erosion (Figure 12.4). The capacity building of local communities in this initiative has not only enhanced their adaptive strength to extreme weather but also provided economic opportunities through eco-tourism, sustainable fishing practices, and carbon credit markets. By drawing on local knowledge and traditional environmental stewardship practices, these communities have transformed mangrove restoration into a sustainable livelihood model while simultaneously enhancing disaster preparedness and reducing the economic impact of cyclones (Sivaramakrishnan & Sreeram, 2015).

FIGURE 12.4 Women engage in mangrove restoration. (Source: Grow Billion Trees.)

The Kuthambakkam Village: Water Management and Climate Resilience

Location: Kuthambakkam, Tamil Nadu

Issue: Water Scarcity and Drought

Community Involved: Rural Farmers and Local Leaders

Kuthambakkam, a village in Tamil Nadu, faced severe water scarcity and agricultural challenges due to prolonged droughts, which worsened in the wake of global climate shifts. The village, once dependent on unreliable water sources and conventional farming methods, struggled to sustain its community, particularly during dry seasons. In response, the village undertook a community-led water management program that involved the entire community in water conservation, rainwater harvesting, and sustainable agricultural practices. Local farmers were trained in water-efficient irrigation methods, and the village built its own network of small check dams and ponds to capture and store rainwater. The initiative also involved the construction of recharge wells and the promotion of organic farming techniques to reduce water consumption. Through community capacity building, Kuthambakkam transformed into a model of sustainable water management (Figure 12.5). The village not only ensured a more reliable water supply but also improved agricultural productivity and food security. Furthermore, the community's collective efforts led to a reduction in waterborne diseases, which were prevalent during droughts, thus improving community well-being outcomes (Ghosh & Mukherjee, 2018).

The Himalayas: Community-Led Disaster Risk Reduction in Uttarakhand

Location: Uttarakhand, Northern India

Issue: Landslides, Flash Floods, and Extreme Weather Events

Community Involved: Mountain Communities and Local Authorities

Uttarakhand, a mountainous state in Northern India, has long faced challenges related to landslides, flash floods, and extreme weather events exacerbated by global climate shifts. In 2013,

FIGURE 12.5 Water conservation. (Source: Vikalp Sangam.)

devastating floods and landslides displaced thousands of people and caused significant loss of life and property. Recognizing the need for disaster risk reduction, the state government and various NGOs initiated a community-based approach to disaster preparedness and risk reduction (Figure 12.6). Local communities were trained to identify hazards, create early warning systems, and implement simple yet effective risk reduction approaches such as better drainage systems, slope stabilization,

FIGURE 12.6 Local authorities and mountain communities involved in disaster risk reduction. (Source: SANDRP.)

and the construction of protective structures. Moreover, the communities were involved in emergency response planning, including evacuation routes and safe shelters. This community-driven approach to disaster risk reduction has not only increased the adaptive strength of mountain communities but also enhanced their capacity to cope with future disasters. The active participation of local communities has led to the adoption of localized solutions that are culturally relevant and practical, ensuring better outcomes during extreme weather events (Ministry of Environment, Forest and Climate Change, Government of India, 2018).

The Self-Help Groups in Odisha: Empowering Women for Climate Resilience

Location: Odisha, Eastern India

Issue: Cyclones, Flooding, and Livelihood Protection

Community Involved: Women's Self-Help Groups (SHGs)

Odisha, a coastal state in India, is highly at risk to cyclones and flooding. Women, especially in rural areas, are often disproportionately affected by these disasters due to limited access to resources, low decision-making power, and economic vulnerability. In Odisha, women's SHGs have played a crucial role in building community adaptive strength to global climate shifts (Figure 12.7a). These groups, which are often formed at the grassroots level, focus on empowering women through financial inclusion, skill development, and disaster preparedness training. They have been instrumental in promoting community-based disaster response plans, establishing early warning systems, and providing mutual aid in the aftermath of cyclones and floods. One key initiative was the establishment of disaster-resilient livelihoods for women in rural areas. SHGs have encouraged women to engage in sustainable livelihood activities such as farming drought-resistant crops, setting up small businesses, and participating in climate-smart agriculture projects (Figure 12.7b). Additionally, these

FIGURE 12.7 (a) Self-help group in Odisha; (b) women involved in livelihood protection (*World Bank*).

groups have been involved in advocacy efforts for better disaster relief policies and climate adjustment approaches. By providing women with the knowledge and resources to become climate adaptive strength advocates, SHGs have not only enhanced the social and economic status of women but also strengthened the overall adaptive strength of their communities.

The Jhabua District: Forest Rights and Climate Change Adaptation

Location: Jhabua District, Madhya Pradesh

Issue: Forest Degradation, Livelihoods, and Climate Vulnerability

Community Involved: Tribal Communities

In Jhabua, a tribal district in Madhya Pradesh, forest-dependent communities have faced the dual challenges of forest degradation and changing weather patterns that threaten their traditional livelihoods. With the rise of climate-related issues such as erratic rainfall and increased temperature, tribal farmers have found it harder to rely on forest resources for sustenance. In response, local tribal communities have taken the lead in protecting and restoring their forests. Through the Forest Rights Act, these communities have asserted their legal rights to manage forests, promoting sustainable forest management practices and engaging in tree-planting initiatives to combat soil erosion, improve water retention, and create carbon sinks (Figure 12.8). Empowered by these rights, tribal communities have also embraced alternative livelihoods, including non-timber forest product (NTFP) collection, eco-tourism, and community-based forest management. These efforts have helped strengthen their adaptive strength to the effects of global climate shifts, improve food security, and protect their cultural heritage.

IMPACT OF CLIMATE CHANGE ON WOMEN IN INDIA

India is particularly at risk to global climate shifts due to its geographical location, heavy dependence on agriculture, and high poverty levels (Menon & Sharma, 2017). For women, especially in rural areas, the consequences of global climate shifts are multifaceted:

1. **Increased Workload:** Women in rural India are primarily responsible for household chores, water collection, food production, and caring for family members. Climate change disrupts these activities by exacerbating water scarcity, reducing agricultural productivity, and increasing the frequency and severity of natural disasters. As a result, women face an increased burden of work, often with little support or resources to adapt to these challenges (Banerjee, 2017).

FIGURE 12.8 Locals of Jhabua involved in protection of forest and livelihood. (Source: Citizens for Justice and Peace.)

2. **Vulnerability in Agriculture:** Women in India are heavily involved in agriculture, often in informal roles such as farming, weeding, and harvesting, even though they may not have formal land ownership rights. Climate change-induced droughts, floods, and changing rainfall patterns threaten agricultural productivity and food security. This results in greater poverty and food insecurity, which disproportionately affects women, particularly those from poor, rural communities (IPCC, 2022).
3. **Health Impacts:** Climate change exacerbates community well-being issues, especially for women and children. Extreme heat waves and erratic weather patterns increase the risks of heat-related illnesses, respiratory diseases, and waterborne diseases. Women are also more susceptible to malnutrition, as they often prioritize their family's nutritional needs over their own. These health challenges compound the socio-economic vulnerabilities that women already face.
4. **Displacement and Migration:** Women are particularly at risk of displacement caused by climate-induced disasters such as floods and cyclones. In many cases, displaced women face additional challenges, such as lack of access to safe sanitation, sexual violence, and loss of livelihoods. Migrant women often face exploitation, limited access to healthcare, and social marginalization (Sengupta & Bhatnagar, 2019). These vulnerabilities are compounded by the fact that women's voices are often underrepresented in migration and displacement decision-making processes.

WOMEN EMPOWERMENT AS A TOOL FOR CLIMATE CHANGE MITIGATION AND ADAPTATION

Empowering women to be leaders in climate action is not only a matter of gender equality but also an effective strategy for enhancing adaptive strength and building sustainable communities. Here are some key aspects of how women's capacity building is essential in the context of global climate shifts in India:

1. **Women as Key Agents of Change in Agriculture:** Women in rural India are crucial to the agricultural sector. By empowering women with the skills and knowledge needed to adopt sustainable farming practices, climate-resilient crops, and eco-friendly agricultural techniques, they can significantly contribute to mitigating the effects of global climate shifts. Women can also be instrumental in promoting organic farming, water conservation, and soil health practices that reduce the environmental impact of agriculture. For example, the Bhoomi Sena program in Maharashtra trains women farmers in sustainable agricultural techniques that help conserve water, prevent soil erosion, and improve productivity (Shukla, 2020). When women farmers are empowered with knowledge about climate-resilient agriculture, they can enhance food security and increase community adaptive strength to climate-related shocks.
2. **Access to Financial Resources and Climate Resilience:** One of the key elements of women's capacity building in the face of global climate shifts is economic independence. Programs like SHGs, microfinance schemes, and cooperative models provide women with financial resources, access to loans, and support for small-scale businesses. These programs enable women to invest in climate-resilient technologies, sustainable livelihoods, and disaster preparedness. In many parts of India, rural women are using microfinance to create climate-resilient enterprises, such as solar-powered cooking stoves, biogas plants, and sustainable agro-processing ventures. Financial capacity building enables women to become more self-reliant and reduces their vulnerability to the economic effects of global climate shifts.
3. **Participation in Decision-Making and Climate Policies:** Women's active participation in climate decision-making processes is critical to creating comprehensive policies that address their specific needs. Gender-responsive climate policies can ensure that women's perspectives

and priorities are integrated into national and local climate approaches. Women's involvement in policy discussions can lead to more equitable distribution of resources, better disaster preparedness approaches, and effective implementation of climate adjustment projects. At the grassroots level, the involvement of women in community-based organizations and local governance (e.g., panchayats) ensures that women have a voice in decision-making related to natural resource management, disaster risk reduction, and global climate shifts adjustment (Bharati & Sahoo, 2018).

4. **Climate Education and Awareness:** Empowering women through education and awareness-raising activities is essential for building a more climate-conscious society. Programs that teach women about climate science, the effects of global climate shifts, and sustainable practices help them become better advocates for climate adaptive strength. Women's education on global climate shifts also enhances their ability to adopt climate-smart solutions and share this knowledge with their communities. Various NGOs and local initiatives are offering training programs on global climate shift adjustment for women, such as capacity-building workshops on water conservation, energy-efficient technologies, and disaster preparedness. These programs enable women to not only protect their families but also lead collective actions in their communities to mitigate climate risks.
5. **Leadership and Advocacy:** Women leaders, particularly from rural areas, are already driving change in many parts of India. Whether it's advocating for better water management, promoting renewable energy, or leading climate action at the local level, women are at the forefront of adaptive strength-building efforts. The leadership of women in climate adjustment is essential for creating long-term, sustainable solutions. By recognizing and amplifying women's leadership in climate action, communities can develop more innovative and comprehensive approaches to climate adaptive strength. For instance, in the Kutch region of Gujarat, women have been actively involved in water conservation initiatives, including the creation of check dams and rainwater harvesting systems. Their leadership in these initiatives has led to improved water access for entire communities, demonstrating how women's participation can bring about effective climate solutions.

GOVERNMENT AND NGO INITIATIVES SUPPORTING WOMEN AND CLIMATE CHANGE

Several government programs and non-governmental initiatives in India focus on empowering women in the context of global climate shifts: With an emphasis on empowering women via sustainable agriculture, water management, and disaster preparedness, many government and NGO projects have been put into place in India to assist women in tackling climate change. For example, the Self-Employed Women's Association (SEWA) has trained women in drought-resistant farming and water conservation methods as part of its climate change adaptation initiatives in Gujarat. According to a survey, 70% of women who participated in these programs reported having better access to food and water, and over three years, agricultural yields increased by 20%. In a similar vein, 60% of women now take part in local climate resilience initiatives thanks to the Indian government's National Action Plan on Climate Change (NAPCC) and state-level climate action plans like Uttarakhand's.

By empowering rural women through SHGs, the JEEViKA initiative in Bihar has increased agricultural productivity by 50% and assisted women in mitigating the consequences of drought by adopting climate-resilient practices. Furthermore, gender-sensitive disaster management practices have been incorporated by the National Disaster Management Authority (NDMA), particularly in areas like Kerala and Odisha, where 50% of women participated in disaster response teams, resulting in faster recovery periods and fewer casualties. Women farmers have also benefited from the Pradhan Mantri Fasal Bima Yojana (PMFBY), a crop insurance program, especially in Madhya Pradesh and Uttar Pradesh, where 75% of women were compensated for crop losses brought on by climate events. Finally, by empowering women to manage water resources, WaterAid India's Women and Water Program in Rajasthan has improved community resilience and cut down on water collection time by 40%. There are also obstacles in scaling these efforts and making sure that women, particularly in isolated rural areas, fully benefit from such programs, even if these projects have demonstrated favorable effects, such as greater food security, water access, and disaster preparedness. Improving the long-term efficacy of these efforts will need sustained investment in gender-responsive climate policy and the removal of structural obstacles.

- **Beti Bachao Beti Padhao:** A program to promote gender equality and improve the status of girls in India, which indirectly addresses the capacity building of women in climate adaptive strength through education and community engagement.
- **Mahatma Gandhi National Rural Employment Guarantee Act (MGNREGA):** This government initiative provides women with employment opportunities in rural areas, helping them build adaptive strength to climate effects while ensuring a steady income.
- **National Action Plan on Climate Change:** The NAPCC includes various climate action plans that promote women's involvement in global climate shift reduction and adjustment approaches, particularly in rural areas.

Many NGOs, such as SEWA and the Barefoot College, have empowered women through skill-building programs in renewable energy, eco-friendly agriculture, and disaster management. Empowering women is not only a matter of justice but also a strategic approach to building climate adaptive strength. By integrating women into the global climate shifts dialogue and ensuring that they have the resources, knowledge, and platforms to lead, India can better mitigate the effects of global climate shifts. A gender-sensitive approach to climate action ensures that women are not only recipients of aid but also active contributors to solutions that enhance community adaptive strength, promote sustainable development, and build a more equitable society. The intersection of women's capacity building and global climate shifts offers a powerful opportunity to create lasting, sustainable change in India (Sathish, 2015).

CONCLUSION

The well-being of communities is urgently threatened by climate change, which disproportionately affects vulnerable groups. Existing social and economic disparities are made worse by rising temperatures, harsh weather, and changing disease patterns. Building community capacity, however, provides an essential route to adaptability. By involving local communities in decision-making, encouraging education, and guaranteeing resource accessibility, communities can create flexible plans to reduce health hazards. Case studies from India and the Philippines show how grassroots activities, such as disaster preparedness and women-led climate adaptation projects, may successfully tackle climate issues. The significance of community-driven solutions that combine traditional knowledge, sustainable development strategies, and technical innovations is highlighted by these examples. Building long-term adaptive strength also requires empowering women, especially in the fields of agriculture, water management, and disaster relief. More fair and successful climate adaptation strategies can be fueled by gender-sensitive climate policy and a rise in the number of women in leadership positions. These findings have a number of policy ramifications. First, governments and international organizations ought to give community capacity building top priority in order to provide local populations with the tools and information they need to respond to the effects of climate change. Second, gender-responsive policies are necessary to guarantee that women's contributions to climate action are acknowledged and encouraged. Furthermore, combining traditional wisdom with contemporary advancements can result in more comprehensive and long-lasting climate adaptation plans. In order to respond to particular regional threats, governments should also concentrate on promoting localized disaster preparedness. The long-term effects of community-driven adaptation initiatives should be further investigated in future studies, especially with regard to sustainability and scalability. Along with examining how gender-sensitive policies may better address climate injustices, it should also look into how gender and climate change intersect and how technology can improve community resilience. Research on how traditional knowledge might be incorporated into policymaking and the investigation of cross-sector partnerships to support community adaptation initiatives are also necessary. In the end, societies may more effectively handle the changing climate crises while preserving the welfare of current and future generations by giving priority to adaptive strength, equality, and sustainability and filling in these research gaps.

REFERENCES

Adger, W. N., Hughes, T. P., Folke, C., Carpenter, S. R., & Rockström, J. (2005). Social-ecological resilience to coastal disasters. *Science*, 309(5737), 1036–1039.

Banerjee, B. (2017). Gender and climate change adaptation in India: An analysis of local adaptation strategies and gender-responsive policies. *Development Policy Review*, 35(6), 825–849.

Berkes, F., Colding, J., & Folke, C. (2003). *Navigating Social-Ecological Systems: Building Resilience for Complexity and Change*. Cambridge University Press.

Bharati, L., & Sahoo, P. (2018). Climate change and gender inequality: Addressing women's needs in the context of rural India. *Global Environmental Change*, 51, 16–23.

Bina, A. (1992). The gender and environment debate: Lessons from India. *Feminist Studies*, 18(1), 119–158.

Cannon, T., & Müller-Mahn, D. (2010). Community-based disaster risk reduction and resilience: A new approach to climate change adaptation. *Development Policy Review*, 28(3), 303–318.

Costello, A., Abbas, M., Allen, A., Ball, S., Bell, S., & Bellamy, R. (2009). Managing the health effects of global climate shifts. *The Lancet*, 373(9676), 1693–1733.

Ghosh, D., & Mukherjee, P. (2018). Women's role in disaster risk reduction and climate change adaptation: Case studies from coastal India. *Disasters*, 42(3), 480–498.

Haines, A., & Ebi, K. L. (2019). The imperative of climate change and health adaptation. *The Lancet Planetary Health*, 3(7), e309–e310.

Horton, R., & Lo, S. (2015). Climate change and health: A call to action. *The Lancet*, 386(10006), 1974–1976.

India Climate Dialogue (2020). The role of women in India's climate policy and adaptation efforts. India Climate Dialogue. Retrieved from https://indiaclimatedialogue.net/.

IPCC (2022). Climate change 2022: impacts, adaptation, and vulnerability. *Intergovernmental Panel on Climate Change (IPCC)*. Retrieved from https://www.ipcc.ch/report/ar6/wg2/.

Lloyd, S. J., Kovats, S. R., & Chalabi, Z. (2011). Climate change, cholera, and other waterborne diseases: A review. *Environmental Health Perspectives*, 119(7), 999–1005.

Luber, G., & McGeehin, M. A. (2008). Climate change and extreme heat events. *American Journal of Preventive Medicine*, 35(5), 429–435.

Menon, A., & Sharma, A. (2017). Women and water scarcity in rural India: A gendered perspective on climate change adaptation. *Climate and Development*, 9(2), 148–156.

Ministry of Environment, Forest and Climate Change, Government of India (2018). Gender and climate change: A policy framework for the Indian context. Government of India.

Moseley, M., & Kanter, L. (2020). Community empowerment for climate change mitigation and health promotion: Challenges and opportunities. *Global Environmental Change*, 60, 102025.

Patel, V., & Kleinman, A. (2018). The social and public health implications of climate change. *American Journal of Public Health*, 108(1), 33–34.

Patz, J. A., & Balbus, J. M. (2014). Health and climate change: A research agenda for the U.S. *Global Change Research Program. Environmental Health Perspectives*, 122(7), 1–9. Here are 16 additional references that provide valuable insights into global climate shifts and women capacity building in India. These sources cover topics related to gender and climate vulnerability, women's leadership, and adjustment approaches.

Rao, K. D., & Berman, P. (2019). Climate change and public health: A global overview of the threats and opportunities. *Global Health Action*, 12(1), 1686–1695.

Reddy, B. S. (2016). Empowering women in agriculture: Impact of climate change on rural women's livelihoods in India. *Indian Journal of Rural Development*, 33(2), 48–63.

Sathish, A. (2015). Women and climate change: Empowering women in the face of environmental challenges. *Indian Journal of Environmental Sciences*, 33(1), 59–69.

Sengupta, R., & Bhatnagar, S. (2019). Climate change adaptation strategies in India: Empowering women to lead change. *Economic and Political Weekly*, 54(31), 64–72.

Shah, P., & Desai, A. (2014). Gender and climate change adaptation: The case of coastal areas in India. *Indian Journal of Gender Studies*, 21(2), 219–239.

Shukla, S. (2020). Women's role in disaster risk reduction and climate adaptation: A case study of rural women in Bihar. *Indian Journal of Environment & Ecological Studies*, 29(4), 47–58.

Sivaramakrishnan, K., & Sreeram, V. (2015). Women in the climate change era: Impact, adaptation, and empowerment strategies in India. *International Journal of Environmental Studies*, 22(3), 89–106.

Smith, K. R., & Bradley, D. (2016). Public health and climate change: Challenges and opportunities. *Annual Review of Public Health*, 37(1), 115–131.

Srinivasan, V. (2017). Climate change and health: Impacts and adaptation strategies for low-income communities. *International Journal of Environmental Health Research*, 27(5), 472–482.

Swaminathan, M. (2019). The role of women in climate-resilient agriculture in India: Lessons from South India. *World Journal of Agricultural Research*, 7(5), 123–135.

UNDP (2020). Gender equality and climate change: India's path to resilient development. United Nations Development Programme. https://www.undp.org/publications/undp-gender-equality-strategy-2020-annual-report

Watts, N., Amann, M., Arnell, N., Ayeb-Karlsson, S., Belesova, K., Bouley, T.& Campbell-Lendrum, D. (2018). The 2018 report of the Lancet Countdown on health and global climate shifts: shaping the health of nations for centuries to come. *The Lancet*, 392(10163), 2479–2514.

WorldBank.(2020).LightingupAfrica:Bringingrenewableoff-gridenergytocommunities.https://www.worldbank.org/en/news/feature/2020/08/13/lighting-up-africa-bringing-renewable-off-grid-energy-to-communities

World Bank. (2022). World Bank Group announces major initiative to electrify Sub-Saharan Africa with distributed renewable energy. https://www.worldbank.org/en/news/press-release/2022/11/09/world-bank-group-announces-major-initiative-to-electrify-sub-saharan-africa-with-distributed-renewable-energy

World Health Organization (2015). *Climate Change and Health: Risk and Response*. World Health Organization.

13 Assessing the Impact of Indoor Air Pollution on Self-Reported Health among Older Adults in India

Evidence from LASI Wave 1 Data

Md Mohsin and Syed Naushad Ahmad

INTRODUCTION

The UN Climate Change Conference has brought attention to the vital requirement for international action to tackle climate change, with rising greenhouse gas emissions mostly causing air pollution (Bollen et al., 2009; Chowdhury et al., 2019; Nadeau et al., 2022). In addition to its environmental impact, air pollution presents serious health hazards, contributing to respiratory disease, lung cancer, heart disease, and neurological disorders (Burslem et al., 2000; Deng et al., 2015; Deng et al., 2016; Jiang et al., 2022; Zha et al., 2022). Research indicates that harmful pollutants can breach the blood–brain barrier, potentially leading to psychiatric conditions such as neurodegenerative diseases, sleep disturbances, depression, and cognitive impairment. Older individuals are particularly at risk, as prolonged exposure to air pollution heightens their vulnerability to illnesses like Parkinson's and Alzheimer's disease. Particulate matter, sulfur dioxide, and nitrogen dioxide are among the most harmful pollutants, significantly affecting countries like India, which faces challenges with both outdoor and indoor air pollution (IAP) (Chattopadhyay & Shaw, 2021). A recent study estimates that air pollution has led to roughly 7 million premature deaths worldwide, with indoor pollution responsible for 40% of these fatalities (WHO, 2021). Key sources of IAP include burning fuelwood, agricultural residues, coal, dung cakes, incense sticks, mosquito coils, and exposure to secondhand smoke inside the house (Wong et al., 2020; Saenz et al., 2021). In 2019, IAP was responsible for 4.5% of disability-adjusted life years (DALY) and 0.61 million fatalities in India (Pandey et al., 2021). Residential solid fuel combustion, particularly widespread in South Asia, releases hazardous substances such as black carbon and organic carbon, further worsening health risks (Bond et al., 2004). About 4 million fatalities worldwide are attributed to IAP, which is a prominent cause of lung cancer (8%), ischemic cardiac anomaly (27%), chronic obstructive pulmonary disease (20%), pneumonia (27%), and stroke (18%) (WHO, 2022). Furthermore, exposure to indoor pollutants has been linked to cognitive decline and mental health disorders (Luo et al., 2021; Shaw et al., 2024). In India, major contributors to indoor pollution include combustion activities, building materials, and bioaerosols, with the residential sector accounting for 20%–50% of IAP (Rao et al., 2021). In rural and low-income families in India, unsafe cooking methods utilizing firewood (45%), cow dung cake, coal, charcoal (10.4%), and kerosene (2.9%) are common (Kankaria et al., 2014). Urban populations, on the other hand, are exposed to vehicular and industrial emissions, further compounding public health risks (Elf et al., 2018; Rao et al., 2021). The use of biomass solid fuels in developing nations with low and moderate incomes, particularly for cooking and heating, significantly impacts respiratory health,

 DOI: 10.1201/9781003620747-15

cardiovascular diseases, birth defects, and cognitive impairments (Saenz et al., 2021; Shaw et al., 2024). Poor ventilation exacerbates these effects, allowing polluted air to circulate within households, leading to inflammation and neuronal dysfunction (Peters et al., 2006). The World Health Organization (WHO) report published in 2019 indicated that nearly 90% of the global population breathes polluted air, with IAP being a major contributor to disease burden (Murray et al., 2020). The issue is more severe in developing countries, where dependence on traditional biomass fuels is high (James et al., 2020; Krishnamoorthy et al., 2022). Older adults and women who spend more time in domestic settings are disproportionately affected by the health consequences of unclean cooking fuels (Sehgal et al., 2014; James et al., 2020). India's aging population is currently at 8.6%, which is predicted to rise by 19.1% by 2050 (Census Commissioner, India, 2011). This demographic shift raises concerns about the increasing reliance on traditional biomass fuels and their detrimental effects on older adults (Sigsgaard et al., 2015). The WHO defines health as the state of being physically, mentally, and socially well. Drawing on extant literature, this investigation explores the repercussions of IAP on health issues, with a particular emphasis on fuel sources and cooking methods (Sinha & Ray, 2015; Mocumbi et al., 2019). Additionally, it explores socioeconomic and environmental factors influencing older adults' well-being (Agrawal & Yamamoto, 2015; Belkin, 2018; Dakua et al., 2022). While previous studies on IAP have primarily focused on particular areas/regions and specific demographic groups, limited research has explored its direct impact on the health of older adults in India reported by themselves. This chapter seeks to bridge that gap by utilizing data from Wave 1 of the longitudinal aging study in India (LASI). This chapter's primary objective is to evaluate the connection between IAP and health outcomes in India's older adults and provide a comprehensive analysis of how IAP affects older individuals' respiratory, cardiovascular, and mental health. This research is crucial for policymakers, environmentalists, and healthcare professionals, offering evidence-based insights that can guide interventions to improve indoor air quality and mitigate associated health risks. This chapter makes a unique contribution to our understanding of the impact of IAP on India's aging population by looking at a nationally representative dataset. It suggests that long-term exposure to IAP raises older persons' risk of depression, respiratory disorders, cardiovascular diseases, and chronic illnesses. With the use of its findings, the research hopes to improve the health of vulnerable people by informing focused policy measures, including encouraging cleaner cooking technology, increasing ventilation, and lowering family dependency on solid fuels. This study is crucial and relevant because it offers factual support for a significant public health concern impacting India's aging population. Reducing health inequities linked to environmental pollution, improving housing and cooking practices, and establishing future health policies would all benefit from an understanding of the long-term effects of IAP on older people. The results of this study will add to the current conversation on aging and air quality by highlighting the necessity of long-term solutions to reduce the risks associated with indoor pollution.

SOURCE OF DATA

This study relies on data from Wave 1 of the LASI, conducted from April 2017 to December 2018 (IIPS, 2020). The goal of the extensive, nationally representative LASI survey is to comprehend the social, demographic, economic, and health facets of aging in India (IIPS, 2020). The study was designed following the framework of the *Global Health and Retirement Study (HRS)* and is supported by the *Ministry of Health and Family Welfare, Government of India* (IIPS, 2020). It gathers detailed data on the economic well-being, family dynamics, social security, and physical and mental health of people aged 45 years and above (IIPS, 2020). Three-stage sampling is employed for rural areas, while four-stage sampling is utilized for urban areas as part of LASI's multistage stratified probability cluster sampling technique (IIPS, 2020). The publicly available dataset consists of 72,250 respondents aged 45 years and above, including 40,786 individuals aged 45–59 years and 31,464 individuals aged 60 and above (IIPS, 2020).

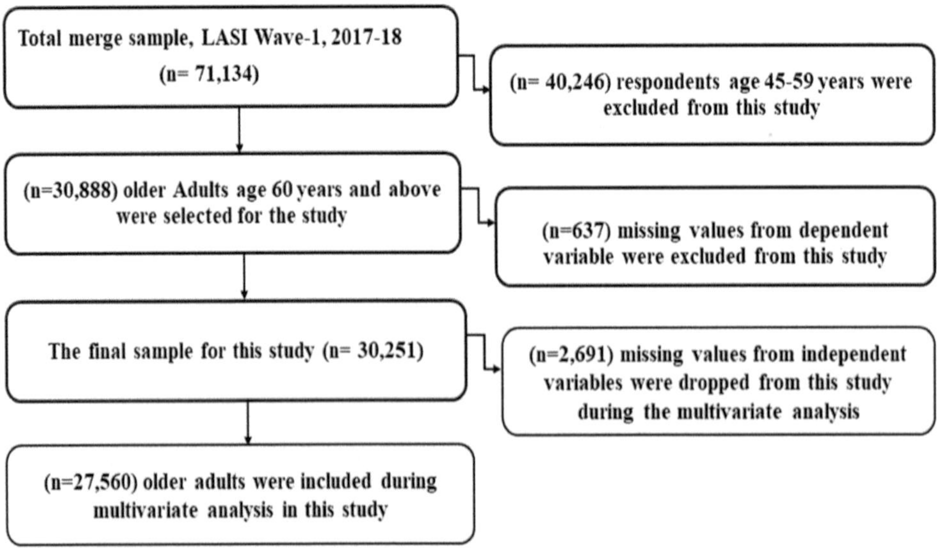

FIGURE 13.1 Flow diagram depicting inclusion of older adults (aged 60 years and above) for analysis from the LASI Wave-1 Data, India (2017–2018).

STUDY SAMPLE

The flow diagram, Figure 13.1, illustrates the stepwise inclusion process of older adults (aged 60 and above) for analysis from LASI Wave 1 data. The initial merged sample of individual and household datasets included 71,134 respondents from the LASI Wave 1 study (IIPS, 2020). Individuals aged 45–59 years (n=40,246) were excluded to focus solely on older adults (60+ years). After exclusion, 30,888 older adults (60+ years) were identified for the study. Six hundred thirty-seven respondents with missing dependent variable values were removed, resulting in a final sample of 30,251 older adults. During multivariate analysis, 2,691 respondents with missing values in independent variables were excluded. The final analytical sample included 27,560 older adults. This structured sample selection ensures a robust and reliable dataset for analyzing factors influencing older adults' health and well-being in India.

DEPENDENT VARIABLE

Self-reported health (SRH), a binary measure that reflects participants' subjective evaluation of their general well-being, is the dependent variable in this study. SRH is shaped by various socioeconomic, biological, and psychological factors. As part of the LASI survey, respondents were asked: "Now I want to ask you about your general health (IIPS, 2020). Overall, how is your health in general? Would you say it is very good, good, fair, poor, or very poor?" (IIPS, 2020). Following the approach used in previous research, SRH was categorized into two groups: "Good" SRH, which includes responses of good, very good, and fair. "Poor" SRH contains responses from the poor and the very poor. This classification provides a more apparent distinction in analyzing health outcomes among older adults (Blazer & Houpt, 1979; Rani et al., 2023).

INDEPENDENT VARIABLES

In this study, IAP was assessed based on cooking fuel type, kitchen setup, and cooking practices within households (Dakua et al., 2022; Fatmi et al., 2014). Cooking fuels were categorized into biogas, LPG, electricity, kerosene, coal, wood, and dung cake to examine their impact on health.

Cooking practices were classified into four categories: use of exhaust fans (modern ventilation), chimneys (traditional ventilation), cooking near a window or door (natural ventilation), and no ventilation or other methods. The study also considered the presence of a separate kitchen, distinguishing between households with and without a dedicated cooking space. The investigation covered a number of socioeconomic, health, and demographic characteristics. Age group (60–64, 65–69, 70–74, and 75 years and above), sex (male, female), social category (Other Backward Class, Scheduled Tribe, Scheduled Caste, and General), and religion (Hindu, Muslim, Sikh, Christian, and others) were among the demographic characteristics. Living arrangements (living with spouse and children, living alone or living with spouse only), education (having formal education and no education), and monthly per capita consumption and expenditure (rich, middle, and poor) were socioeconomic factors. Health-related factors examined the presence of lung disease, chronic cough, diabetes, heart disease, depression, and hypertension. By incorporating these variables, the study provides a comprehensive understanding of the factors influencing SRH among older adults in India (Ortiz et al., 2020; Hamplová et al., 2022; Singh & Chattopadhyay, 2023; Halder et al., 2024). The definitions and coding of these variables are presented in Table 13.1.

STATISTICAL ANALYSIS

The variables impacting poor SRH among older individuals in India were examined using a number of statistical studies. Initially, the frequency of poor SRH in various demographic categories was determined using descriptive statistics. A multivariate binary logistic regression model has been used to examine the association between SRH and other explanatory variables after adjusting for pertinent confounders. Unadjusted odds ratios were computed for every predictor variable before the adjusted model was built. Appropriate individual weights were added to the estimations to guarantee national representativeness. STATA software was used for all statistical analyses. Regression models only contained the statistically significant variables (p-value <0.05) in the adjusted and unadjusted analyses.

TABLE 13.1
Definition/Codes of Variables

Variables	Code/Definition
Cooking fuel	0=Biogas, 1=LPG, 2=Electric, 3=Kerosene, 4=Coal, 5=Wood, 6=Dung cake
Separate kitchen	0=Yes, 1=No
Cooking practice	0=Exhaust fan, 1=Chimney, 2=Near window or door, 3=None or others
Age group	0=60–64, 1=65–69, 2=70–74, 3=75 and above
Sex	0=Male, 1=Female
Social group	0=Schedule Caste (SC), 1=Schedule Tribe (ST), 2=Other Backward Class (OBC), 3=General
Religious group	0=Hindu, 1=Muslim, 2=Christian, 3=Sikh, 4=Others
Living arrangement	0=Living with children and spouse, 1=Living alone or with spouse
Schooling	0=Yes, 1= No
MPCE	0=Middle, 1=Rich, 2=Poor
Lungs disease	0=No, 1=Yes
Cough issue	0=No, 1=Yes
Diabetes	0=No, 1=Yes
Heart disease	0=No, 1=Yes
Hypertension	0=No, 1=Yes
Feeling depressed	0=No, 1=Yes

RESULTS

Background Characteristics

Table 13.2 presents the background characteristics of 30,251 older adults in India aged 60 and above. The gender distribution is nearly balanced, with 51.9% female and 48.1% male. Most of the responders (32.67%), those between the ages of 60 and 64, and those between the ages of 65 and 69 (28.36%) make up the next largest age group. Hindus constitute the largest religious group (73.3%),

TABLE 13.2
Background Characteristics of the Target Population and Prevalence of Self-Reported Health among Older Adults in India, 2017–2018

Variables & Categories	Self-Reported Health Good (%)	Self-Reported Health Poor (%)	Sample *N* (%)
	Cooking Fuel Type		
Biogas	87 (82.08)	19 (17.92)	106 (0.35)
LPG	12,095 (77.95)	3,421 (22.05)	15,516 (51.65)
Electric	143 (80.34)	35 (19.66)	178 (0.59)
Kerosene	167 (74.89)	56 (25.11)	223 (0.74)
Coal	472 (75.88)	150 (24.12)	622 (2.07)
Crops & wood	9,275 (75.92)	2,942 (24.08)	12,217 (40.67)
Dung cake	888 (75.38)	290 (24.62)	1,178 (3.92)
	Separate Kitchen		
Yes	15,494 (77.30)	4,550 (22.70)	20,044 (66.48)
No	7,687 (76.06)	2,420 (23.94)	10,107 (33.52)
	Cooking Practice		
Exhaust fan	1,362 (80.93)	321 (19.07)	1,683 (6.04)
Chimney	2,555 (76.91)	767 (23.09)	3,322 (11.92)
Near the window or door	13,264 (77.79)	3,787 (22.21)	17,051 (61.18)
None or others	4,383 (75.41)	1,429 (24.59)	5,812 (20.86)
	Age Group		
60–64	8,134 (82.30)	1,749 (17.70)	9,883 (32.67)
65–69	6,792 (79.17)	1,787 (20.83)	8,579 (28.36)
70–74	4,120 (74.81)	1,387 (25.19)	5,507 (18.20)
75 and Above	4,214 (67.08)	2,068 (32.92)	6,282 (20.77)
	Sex Group		
Male	11,507 (79.12)	3,036 (20.88)	14,543 (48.07)
Female	11,753 (74.82)	3,955 (25.18)	15,708 (51.93)
	Social Group		
General	6,841 (77.63)	1,971 (22.37)	8,812 (29.18)
OBC	8,506 (74.12)	2,970 (25.88)	11,476 (38.00)
SC	3,694 (74.45)	1,268 (25.55)	4,962 (16.43)
ST	4,179 (84.46)	769 (15.54)	4,948 (16.39)
	Religious Group		
Hindu	16,965 (76.47)	5,221 (23.53)	22,186 (73.34)
Muslim	2,639 (74.32)	912 (25.68)	3,551 (11.74)
Christian	2,447 (81.08)	571 (18.92)	3,018 (9.98)

(Continued)

TABLE 13.2 (*Continued*)
Background Characteristics of the Target Population and Prevalence of Self-Reported Health among Older Adults in India, 2017–2018

	Self-Reported Health		
Variables & Categories	**Good (%)**	**Poor (%)**	**Sample *N* (%)**
Sikh	749 (79.43)	194 (20.57)	943 (3.12)
Others	460 (83.18)	93 (16.82)	553 (1.83)
	Living Arrangement		
Living with children & spouse	16,318 (77.62)	4,704 (22.38)	21,022 (69.49)
Living alone or with a spouse	6,942 (75.22)	2,287 (24.78)	9,229 (30.51)
	Schooling		
Yes	11,061 (78.65)	3,002 (21.35)	14,063 (46.49)
No	12,199 (75.36)	3,989 (24.64)	16,188 (53.51)
	MPCE		
Rich	8,904 (76.30)	2,765 (23.70)	11,669 (38.57)
Middle	4,866 (78.83)	1,307 (21.17)	6,173 (20.41)
Poor	9,490 (76.48)	2,919 (23.52)	12,409 (41.02)
	Lungs Disease		
No	21,960 (78.48)	6,021 (21.52)	27,981 (92.53)
Yes	1,294 (57.28)	965 (42.72)	2,259 (7.47)
	Cough Issue		
No	20,698 (78.26)	5,749 (21.74)	26,447 (87.72)
Yes	2,475 (66.87)	1,226 (33.13)	3,701 (12.28)
	Diabetes		
No	20,177 (78.89)	5,399 (21.11)	25,576 (84.58)
Yes	3,076 (65.98)	1,586 (34.02)	4,662 (15.42)
	Heart Disease		
No	22,372 (77.91)	6,345 (22.09)	28,717 (94.96)
Yes	882 (57.91)	641 (42.09)	1,523 (5.04)
	Hypertension		
No	15,922 (80.86)	3,769 (19.14)	19,691 (65.12)
Yes	7,329 (69.48)	3,219 (30.52)	10,548 (34.88)
	Feeling Depressed		
No	19,929 (80.00)	4,982 (20.00)	24,911 (82.76)
Yes	3,217 (62.01)	1,971 (37.99)	5,188 (17.24)
Total	23,260 (76.89)	6,991 (23.11)	30,251 (100.00)

Note: Sample weights were used to calculate percentages using LASI Wave 1 data from India in 2017–2018. OBC: Other Backward Class, MPCE: Monthly Per Capita Consumption and Expenditure, SC: Schedule Caste, ST: Schedule Tribe.

followed by Muslims (11.7%), Christians (10%), Sikhs (3.1%), and others (1.8%). The social composition includes respondents from the General category (29.2%), Other Backward Class (OBC) (38%), Scheduled Tribe (ST) (16.4%), and Scheduled Caste (SC) (16.4%). Education levels are low, with 53.5% of respondents never attending school.

In terms of economic status, 38.6% belong to the rich category, 20.4% to the middle-income group, and 41% to the poor category. Health conditions are prevalent among the respondents, with 7.5% reporting lung disease, 12.3% experiencing a persistent cough, 15.4% having diabetes, 5% suffering from heart disease, 34.9% diagnosed with hypertension, and 17.2% experiencing depression. The data provides an overview of the demographic, social, and health characteristics of India's elderly population. Regarding household characteristics, cooking fuel type varies, with the majority using LPG (51.7%), followed by crop residue wood (40.7%), dung cakes (3.9%), coal (2.1%), kerosene (0.7%), electricity (0.6%), and biogas (0.3%). A separate kitchen is present in 66.5% of households, while 33.5% cook in shared or open spaces. Cooking practices also differ, with 61.2% cooking near a window or door, 11.9% using a chimney, 6% using an exhaust fan, and 20.9% having no proper ventilation.

Usage of Harmful Products inside Indian Households

Figure 13.2 presents the percentage of Indian households using various harmful indoor products daily, providing insights into household practices and their potential health implications. These products range from traditional incense sticks to modern mosquito repellents and smoking habits, shaping the indoor air quality and living environment. Incense sticks emerge as the most commonly used product, with 53.1% of households reporting daily usage. This highlights their cultural and religious significance in Indian homes, where they are frequently used for worship and aromatic purposes. Because of worries about mosquito-borne diseases, including dengue and malaria, 7.4% of homes utilize mosquito coils. Similarly, 10.92% of families employ electric insect repellents or liquid vaporizers, suggesting a moderate preference for contemporary pest management techniques. A relatively smaller proportion (1.47%) of households use fast cards, sticks, or cakes, suggesting their limited cultural or personal preference compared to other mosquito repellents. Due to IAP, smoking inside is a substantial health concern, as reported by 24.3% of families. Both smokers and non-smokers in the home are impacted when tobacco smoke is present indoors because it increases exposure to dangerous contaminants.

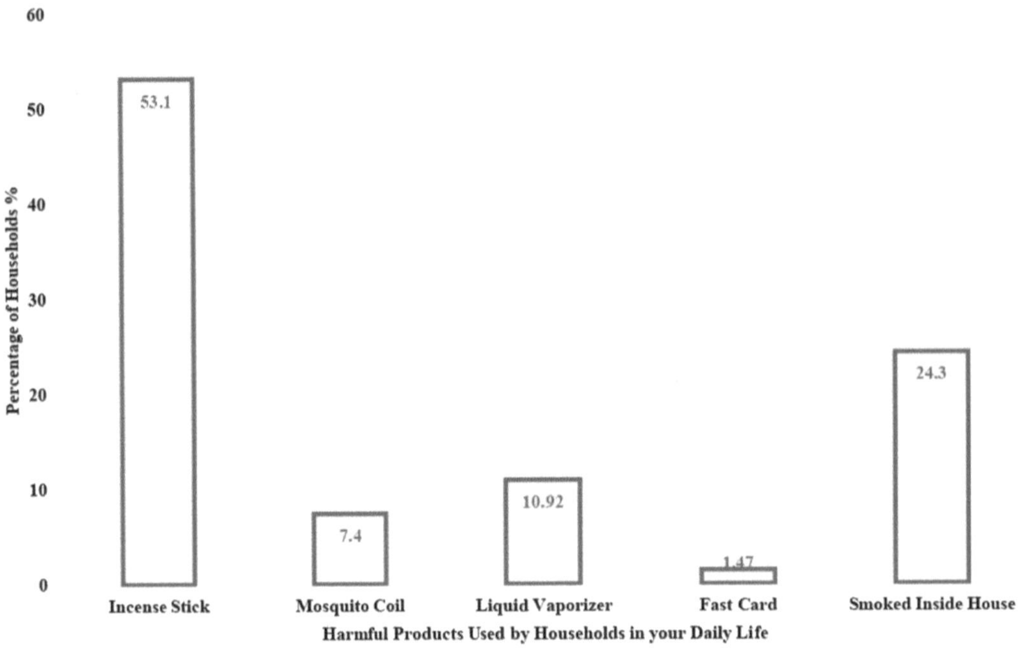

FIGURE 13.2 Percentage of Indian households that utilize different harmful products regularly.

All things considered, Figure 13.2 shows how traditional and contemporary customs coexist in Indian homes. Due to cultural reasons, incense sticks continue to be widely used, although the use of mosquito repellents, both conventional (mosquito coils) and contemporary (liquid vaporizers), reflects health-conscious practices. However, the high frequency of indoor smoking highlights the need for awareness and initiatives to lessen its impact and raises concerns about air quality and possible respiratory health consequences.

The Percentage of Poor Self-Reported Health among Older Adults in India

The prevalence of poor SRH varied based on cooking fuel type, household characteristics, and demographic factors. A higher proportion of Indian older adults using solid biomass fuels such as wood and crop residue (24.1%), coal (24.1%), and dung cakes (24.6%) reported poor SRH compared to those using clean fuels like LPG (22%) or biogas (17.9%). Cooking methods also influenced health outcomes, with a more significant proportion of individuals cooking near a window or door (22.2%) and those using traditional chimneys (23.1%) reporting poor SRH than those using exhaust fans (19.1%). Furthermore, respondents without proper ventilation (24.6%) experienced the highest proportion of poor SRH (Table 13.2). The proportion of poor SRH increased with age, with the highest prevalence among those aged 75 and above (32.9%), followed by the 70–74 years group (25.2%). A significant proportion of women (25.2%) reported poor health compared to men (20.9%). Social disparities were evident, with a more substantial proportion of OBC (25.9%) and SC (25.5%) groups reporting poor SRH compared to the General category (22.4%). Among religious groups, Muslims (25.7%) had the highest proportion, followed by Hindus (23.5%), while Christians and Sikhs reported slightly lower proportions. Economic status, education, and living arrangements also played a role. A more significant proportion of older adults in the poorest category (23.5%) reported poor SRH compared to those in the wealthiest category (23.7%). Individuals with no schooling (24.6%) had a higher proportion of poor SRH than those with education (21.3%). Additionally, a more significant proportion of those living alone or only with a spouse (24.8%) reported poor health compared to those living with their families (22.4%). Health conditions significantly impacted SRH, with the highest proportion of poor health observed among respondents reporting lung disease (42.7%), diabetes (34%), and depression (38%). Those with hypertension (30.5%) and heart disease (42.1%) also had a more significant proportion of poor SRH than those without these conditions. The findings highlight the disparities in health outcomes among older adults, influenced by environmental, social, and economic factors.

The Percentage of Older Adults in Indian States Who Have Poor Self-Reported Health

As shown in Table 13.3 and Figure 13.3, the incidence of poor SRH among older individuals in India varies considerably by state. Based on the percentage of older persons reporting poor SRH, the states are divided into three groups: low prevalence (5.12%–20.40%), moderate prevalence (20.41%–35.69%), and high prevalence (35.70%–50.98%). States in this category, marked in yellow on the map, exhibit a relatively lower proportion of older adults experiencing poor SRH.

These states include Nagaland (8.1%), Meghalaya (5.1%), Arunachal Pradesh (9.5%), Chhattisgarh (10.1%), Delhi (12.1%), Mizoram (12.6%), Daman & Diu (11.8%), and Karnataka (10.9%). The northeastern states, parts of Central India, and some Union Territories fall under this category, indicating better SRH status among the elderly. States represented in green on the map show a moderate proportion of older adults reporting poor SRH. . These states include Northern, Eastern, and Central India. States highlighted in the blue report the highest proportion of older adults experiencing poor SRH. These include Kerala (51.0%), Tamil Nadu (50.2%), and Puducherry (41.2%). The Southern region exhibits the highest prevalence of poor SRH among older adults, indicating significant health concerns in these areas. The variation in the prevalence of poor SRH across

TABLE 13.3
Prevalence of Self-Reported Health across the State among Older Adults in India, 2017–2018

State	Self-Reported Health	
	Good (%)	Poor (%)
Arunachal Pradesh	90.48	9.52
Nagaland	91.86	8.14
Madhya Pradesh	76.95	23.05
Tamil Nadu	49.83	50.17
Puducherry	58.81	41.19
Himachal Pradesh	74.16	25.84
Punjab	77.92	22.08
Delhi	87.93	12.07
Rajasthan	79.55	20.45
Mizoram	87.42	12.58
Tripura	75.72	24.28
Jharkhand	80.60	19.40
Odisha	77.11	22.89
Chhattisgarh	89.95	10.05
Daman & Diu	88.17	11.83
Dadra & Nagar Haveli	83.72	16.28
Kerala	49.02	50.98
Andaman & Nicobar Islands	90.18	9.82
Jammu & Kashmir	73.25	26.75
Chandigarh	82.98	17.02
Bihar	78.62	21.38
Manipur	84.55	15.45
Meghalaya	94.88	5.12
Assam	78.92	21.08
West Bengal	68.36	31.64
Gujarat	85.70	14.30
Andhra Pradesh	76.91	23.09
Karnataka	89.14	10.86
Maharashtra	84.43	15.57
Goa	67.22	32.78
Lakshadweep	79.19	20.81
Uttarakhand	81.04	18.96
Haryana	75.61	24.39
Uttar Pradesh	74.79	25.21
Telangana	78.14	21.86
India	**76.89**	**23.11**

Indian states underscores the need for region-specific healthcare policies and interventions aimed at addressing elderly health concerns, particularly in states with high prevalence.

Factors Influencing Poor Self-Reported Health among Older Adults in India

The findings of a multivariate logistic regression study examining the association between IAP factors, such as cooking fuel type, cooking methods, and other socioeconomic and health-related variables, and poor SRH are shown in Table 13.4. The results show that older persons who are

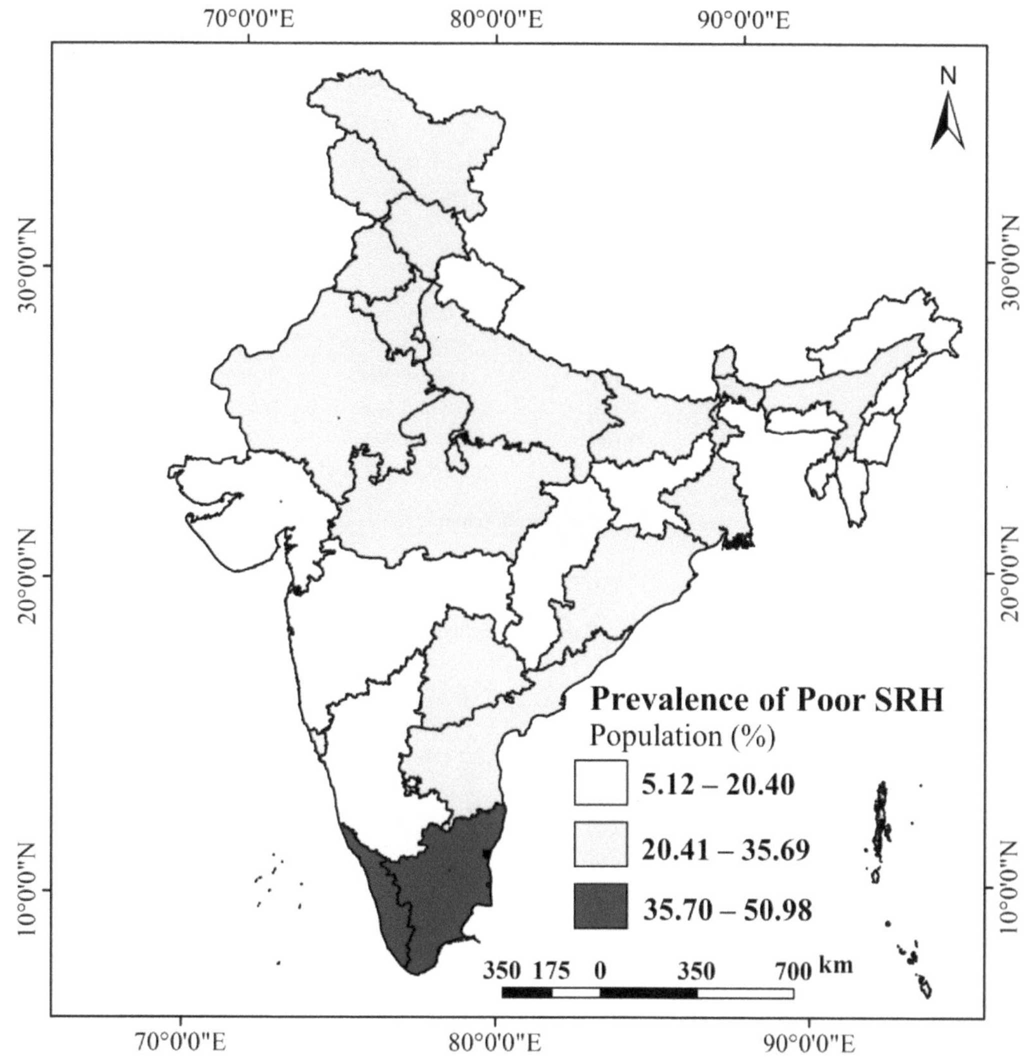

FIGURE 13.3 The percentage of older adults in Indian states who report having poor health.

exposed to specific household environmental circumstances are more likely to have poor SRH. The study discovered that cooking fuel type significantly affected SRH after controlling for demographic, socioeconomic, and health-related characteristics (such as age, sex, social group, religion, MPCE, housing arrangements, and education). People who used kerosene (AOR = 2.859, 95% CI = 1.213–6.735), wood and agricultural residue (AOR = 2.324, 95% CI = 1.215–4.447), and dung cakes (AOR = 2.297, 95 % CI = 1.138–4.637) were at a considerably greater risk than those who used biogas. The presence of a separate kitchen was associated with better SRH, whereas those without one had greater odds of reporting poor health (AOR = 1.112, 95% CI = 0.993–1.244), suggesting that cooking in shared or poorly ventilated spaces may be detrimental to health. Additionally, poor ventilation while cooking increased the risk of poor SRH, with those cooking near a window or door without an exhaust fan (AOR = 1.809, 95% CI = 1.111–2.943) and those cooking without any ventilation (AOR = 1.950, 95% CI = 1.191–3.191) being more vulnerable. Sociodemographic variables were also very important in determining the results of SRH. Compared to those aged 60–64, older persons 75 years of age and older were more likely to report poor health (AOR = 2.105, 95%

TABLE 13.4
Indoor Air Pollution and Other Contributing Factors to Older Adults' Poor Self-Reported Health in India, 2017–2018

Variable	Model 1 UOR	Model 1 [95% CI]	Model 2 AOR	Model 2 [95% CI]
Cooking Fuel				
Biogas (Reference)	1	.	1	.
LPG	1.918**	1.038–3.542	1.833**	0.963–3.489
Electric	1.578	0.705–3.535	1.463	0.625–3.425
Kerosene	3.122***	1.399–6.968	2.859***	1.213–6.735
Coal	1.733	0.887–3.386	1.773*	0.868–3.620
Crop residue & wood	2.166***	1.171–4.008	2.324***	1.215–4.447
Dung cake	2.171**	1.119–4.213	2.297***	1.138–4.637
Separate Kitchen				
Yes (Reference)	1	.	1	.
No	1.131**	1.019–1.256	1.112**	0.993–1.244
Cooking Practice				
Exhaust fan (Reference)	1	.	1	.
Chimney	1.302*	0.920–1.842	1.428	0.832–2.450
Near the window or door	1.480***	1.109–1.975	1.809***	1.111–2.943
None or others	1.583***	1.169–2.143	1.950***	1.191–3.191
Age Group				
60–64 (Reference)			1	.
65–69			1.115*	0.982–1.265
70–74			1.359***	1.156–1.597
75 and above			2.105***	1.834–2.417
Sex				
Male (Reference)			1	.
Female			1.166**	1.037–1.311
Social Group				
General (Reference)			1	.
OBC			1.189***	1.044–1.353
SC			1.266***	1.087–1.475
ST			0.755***	0.618–0.924
Religious Group				
Hindu (Reference)			1	.
Muslim			1.215***	1.048–1.407
Christian			2.082***	1.641–2.642
Sikh			0.929	0.729–1.184
Others			0.827	0.538–1.271
Living Arrangement				
Living with children & spouse (Ref)			1	.
Living alone or with a spouse			1.191***	1.065–1.332
Schooling				

(Continued)

TABLE 13.4 (*Continued*)
Indoor Air Pollution and Other Contributing Factors to Older Adults' Poor Self-Reported Health in India, 2017–2018

	Model 1		Model 2	
Variable	**UOR**	**[95% CI]**	**AOR**	**[95% CI]**
Yes (Reference)			1	.
No			1.182***	1.053–1.327
MPCE				
Middle (Reference)			1	.
Rich			1.110*	0.960–1.282
Poor			1.210***	1.054–1.382
Lung Disease				
No (Reference)			1	.
Yes			1.947***	1.577–2.404
Cough				
No (Reference)			1	.
Yes			1.776***	1.554–2.029
Diabetes				
No (Reference)			1	.
Yes			1.810***	1.559–2.102
Heart Disease				
No (Reference)			1	.
Yes			1.481***	1.084–2.022
Hypertension				
No (Reference)			1	.
Yes			1.534***	1.381–1.703
Feeling Depressed				
No (Reference)			1	.
Yes			2.123***	1.894–2.379

Significance Level: *** $p<0.01$, ** $p<0.05$, * $p<0.1$.

Note: Model 1 Unadjusted Figure of indoor air pollution (UOR= unadjusted odds ratio), Model 2 adjusted figure for sociodemographic, economic, and health variables (AOR= adjusted odds ratio), CI= confidence interval.

CI = 1.834–2.417). Poor SRH was reported by women more frequently than by males (AOR = 1.166, 95% CI = 1.037–1.311).

While Scheduled Tribes (ST) had lower odds of reporting poor health (AOR = 0.755, 95% CI = 0.618–0.924), membership in socially disadvantaged groups, such as Scheduled Castes (SC) (AOR = 1.266, 95% CI = 1.087–1.475) and Other Backward Classes (OBC) (AOR = 1.189, 95% CI = 1.044–1.353), was linked to poorer SRH. Compared to Hindus, Muslims (AOR = 1.215, 95% CI = 1.048–1.407) and Christians (AOR = 2.082, 95% CI = 1.641–2.642) were more likely to report poor health. Living arrangements and economic circumstances also have an impact on SRH results. Compared to older persons who lived with their spouse and children, those who lived alone or with their spouse alone were more likely to be in poor health (AOR = 1.191, 95% CI = 1.065–1.332). People from poorer socioeconomic origins (AOR = 1.210, 95% CI = 1.054–1.382) and those without formal education (AOR=1.182, 95% CI=1.053–1.327) similarly reported worse health. Chronic health problems were substantially associated with poor SRH. Heart disease (AOR=1.481, 95% CI=1.084–2.022), diabetes (AOR=1.810, 95% CI=1.559–2.102), persistent cough (AOR=1.776, 95% CI=1.554–2.029), lung disease (AOR=1.947, 95% CI=1.577–2.404), and hypertension (AOR=1.534, 95% CI=1.381–1.703) were among the conditions that older adults were more likely to report poor health. Furthermore, a significant predictor of poor SRH was depression (AOR=2.123, 95% CI=1.894–2.379) (see Table 13.4).

DISCUSSION

The complex relationship between IAP from domestic cooking fuel and its detrimental effects on the SRH of senior citizens in India is examined in this study. Our results, based on nationally representative cross-sectional type data from LASI Wave 1, show a strong correlation between older persons' exposure to IAP and a higher chance of having poor SRH.

Use of Biomass Solid Fuel and Its Effects on Health

A critical worldwide and public health concern is the ongoing reliance on solid fuels, such as coal, kerosene, crop waste, wood, shrubs, and animal manure. Approximately 3 billion individuals worldwide still depend on these fuels for household energy needs, with India being one of the most affected nations. Using these conventional fuels increases the risk of serious health problems, especially for older persons who are more vulnerable because of lifetime exposure. In India's north and east, where poor health problems are more common, IAP is noticeably more significant. In rural areas of these countries, traditional cooking practices remain widespread, particularly the use of solid biomass fuels such as coal, firewood, cow dung, and crop residue. Additionally, large-scale agricultural biomass burning in the northern Indo-Gangetic plains and deliberate forest fires in the northeastern states, where land is cleared and burned for human activities, are major contributors to air pollution. (Singh et al., 2021). Given these several factors contributing to both outdoor and IAP, India ranks third globally in terms of carbon emissions, after the United States and China (Timperley, 2019). In India, 57% of households continue to depend on solid biomass fuels for cooking and heating (IIPS, 2020). The Indian population experiences IAP at three times the level of ambient air pollution, which is the second-leading cause of mortality in the country (Dandona et al., 2020). Older adults are more susceptible to IAP because of their decreased mobility, longer exposure times, and deteriorating ability to expel toxins from their bodies as they age (Mangoni & Jackson, 2004; Ferrucci et al., 2016). The findings of the current study reinforce previous research, which has consistently linked solid fuel use with deteriorating health conditions, including respiratory and cardiovascular diseases. Economic constraints force households to resort to non-renewable, high-emission fuels, which further complicates efforts to transition toward cleaner and renewable alternatives. Research from Southeast Asia, Latin America, and Sub-Saharan Africa has also shown how burning biomass fuels negatively affects both individual and community health (Belkin, 2018; James et al., 2020; Luo et al., 2021; Saenz et al., 2021; Zha et al., 2022; Oluwatosin et al., 2022). These results support Sustainable Development Goal 7, which highlights the need for ecologically friendly cooking methods and widespread access to renewable energy sources. Many researchers have also emphasized the importance of these actions and the benefits of switching to renewable energy sources in reducing environmental and health risks (Parra et al., 2020; Čukić et al., 2021; Carlsen & Bruggemann, 2022).

Impact of Cooking Practices on Health Consequences

How cooking is conducted significantly influences health outcomes. This study reveals that older adults residing in households without adequate ventilation, such as electric chimneys or exhaust fans, are at an 82% greater risk of experiencing poor SRH. The lack of proper ventilation exacerbates exposure to harmful pollutants, increasing susceptibility to chronic illnesses. Additionally, the use of kerosene as a cooking fuel, as well as exposure to secondhand and neighborhood smoke, further heightens health risks among older adults. Inadequate ventilation is often a by-product of socioeconomic limitations, where low-income households struggle to afford necessary infrastructure improvements. These findings are consistent with research conducted in Mexico and China, which has demonstrated strong correlations between cooking practices, household air quality, and adverse health effects in older populations (Hou et al., 2018; Saenz et al., 2021; Du et al., 2022). Given the global nature of this issue, targeted interventions, such as those facilitated by India's Ujjwala Yojana scheme, should be reinforced and expanded to improve access to cleaner fuels and enhanced ventilation solutions.

Association with Sociodemographic and Economic Aspects

The sociodemographic and economic circumstances of older persons have an impact on the health effects of IAP. Long-term exposure to biomass fuels disproportionately increases the risk of health deterioration for women and the oldest old, who are more likely to be involved in cooking activities (Borooah et al., 2013; Sharma et al., 2013; MOSPI, 2019). Education and income levels also play an essential role in determining access to cleaner fuel alternatives (Cagney & Lauderdale, 2002; Wu et al., 2016). Households with lower educational attainment are less likely to adopt cleaner cooking technologies, often due to a lack of awareness regarding their health benefits. Similarly, financial constraints prevent many families from transitioning to cleaner fuels, leading to continued reliance on solid fuels despite their known adverse health effects (Cagney & Lauderdale, 2002; Wu et al., 2016; Chafe & Chowdhury, 2021; Krishnamoorthy et al., 2022). Additionally, socio-cultural aspects such as caste and religion also influence health disparities among older adults. Our findings indicate that older adults from SC (Scheduled Castes) and OBC (Other Backward Classes) are at a greater risk of poor SRH and cognitive health compared to those from the General category. This may be attributed to differences in socioeconomic status, living conditions, and access to healthcare services. Previous research has also highlighted similar disparities, emphasizing the role of caste-based social stratification in shaping health outcomes (Borooah et al., 2013; Sharma et al., 2013; Padmanaban, 2024). Similarly, religious affiliation plays a role in health disparities, with Muslims and Christians exhibiting a greater likelihood of reporting poor SRH compared to Hindus. These variations highlight the need for inclusive health policies that cater to the unique challenges faced by marginalized communities. Previous studies have also reported comparable findings, demonstrating that religious minority groups often experience more significant health disadvantages due to socioeconomic and institutional factors (Padmanaban, 2024).

Association with Particular Health Conditions

This study emphasizes how older persons with pre-existing medical issues are more susceptible to the negative consequences of IAP. Poor SRH is substantially more likely to be reported by those with diabetes, heart disease, lung diseases, chronic cough, depression, and hypertension. It is noteworthy that people with lung diseases are 1.95 times more likely to experience adverse health outcomes than people in good health. These findings are consistent with various international research highlighting the significant influence of IAP on illnesses of the heart, lungs, and mind. International studies conducted in the US, Canada, and China have repeatedly shown that indoor pollution exposure is a substantial risk factor for depression and other long-term illnesses (Mu et al., 2013; Chafe et al., 2014; Jeuland et al., 2015; Eze et al., 2015; Ghorani-Azam et al., 2016; Szyszkowicz et al., 2016; Kioumourtzoglou et al., 2017; Wang & Yang, 2018).

Climate Change, Indoor Air Pollution, and Human Health

In addition to posing serious health risks, IAP accelerates climate change by releasing greenhouse gases and particulate matter. Burning biomass fuels like coal, firewood, and cow dung generates black carbon, a transient climatic pollutant that absorbs sunlight and warms the atmosphere, greatly accelerating global warming (Bond et al., 2013; Ramanathan & Carmichael, 2008). Furthermore, methane and carbon monoxide are produced when these fuels are not completely burned, which speeds up climate change (Shindell et al., 2012). By increasing the accumulation of pollutants in houses with inadequate ventilation and escalating the dangers to respiratory and cardiovascular health, rising global temperatures and extreme weather events can worsen IAP (Ebi & McGregor, 2008; Patz et al., 2014). As the negative impacts of indoor pollution are exacerbated by temperature changes brought on by climate change and declining air quality, older adults, especially those living in low-income households, face increased vulnerability (IPCC, 2021). As a result, reducing IAP is essential for both enhancing public health and lessening its broader environmental effects. Reducing dependency on solid fuels, improving ventilation in the home, and implementing sustainable energy solutions may all be combined to fight climate change and diseases linked to indoor pollution (Smith et al., 2009; Wilkinson et al., 2009).

Policy Implications and Public Health Interventions

The findings of this study underscore the pressing need for comprehensive public health interventions aimed at reducing IAP and promoting cleaner cooking alternatives. The expansion of government-led initiatives, such as the Pradhan Mantri Ujjwala Yojana, is crucial in enhancing access to affordable clean fuel solutions, particularly in rural and low-income households (Jagadale & Kemper, 2022). Additionally, public health campaigns should emphasize the importance of improving ventilation through the adoption of electric chimneys, exhaust fans, and other air-purification technologies. Strengthening rural healthcare infrastructure and increasing awareness about the health risks associated with solid fuel use are essential steps in mitigating the burden of IAP (Rehfuess, 2006). Policymakers should also consider integrating household air quality assessments into routine health screenings, ensuring that vulnerable populations receive timely interventions and support. The broader implementation of policies that promote sustainable and eco-friendly energy alternatives will contribute to long-term improvements in public health and environmental sustainability (Ezzati & Kammen, 2002).

Building on Global Insights

This study contributes to the growing body of international research on IAP and its profound effects on aging populations. Findings from studies conducted in Mexico, China, and developed nations provide valuable insights into the diverse health implications of prolonged exposure to indoor pollutants. Aligning with global discourse, the present chapter reinforces the necessity for cross-country collaborations in addressing IAP-related health risks. The collective evidence highlights the need for international knowledge exchange and shared policy frameworks aimed at advancing clean energy transitions and public health strategies (Buyinza Faisal & Vicent, 2021; Dida et al., 2022; Saenz et al., 2018; Yeatts et al., 2012). Overall, this study makes a significant contribution to the discourse on IAP and its detrimental effects on Indian older adults. The findings emphasize the essential need for targeted health interventions, improved household energy policies, and strengthened global efforts to reduce the health burdens associated with IAP. By integrating sustainable practices and fostering policy-driven clean energy transitions, significant progress can be achieved in ensuring a healthier aging population and promoting environmental well-being (Grazuleviciute-Vileniske et al., 2020).

Strengths and Limitations of the Study

Using recently published 2021 data, this study offers a well-representative sample from every state and an updated understanding of older adults' health. However, there are a few limitations to take into account. First, because people's judgments of their own health might differ, relying solely on SRH evaluations may create bias. Second, the cross-sectional form of the study makes it more challenging to identify the factors that influence the variables. Therefore, the results must be carefully interpreted. Third, an extensive evaluation of how IAP affects health outcomes is hampered by the lack of ventilation data. Finally, omitted variable bias may result from the analysis's limitations due to the variables included in the dataset.

CONCLUSION

This study highlights the strong correlation between IAP and poor SRH among older adults in India. The health risks are significantly exacerbated by the ongoing use of solid fuels, including wood, dung cakes, and kerosene, in conjunction with conventional cooking techniques in areas with inadequate ventilation, highlighting the urgent need for focused interventions. Promoting clean cooking options, increasing access to alternative energy sources like biogas, electricity, and LPG, and improving home ventilation with cutting-edge technology like electric chimneys and exhaust systems should be top priorities for policymakers. Strengthening initiatives like the Ujjwala Yojana can further support the widespread adoption of clean fuels, improving indoor air quality and reducing health inequalities. This chapter also highlights the need for tailored public health treatments for older persons, especially those with pre-existing ailments, including respiratory diseases, cardiovascular problems, and mental health disorders that are made worse by IAP. To lessen the negative consequences of IAP, a comprehensive, multi-sectoral strategy combining clean energy regulations, infrastructural upgrades, and health awareness programs is necessary. Policymakers can significantly improve elderly people's quality of life by tackling the root causes of IAP and how it affects vulnerable populations. A healthy aging population and sustainable environmental advancement in India will result from fortifying policy frameworks, giving clean energy transitions top priority, and encouraging community-driven solutions.

LIST OF ABBREVIATIONS

AOR: Adjusted odd ratio
CI: Coefficient of interval
IAP: Indoor air pollution
LASI: Longitudinal aging study in India
LPG: Liquified petroleum gas
MPCE: Monthly per capita consumption and expenditure
SRH: Self-reported health
UOR: Unadjusted odd ratio

ACKNOWLEDGMENT

The authors express their gratitude to Aligarh Muslim University, India, and the Department of Geography for their assistance and provision of the necessary infrastructure to enable the completion of this work.

DATA AVAILABILITY

Openly available data from the longitudinal aging study in India (LASI) was used in this study: https://www.iipsindia.ac.in/content/LASI-data.

DECLARATIONS

Declarations and Statements: We certify that no one else has met the requirements to be listed as an author and that all of the previously specified writers have reviewed and approved the work. Additionally, we attest that we have all given our approval to the author's order as specified in the article.

Funding: We certify that no funding, money, or other assistance was obtained in order to prepare this manuscript. The authors stated that they had no competing interests, either financial or non-financial.

Conflict of Interest: The authors clearly declare that there is no conflict of interest between them.

AUTHOR'S CONTRIBUTION

Conceptualization: Md Mohsin and Syed Naushad Ahmad; **Methodology:** Md Mohsin; **Data Analysis:** Md Mohsin; **Mapping:** Md Mohsin; **Formal analysis:** Md Mohsin, Syed Naushad Ahmad; **Final Review:** Md Mohsin, Syed Naushad Ahmad; **Supervision:** Professor Syed Naushad Ahmad

REFERENCES

Agrawal, S., & Yamamoto, S. (2015). Effect of Indoor air pollution from biomass and solid fuel combustion on symptoms of preeclampsia/eclampsia in Indian women. *Indoor Air*, *25*(3), 341–352. https://doi.org/10.1111/ina.12144

Belkin, H. E. (2018). Chapter 16—Environmental human health issues related to indoor air pollution from domestic biomass use in rural China: A review. In B. D. Vivo, H. E. Belkin, & A. Lima (Eds.), *Environmental Geochemistry* (Second Edition, pp. 417–434). Elsevier. https://doi.org/10.1016/B978-0-444-63763-5.00017-3

Blazer, D. G., & Houpt, J. L. (1979). Perception of poor health in the healthy older adult. *Journal of the American Geriatrics Society*, *27*(7), 330–334. https://doi.org/10.1111/j.1532-5415.1979.tb06051.x

Bollen, J., Zwaan, B. van der, Brink, C., & Eerens, H. (2009). Local air pollution and global climate change: A combined cost-benefit analysis. *Resource and Energy Economics*, *31*(3), 161–181. https://doi.org/10.1016/j.reseneeco.2009.03.001

Bond, T. C., Doherty, S. J., Fahey, D. W., Forster, P. M., Berntsen, T., DeAngelo, B. J., … Zender, C. S. (2013). Bounding the role of black carbon in the climate system: A scientific assessment. *Journal of Geophysical Research: Atmospheres*, *118*(11), 5380–5552. https://doi.org/10.1002/jgrd.50171

Bond, T., Venkataraman, C., & Masera, O. (2004). Global atmospheric impacts of residential fuels. *Energy for Sustainable Development*, *8*(3), 20–32. https://doi.org/10.1016/S0973-0826(08)60464-0

Borooah, V. K., Sabharwal, N. S., & Thorat, S. (2013). *Gender and Caste-Based Inequality in Health Outcomes in India*. Retrieved from https://api.semanticscholar.org/CorpusID:53322228

Burslem, D. F. R. P., Whitmore, T. C., & Brown, G. C. (2000). Short-term effects of cyclone impact and long-term recovery of tropical rain forest on Kolombangara, Solomon Islands. *Journal of Ecology*, *88*(6), 1063–1078. https://doi.org/10.1046/j.1365-2745.2000.00517.x

Buyinza Faisal, J. K., & Vicent, S. (2021). Household air pollution and household health in Uganda. *Development Southern Africa*, *38*(3), 437–453. https://doi.org/10.1080/0376835X.2020.1856039

Cagney, K. A., & Lauderdale, D. S. (2002). Education, wealth, and cognitive function in later life. *The Journals of Gerontology: Series B*, *57*(2), P163–P172. https://doi.org/10.1093/geronb/57.2.P163

Carlsen, L., & Bruggemann, R. (2022). The 17 United Nations' sustainable development goals: A status by 2020. *International Journal of Sustainable Development & World Ecology*, *29*(3), 219–229. https://doi.org/10.1080/13504509.2021.1948456

Chafe, Z., & Chowdhury, S. (2021). A deadly double dose for India's poor. *Nature Sustainability*, *4*(10), 835–836. https://doi.org/10.1038/s41893-021-00752-0

Chafe Zoë A., Brauer, M., Klimont Z., Van Dingenen, R., Mehta, S., Shilpa, R., … Smith Kirk, R. (2014). Household cooking with solid fuels contributes to ambient PM2.5 air pollution and the burden of Disease. *Environmental Health Perspectives*, *122*(12), 1314–1320. https://doi.org/10.1289/ehp.1206340

Chattopadhyay, A., & Shaw, S. (2021). Association between air pollution and COVID-19 pandemic: An investigation in Mumbai, India. *GeoHealth*, *5*(7), e2021GH000383. https://doi.org/10.1029/2021GH000383

Chowdhury, S., Dey, S., Guttikunda, S., Pillarisetti, A., Smith, K. R., & Girolamo, L. D. (2019). Indian annual ambient air quality standard is achievable by completely mitigating emissions from household sources.

Proceedings of the National Academy of Sciences, *116*(22), 10711–10716. https://doi.org/10.1073/pnas.1900888116

Čukić, I., Kypridemos, C., Evans, A. W., Pope, D., & Puzzolo, E. (2021). Towards sustainable development goal 7 "Universal Access to Clean Modern Energy": National strategy in rwanda to scale clean cooking with bottled gas. *Energies*, *14*(15), 4582. https://doi.org/10.3390/en14154582

Dakua, M., Karmakar, R., & Barman, P. (2022). Exposure to indoor air pollution and the cognitive functioning of elderly rural women: A cross-sectional study using LASI data, India. *BMC Public Health*, *22*(1), 2272. https://doi.org/10.1186/s12889-022-14749-7

Dandona, R., Kumar, G. A., Henry, N. J., Joshua, V., Ramji, S., Gupta, S. S., … Dandona, L. (2020). Subnational mapping of under-5 and neonatal mortality trends in India: The Global Burden of Disease Study 2000–17. *The Lancet*, *395*(10237), 1640–1658. https://doi.org/10.1016/S0140-6736(20)30471-2

Deng, Q., Lu, C., Ou, C., Chen, L., & Yuan, H. (2016). Preconception, prenatal, and postnatal exposure to outdoor and indoor environmental factors on allergic diseases/symptoms in preschool children. *Chemosphere*, *152*, 459–467. https://doi.org/10.1016/j.chemosphere.2016.03.032

Deng, Q., Lu, C., Ou, C., & Liu, W. (2015). Effects of early life exposure to outdoor air pollution and indoor renovation on childhood asthma in China. *Building and Environment*, *93*, 84–91. https://doi.org/10.1016/j.buildenv.2015.01.019

Dida, G. O., Lutta, P. O., Abuom, P. O., Mestrovic, T., & Anyona, D. N. (2022). Factors predisposing women and children to indoor air pollution in rural villages, Western Kenya. https://doi.org/10.1186/s13690-022-00791-9

Du, X., Li, X., Qian, P., & Wu, H. (2022). Indoor air pollution from solid fuels use, inflammation, depression and cognitive function in middle-aged and older Chinese adults. *Journal of Affective Disorders*, *319*, 370–376. https://doi.org/10.1016/j.jad.2022.09.103

Ebi, K. L., & McGregor, G. (2008). Climate change, tropospheric ozone and particulate matter, and health impacts. *Environmental Health Perspectives*, *116*(11), 1449–1455. https://doi.org/10.1289/ehp.11463

Elf, J. L., Kinikar, A., Khadse, S., Mave, V., Suryavanshi, N., Gupte, N., … Golub, J. E. (2018). Sources of household air pollution and their association with fine particulate matter in low-income urban homes in India. *Journal of Exposure Science & Environmental Epidemiology*, *28*(4), 400–410. https://doi.org/10.1038/s41370-018-0024-2

Eze Ikenna C., Hemkens Lars G., Bucher Heiner C., Hoffmann Barbara, Schindler Christian, Künzli Nino, … Probst-Hensch Nicole M. (2015). Association between ambient air pollution and diabetes mellitus in Europe and North America: Systematic review and meta-analysis. *Environmental Health Perspectives*, *123*(5), 381–389. https://doi.org/10.1289/ehp.1307823

Ezzati, M., & Kammen, D. M. (2002). Household energy, indoor air pollution, and health in developing countries: Knowledge base for effective interventions. *Annual Reviews*. https://doi.org/10.1146/annurev.energy.27.122001.083440

Fatmi, Z., Coggon, D., Kazi, A., Naeem, I., Kadir, M. M., & Sathiakumar, N. (2014). Solid fuel use is a major risk factor for acute coronary syndromes among rural women: A matched case-control study. *Public Health*, *128*(1), 77–82. https://doi.org/10.1016/j.puhe.2013.09.005

Ferrucci, L., Cooper, R., Shardell, M., Simonsick, E. M., Schrack, J. A., & Kuh, D. (2016). Age-related change in mobility: Perspectives from life course epidemiology and geroscience. *The Journals of Gerontology: Series A, 71*(9), 1184–1194. https://doi.org/10.1093/gerona/glw043

Ghorani-Azam, A., Riahi-Zanjani, B., & Balali-Mood, M. (2016). Effects of air pollution on human health and practical measures for prevention in Iran. *Journal of Research in Medical Sciences*, *21*(1). Retrieved from https://journals.lww.com/jrms/fulltext/2016/21000/effects_of_air_pollution_on_human_health_and.65.aspx

Grazuleviciute-Vileniske, I., Seduikyte, L., Teixeira-Gomes, A., Mendes, A., Borodinecs, A., & Buzinskaite, D. (2020). Aging, living environment, and sustainability: What should be taken into account? *Sustainability*, *12*(5). https://doi.org/10.3390/su12051853

Halder, M., Saha, J., Roy, A., Roy, D., & Chouhan, P. (2024). Functional disability and its associated factors among the elderly in rural India using LASI Wave 1 data. *Journal of Public Health*, *32*(6), 1069–1081. https://doi.org/10.1007/s10389-023-01890-9

Hamplová, D., Klusáček, J., & Mráček, T. (2022). Assessment of self-rated health: The relative importance of physiological, mental, and socioeconomic factors. *Plos One*, *17*(4), e0267115. https://doi.org/10.1371/journal.pone.0267115

Hou, B., Liao, H., & Huang, J. (2018). Household cooking fuel choice and economic poverty: Evidence from a nationwide survey in China. *Energy and Buildings*, *166*, 319–329. https://doi.org/10.1016/j.enbuild.2018.02.012

International Institute for Population Sciences (IIPS), NPHCE, Mo HFW, Harvard T. H. Chan School of Public Health (HSPH), and the University of Southern California (USC). (2020). *Longitudinal Ageing Study in India (LASI) Wave 1,* 2017–18. India Report: (International Institute for Population Sciences, 2020).

IPCC (2021). (n.d.). *Climate Change 2021: The Physical Science Basis. Contribution of Working Group I to the Sixth Assessment Report of the Intergovernmental Panel on Climate Change.* Cambridge University Press.

Jagadale, S. R., & Kemper, J. (2022). 'Give It Up!': A macro-social marketing approach to India's clean cooking fuel access. *Journal of Macromarketing*, *42*(3), 433–453. https://doi.org/10.1177/02761467221107556

James, B. S., Shetty, R. S., Kamath, A., & Shetty, A. (2020). Household cooking fuel use and its health effects among rural women in southern India—A cross-sectional study. *Plos One*, *15*(4), e0231757. https://doi.org/10.1371/journal.pone.0231757

Jeuland, M., Pattanayak, S. K., & Bluffstone, R. (2015). The economics of household air pollution. *Annual Reviews*. https://doi.org/10.1146/annurev-resource-100814-125048

Jiang, W., Chen, H., Liao, J., Yang, X., Yang, B., Zhang, Y., … Yang, L. (2022). The short-term effects and burden of particle air pollution on hospitalization for coronary heart disease: A time-stratified case-crossover study in Sichuan, China. *Environmental Health*, *21*(1), 19. https://doi.org/10.1186/s12940-022-00832-4

Kankaria, A., Nongkynrih, B., & Gupta, S. K. (2014). Indoor air pollution in India: Implications on health and its control. *Indian Journal of Community Medicine*, *39*(4). Retrieved from https://journals.lww.com/ijcm/fulltext/2014/39040/indoor_air_pollution_in_india__implications_on.4.aspx

Kioumourtzoglou, M.-A., Power, M. C., Hart, J. E., Okereke, O. I., Coull, B. A., Laden, F., & Weisskopf, M. G. (2017). The association between air pollution and onset of depression among middle-aged and older women. *American Journal of Epidemiology*, *185*(9), 801–809. https://doi.org/10.1093/aje/kww163

Krishnamoorthy, Y., Rajaa, S., Ramasubramani, P., & Saya, G. K. (2022). Association between indoor air pollution and cognitive function among a nationally representative sample of middle-aged and older adults in India—A multilevel modeling approach. *Indoor Air*, *32*(1), e12929. https://doi.org/10.1111/ina.12929

Luo, Y., Zhong, Y., Pang, L., Zhao, Y., Liang, R., & Zheng, X. (2021). The effects of indoor air pollution from solid fuel use on cognitive function among middle-aged and older population in China. *Science of the Total Environment*, *754*, 142460. https://doi.org/10.1016/j.scitotenv.2020.142460

Mangoni, A. A., & Jackson, S. H. D. (2004). Age-related changes in pharmacokinetics and pharmacodynamics: Basic principles and practical applications. *British Journal of Clinical Pharmacology*, *57*(1), 6–14. https://doi.org/10.1046/j.1365-2125.2003.02007.x

Mocumbi, A. O., Stewart, S., Patel, S., & Al-Delaimy, W. K. (2019). Cardiovascular effects of indoor air pollution from solid fuel: Relevance to Sub-Saharan Africa. *Current Environmental Health Reports*, *6*(3), 116–126. https://doi.org/10.1007/s40572-019-00234-8

MOSPI. (2019). *Time Use in India Report, Time Use Survey (Jan–Dec 2019).* Ministry of Statistics and Programme Implementation, National Statistical Office, Government of India. Retrieved from Ministry of Statistics and Programme Implementation, National Statistical Office, Government of India website: www.mospi.gov.in

Mu, L., Liu, L., Niu, R., Zhao, B., Shi, J., Li, Y., … Zhang, Z.-F. (2013). Indoor air pollution and risk of lung cancer among Chinese female non-smokers. *Cancer Causes & Control:* CCC, *24*(3), 439—450. https://doi.org/10.1007/s10552-012-0130-8

Murray, C. J. L., Aravkin, A. Y., Zheng, P., Abbafati, C., Abbas, K. M., Abbasi-Kangevari, M., … Lim, S. S. (2020). Global burden of 87 risk factors in 204 countries and territories, 1990–2019: A systematic analysis for the Global Burden of Disease Study 2019. *The Lancet*, *396*(10258), 1223–1249. https://doi.org/10.1016/S0140-6736(20)30752-2

Nadeau, K. C., Agache, I., Jutel, M., Annesi Maesano, I., Akdis, M., Sampath, V., … Akdis, C. A. (2022). Climate change: A call to action for the United Nations. *Allergy*, *77*(4), 1087–1090. https://doi.org/10.1111/all.15079

Office of the Registrar & General & Census Commissioner, India. (2011). *Census of India*. New Delhi. Retrieved from https://censusindia.gov.in/2011census/population_enumeration.html

Oluwatosin, D., Ola, K., Olayinka, A., et al. (2022). *The Impacts of the Use of Biomass Solid Fuels for Household Cooking in Sub-Saharan Africa—A Review*. Retrieved from https://doi.org/10.21203/rs.3.rs-1840692/v1

Ortiz, M., Itard, L., & Bluyssen, P. M. (2020). Indoor environmental quality related risk factors with energy-efficient retrofitting of housing: A literature review. *Energy and Buildings*, *221*, 110102. https://doi.org/10.1016/j.enbuild.2020.110102

Padmanaban, D. (2024). *India's Disadvantaged Groups Face more Air Pollution* (No. 105). Eos. Retrieved from Eos website: https://doi.org/10.1029/2024EO240007

Pandey, A., Brauer, M., Cropper, M. L., Balakrishnan, K., Mathur, P., Dey, S., … Dandona, L. (2021). Health and economic impact of air pollution in the states of India: The Global Burden of Disease Study 2019. *The Lancet Planetary Health*, *5*(1), e25–e38. https://doi.org/10.1016/S2542-5196(20)30298-9

Parra, C., Kirschke, J., & Ali, S. H. (2020). Ensure access to affordable, reliable, sustainable, and modern energy for all. In *Mining, Materials, and the Sustainable Development Goals (SDGs)* (pp. 61–68). Taylor & Francis. Retrieved from https://doi.org/10.1201/9780367814960-7

Patz, J. A., Frumkin, H., Holloway, T., Vimont, D. J., & Haines, A. (2014). Climate change: Challenges and opportunities for global health. *JAMA*, *312*(15), 1565–1580. https://doi.org/10.1001/jama.2014.13186

Peters, A., Veronesi, B., Calderón-Garcidueñas, L., Gehr, P., Chen, L. C., Geiser, M., … Schulz, H. (2006). Translocation and potential neurological effects of fine and ultrafine particles are a critical update. *Particle and Fibre Toxicology*, *3*(1), 13. https://doi.org/10.1186/1743-8977-3-13

Ramanathan, V., & Carmichael, G. (2008). Global and regional climate changes due to black carbon. *Nature Geoscience*, *1*(4), 221–227. https://doi.org/10.1038/ngeo156

Rani, R., Arokiasamy, P., Meitei, W. B., & Sikarwar, A. (2023). Association between indoor air pollution and cognitive function of older adults in India: A cross-sectional multilevel analysis. *Journal of Public Health*, *31*(3), 369–379. https://doi.org/10.1007/s10389-021-01521-1

Rao, N. D., Kiesewetter, G., Min, J., Pachauri, S., & Wagner, F. (2021). Household contributions to and impacts from air pollution in India. *Nature Sustainability*, *4*(10), 859–867. https://doi.org/10.1038/s41893-021-00744-0

Rehfuess, E. (2006). *Fuel for Life: Household Energy and Health*. World Health Organization. Retrieved from World Health Organization website: https://iris.who.int/bitstream/handle/10665/43421/9241563168_eng.pdf

Saenz, J. L., Adar, S. D., Zhang, Y. S., Wilkens, J., Chattopadhyay, A., Lee, J., & Wong, R. (2021). Household use of polluting cooking fuels and late-life cognitive function: A harmonized analysis of India, Mexico, and China. *Environment International*, *156*, 106722. https://doi.org/10.1016/j.envint.2021.106722

Saenz, J. L., Wong, R., & Ailshire, J. A. (2018). Indoor air pollution and cognitive function among older Mexican adults. *Journal of Epidemiology & Community Health*, *72*(1), 21–26. https://doi.org/10.1136/jech-2017-209704

Sehgal, M., Rizwan, S. A., & Krishnan, A. (2014). Disease burden due to biomass cooking-fuel-related household air pollution among women in India. *Global Health Action*, *7*(1), 25326. https://doi.org/10.3402/gha.v7.25326

Sharma, D., Mazta, S. R., & Parashar, A. (2013). Prevalence of cognitive impairment and related factors among elderly: A population-based study. *Journal of Dr. NTR University of Health Sciences*, *2*, 171.

Shaw, S., Kundu, S., Chattopadhyay, A., & Rao, S. (2024). Indoor air pollution and cognitive function among older adults in India: A multiple mediation approach through depression and sleep disorders. *BMC Geriatrics*, *24*(1), 81. https://doi.org/10.1186/s12877-024-04662-6

Shindell, D., Kuylenstierna, J. C. I., Vignati, E., Dingenen, R. van, Amann, M., Klimont, Z., … Fowler, D. (2012). Simultaneously mitigating near-term climate change and improving human health and food security. *Science*, *335*(6065), 183–189. https://doi.org/10.1126/science.1210026

Sigsgaard, T., Forsberg, B., Annesi-Maesano, I., Blomberg, A., Bølling, A., Boman, C., … Brunekreef, B. (2015). Health impacts of anthropogenic biomass burning in the developed world. *European Respiratory Journal*, *46*(6), 1577–1588. https://doi.org/10.1183/13993003.01865-2014

Singh, A., & Chattopadhyay, A. (2023). Malnutrition among older adults in India: Does gender play a role? *Aging and Health Research*, *3*(2), 100143. https://doi.org/10.1016/j.ahr.2023.100143

Singh, P., Roy, A., Bhasin, D., Kapoor, M., Ravi, S., & Dey, S. (2021). Crop fires and cardiovascular health – A study from North India. *SSM – Population Health*, *14*, 100757. https://doi.org/10.1016/j.ssmph.2021.100757

Sinha, D., & Ray, M. R. (2015). Health effects of indoor air pollution due to cooking with biomass fuel. In S. M. Roberts, J. P. Kehrer, & L.-O. Klotz (Eds.), *Studies on Experimental Toxicology and Pharmacology* (pp. 267–302). Cham: Springer International Publishing. https://doi.org/10.1007/978-3-319-19096-9_14

Smith, K. R., Jerrett, M., Anderson, H. R., Burnett, R. T., Stone, V., Derwent, R., … Thurston, G. (2009). Public health benefits of strategies to reduce greenhouse-gas emissions: Health implications of short-lived greenhouse pollutants. *The Lancet*, *374*(9707), 2091–2103. https://doi.org/10.1016/S0140-6736(09)61716-5

Szyszkowicz, M., Kousha, T., Kingsbury, M., & Colman, I. (2016). Air Pollution and Emergency Department Visits for Depression: A Multicity Case-Crossover Study. *Environmental Health Insights*, *10*, EHI.S40493. https://doi.org/10.4137/EHI.S40493

Timperley J. (2019). *The Carbon Brief Profile: India*. Retrieved from https:// www.carbonbrief.org/the-carbon-brief-profile-india/

Wang, Q., & Yang, Z. (2018). Does chronic disease influence susceptibility to the effects of air pollution on depressive symptoms in China? *International Journal of Mental Health Systems*, *12*(1), 33. https://doi.org/10.1186/s13033-018-0212-4

WHO. (2021). *Air Pollution*. Retrieved from https://www.who.int/health-topics/a ir-pollution#tab=tab_1A

WHO. (2022). *Air Quality Database*. Retrieved from https://www.who.int/data/gho/data/themes/air-pollution/who-air-quality-database

Wilkinson, P., Smith, K. R., Davies, M., Adair, H., Armstrong, B. G., Barrett, M., … Chalabi, Z. (2009). Public health benefits of strategies to reduce greenhouse-gas emissions: Household energy. *The Lancet*, *374*(9705), 1917–1929. https://doi.org/10.1016/S0140-6736(09)61713-X

Wong, A., Lou, W., Ho, K., Yiu, B. K., Lin, S., Chu, W. C., … Mok, V. C. (2020). Indoor incense burning impacts cognitive functions and brain functional connectivity in community older adults. *Scientific Reports*, *10*(1), 7090. https://doi.org/10.1038/s41598-020-63568-6

Wu, F., Guo, Y., Zheng, Y., Ma, W., Kowal, P., Chatterji, S., & Wang, L. (2016). Social-economic status and cognitive performance among Chinese aged 50 years and older. *Plos One*, *11*(11), e0166986. https://doi.org/10.1371/journal.pone.0166986

Yeatts, K. B., El-Sadig, M., Leith, D., Kalsbeek, W., Al-Maskari, F., Couper, D., … Olshan, A. F. (2012). Indoor air pollutants and health in the United Arab Emirates. *Environmental Health Perspectives*, *120*(5), 687–694. https://doi.org/10.1289/ehp.1104090

Zha, Z., Li, G., Lv, Y., Liu, L., He, J., Xu, W., … Huang, F. (2022). The effects of air pollution on the lung cancer mortality in rural areas of eastern China: A multi-region study. *Environmental Science and Pollution Research*, *29*(30), 45716–45729. https://doi.org/10.1007/s11356-022-19027-y

14 Climate Change and Health Vulnerability

Examining the Impacts on Flood-Affected Marginalized Communities in Nirmal Char

Snehashis Alam and Mohan Kumar Bera

INTRODUCTION

Climate change is increasing the frequency and severity of natural hazards, threatening people's lives and livelihoods, and disrupting services (Bowen et al., 2024; Codjoe et al., 2020). Some of the direct impacts of climate change are sea-level rise, extreme weather events, including drought and flooding, increases in temperature leading to wildfires, and changes in rainfall patterns (Klipper et al., 2021). Extreme weather events cause damage to physical infrastructures, injury, loss of life, and reduced supply of food and safe drinking water (Codjoe et al., 2020). Floods are the most disastrous natural hazards, mainly caused by accidental discharges from water reservoirs such as dams, snowmelt, tidal influences, or, primarily, extreme rainfall (Bowen et al., 2024; Mandal, Dharanirajan, et al., 2023). The low-lying areas are highly prone to floods (Mandal, Sarkar, et al., 2023), and a large number of people live along the floodplains, deltas, coastal areas, and chars. According to the Intergovernmental Panel on Climate Change (IPCC), low-lying areas, including coastal regions and small islands, are the most vulnerable areas to climate change (IPCC, 2022) Which is caused by sea-level rise, storm surges, floods, and erosion (Bindoff et al., 2022; IPCC, 2022; Collins et al., 2022; Oppenheimer et al., 2022).

Char lands are unique low-lying areas in major rivers' floodplains, particularly in South and Southeast Asia. These areas are characterized by their dynamic nature, formed and re-formed by shifting river courses and silt deposits (Mukherjee, 2011; Rudra, 2003). Chars, which are poorly connected to the mainland, are highly vulnerable to floods and river erosion. Despite this, they have become permanent homes for thousands from low-income or ultra-poor communities (Alam & Bera, 2024; Islam, 2021; Paul & Islam, 2015). Living and working conditions in char lands are poor, marked by issues such as illiteracy, lack of social awareness, and inadequate communication, and transportation (Das & Samanta, 2022; Mollah, 2011). These factors limit access to local government services, including legal aid, healthcare, education, banking, and policing (Mollah, 2011). Climate change-induced hydro-climatic hazards have increased the vulnerability to accessing adequate livelihood and health services in these areas (Barua & Hazarika, 2008). Frequent floods cause poor mental health issues, contagious diseases, and skin infections (Mandal et al., 2023). Floods cause significant economic losses and unemployment, adversely affecting communities' ability to access health services (Hossain et al., 2024). They also lead to significant damage to physical and social infrastructures (Mandal, Dharanirajan, et al., 2023), leads to a lack of health services, insufficient food supply, displacement of populations, overcrowding in shelters, poor sanitation, and an increased incidence of water- and vector-borne diseases (Alam & Bera, 2024; Islam, 2021). The collapse of the transportation system has caused extreme suffering to the people of the char areas. It has

DOI: 10.1201/9781003620747-16

often become impossible to reach the hospital or collect necessary medicines in times of emergency (Alam & Bera, 2024). In such a situation, uncertainty has arisen in the daily lives of those areas, making them more physically and mentally vulnerable. Children and the elderly are especially at risk in this situation, as they cannot walk without the help of others, and even the slightest lack of medical care increases their risk of death (Ghosh & Ghosal, 2021).

Elderly people often face difficulties in accessing emergency services and medical facilities (Krongthaeo et al., 2021). They are at increased risk from climate change-induced floods, as their existing health problems are exacerbated (Chan, 2019). Social isolation and limited access to medical care during disasters exacerbate health risks. In addition, flood-related physical stress can exacerbate chronic diseases (Bei et al., 2013). Many elderly people rely on their families for daily care, which is particularly noticeable in female-headed households, where male members are often busy working outside the home, making it difficult to receive adequate support during a crisis (L. Wang et al., 2020).

Children aged 6–12 are particularly vulnerable to climate-induced disasters because they are at a sensitive stage of development (Baez & Santos, 2007; Sajid & Bevis, 2021). Participation in regular activities such as school and sports makes them more vulnerable to the effects of climate change-induced floods (Suri et al., 2014). The floods have caused widespread disruption to the education system, food and nutrition shortages, and access to healthcare, which have had long-term negative impacts on children's development. These challenges further limit children's future opportunities. At the same time, the World Health Organization warns that such situations can lead to psychological distress, such as anxiety, behavioural changes, or lack of self-confidence (WHO, 2023). In this context, the implementation of child-centred psychosocial support and protection programmes is essential.

UNDERSTANDING THE HEALTH VULNERABILITY OF CHILDREN AND ELDERLY PEOPLE IN THE CONTEXT OF CLIMATE-INDUCED DISASTERS

Floods are one of the world's most frequent and catastrophic climate-induced disasters, resulting in increased mortality, economic losses, and environmental damage, especially in developing and underdeveloped countries (Alam & Bera, 2024; Srikuta et al., 2015). Floods cause physical injuries, worsen existing medical conditions and mental health problems, and spread infectious diseases among children, the elderly, and other vulnerable people in society (Table 14.1).

Children are highly vulnerable to floods due to their physical limitations and reliance on family members, which poses significant health risks such as drowning, infections, malnutrition, and psychological trauma. Their underdeveloped immune systems make them particularly susceptible to the health effects of floods. Flooding disrupts food supplies and livelihoods, leading to severe nutritional deficiencies and diseases among children (Agabiirwe et al., 2022a; Mirzabaev et al., 2023; J. Rodriguez-Llanes et al., 2016; J. M. Rodriguez-Llanes et al., 2011). They are also at higher risk for waterborne and vector-borne illnesses in affected areas, with a noticeable increase in illnesses during the flood season caused by contaminated drinking water sources (Mirzabaev et al., 2023; P. Wang et al., 2023). Socio-economic factors further exacerbate their vulnerability; households in lower economic strata often lack adequate housing, healthcare access, clean water, and proper sanitation, increasing health risks for children in flood-prone regions (Krieger & Higgins, 2002).

Research shows that disasters affect the physical well-being of elderly people. Their vulnerability depends on physical abilities, chronic health issues, and socio-economic support systems. Disrupted health facilities create major challenges due to chronic conditions such as diabetes, hypertension, and cardiovascular diseases (Alam & Bera, 2024; Okaka & Odhiambo, 2019; Srikuta et al., 2015). A lack of health facilities and restricted medications after disasters can increase illness and mortality (Saulnier et al., 2018). Evacuating elderly people during disasters becomes challenging because of restricted mobility and caregiver dependency (Krongthaeo et al., 2021). Many elderly people are

TABLE 14.1
Health Effects of Floods on Children and the Elderly

Impact of Flood	Description	Health Effects	References
Water contamination	Flooding can contaminate drinking water sources with sewage, agricultural runoff, and industrial pollutants, leading to increased levels of harmful bacteria, pathogens, and chemicals.	Contaminated water and poor sanitation contribute to the spread of diseases such as cholera, diarrhoea, dysentery, hepatitis A, typhoid, and various gastrointestinal illnesses.	Andrade et al. (2018) and P. Wang et al. (2023)
Injury and drowning	Floods can lead to physical injuries caused by debris, falls, accidents during evacuation or clean-up, and the risk of drowning.	These injuries may include cuts, bruises, fractures, head injuries, and drowning.	Mandal, Dharanirajan, et al. (2023) and Okaka and Odhiambo (2019)
Vector-borne diseases	Flooding increases viable vector breeding areas and accelerates breeding due to faster egg hatching in hotter climates.	That causes malaria, dengue, yellow fever, chikungunya, Zika, and lymphatic filariasis.	Bennett and Friel (2014) and Srikuta et al. (2015)
Respiratory issues	Flooding can worsen respiratory issues due to housing damage and compromised health systems, increasing exposure and vulnerability to viral respiratory infections.	Asthma, bronchitis, allergies, and other respiratory infections.	Mirsaeidi et al. (2016)
Food security	Flooding destroys crops, disrupts food supplies, and contaminates food stores.	Malnutrition, foodborne illnesses, and hunger-related health issues.	Echendu (2022)
Malnutrition	Floods can exacerbate undernutrition, affecting food security, leading to inadequate childcare practices, and compromising water and sanitation systems.	Undernutrition, lack of essential micronutrients (such as vitamin A and iron), and a weakened immune system.	Agabiirwe et al. (2022b) and Bennett and Friel (2014)
Healthcare access	Damage to healthcare facilities and infrastructure can impede the accessibility and availability of medical services.	Exacerbation of long-standing medical conditions, limited access to emergency medical treatment, and disrupted medication distribution.	Alam and Bera (2024) and Srikuta et al. (2015)

poor and lack the resources for recovery. Social isolation exacerbates psychological impacts, including anxiety, depression, and post-traumatic stress disorder, while a disrupted healthcare system limits access to necessary medical and psychological support (Bei et al., 2013; Sarkar et al., 2023).

NEED FOR THE STUDY

The studies available on flooding, climate change, and health vulnerabilities indicate a significant gap in understanding the unique circumstances of chars (riverine islands) (Agabiirwe et al., 2022a; Bei et al., 2013; Krongthaeo et al., 2021; Mirzabaev et al., 2023; Phraknoi et al., 2023; J. Rodriguez-Llanes et al., 2016; J. M. Rodriguez-Llanes et al., 2011). These areas are very unstable, often experience flooding, and are socially and economically disadvantaged (Mukherjee, 2011; Rudra, 2003). Climate change has increased the frequency and severity of floods in the char region, heightening health risks associated with vector-borne and waterborne diseases. This situation has further exacerbated food insecurity and limited access to essential healthcare services.

The ecological characteristics and extent of damage in char areas differ from other flood-prone regions due to geographical isolation, lack of permanent infrastructure, and limited access to healthcare, drinking water, and sanitation facilities. This unique situation increases health risks for char dwellers, especially school-aged children and the elderly. However, these risks remain unexplored in the existing literature.

OBJECTIVES

This chapter analyses the impact of climate change-induced floods on children aged 6–12 years and elderly people aged 60 years and above in Nirmal Char, Murshidabad district, India. Basically, it assesses how and to what extent the health risks of these two groups become evident during and after floods. In addition, it includes an analysis of how water and sanitation systems become compromised and what kind of impact this has on the health of these populations. The main objective of the study is to recommend effective policies and targeted interventions for vulnerable populations in flood-prone areas so that health risks can be reduced and resilience can be increased. The objectives of this study are as follows:

i. To assess the level of health risks of school-going children (6–12 years) and elderly people (60 years and above) during and after climate-induced floods.
ii. To analyse the impact of water and sanitation system degradation on the health of children and elderly people during crises.

THEORETICAL FRAMEWORK

This chapter uses the IPCC vulnerability framework and the Social Determinants of Health (SDOH) framework to analyse the nature and extent of health risks to children aged 6–12 years and the elderly population in the char region under climate change-affected flood conditions. This analytical framework allows us to understand that the context is not only environmental hazards but also multidimensional and interrelated processes.

The IPCC framework captures vulnerability in terms of how and to what extent a system is affected by the impact of natural hazards. This framework considers three main components: exposure, or direct contact with the hazard; sensitivity, that is, who is most affected by the response to the hazard; and adaptive capacity, which refers to the ability of individuals or societies to adapt to the crisis (IPCC, 2001a; IPCC, 2001b).

In this context, exposure is defined as the geographical and social context in which people and infrastructure are exposed to risk (Tate et al., 2021). Sensitivity analysis can help to understand how health systems, nutritional status, and psychosocial stressors exacerbate risk levels among these groups. Adaptation is constrained when health infrastructure, financial resources, or social support systems are inadequate. An interrelated analysis of this framework sheds light on the complexities of the lives of nomads and the underlying factors that underlie their daily disaster resilience. This chapter is not only concerned with identifying risks but also with understanding why some groups are repeatedly pushed to the margins and how policy interventions, social supports, and planned adaptation measures can provide practical pathways to recovery.

SDOH are non-medical factors that influence an individual's health outcomes and overall well-being (WHO, 2013). SDOH are defined as the structural determinants and conditions of how people are born, grow up, live, work, and age (Marmot et al., 2008), and are recognized worldwide as important determinants of health outcomes. These factors have a direct impact on people's health, well-being, and quality of life (Nomura et al., 2016). SDOH can generally be divided into five main categories: economic stability, educational access and quality, healthcare access and quality, neighbourhood and built environment, and social and community context (WHO, 2013). Each of these factors can have a significant impact on health outcomes for residents of flooded char areas. Therefore, this chapter uses the SDOH framework to analyse how multiple social determinants are associated with health risks in flood-affected char dwellers and how the interconnections between these risks operate in the context of communities living in char areas. This analysis can help determine specific intervention plans and sustainable adaptation strategies.

This chapter comprises five sections: Introduction, Materials and Methods, Results, Discussion, and Conclusion. The introduction covers chars' characteristics, the hazards' impacts on char dwellers, the research's aim and objective, and the understanding of vulnerability assessment frameworks. The "Materials and Methods" section outlines the research methodology. The "Results" section presents the research results. The "Discussion" section discusses them, and the "Conclusion" section concludes the chapter.

MATERIALS AND METHODS

STUDY AREA

The sandy areas that emerge as islands in river channels are chars. These land formations now divide India and Bangladesh in Murshidabad. Within the Indian territory of Murshidabad, several chars have formed in the Padma River, specifically in Bhagwangola II, Raghunathganj II, Raninagar-II, and Jalangi blocks. The largest among them is Nirmal Char, located roughly between 24° 19′ N to 24° 22′ N latitude and 88° 22′ E to 88° 31′ E longitude of Bhagwangola II block in Murshidabad (Figure 14.1). Nirmal Char encompasses roughly 61 square kilometers and comprises villages like, Patibona, Char Krishnapur, Ghoshpara, Machchar, Mansurpur, Paikmari, and Natun Rajapur. The population of this region around 10000.

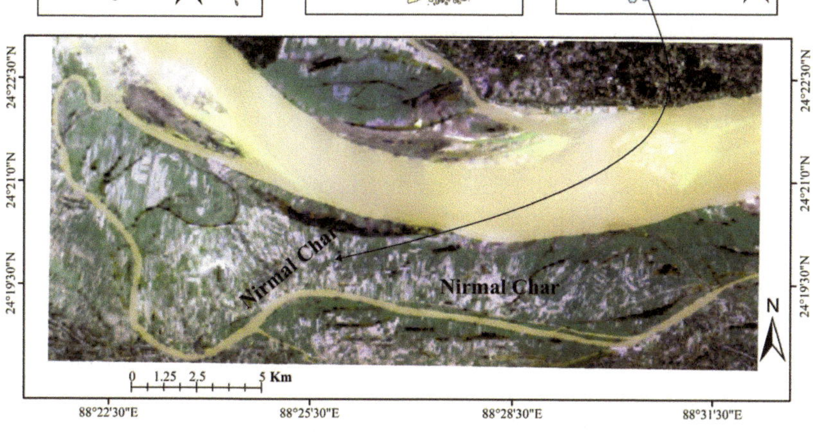

FIGURE 14.1 Nirmal Char in Murshidabad District.

Methodology

This chapter employed a mixed-method approach to assess the health vulnerability of school-aged children and the elderly in the flood-affected Nirmal Char in Murshidabad. This methodology integrates both quantitative and qualitative data collection methods to provide a comprehensive understanding of the health challenges faced by char dwellers.

METHODS

Quantitative data were collected through household surveys. A total of 133 households from eight villages of Nirmal Char were randomly selected for the semi-structured questionnaire survey. The sample size was determined using a sample size calculator on the website https://surveysystem.com/sscalc.htm. For the Nirmal Char sample size at a 95% confidence level, the confidence interval was determined to be 8.44. The survey targeted the heads of households, as they were responsible for their families' well-being. Socio-demographic data, including age, gender, occupation, education, marital status, number of household members, number of male household members aged 60 years and above living in the household, economic status, and lifestyle, were collected. Information was also collected on health care services, water and sanitation, prevalence of waterborne diseases, use of sanitation facilities, quality of drinking water, disease reports, methods of managing health problems, and government assistance during floods.

Qualitative data were collected through in-depth interviews with local government officials, elected panchayat members, and affected households to investigate the impact on health, coping strategies, and effectiveness of healthcare during floods. This approach provides insights into the socio-cultural factors influencing health vulnerabilities related to climate change and socio-economic deprivation.

Quantitative data were analysed using Microsoft Excel 2021, while qualitative data obtained from interviews were transcribed and thematically analysed to identify challenges and community perceptions about health vulnerabilities. Ethical clearance was obtained from the Human Research Ethics Committee (*HEC/BPGC/2023/025*) of Birla Institute of Technology and Science, Pilani K.K. Birla Goa Campus.

RESULTS

Socio-Economic Profile of the Study Households

The demographic and socio-economic profile analysis of the 133 respondents from Nirmal Char (Table 14.2) indicates that 67.67% are male and 32.33% are female (Figure 14.2a). The age distribution of the respondents reveals that 15.79% are between 18 and 30 years, 24.06% are 31–40 years, 18.05% are 41–50 years, 14.29% are 51–59 years, and 27.82% are above 60 years (Figure 14.2b). While most household heads are male (90.23%), there are only 9.77% female-headed households. On average, families in the study area have approximately 3.73, with 42.11% having 1–3 members, 48.87% having 4–5 members, and 9.02% having more than 6 members (Figure 14.2c). Educational levels are relatively modest, with 33.08% having education up to Class 1–5, 22.56% up to Class 5–8, 18.80% up to secondary level, and 9.77% up to higher secondary. In comparison, 15.79% of the population is illiterate (Figure 14.2d). The majority of them owned small farmlands, which were less than four bighas in size. However, due to frequent floods and river-bank erosion, some portion of their land was lost to the river every year. All of the respondents were married, and none of them were divorced or unmarried. As for occupations, 51.13% are involved in agriculture, 36.09% in wage labour, 3.76% in fishing, and 9.02% in other occupations (Figure 14.2e). Housing conditions indicate that 29.32% live in semi-pucca houses, while 70.68% reside in huts (Figure 14.2f). Most households (96.99%) have lived there for 10 years or more. Moreover, 44.36% of households have school-aged children (6–12 years), 27.82% have elderly members above 60 years, and 64.86% of the

TABLE 14.2
Socio-Demographic Characteristics of Respondents

Variables	Categories	Respondents (N = 133)	Percentage (%)
Sex	Male	90	67.67
	Female	43	32.33
Age	18–30	21	15.79
	31–40	32	24.06
	41–50	24	18.05
	51–59	19	14.29
	Above 60	37	27.82
Gender of the household head	Male	120	90.23
	Female	13	9.77
Family size	1–3	56	42.11
	4–5	65	48.87
	Above 6	12	9.02
Education	Class 1–5	44	33.08
	Class 5–8	30	22.56
	Secondary	25	18.80
	Higher secondary	13	9.77
	Illiterate	21	15.79
Occupations	Agriculture	68	51.13
	Wage labour	48	36.09
	Fishing	5	3.76
	Others	12	9.02
Types of houses	Semi-pucca	39	29.32
	Hut	94	70.68
Length of residence (years)	Below 10	4	3.01
	10 and above	129	96.99
Households with school-aged children (6–12 years)	Yes	59	44.36
	No	74	55.64
Households with the elderly above 60 years	Yes	37	27.82
	No	96	72.18
The elderly with chronic illness	Yes (out of 37)	24	64.86
	No (out of 37)	13	35.14
Distance from the nearest health facility	Less than 5 km	15	11.28
	6–10 km	101	75.94
	Above 10 km	17	12.78
Health facilities impacted by the recent flood	Yes	133	100.00
	No	0	0.00
Damaged drinking water supply infrastructure	Yes	118	88.72
	No	15	11.28

elderly suffer from chronic illnesses. Access to health facilities is challenging, with 75.94% living 6–10 km from the nearest facility. Furthermore, 88.72% of households experienced damage to their drinking water supply infrastructure due to recent floods.

Health Vulnerabilities of Children during the Flood Season

Waterborne diseases are prevalent among children in flood-affected areas, with over 88% of households reporting diarrhoea. More than 90% experienced common health issues, including fevers, coughs, malaria, and diarrhoea, in the month following the flood. One mother noted, "Malaria

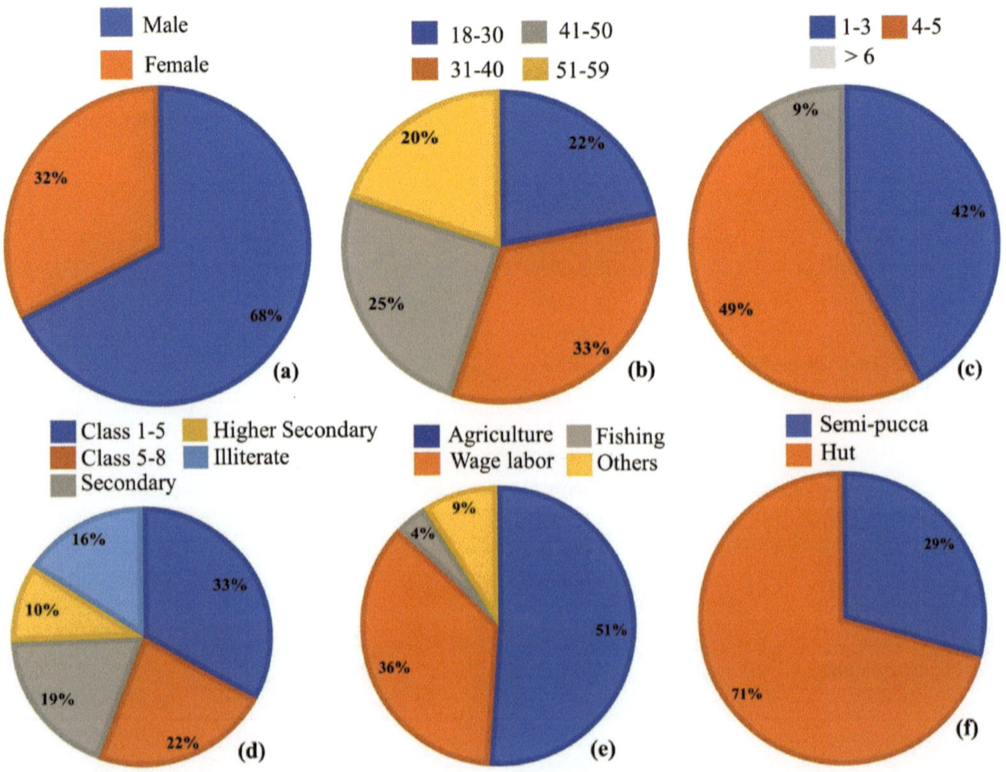

FIGURE 14.2 (a) Sex of the participants, (b) age of the participants, (c) family size of the participants, (d) education level of the participants, (e) occupation types of the participants, and (f) types of houses of the participants.

and diarrhoea are the primary diseases in Nirmal Char during the flood season, primarily caused by unsafe drinking water. Vector-borne diseases also become prominent due to prolonged waterlogging."

The survey indicates that all households rely on tube wells for drinking and household water, yet 94.74% do not purify or boil the water before consumption. Only 6% utilize boiled water or halogen tablets for purification. Additionally, 79.70% use temporary pit latrines, while just 20.30% have access to permanent latrines. Over 85% reported using ash for handwashing after toilet use, with only 5% using soap. During floods, 91.74% had access to unpurified drinking water, and 86.44% practiced open defecation due to a lack of sanitation facilities.

Health-seeking behaviour heavily depends on access to health facilities. The survey found that over 73% of respondents prefer allopathic treatment, while the rest favour homeopathy. However, assessing health vulnerability is difficult due to long distances to medical care and inadequate transportation. In Nirmal Char, only 11.28% of households live within a 5-km radius of the Block Public Health Centre (BPHC), while 88.72% reside 6–15 km away. Poor road conditions necessitate reliance on boats for transport, with tractors, motorbikes, and vans used seasonally. The lack of proper transportation complicates healthcare access, creating a significant burden for char dwellers. This, combined with limited healthcare services, worsens the immediate health impacts and long-term consequences on children. The study found that educational continuity was severely disrupted, with children missing school for months, particularly during floods. Over 55% of households reported changing primary income sources multiple times in the last decade, with the agricultural sector significantly affected by crop loss, leading to food shortages.

TABLE 14.3
Health Vulnerabilities of Children during the Flood Season

Health Issue	Findings
Waterborne diseases	88.14% of households reported diarrhoea; malaria and diarrhoea were prevalent due to unsafe drinking water.
Vector-borne diseases	Malaria cases are rising due to stagnant water and poor sanitation.
Unsafe drinking water	91.74% of households accessed unpurified water; only 6% purified it.
Sanitation & hygiene	86.44% practiced open defecation; 85% used ash for handwashing, with only 5% using soap.
Healthcare access	About 88.72% lived 6–15 km from health centres, facing access issues due to poor transportation.
Education disruptions	30% drop-out rate, leading to child labour and migration.
Economic impact on children	55% of households changed their sources of income, and children are engaging in labour due to financial struggles.
Food & nutrition	65% of households faced food shortages, resulting in a 36% increase in child malnutrition.

Economic instability has increased socio-economic vulnerability, prompting families to require children to contribute to financial needs, contributing to a 30% school drop-out rate. Many dropouts become child labourers or migrate with family members in search of work. As male members seek jobs, women (9.77% of the sample) take on multiple roles, managing family needs amidst unstable financial support. Reports of physical abuse towards children in these areas have also increased. Floods adversely affect both physical and mental health. Moreover, over 65% of households reported inadequate food supply, leading to a 36% increase in child malnutrition in Nirmal Char. Programmes like the mid-day meal initiative fail to function effectively during floods, resulting in reduced food consumption and inadequate support for children. This inadequate nutrition, coupled with physical and mental stress and harsh living conditions, significantly impacts the growth and development of children on char land (Table 14.3).

Health Vulnerabilities of Elderly People during the Flood Season

A total of 64.86% of elderly people reported chronic health issues such as diabetes, hypertension, and arthritis, facing challenges in receiving treatment due to limited access to healthcare. This issue affects 75.94% of those who live 6–10 km from the nearest hospital, particularly during the flood season. A 72-year-old widow explained, "I cannot afford medical treatment for myself and my family. We struggle with unpaved roads and limited financial resources when floods hit. Floods destroy everything, making clean water hard to access, and although unsafe water causes illness, we have no choice." Food insecurity affects 65% of elderly respondents, stemming from financial instability and inadequate pensions. A 65-year-old shared, "Floods and river erosion have washed away my house multiple times over the last 30 years. I've relocated due to erosion, leaving me with few resources. I live in a hut made of bamboo and cannot afford proper medical treatment when I am ill."

Elderly people also face risks from inadequate sanitation and limited access to safe drinking water. According to the survey, 91.53% of respondents consume unsafe water during floods, and 86.44% practice open defecation, increasing the likelihood of disease. Many compromises hygiene due to these circumstances. With women often taking on additional responsibilities after the migration of male family members, elderly people struggle to manage their health issues effectively.

TABLE 14.4
Health Vulnerabilities of Elderly People during the Flood Season

Health Issue	Findings
Chronic diseases	64.86% reported diabetes, hypertension, and arthritis; limited access to healthcare.
Distance to healthcare	With poor transportation facilities, 75.94% lived 6–10 km from the nearest hospital.
Food & nutrition	65% faced food insecurity due to financial instability and inadequate pensions.
Unsafe drinking water	91.53% consumed unpurified water during floods.
Sanitation & hygiene	86.44% practiced open defecation, increasing the risk of waterborne diseases.

Social isolation and unstable economic conditions further worsen their health risks. Many elderly people live alone or receive minimal assistance, making them more vulnerable to the impacts of flooding and agricultural loss (Table 14.4).

DISCUSSION

The study found that the living environment in Nirmal Char is very poor. The housing, sanitation, and environmental conditions here are not at all hygienic. Especially during floods, the elderly and school-going children face serious health risks. Most families live in semi-permanent houses made of tin or bamboo, while some families spend their days in temporary shelters made of polythene or plastic. Due to the lack of a proper drainage system, water accumulates during rains or floods, which creates an environment conducive to the spread of waterborne diseases. As a result, the risk of diseases like diarrhoea, cholera, and malaria increases significantly. Many families who participated in the survey clearly mentioned these health problems and risks.

The current research revealed that 88.14% of households reported diarrheal illness during the recent major floods, which also aligns with the outcomes of similar studies conducted in flood-vulnerable informal settlements. For instance, Okaka and Odhiambo (2019) indicated in a health risk study in Mombasa, Kenya, that household factors, such as water, sanitation, and environmental risk factors, had a substantial impact on health risks during floods. Also, a study by Sakijege et al. (2012) reported that 90% of families in the Keko Machungwa informal settlement in Dar es Salaam utilized pit latrines for sanitation and resided in places with inadequate or no drainage facilities, thereby enhancing health hazards. It also reported that approximately 70.68% of homes were constructed of mud and bamboo and thus were very vulnerable to destruction by floods. Fatemi et al. (2020) demonstrate that poor housing raises health threats in emergency situations, with emphasis on the relationship between housing conditions and health consequences in times of crisis and the necessity for intervention. Sanitation conditions are extremely poor, and there is a high rate of open defecation (86.44%).

The study found that the resilience of people in Nirmal Char is quite limited. The main reasons for this are inadequate sanitation facilities, economic uncertainty, and lack of adequate healthcare. Only 11.28% of households in the area live within 5 km of the nearest health centre, which becomes a major problem during floods, especially for female-headed households or households with elderly members. Health care limitations play a significant role in contributing to the risk of chronic diseases. Food insecurity is another pressing issue, particularly during and after floods. Nearly 36% of children are malnourished, and more than 65% of households report that they do not have enough food. These data show that natural disasters damage homes and assets, and reduce people's access to food and health care. This increases the risk of chronic diseases (Bhutta et al., 2013). Floods and poverty have forced many families to change their sources of income to survive. This experience illustrates how economic instability weakens the adaptive capacity of families.

Elderly people about 64.86% in Nirmal Char were already suffering from pre-existing health problems. This data suggests that there is a close relationship between people's physical frailty and their socio-economic status. The study also shows that health problems, low literacy rates, and poverty increase people's vulnerability after floods. Frequent floods severely disrupt not only homes but also children's education. Not all children (100%) were able to go to school for several months. As a result, their educational progress was halted, which could become a major obstacle to good jobs or a better life in the future. At the same time, families are so stressed economically that many are forced to look for new ways of earning a living. Some even involve children in income generation, which is even more worrisome. This situation is also consistent with the results of other studies (Dohaney et al., 2020; Lowe et al., 2013), which state that poverty, health problems, and lack of education influence each other and reduce post-disaster resilience.

The findings of this research indicate that the elderly people in Nirmal Char are also facing physical issues, but in addition to that, they are also facing social isolation and economic insecurity, which make their lives complicated. This is consistent with research conducted by Heagele and Pacquiao (2019) and Rafiey et al. (2016), which reported that when health care access is compromised following a disaster, existing illness in the elderly deteriorates. There is no strong pension or social security system in the region, so many struggle to provide basic essentials like medical care or healthy food (Berkman et al., 2014). This is added to by isolation. Older adults are left isolated after floods, without the assistance of families or neighbours (Lowe et al., 2013; Morrow, 1999). This social loneliness is enhanced even further when younger generations of family members travel elsewhere to find work. Due to this, elderly individuals postpone receiving emergency medical care. Even the community support systems during the flood have been broken down, leaving them even more isolated. This research is in support of Galea et al.'s (2005) observation that a high level of social support can greatly reduce the impact on individuals following a disaster.

The people living in Nirmal Char have developed different methods to cope with recurring floods. These include constructing fences around their houses using sandbags or mud, constructing elevated platforms, raising tube well pipes, and shifting locations to higher altitudes. Though these steps give temporary comfort, they are mostly reactionary. This is in contrast to the strategy by Taş et al. (2013), where preparedness and risk reduction in management are stressed as critical over evacuation and recovery. In Nirmal Char, however, planning is hindered by such obstacles as poverty, the absence of institutional backing, and poor infrastructure. Consequently, residents are left leaning on temporary coping mechanisms instead of enduring long-term solutions. Health systems must be capacitated to deal with the risks posed by climate-related flooding. Interventions of primary concern include putting in place early warning systems, specific health campaigns, flood health action plans, and integrated response systems. Additionally, inter-sectoral coordination between the health sector and non-health sectors is essential. Based on Srikuta et al. (2015), the effectiveness of long-term disaster prevention and preparation lies in multiple stakeholders working in a coordinated manner, highlighting the significance of community involvement.

CONCLUSION

This study analyses the health risks of school-going children and elderly people living in the flood-prone Nirmal Char area of Murshidabad district, West Bengal, India, in the context of climate change. This evidence-based study provides a realistic picture of how the geographical isolation and recurrent floods of the char area have a profound impact on people's livelihoods. The observations showed that about 98% of households face annual or recurrent floods, with most people living in poorly constructed mud houses, temporary shelters made mainly of mud and bamboo. This fragile infrastructure not only makes them uninhabitable but also multiplies the risk of life during disasters.

Children are the most affected in such situations. The lack of safe drinking water and adequate sanitation leads to high rates of vector-borne diseases, including diarrhoea, typhoid, and malaria. More than 88% of households reported such health problems. In addition, 30% of children have been forced to drop out of school due to disruptions in school continuity, and many have been forced into child labour due to financial stress. Disruptions in mid-day meal programmes and chronic food shortages are hampering children's malnutrition and normal mental and physical development.

The situation of the elderly is no different. Limited mobility, social isolation, and existing health complications have made them even more vulnerable. Many are forced to live with inadequate medical care due to difficulties in accessing healthcare in times of emergency. About 65% of the elderly who participated in the study experienced food insecurity. In this context, relief or emergency assistance alone is not enough as an effective strategy for disaster management. Families need to be made capable of coping with disasters through increased awareness, preparedness, and self-reliance. At the same time, ensuring housing stability, uninterrupted flow of clean water, improved sanitation, and strengthening the effectiveness of local health centres are essential. Most importantly, realistic and coordinated efforts between government and non-government organizations must be ensured. Health risks can be significantly reduced by taking the right steps, making informed decisions, and implementing plans tailored to local realities before, during, and after a flood.

LIMITATIONS

This study has revealed some important aspects of the health risks of children and the elderly in the Nirmal Char area, but it has some natural limitations that need to be considered.

First, the data collection was based on only 133 households and was limited to a specific char area. As a result, it is difficult to say whether the findings here will apply to other regions in the same way. In addition, since the participants described their experiences in their own way, there is a risk of personal memory errors or misinterpretations, especially for older people.

Another issue is that, as much as the study was able to capture local realities, it was not possible to consider complex natural or administrative factors such as river control systems or upstream dam policies, which often influence the type of flooding and the extent of human damage.

In addition, because the study focused only on children and the elderly, other vital groups such as pregnant women, people with disabilities, or single-female-headed households were left out of the analysis, even though in reality they are also heavily affected by floods.

Given these limitations, future work with larger, more diverse populations may provide a clearer picture of the nature of health risks. At the same time, understanding both local and institutional influences together may provide more effective guidance for addressing health risks during disasters.

REFERENCES

Agabiirwe, C. N., Dambach, P., Methula, T. C., & Phalkey, R. K. (2022a). Impact of floods on undernutrition among children under five years of age in low- and middle-income countries: a systematic review. *Environmental Health: A Global Access Science Source*, *21*(1), 98. https://doi.org/10.1186/s12940-022-00910-7

Agabiirwe, C. N., Dambach, P., Methula, T. C., & Phalkey, R. K. (2022b). Impact of floods on undernutrition among children under five years of age in low- and middle-income countries: a systematic review. *Environmental Health*, *21*(1), 98. https://doi.org/10.1186/s12940-022-00910-7

Alam, S., & Bera, M. K. (2024). Flood-Induced Health Vulnerability: A Case Study of Nirmal Char in Murshidabad District, West Bengal. *Disaster Advances*, *17*(6), 9–16. https://doi.org/10.25303/176da09016

Andrade, L., O'Dwyer, J., O'Neill, E., & Hynds, P. (2018). Surface water flooding, groundwater contamination, and enteric disease in developed countries: A scoping review of connections and consequences. *Environmental Pollution*, *236*, 540–549. https://doi.org/10.1016/j.envpol.2018.01.104

Baez, J., & Santos, I. (2007). Children's vulnerability to weather shocks: A natural disaster as a natural experiment. *Social Science Research Network,* New York, 1–28. https://alnap.org/help-library/resources/childrens-vulnerability-to-weather-shocks-a-natural-disaster-as-a-natural-experiment/

Barua, P., & Hazarika, J. (2008). A study on health awareness in Chars of Assam. *Journal of Health Management*, *10*(3), 345–351. https://doi.org/10.1177/097206340801000304

Bei, B., Bryant, C., Gilson, K.-M., Koh, J., Gibson, P., Komiti, A., Jackson, H., & Judd, F. (2013). A prospective study of the impact of floods on the mental and physical health of older adults. *Aging & Mental Health*, *17*(8), 992–1002. https://doi.org/10.1080/13607863.2013.799119

Bennett, C., & Friel, S. (2014). Impacts of climate change on inequities in child health. *Children*, *1*(3), 461–473. https://doi.org/10.3390/children1030461

Berkman, L. F., Kawachi, I., & Glymour, M. M. (2014). *Social Epidemiology* (L. F. Berkman, I. Kawachi, & M. M. Glymour (eds.)). Oxford University Press. https://doi.org/10.1093/med/9780195377903.001.0001

Bhutta, Z. A., Das, J. K., Rizvi, A., Gaffey, M. F., Walker, N., Horton, S., Webb, P., Lartey, A., & Black, R. E. (2013). Evidence-based interventions for improvement of maternal and child nutrition: What can be done and at what cost? *The Lancet*, *382*(9890), 452–477. https://doi.org/10.1016/S0140-6736(13)60996-4

Bindoff, N. L., Cheung, W. W. L., Kairo, J. G., Arístegui, J., Guinder, V. A., Hallberg, R., Hilmi, N., Jiao, N., Karim, M. S., Levin, L., O'Donoghue, S., Cuicapusa, S. R. P., Rinkevich, B., Suga, T., Tagliabue, A., Bindoff, P. W., N.L., Cheung, W. W. L., Kairo, J. G., … Williamson, P. (2022). Changing ocean, marine ecosystems, and dependent communities. In *The Ocean and Cryosphere in a Changing Climate* (pp. 447–588). Cambridge University Press. https://doi.org/10.1017/9781009157964.007

Bowen, K. J., Ebi, K. L., Woodward, A., McIver, L., Tukuitonga, C., & Nayna Schwerdtle, P. (2024). Human health and climate change in the Pacific: A review of current knowledge. *Climate and Development*, *16*(2), 119–133. https://doi.org/10.1080/17565529.2023.2185479

Chan, E. Y. Y. (2019). *Disaster Public Health and Older People* (1st ed.). Routledge. https://doi.org/10.4324/9781351127622

Codjoe, S. N. A., Gough, K. V., Wilby, R. L., Kasei, R., Yankson, P. W. K., Amankwaa, E. F., Abarike, M. A., Atiglo, D. Y., Kayaga, S., Mensah, P., Nabilse, C. K., & Griffiths, P. L. (2020). Impact of extreme weather conditions on healthcare provision in urban Ghana. *Social Science & Medicine*, *258*, 113072. https://doi.org/10.1016/j.socscimed.2020.113072

Collins, M., Sutherland, M., Bouwer, L., Cheong, S.-M., Frölicher, T., Combes, H. J. Des, Roxy, M. K., Losada, I., McInnes, K., Ratter, B., Rivera-Arriaga, E., Susanto, R. D., Swingedouw, D., & Tibig, L. (2022). Extremes, abrupt changes and managing risks. In H.-O. Pörtner, D. C. Roberts, V. Masson-Delmotte, P. Zhai, M. Tignor, E. Poloczanska, K. Mintenbeck, A. Alegría, M. Nicolai, A. Okem, J. Petzold, B. Rama, & N. M. Weyer (Eds.), *The Ocean and Cryosphere in a Changing Climate* (pp. 589–656). Cambridge University Press. https://doi.org/10.1017/9781009157964.008

Das, R., & Samanta, G. (2022). Impact of floods and river-bank erosion on the riverine people in Manikchak Block of Malda District, West Bengal. *Environment, Development and Sustainability*, *0123456789*. https://doi.org/10.1007/s10668-022-02648-1

Dohaney, J., de Róiste, M., Salmon, R. A., & Sutherland, K. (2020). Benefits, barriers, and incentives for improved resilience to disruption in university teaching. *International Journal of Disaster Risk Reduction*, *50*, 101691. https://doi.org/10.1016/j.ijdrr.2020.101691

Echendu, A. J. (2022). Flooding, food security and the sustainable development goals in Nigeria: An assemblage and systems thinking approach. *Social Sciences*, *11*(2), 59. https://doi.org/10.3390/socsci11020059

Fatemi, M. N., Okyere, S. A., Diko, S. K., Kita, M., Shimoda, M., & Matsubara, S. (2020). Physical vulnerability and local responses to flood damage in peri-urban areas of Dhaka, Bangladesh. *Sustainability*, *12*(10), 3957. https://doi.org/10.3390/su12103957

Galea, S., Nandi, A., & Vlahov, D. (2005). The epidemiology of post-traumatic stress disorder after disasters. *Epidemiologic Reviews*, *27*(1), 78–91. https://doi.org/10.1093/epirev/mxi003

Ghosh, M., & Ghosal, S. (2021). Climate change vulnerability of rural households in flood-prone areas of Himalayan foothills, West Bengal, India. *Environment, Development and Sustainability*, *23*(2), 2570–2595. https://doi.org/10.1007/s10668-020-00687-0

Heagele, T., & Pacquiao, D. (2019). Disaster vulnerability of elderly and medically frail populations. *Health Emergency and Disaster Nursing*, *6*(1), 50–61. https://doi.org/10.24298/hedn.2016-0009

Hossain, A., Chowdhury, A. T., Mahbub, M., Khan, M., Rahman, T., Sharif, A. Bin, Hijazi, H., & Alameddine, M. (2024). Natural disasters, livelihood, and healthcare challenges of the people of a riverine island in Bangladesh: A mixed-method exploration. *PloS One*, *19*(3), e0298854. https://doi.org/10.1371/journal.pone.0298854

IPCC. (2001a). *Climate Change 2001: Impacts, Adaptation, and Vulnerability: Contribution of Working Group II to the Third Assessment Report of the Intergovernmental Panel on Climate Change* (I. J. J. McCarthy, O. F. Canziani, N. A. Leary, D. J. Dokken, & K. S. White (eds.)). Cambridge University Press.

IPCC. (2001b). *Climate Change 2001: Impacts, Adaptation, and Vulnerability* (J. J. McCarthy, O. F. Canziani, N. A. Leary, D. J. Dokken, & K. S. White (eds.)). Cambridge University Press. https://www.ipcc.ch/site/assets/uploads/2018/03/WGII_TAR_full_report-2.pdf

IPCC. (2022). The ocean and cryosphere in a changing climate: special report of the intergovernmental panel on climate change. In *The Ocean and Cryosphere in a Changing Climate*. https://doi.org/10.1017/9781009157964

Islam, M. R. (2021). Water, sanitation and hygiene practices among disaster-affected char land people: Bangladesh experience. *Natural Hazards*, *107*(2), 1167–1190. https://doi.org/10.1007/s11069-021-04623-x

Klipper, I. G., Zipf, A., & Lautenbach, S. (2021). Flood impact assessment on road network and healthcare access at the example of Jakarta, Indonesia. *AGILE: GIScience Series*, *2*, 1–11. https://doi.org/10.5194/agile-giss-2-4-2021

Krieger, J., & Higgins, D. L. (2002). Housing and health: Time again for public health action. *American Journal of Public Health*, *92*(5), 758–768. https://doi.org/10.2105/AJPH.92.5.758

Krongthaeo, S., Piaseu, N., Junda, T., & Wall, B. M. (2021). Community-based flood preparedness for Thai dependent older adults. *International Journal of Disaster Risk Reduction*, *63*, 102460. https://doi.org/10.1016/j.ijdrr.2021.102460

Lowe, D., Ebi, K., & Forsberg, B. (2013). Factors increasing vulnerability to health effects before, during and after floods. *International Journal of Environmental Research and Public Health*, *10*(12), 7015–7067. https://doi.org/10.3390/ijerph10127015

Mandal, K. K., Dharanirajan, K., Meena, M. L., Jaman, T., & Rana, S. (2023). Application of geospatial tools in the assessment of flood hazard impact on social vulnerability of Malda district, West Bengal, India. *Natural Hazards Research*. https://doi.org/10.1016/j.nhres.2023.11.008

Mandal, K. K., Sarkar, T., Alam, S., Dharanirajan, K., & Sharma S. V., S. (2023). Application of Sentinel-1A SAR data for village level flood inundation mapping in Malda District, West Bengal, India. *Journal of Geographical Studies*, *7*(1), 1–13. https://doi.org/10.21523/gcj5.22070101

Marmot, M., Friel, S., Bell, R., Houweling, T. A., & Taylor, S. (2008). Closing the gap in a generation: Health equity through action on the social determinants of health. *The Lancet*, *372*(9650), 1661–1669. https://doi.org/10.1016/S0140-6736(08)61690-6

Mirsaeidi, M., Motahari, H., Taghizadeh Khamesi, M., Sharifi, A., Campos, M., & Schraufnagel, D. E. (2016). Climate change and respiratory infections. *Annals of the American Thoracic Society*, *13*(8), 1223–1230. https://doi.org/10.1513/AnnalsATS.201511-729PS

Mirzabaev, A., Bezner Kerr, R., Hasegawa, T., Pradhan, P., Wreford, A., Cristina Tirado von der Pahlen, M., & Gurney-Smith, H. (2023). Severe climate change risks to food security and nutrition. *Climate Risk Management*, *39*, 100473. https://doi.org/10.1016/j.crm.2022.100473

Mollah, S. (2011). Vulnerability of Char Dwellers to flood hazards – A case study in two chars in Murshidabad District, West Bengal. *Indian Journal of Applied Research*, *3*(3), 170–172. https://doi.org/10.15373/2249555X/MAR2013/53

Morrow, B. H. (1999). Identifying and mapping community vulnerability. *Disasters*, *23*(1), 1–18. https://doi.org/10.1111/1467-7717.00102

Mukherjee, J. (2011). No voice, No choice: Riverine changes and human vulnerability in the 'Chars' of Malda and Murshidabad. *Occasional Paper No. 28, Institute of Development Studies*, Kolkata, India, 91(July). https://idsk.edu.in/wp-content/uploads/2015/07/OP-28.pdf

Nomura, S., Parsons, A. J. Q., Hirabayashi, M., Kinoshita, R., Liao, Y., & Hodgson, S. (2016). Social determinants of mid- to long-term disaster impacts on health: A systematic review. *International Journal of Disaster Risk Reduction*, *16*, 53–67. https://doi.org/10.1016/j.ijdrr.2016.01.013

Okaka, F. O., & Odhiambo, B. D. O. (2019). Health vulnerability to flood-induced risks of households in flood-prone informal settlements in the Coastal City of Mombasa, Kenya. *Natural Hazards*, *99*(2), 1007–1029. https://doi.org/10.1007/s11069-019-03792-0

Oppenheimer, M., Glavovic, B. C., Hinkel, J., Wal, R. van de, Magnan, A. K., Abd-Elgawad, A., Cai, R., CifuentesJara, M., DeConto, R. M., Ghosh, T., Hay, J., Isla, F., Marzeion, B., Meyssignac, B., & Sebesvari, Z. (2022). Sea level rise and implications for low-lying islands, coasts and communities. In H.-O. Pörtner, D. C. Roberts, V. Masson-Delmotte, P. Zhai, M. Tignor, E. Poloczanska, K. Mintenbeck, A. Alegría, M. Nicolai, A. Okem, J. Petzold, B. Rama, N. M. Weyer (Eds.), *The Ocean and Cryosphere in a Changing Climate* (pp. 321–446). Cambridge University Press. https://doi.org/10.1017/9781009157964.006

Paul, S., & Islam, M. R. (2015). Ultra-poor char people's rights to development and accessibility to public services: A case of Bangladesh. *Habitat International*, *48*, 113–121. https://doi.org/10.1016/J.HABITATINT.2015.03.018

Phraknoi, N., Sutanto, J., Hu, Y., Goh, Y. S., & Lee, C. E. C. (2023). Older people's needs in urban disaster response: A systematic literature review. *International Journal of Disaster Risk Reduction*, *96*, 103809. https://doi.org/10.1016/j.ijdrr.2023.103809

Rafiey, H., Momtaz, Y., Alipour, F., Khankeh, H., Ahmadi, S., Sabzi Khoshnami, M., & Haron, S. A. (2016). Are older people more vulnerable to long-term impacts of disasters? *Clinical Interventions in Aging*, *11*, 1791–1795. https://doi.org/10.2147/CIA.S122122

Rodriguez-Llanes, J. M., Ranjan-Dash, S., Degomme, O., Mukhopadhyay, A., & Guha-Sapir, D. (2011). Child malnutrition and recurrent flooding in rural eastern India: A community-based survey. *BMJ Open*, *1*(2), 1–8. https://doi.org/10.1136/bmjopen-2011-000109

Rodriguez-Llanes, J., Ranjan-Dash, S., Mukhopadhyay, A., & Guha-Sapir, D. (2016). Flood-exposure is associated with higher prevalence of child undernutrition in rural eastern India. *International Journal of Environmental Research and Public Health*, *13*(2), 210. https://doi.org/10.3390/ijerph13020210

Rudra, K. (2003). *The Encroaching Ganga and Social Conflicts: The Case of West Bengal, India*. Kolkata. https://www.gangawaterway.in/assets/02Rudra.pdf

Sajid, O., & Bevis, L. E. M. (2021). Flooding and child health: Evidence from Pakistan. *World Development*, *146*, 105477. https://doi.org/10.1016/j.worlddev.2021.105477

Sakijege, T., Lupala, J., & Sheuya, S. (2012). Flooding, flood risks and coping strategies in urban informal residential areas: The case of Keko Machungwa, Dar es Salaam, Tanzania. *Jàmbá: Journal of Disaster Risk Studies*, *4*(1). https://doi.org/10.4102/jamba.v4i1.46

Sarkar, S. M., Dhar, B. K., Fahlevi, M., Ahmed, S., Hossain, M. J., Rahman, M. M., Gazi, M. A. I., & Rajamani, R. (2023). Climate change and aging health in developing countries. *Global Challenges*, *7*(8). https://doi.org/10.1002/gch2.202200246

Saulnier, D., Hanson, C., Ir, P., Mölsted Alvesson, H., & Von Schreeb, J. (2018). The effect of seasonal floods on health: Analysis of six years of national health data and flood maps. *International Journal of Environmental Research and Public Health*, *15*(4), 665. https://doi.org/10.3390/ijerph15040665

Srikuta, P., Inmuong, U., Inmuong, Y., & Bradshaw, P. (2015). Health vulnerability of households in flooded communities and their adaptation measures. *Asia-Pacific Journal of Public Health*, *27*(7), 743–755. https://doi.org/10.1177/1010539514568709

Suri, S., Ahmad, F., Yahaya, A. S., Mokhtar, Z. A., & Halim, M. H. (2014). Climate change impact on water level in Peninsular Malaysia. *Journal of Civil Engineering Research*, *4*(3A), 228–232.

Taş, M., Taş, N., Durak, S., & Atanur, G. (2013). Flood disaster vulnerability in informal settlements in Bursa, Turkey. *Environment and Urbanization*, *25*(2), 443–463. https://doi.org/10.1177/0956247813501134

Tate, E., Rahman, M. A., Emrich, C. T., & Sampson, C. C. (2021). Flood exposure and social vulnerability in the United States. *Natural Hazards*, *106*(1), 435–457. https://doi.org/10.1007/s11069-020-04470-2

Wang, L., Yang, L., Di, X., & Dai, X. (2020). Family support, multidimensional health, and living satisfaction among the elderly: A case from Shaanxi Province, China. *International Journal of Environmental Research and Public Health*, *17*(22), 8434. https://doi.org/10.3390/ijerph17228434

Wang, P., Asare, E. O., Pitzer, V. E., Dubrow, R., & Chen, K. (2023). Floods and diarrhea risk in young children in low- and middle-income countries. *JAMA Pediatrics*, *177*(11), 1206. https://doi.org/10.1001/jamapediatrics.2023.3964

WHO. (2013). *Social Determinants of Health*. World Health Organization. https://www.who.int/health-topics/social-determinants-of-health#tab=tab_3

WHO. (2023). *Anxiety Disorders*. World Health Organization. https://www.who.int/news-room/fact-sheets/detail/anxiety-disorders.

Part 3

Natural Resources, Ecosystems, and Climate Impact Assessments

Modeling, Mapping, and Risk Assessment

15 Renewable Energy Transition
A Multilevel Perspective on Climate Change and Geopolitical Impacts

Farhana Parvin

INTRODUCTION

The global pursuit of carbon neutrality is not only transforming national energy infrastructures but also reshaping geopolitical landscapes, as the transition from fossil fuels to renewable energy sources such as solar, wind, and hydrogen alters the strategic importance of resources, supply chains, and technological control in the emerging low-carbon global order (Li et al., 2022; Mahmood et al., 2022; Zhang & Chen, 2022). In the face of the escalating climate crisis, the global community is positioned at a critical crossroads, where decisive policy actions and strategic energy interventions will shape the environmental, economic, and geopolitical trajectory of the 21st century. The International Renewable Energy Agency (IRENA), in its report Global Energy Transformation: A roadmap to 2050 (IRENA, 2019a), emphasizes that meeting the objectives of the Paris Agreement requires a fundamental overhaul of the global energy architecture. To align with the decarbonization trajectory, renewable energy must constitute at least two-thirds of the total final energy supply by 2050, underscoring the critical need for accelerated policy interventions, technological innovation, and large-scale investment in sustainable energy infrastructure (IRENA, 2019b). This transformation entails a substantial reduction in total energy demand through aggressive energy efficiency measures, widespread electrification across the industrial, transportation, and residential sectors, and a significant increase in the contribution of renewable energy sources, including solar, wind, and sustainable biofuels, to the global energy mix. The United Nations Climate Change Conference (COP28) represents a critical inflection point, serving as a high-level platform for multilateral engagement and coordinated climate action (UN, 2023). Beyond a mere diplomatic assembly, COP28 embodies the collective commitment of nations to advancing comprehensive and enforceable climate strategies, fostering resilience, and accelerating the transition toward a sustainable global economy (Yu et al., 2024). These shifts not only support climate goals but also catalyze broader socioeconomic and geopolitical changes, redefining the contours of power, influence, and resource dependencies in a rapidly decarbonizing world.

There is now a growing international consensus on the urgent need for climate action, particularly within the energy sector, which accounts for nearly two-thirds of global greenhouse gas (GHG) emissions (IEA, 2022). This recognition has catalyzed the formulation of ambitious policy frameworks and long-term decarbonization commitments by major economies. At the 2021 Leaders Climate Summit, the US President Joe Biden pledged to reduce the country's carbon emissions by 50% by 2030 (relative to 2005 levels) and to achieve full carbon neutrality by 2050. The European Union (EU) has similarly committed to a net-zero emissions target by 2050, while Germany has set an even more accelerated timeline, aiming to reach carbon neutrality by 2045. Japan, under then-Prime Minister Yoshihide Suga, also reaffirmed its target of carbon neutrality by 2050 (Wang et al., 2023). These national pledges underscore a unified global ambition to transition away from fossil-fuel-based energy systems and toward sustainable, resilient, and low-carbon alternatives, laying the groundwork for a new era of energy geopolitics.

DOI: 10.1201/9781003620747-18

As the global energy transition gathers momentum, the traditional hydrocarbon-centered geopolitical framework is undergoing a significant transformation. While the primary objective of the transition from nonrenewable to renewable energy is to reduce carbon emissions to net zero as outlined in the Paris Agreement, where 195 nations committed to limiting environmental impacts and enhancing climate affordability, it also signals a deeper structural transformation in global energy production and consumption (IPCC, 2021; IEA, 2021; IEA, 2023b; IRENA, 2023). The shift toward renewables, critical minerals, and decentralized energy systems (DES) is giving rise to new geopolitics of energy, one defined not just by access to oil and gas, but by control over technological infrastructure, rare earth elements (REEs), and renewable energy supply chains (IRENA, 2019b; Scholten et al., 2020). Geopolitical attention has traditionally centered on oil-rich regions such as the Persian Gulf, the Caspian Basin, and the Arctic, as well as strategic maritime chokepoints like the Strait of Hormuz and the Suez Canal. These areas are seen as critical nodes in the global energy supply chain, and any disruption there can have cascading effects on global energy markets (Bouzarovski & Konieczny, 2010; Johnson, 2017). More recently, tensions between the EU and Russia, especially following disputes over natural gas supply routes and pricing have brought natural gas geopolitics to the forefront, particularly in the context of European energy dependence and diversification strategies (Overland, 2015; Overland et al., 2022; Goldthau & Sitter, 2021). Historical events have consistently illustrated how geopolitical dynamics can significantly disrupt energy markets.

During World War I, geopolitical analysis was largely framed around competition for access to energy fuels, particularly oil, an idea that popularized the phrase "it's all about oil" (Hirsch et al., 2005; Klare, 2012; Bradshaw, 2009). This framing persisted through the 20th century until the oil crises of 1973 and 1979 fundamentally reshaped the discourse, transitioning the concept of oil security from a military concern to one of economic stability and resilience (Santos, 2020; Luft & Korin, 2009). However, the green energy transition is a highly complex topic, fraught with significant social, infrastructure, economic, and political issues (Rubin et al., 2021; Ali et al., 2022; Owen et al., 2023). One of the anticipated geopolitical consequences of the green energy transition is the diminishing significance of oil-producing nations in global affairs. Countries heavily reliant on oil exports, such as Saudi Arabia and Russia, may experience a decline in their geopolitical influence, particularly in developed economies with advanced electrification and reduced fossil fuel dependence (Costa et al., 2022; Ghorbani et al., 2024). This shift could destabilize or erode longstanding international relationships that have been historically shaped by conventional energy trade. Furthermore, the deep financialization of the oil market—characterized by extensive interconnections between global financial institutions and fossil fuel assets—suggests that volatility in oil prices could have far-reaching economic repercussions. A sustained decline in oil demand may not only undermine the economic stability of petroleum-dependent states but also introduce systemic risks to global financial markets, potentially exacerbating financial instability on a broader scale (Gkillas et al., 2021). One likely consequence is a transfer of power from countries that supply fossil fuels to those that supply green energy minerals and metals; this could have a variety of effects and, through feedback, change the energy transition process and international relations (Chishti et al., 2023; Ghorbani et al., 2024).

The historical dependence on fossil fuels has not only shaped geopolitical dynamics but has also been a significant driver of political tensions, socioeconomic instability, and military conflicts, as states have competed over access to these geographically concentrated resources. In contrast, renewable energy sources, such as wind and solar power, are more evenly distributed across regions, enabling greater energy autonomy, reducing reliance on costly imports, and mitigating economic vulnerabilities associated with fossil fuel dependence. Hence, another impact of the transition to a low-carbon economy presents distinct geopolitical and economic challenges between nations (Jacobsen, 2019; Sattich et al., 2021; Dall-Orsoletta et al., 2022; Ghorbani et al., 2024). The increasing demand for critical minerals and REEs, which are indispensable for renewable energy technologies, may give rise to new forms of resource competition. The International Energy Agency (IEA) projects that by 2040, the share of oil in the global energy mix will decline significantly, decreasing from its current level of approximately 32%–20% (IEA, 2020b). Simultaneously, the risk of stranded fossil fuel assets threatens the economic stability of hydrocarbon-dependent states, potentially leading to revenue declines and sociopolitical unrest. Moreover, the uneven pace of the global energy transition exacerbates structural inequalities, as some nations may lack the technological, financial, or infrastructural capacity to integrate into the emerging energy paradigm. Scholars have further argued that the proliferation of clean technologies may perpetuate a novel form of extractive capitalism, wherein resource exploitation persists under a sustainable energy framework. Consequently, while the shift toward renewable energy presents opportunities for enhanced energy security and geopolitical stability, it necessitates a nuanced approach to managing emerging economic asymmetries and geopolitical reconfigurations.

Energy transition and geopolitical risks are inextricably linked, as energy systems are often vulnerable to global political instabilities and conflicts, often lead to war. This study seeks to critically examine the global renewable energy transition amid shifting geopolitical dynamics, by adopting the multilevel perspective (MLP) framework as its analytical foundation. Rooted in transition theory, MLP offers a nuanced understanding of how socio-technical systems evolve over time through the interplay of processes at three analytical levels: the broader socio-technical landscape (macro), socio-technical regimes (meso), and niche innovations (micro) (Geels, 2002; Geels & Schot, 2007). In the context of renewable energy, the landscape level was structured by the declining relevance of fossil fuels and the accelerating rise of renewable energy technologies. As the geopolitical weight of hydrocarbons diminishes, states traditionally empowered by fossil fuel exports face growing structural vulnerabilities, including the risk of economic destabilization due to the devaluation of carbon-intensive assets. At the regime level, traditional energy superpowers face structural challenges, as fossil fuel-dependent economies risk economic destabilization due to the devaluation of hydrocarbon assets, while new energy actors, such as China and the EU, leverage their dominance in renewable technology and supply chains to gain geopolitical advantage. Simultaneously, at the niche level, advancements in battery storage, hydrogen energy, and smart grids are fostering a more decentralized and localized energy landscape, enhancing energy security but also introducing new geopolitical dependencies on critical materials such as lithium and REEs. By applying the MLP framework to the study of renewable energy geopolitics, researchers can systematically examine how technological progress, policy shifts, and market disruptions are reconfiguring global power structures, ultimately shaping the trajectory of the energy transition and its geopolitical consequences.

MULTILEVEL PERSPECTIVE: THEORETICAL FRAMEWORK

The MLP framework provides a valuable theoretical lens for analyzing the evolving geopolitics of renewable energy, offering a structured approach to understanding the interactions between technological innovations, institutional shifts, and broader structural transformations. Within the MLP framework, transitions occur through the dynamic interplay of three levels: *the landscape (macro-level)*, which encompasses external pressures such as climate change, resource depletion, and international climate policies; *the regime (meso level)*, which represents the dominant energy system characterized by entrenched fossil fuel infrastructures, geopolitical dependencies, and institutional frameworks; and *the niche (micro level)*, where emerging renewable technologies, DES and policy innovations challenge the status quo (Geels, 2002). The geopolitical implications of the renewable energy transition are deeply embedded within these levels of interaction (Figure 15.1).

The *landscape level* encompasses broad, exogenous forces such as climate change, market forces and technological advancements, geopolitical tensions, public pressure, social movements, and corporate commitments that influence both regime stability and niche development without being directly altered by short-term actions of actors within the system (Brunet et al., 2021). These exogenous shocks are accelerating the momentum for renewable energy deployment by disrupting

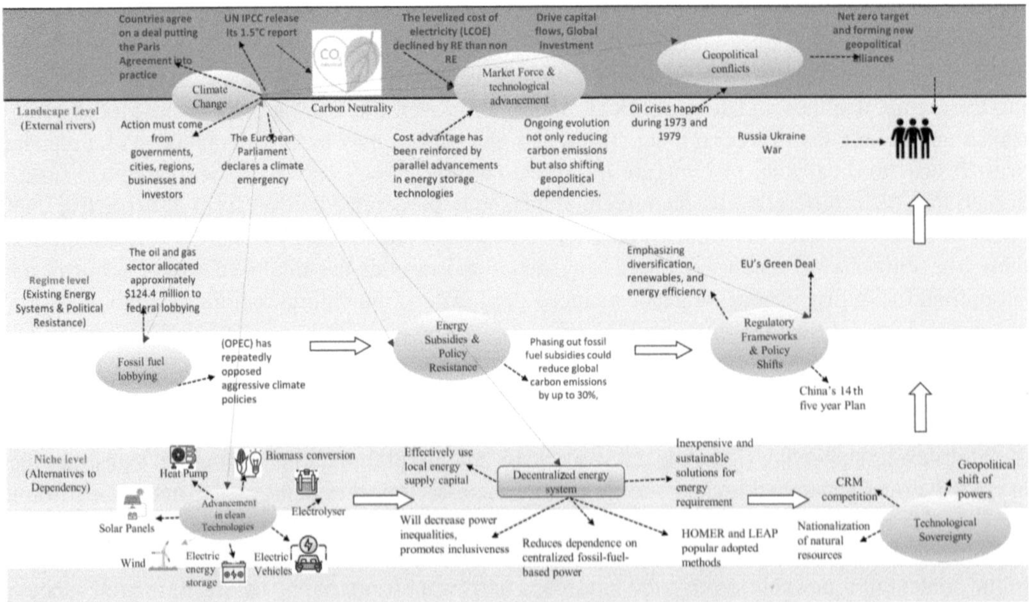

FIGURE 15.1 Interaction between the levels and the phases of the MLP framework. (Source: Prepared by author.)

the fossil fuel regime's hegemony and amplifying the urgency of diversification and decarbonization. Landscape-level pressures, including international climate commitments, global civil society demands, and macroeconomic shifts toward green investments, are converging to create a structural opening for renewable technologies to gain traction. In line with the MLP framework, these macro-level disruptions are not merely background noise but active catalysts that enable niche innovations—such as solar PV, offshore wind, and green hydrogen—to scale up and potentially reconfigure the existing socio-technical energy regime.

At the *regime level* (meso level) of the MLP framework, fossil fuel lobbying continues to wield considerable political and economic influence. Fossil fuel interests actively shape national and international policy agendas, often undermining climate action by promoting regulatory stagnation or diluting environmental standards. This lobbying power is closely tied to energy subsidies and policy resistance, where entrenched fiscal incentives amounting to $7 trillion globally in 2022 artificially lower the cost of fossil energy and distort market dynamics in favor of the status quo (IMF, 2023). Such subsidies not only perpetuate dependence on carbon-intensive fuels but also constrain the financial space for scaling up clean energy alternatives. Despite this resistance, regulatory frameworks and policy shifts are beginning to disrupt the regime's dominance. Progressive policies such as the EU Green Deal and China's 14th Five-Year Plan reflect strategic efforts to realign institutional structures, investment priorities, and legal mandates in favor of renewables. These initiatives aim to institutionalize decarbonization by integrating carbon pricing, emissions caps, and binding renewable targets into national governance systems. However, the regime's transition remains uneven and contested, as legacy infrastructures, vested interests, and institutional inertia continue to buffer the fossil fuel paradigm against rapid displacement. Thus, the regime level embodies a battleground where the inertia of the incumbent system confronts growing pressures from landscape-level disruptions and emerging niche innovations, shaping the trajectory and pace of the global energy transition.

At the *niche level* (micro level), radical innovations and emerging energy alternatives are gradually disrupting the entrenched fossil fuel-based regime and accelerating the transition toward a decarbonized energy system. A core driver of this transformation lies in technological advancements in solar, wind, and hydrogen energy. Solar photovoltaic (PV) technologies have witnessed exponential growth, with global capacity increasing from 30 GW in 2011 to over 920 GW in 2021, making solar PV the fastest-growing renewable energy source due to its modularity, scalability, and cost-effectiveness (IRENA, 2023; Schmela et al., 2023). Similarly, wind power and green hydrogen are expanding their roles, with innovations in offshore wind turbines exceeding 15 MW and electrolyzer capacity scaling rapidly to meet industrial decarbonization goals (IEA, 2022). Concurrently, DES, such as rooftop solar, microgrids, and community-owned renewable infrastructure, are gaining traction, offering resilient, localized solutions that democratize energy access and reduce dependency on centralized fossil-based grids (Sovacool et al., 2020a). These bottom-up energy models support localized energy sovereignty and empower communities to become active participants in the energy transition.

Furthermore, technological sovereignty and rare earth competition have emerged as defining geopolitical elements of the niche space. The global race to secure critical minerals such as lithium, cobalt, and REEs used in batteries, wind turbines, and electrolyzers has intensified, prompting governments to invest in domestic extraction, recycling, and supply chain diversification to reduce dependence on geopolitically sensitive sources like China (Ali et al., 2017; IEA, 2021). These dynamics illustrate how innovation niches, while still emerging, are shaped by broader geopolitical and economic concerns, and under the right landscape conditions, are catalyzed regime shifts. As such, niche-level innovations function as experimental spaces that incubate sustainable technologies and social practices, providing the seeds for systemic transformation toward a resilient, low-carbon global energy future (Geels et al., 2017).

In sum, the MLP has emerged as a widely adopted and influential framework for analyzing socio-technical transitions, particularly within the domain of renewable energy. It has been extensively employed to investigate diverse facets of energy system transformation, including the diffusion of electric vehicles (EVs) (Berkeley et al., 2017), the restructuring of low-carbon power systems (Wainstein & Bumpus, 2016), the deployment of smart grids (Jia-Hai, 2014), and the expansion of offshore wind infrastructure (Flynn, et al., 2019). Within the context of decentralized energy transitions, Gunther and Bernell (2019) utilize the MLP framework to examine the emergence of community choice aggregations (CCAs) in California a model that enables local communities to procure electricity through public agencies, thereby enhancing autonomy and sustainability. Their analysis reveals how CCAs are not only reshaping local governance but also contributing to the broader systemic shift away from centralized, fossil-based regimes toward decentralized, renewable energy models. These applications underscore the analytical strength of MLP in capturing the complex, multilevel dynamics of energy transitions, especially as they intersect with evolving geopolitical imperatives and the reconfiguration of global energy governance.

LANDSCAPE LEVEL: EXTERNAL DRIVING FORCES

Climate Change as a Driving Force to Energy Transition

Climate change has emerged as one of the most pressing global challenges of the 21st century, with far-reaching environmental, socioeconomic, and geopolitical implications. According to the Intergovernmental Panel on Climate Change (IPCC), limiting global warming to 1.5°C above pre-industrial levels is critical to avoiding catastrophic outcomes such as irreversible ecosystem collapse, the submergence of island nations, and severe disruptions to climate systems (IPCC, 2021). Beyond environmental consequences, climate change is projected to have profound economic effects; a 4°C increase in global temperatures could result in an estimated loss of 1%–5% of annual global GDP by 2100 (Hao & Shao, 2021). Empirical evidence further indicates that higher temperatures are correlated with increased social tensions and conflict at both interpersonal and intergroup levels (Hsiang et al., 2013). Central to climate change mitigation is the reduction of GHG emissions, over 73% of which stem from the combustion of fossil fuels for electricity generation, industrial processes, and transportation (IPCC, 2021). While political discourse, particularly in countries like the

United States, where climate skepticism persists among certain policymakers, continues to contest the anthropogenic origins of climate change, a robust scientific consensus affirms human-induced emissions as a dominant driver (Gleick et al., 2013; NRC, 2002). Anthropogenic activities have been linked to a range of observable phenomena, including a rise in global mean surface temperatures, ocean acidification, and reductions in cryospheric coverage. Global climate models project that without significant emissions reductions, average surface temperatures could increase by 1.1°C–6.4°C by 2100, accompanied by rising sea levels, altered precipitation patterns, and a heightened frequency of extreme weather events such as droughts, floods, and tropical cyclones (IPCC, 2007). In response to this escalating threat, nations have adopted ambitious decarbonization targets. The EU, for instance, seeks climate neutrality by 2050 and has already achieved a 22.5% renewable energy share in its 2022 energy mix (IEA, 2023). China, the world's largest emitter, aims to peak its emissions by 2030 and achieve carbon neutrality by 2060, with over half of its newly added power capacity in 2022 sourced from renewables (IEA, 2023). Meanwhile, the United States' Inflation Reduction Act (IRA) allocates $369 billion toward clean energy development, marking a pivotal investment in its energy transition strategy. These policy shifts underscore a growing global consensus that accelerating the deployment of low-carbon technologies is essential not only for climate resilience but also for redefining energy security and economic competitiveness in a rapidly warming world.

Moreover, the transition to clean energy is reshaping the global resource economy. Unlike fossil fuels, renewable energy technologies are material-intensive, requiring REEs and critical minerals such as lithium, cobalt, and nickel. This shift has repositioned countries like the Democratic Republic of Congo, Chile, and Indonesia, rich in such resources, as pivotal players in the global energy transition.

The IEA estimates that demand for critical minerals could increase fourfold by 2040 under a scenario aligned with the Paris Agreement (IEA, 2021). In response, major economies are engaging in strategic resource diplomacy and forming mineral security partnerships to ensure stable supply chains, thereby adding a new layer of complexity to global geopolitics. This competition over technological materials highlights a departure from traditional hydrocarbon geopolitics to a more decentralized, multiresource dynamic, with implications for trade, foreign policy, and industrial strategy.

Geopolitical conflicts significantly impact global energy security, accelerating the shift toward diversified and renewable energy sources. The Russia–Ukraine war, which began in 2022, severely disrupted fossil fuel supply chains, particularly in Europe, as Russia supplied nearly 40% of the EU's natural gas before the conflict (IEA, 2022). In response, the EU launched the REPowerEU plan, aiming to reduce Russian gas imports by two-thirds by the end of 2022 and achieve full independence by 2027 through increased investments in renewables, liquefied natural gas (LNG) imports, and energy efficiency measures. Similarly, Germany restarted coal plants and expanded LNG terminals while simultaneously increasing solar and wind energy deployment to compensate for gas shortages. Meanwhile, China and India ramped up Russian oil imports, altering global energy trade flows. The crisis also accelerated clean energy investments, with global renewable capacity additions reaching 340 GW in 2022, a 25% increase from 2021 (IEA, 2023). These shifts underscore how geopolitical instability is reshaping global energy markets, reinforcing the urgency of energy diversification and security through renewables. Additionally, oil price volatility surged, with Brent crude exceeding $120 per barrel in mid-2022, prompting many nations to accelerate investments in energy independence. The crisis also triggered record investments in renewables, with global clean energy investments surpassing $1.3 trillion in 2022, marking a 10% increase from the previous year (Bloomberg New Energy Finance, 2023). These shifts highlight how geopolitical instability is reshaping energy markets, reinforcing the need for a transition to resilient and sustainable energy systems (SES).

This global shift toward renewable energy is not merely an environmental imperative but also a strategic realignment of energy systems that carries significant geopolitical ramifications.

As countries reduce their reliance on fossil fuels historically concentrated in geopolitically volatile regions such as the Middle East and Russia, the locus of energy production and influence is undergoing a profound transformation. The EU's move to reduce dependence on Russian gas following the 2022 invasion of Ukraine is a prime example of how geopolitical shocks can accelerate renewable energy transitions. The EU has since increased LNG imports from the US and Qatar, invested in offshore wind capacity, and expanded solar installations across southern Europe (IEA, 2023; BP Statistical Review, 2023). Simultaneously, new energy alliances are emerging, particularly between Europe and North African nations, where abundant solar resources are now positioned as strategic assets. For instance, the €2.2 billion Xlinks Morocco-UK Power Project aims to supply 8% of the UK's electricity needs via solar and wind power transmitted through undersea cables (Xlinks, 2023). These developments underscore how decarbonization efforts are not only altering energy markets but also fostering new trade relationships and shifting geopolitical dependencies. In this evolving landscape, national energy strategies are increasingly intertwined with climate policy, technological innovation, and security considerations. Renewable energy not only mitigates emissions but also enhances energy sovereignty by allowing nations to diversify their supply and reduce exposure to external shocks. Countries like India are capitalizing on this potential, aiming for 500 GW of non-fossil fuel capacity by 2030, supported by robust solar manufacturing and green hydrogen initiatives (MNRE, 2023). In parallel, the rapid deployment of DES such as microgrids and off-grid solar solutions in sub-Saharan Africa demonstrates how renewable technologies can simultaneously address energy poverty and resilience to climate-induced disruptions. As such, the decarbonization of energy systems is not a uniform process but rather a globally differentiated trajectory shaped by local capacities, resources, and geopolitical positioning.

Market Forces and Technological Advancements

Market dynamics and technological innovation have fundamentally reshaped the cost structure and competitiveness of renewable energy, establishing it as a viable and increasingly dominant alternative to fossil fuels in global energy markets. The levelized cost of electricity (LCOE)—a critical metric for comparing energy sources—has experienced dramatic reductions for key renewables over the past decade. Since 2010, the LCOE for solar PV and onshore wind has declined by approximately 82% and 61%, respectively, largely due to economies of scale, technological enhancements, streamlined manufacturing, and global supply chain maturation (IRENA, 2023). By 2022, the global average LCOEs for solar PV and onshore wind had dropped to $40/MWh and $33/MWh, respectively—well below the average costs of coal and natural gas in many markets (Bloomberg New Energy Finance, 2023). This cost advantage has been reinforced by parallel advancements in energy storage technologies, particularly lithium-ion batteries, whose prices plummeted from $1,200/kWh in 2010 to just $139/kWh in 2023 an 88% reduction, thereby improving the dispatchability and reliability of intermittent renewable sources (Bloomberg New Energy Finance, 2023). Breakthroughs in materials science, such as the development of perovskite-silicon tandem solar cells achieving over 30% efficiency in laboratory settings, further suggest continued improvements in cost-effectiveness and performance (NREL, 2023). Offshore wind has also undergone a technological leap, with next-generation turbines exceeding 15 MW in capacity, significantly lowering installation and maintenance costs while boosting output. Collectively, these developments have driven unprecedented capital flows into the sector; global investments in renewable energy reached $1.3 trillion in 2022, outpacing fossil fuel investments for the first time in history (IEA, 2023). This convergence of market forces and technological progress not only accelerates the global energy transition but also enhances energy security, system resilience, and long-term economic sustainability—key pillars for achieving net-zero targets and mitigating climate change. The confluence of declining costs and accelerated innovation has positioned renewables not just as environmentally favorable, but as strategically indispensable components of national and global energy strategies. These economic and technical trends are reshaping investment behavior, infrastructure development, and energy planning frameworks across both developed and emerging economies. In particular, utility-scale solar and wind projects are becoming the cornerstone of new energy capacity additions. For instance, in 2022, over 83% of all new power generation capacity globally came from renewables, with solar accounting for more than half of that growth (IRENA, 2023). This trajectory reflects a significant departure from fossil-fuel-dominated systems, as public and private sectors increasingly recognize renewables as the most cost-effective pathway for energy expansion and decarbonization.

Moreover, advancements in digital technologies such as artificial intelligence (AI), blockchain, and Internet of Things (IoT) are further enhancing the efficiency and management of renewable energy systems. Smart grids, predictive maintenance for wind turbines, AI-optimized solar forecasting, and decentralized energy trading platforms are enabling more flexible, reliable, and user-centered power systems. These tools support the integration of high shares of renewables without compromising grid stability, thereby overcoming one of the longstanding challenges of variable renewable energy. Coupled with innovative financing models such as green bonds, power purchase agreements (PPAs), and public–private partnerships these technologies are lowering entry barriers and catalyzing large-scale deployment in regions previously considered high-risk or low-return. From a policy standpoint, governments are increasingly aligning fiscal tools and regulatory frameworks with clean energy goals. The IRA in the US, the European Green Deal, and China's 14th Five-Year Plan all embed renewables as central pillars for climate action and industrial competitiveness. These initiatives provide long-term investment certainty and stimulate domestic manufacturing of clean technologies, further reducing costs and strengthening energy sovereignty. Additionally, international initiatives like the International Solar Alliance and Mission Innovation are fostering knowledge sharing, joint R&D, and cross-border infrastructure that amplify the global impact of national renewable transitions. In summary, the ongoing evolution in renewable energy economics and technologies is catalyzing a fundamental restructuring of the global energy system. This transformation is not only reducing carbon emissions but also shifting geopolitical dependencies, enabling decentralized and democratized energy access, and creating resilient economic structures adapted to 21st-century climate and energy challenges. As costs continue to decline and innovation accelerates, renewables will increasingly define the trajectory of both energy policy and global power relations in the decades to come.

Public Pressure, Social Movements, and Corporate Commitments

Public pressure, social mobilization, and evolving corporate responsibility are emerging as powerful non-state drivers in reshaping the global energy landscape toward sustainability. The widespread influence of civil society and market-based mechanisms is evident in the surge of net-zero pledges now covering over 90% of global GDP, a shift propelled largely by citizen advocacy, youth-led climate movements, and regulatory reforms aimed at curbing GHG emissions (Net Zero Tracker, 2023). One of the most prominent manifestations of this transformation is the exponential growth of Environmental, Social, and Governance (ESG) investing, which reached a staggering $41 trillion in assets under management (AUM) in 2022, reflecting the increasing prioritization of long-term sustainability metrics in financial decision-making (Bloomberg New Energy Finance, 2023). Concurrently, financial institutions have begun to internalize climate risk as a central tenet of portfolio management, with over 550 firms representing $150 trillion in AUM joining the Glasgow Financial Alliance for Net Zero (GFANZ), committing to align their lending and investment strategies with net-zero emissions targets by 2050 (UNFCCC, 2023).

Corporate climate accountability is also gaining momentum through initiatives such as the science-based targets initiative (SBTi), under which more than 4,000 companies worldwide have established verifiable emission reduction pathways consistent with limiting global warming to 1.5°C (SBTi, 2023). Simultaneously, consumer behavior is evolving toward sustainability-conscious preferences 72% of global consumers now favor environmentally friendly products, exerting market pressure on corporations to adopt decarbonization agendas and integrate sustainability into their business models (PwC, 2023). Tech giants such as Apple, Amazon, and Microsoft have responded by committing to net-zero or carbon-negative goals by 2030, thereby setting influential benchmarks across industries. Moreover, the global divestment movement has become a formidable force in destabilizing the fossil fuel economy. As of 2023, institutional investors managing over $40 trillion in assets have pledged to divest from fossil fuels, signaling a structural capital shift toward renewables and low-carbon technologies (DivestInvest, 2023). These investor actions are further complemented by regulatory developments, such as the EU's Corporate Sustainability Reporting Directive (CSRD), which will mandate ESG disclosures from approximately 50,000 companies starting in 2024, and the US Securities and Exchange Commission's (SEC) proposed climate risk disclosure rules for publicly traded firms (EU Commission, 2023; SEC, 2023). Together, these intersecting dynamics of public advocacy, green finance, consumer activism, and regulatory mandates are not only intensifying pressure on corporations and governments but are actively reconfiguring energy governance frameworks and accelerating the global transition toward a low-carbon, climate-resilient economy.

REGIME LEVEL: EXISTING ENERGY SYSTEMS AND POLITICAL RESISTANCE

Fossil Fuel Lobbying & Influence on Policy

The fossil fuel industry persistently resists comprehensive decarbonization efforts (Fatemi & Fooladi, 2013; InfluenceMap, 2019; Trout et al., 2022), actively working to safeguard its political influence and economic hegemony through a range of interrelated mechanisms. These include substantial financial contributions to political action committees and trade associations, as well as strategic lobbying aimed at shaping policy outcomes in their favor. Of particular concern is the significant investment in professional climate lobbyists, despite widespread public support for governmental policies that accelerate the transition to clean energy. The scale of fossil fuel lobbying expenditures is considerable, unrestricted, and often opaque, posing a substantial threat to transparent climate discourse and democratic governance. The concealment of such lobbying activities obstructs informed public decision-making and delays the implementation of critical climate policies (Brulle, 2018). For firms within the fossil fuel sector, achieving deep decarbonization necessitates a fundamental shift toward renewable energy sources, requiring structural overhauls of traditional business models and extensive transformations in core operational strategies. However, in the absence of comprehensive global carbon pricing mechanisms and stringent regulatory frameworks, the pace and effectiveness of this transition will likely vary across firms, leading to divergent adaptation trajectories. Those corporations that have proactively integrated energy transition strategies into their long-term planning are expected to be better insulated from both the physical and transitional risks associated with climate change, whereas laggards may face heightened financial and operational vulnerabilities (Lantushenko & Schellhorn, 2023).

In 2022, the oil and gas sector allocated approximately $124.4 million to federal lobbying, strategically targeting regulatory policies on methane emissions, fossil fuel extraction on federal lands and waters, and financial incentives for contentious carbon capture technologies (OpenSecrets, 2023). Leading industry players such as ExxonMobil ($7.7 million) and Chevron ($3.9 million) significantly contributed to these efforts lobbying efforts to delay climate policies and maintain subsidies, while also maintaining membership in the American Petroleum Institute (API) a formidable lobbying entity historically opposed to clean energy policies (Figure 15.2). The API has actively resisted climate regulations, advocating against the IRA's methane tax, challenging Corporate Average Fuel Economy (CAFE) standards, and obstructing the large-scale transition to EVs. Beyond direct lobbying, fossil fuel interests exert profound political influence, as exemplified in the documentary The White House Effect, which illustrates how fossil fuel lobbying swayed President George H.W. Bush from supporting renewable energy initiatives to casting doubt on climate science. This underscores how industry-backed advocacy extends beyond formal lobbying, shaping political discourse and obstructing decarbonization efforts to protect fossil fuel profitability.

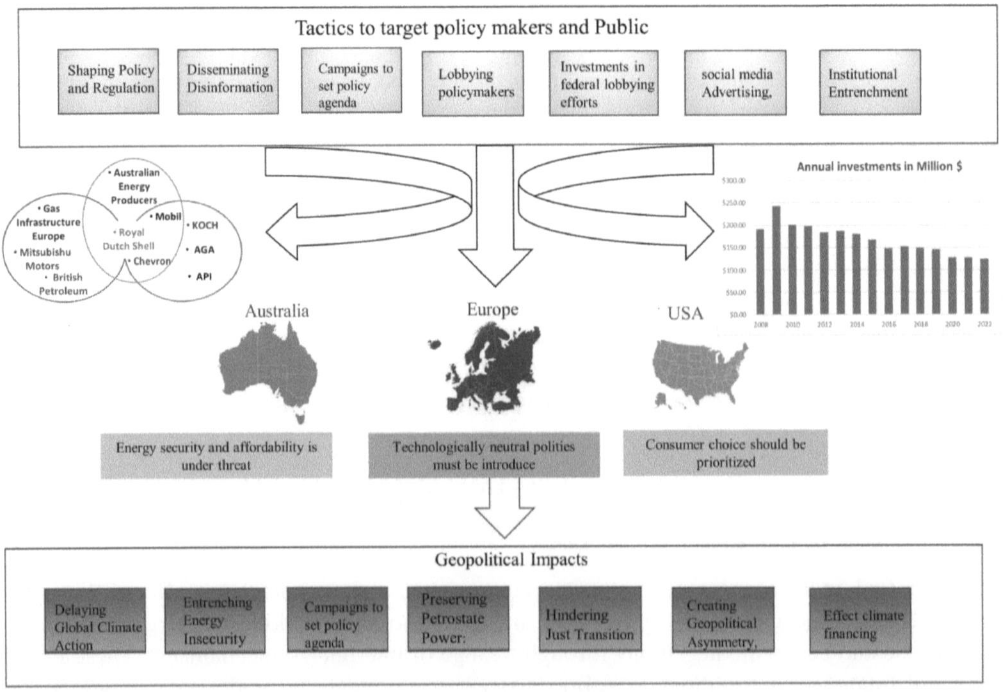

FIGURE 15.2 Fossil fuel lobbying obstructs clean energy transition. (Source: Prepared by author.)

Oil- and gas-exporting countries, such as Saudi Arabia and Russia, prioritize economic stability over decarbonization, as fossil fuels contribute to over 50% of their national revenues (IEA, 2023). The Organization of the Petroleum Exporting Countries (OPEC) has repeatedly opposed aggressive climate policies, arguing that rapid transitions threaten global energy security and economic growth (OPEC, 2023). For example, Saudi Aramco plans to invest $100 billion in new oil and gas projects by 2030, despite global net-zero commitments (Bloomberg, 2023). Russia, which relies on fossil fuels for 45% of its federal budget, has sought to expand Arctic oil exploration and recently increased coal exports to Asia in response to European sanctions (IEA, 2023). Additionally, fossil fuel subsidies remain a significant barrier to energy transition, with global subsidies reaching a record $7 trillion in 2022, primarily in the form of tax breaks and price controls (IMF, 2023). Political resistance also manifests through legal challenges, as seen in the United States, where states like Texas and West Virginia have passed laws restricting ESG-based investment in energy, undermining climate-focused financial regulations (Villegas & Scheyder, 2023). Despite mounting global decarbonization pressures, the persistence of fossil fuel lobbying, economic dependencies, and political resistance at the regime level continues to slow the transition toward a low-carbon energy system.

The fossil fuel industry employs astroturfing to manipulate public perception and obstruct renewable energy progress. In 2019, the Caesar Rodney Institute, funded by the oil industry, launched a deceptive campaign against the Skipjack wind farm, which will power 40,000 homes by 2026. Similarly, FTI Consulting orchestrated fake grassroots campaigns promoting fracking and Arctic drilling. Social media amplifies misinformation, as seen in Exxon's $2 million Facebook ads and the Western States Petroleum Association's 'Levanta Tu Voz' campaign, which misled Latino communities about California's EV transition. Fossil fuel executives, including Exxon's former CEO Rex Tillerson, have also engaged in climate denialism. In response, US states and municipalities, such as Multnomah County, Oregon, are suing fossil fuel firms for fraud and misinformation, increasing legal pressure on the industry.

The fossil fuel industry's use of astroturfing, misinformation, and climate denialism has played a significant role in delaying the transition to clean energy. However, growing legal challenges, such as lawsuits from US states and municipalities, signal increasing accountability for these deceptive practices. As public awareness and regulatory scrutiny intensify, the industry's ability to obstruct climate action may weaken, accelerating the shift toward transparent governance and renewable energy adoption.

Energy Subsidies and Policy Resistance

Despite mounting scientific consensus and international commitments to phase out fossil fuels, government subsidies and entrenched policy resistance continue to hinder the global energy transition. In 2022, fossil fuel subsidies reached an all-time high of $7 trillion globally, equivalent to 7.1% of global GDP (IMF, 2023). These subsidies, through direct payments, tax incentives, and underpriced environmental externalities, create a skewed market structure that artificially reduces the cost of coal, oil, and natural gas. This distorts investment signals and suppresses the competitiveness of renewable technologies, thereby delaying decarbonization. The problem is particularly acute in countries with strong fossil fuel lobbies and existing infrastructure lock-ins. The United States, for instance, allocated $757 billion in fossil fuel subsidies in 2022, despite simultaneously passing the IRA to accelerate clean energy investment (IMF, 2023). Indonesia, while pledging to phase out coal by 2050, allocated over $30 billion in coal and petroleum subsidies in the same year (IEA, 2023), revealing the depth of policy contradictions. Such subsidy regimes not only hinder renewable deployment but also have profound environmental and geopolitical consequences. Research suggests that phasing out fossil fuel subsidies could reduce global carbon emissions by up to 30%, significantly contributing to international net-zero goals (IEA, 2023). Yet, political inertia driven by domestic economic interests, fear of job losses, and powerful fossil fuel lobbies continues to stall meaningful reform. As a result, countries are often caught between ambitious climate goals and the political realities of their energy economies.

The Russia–Ukraine conflict has further exposed the fragility of fossil fuel dependence and highlighted how geopolitical disruptions can force reluctant policy shifts. Europe, historically dependent on Russian natural gas for about 40% of its supply (155 bcm in 2021), was compelled to restructure its energy portfolio following Russia's weaponization of energy exports. The RE Power EU plan, launched in 2022, outlines a roadmap to eliminate this dependency by 2030, emphasizing diversification, renewables, and energy efficiency. This pivot, however, has not been without contradictions. While expanding renewable capacity, some countries such as Germany temporarily returned to coal to manage short-term supply gaps, reflecting the ongoing tension between immediate energy security and long-term decarbonization. Simultaneously, Africa and the MENA region have become central to Europe's new energy diplomacy. Nigeria, for example, supplied 51% of Portugal's LNG imports in 2024, up from negligible levels in 2021, while US exports constituted 40%, displacing Russian imports that dropped to 4.4% (Eurostat, 2024). This strategic shift illustrates how geopolitical crises can overcome institutional inertia and compel a reconfiguration of energy alliances. However, these new trade networks still revolve around fossil-based LNG, which, though cleaner than coal, remains incompatible with long-term climate neutrality. Therefore, while geopolitical shocks can catalyze change, they do not inherently dismantle fossil fuel dependencies. Rather, they often redirect them, entrenching new suppliers instead of enabling a full transition to renewables. Unless fossil fuel subsidies are systematically dismantled and policies aligned with carbon pricing mechanisms and regulatory support for renewables, the global transition will remain hampered by economic and political inertia. This dual challenge of subsidy-induced distortion and policy resistance remains a formidable barrier to the energy transition, even in the face of climate urgency and geopolitical recalibration.

Regulatory Frameworks and Policy Shifts

Regulatory frameworks and targeted policy shifts represent foundational levers in driving the global energy transition, serving as instruments for aligning national energy trajectories with climate objectives. Among the most comprehensive and forward-looking initiatives is the EU's Green Deal, which articulates a legally binding commitment to achieving climate neutrality by 2050. This multifaceted strategy encompasses an integrated approach embedding mechanisms such as carbon pricing through the EU Emissions Trading System (EU ETS), mandatory emissions reduction targets across sectors, and a rapid expansion of renewable energy infrastructure. A cornerstone of this policy suite is the revised Renewable Energy Directive (RED III), which mandates that at least 42.5% of the EU's energy consumption must derive from renewables by 2030, with an aspirational goal of 45% (European Commission, 2023). This legislative commitment is further supported by aligned financial instruments, including the EU Innovation Fund and Just Transition Mechanism, which collectively mobilize billions in public and private investment to ensure equitable decarbonization. These efforts not only institutionalize clean energy adoption but also catalyze cross-sectoral transformation by fostering market confidence, technological innovation, and structural change in energy-intensive industries.

Concurrently, China's 14th Five-Year Plan (2021–2025) reflects a similarly ambitious regulatory realignment, reinforcing the state's role in orchestrating a strategic pivot toward renewable energy dominance. The plan outlines concrete targets to increase the installed capacity of wind and solar energy to 1,200 GW by 2030, the development of green hydrogen production hubs, and the integration of energy storage solutions to stabilize variable generation (NDRC, 2021). Notably, it includes granular policy instruments at the provincial and plant levels, ensuring implementation fidelity through mechanisms such as feed-in tariffs (FiTs), mandatory grid access, and curtailment penalties for coal-based power. These directives are buttressed by carbon trading pilots and regulatory caps on new coal-fired capacity, signaling a tightening policy environment against fossil fuel expansion. China's regulatory model emphasizes top-down coordination, guided investments, and state-driven innovation, showcasing how emerging economies can align developmental imperatives with climate priorities. Together, these regulatory architectures in the EU and China signify a regime-level transformation in global energy governance.

By codifying renewable targets into enforceable legislation and embedding them within broader socioeconomic frameworks, these policies aim not only to decarbonize domestic energy systems but also to exert normative pressure on international climate diplomacy. However, the efficacy of these frameworks hinges on robust policy enforcement, iterative policy learning, and resilience against political and economic headwinds, including fossil fuel lobbying, fluctuating energy prices, and geopolitical instability. Furthermore, successful implementation requires transnational cooperation, particularly in areas such as clean technology transfer, critical mineral supply chains, and coordinated carbon border adjustment mechanisms. In essence, regulatory frameworks are no longer peripheral tools but central mechanisms for orchestrating a systemic transition to a post-carbon economy. Their success will determine the pace and scope of the global shift toward sustainable energy futures.

NICHE LEVEL

Advancements in Solar, Wind and Hydrogen

Within the broader narrative of the global energy transition, *solar energy* has emerged as a central technological pillar at the niche level, exhibiting exponential growth over the past decade. From a marginal contributor constituting merely 1% of the renewable energy mix, solar power expanded to account for approximately 13% of the global renewable energy portfolio by 2021. This remarkable ascent driven by a confluence of technological innovation, policy support, and financial viability positions solar PV as the fastest-growing renewable energy technology, with cumulative installed capacity reaching 920 GW of grid-connected systems and an additional 9 GW of off-grid systems globally by 2021 (BP Statistical Review). The intrinsic modularity, scalability, and declining cost curves of solar technologies have catalyzed their widespread adoption, rendering them increasingly attractive across diverse geographic and socioeconomic contexts.

Solar energy technologies broadly bifurcate into solar PV systems, which directly convert solar irradiance into electricity, and solar thermal systems, which harness solar radiation for heat generation. Among these, solar PV has unequivocally emerged as the dominant configuration, owing to its versatility, ease of deployment, and rapid innovation cycles. Concentrated solar power (CSP), although more limited in deployment, remains relevant for specific applications where dispatchable thermal energy and grid integration flexibility are required. As of 2021, global capacity in solar thermal energy reached 522 GWth, while CSP installations accounted for 6.4 GW, highlighting the technological diversity within the solar domain (IRENA, 2022).

The solar PV market has witnessed a staggering 445% increase in installed capacity between 2011 and 2021, escalating from 30 GW to 163 GW of annual additions (IRENA, 2022). Initially propelled by policy incentives in Europe, the momentum has since shifted toward the Asia-Pacific (APAC) region, which dominated the global market with 57% of annual additions and nearly 59% of total installed PV capacity in 2021. With a compound annual growth rate (CAGR) of 37%, solar PV stands as a well-known renewable energy technology globally. Projections for the near term suggest continued rapid growth: according to various forecasts, annual installations are expected to reach 278 GW by 2025, with total operational capacity ranging from 1.6 TW (IEA) to over 2 TW (IHS Markit; SolarPower Europe). Medium- to long-term projections diverge considerably, with solar PV capacity estimates for 2030 varying between 4.9 TW and 10.2 TW, and as high as 63.4 TW by 2050 in high-ambition scenarios reflecting both technological optimism and the growing indispensability of solar in net-zero energy systems (IEA, 2023; Bloomberg New Energy Finance, 2023).

Despite robust growth in absolute terms, the geographical distribution of solar capacity is anticipated to remain relatively stable. The APAC region is expected to retain its majority share (56%–59%), followed by Europe (maintaining approximately 20%–22%) and the Americas, which may experience marginal growth to 18%–19%. The Middle East and Africa (MEA), while currently accounting for only 3% of global capacity, are projected to double their share to 4% in several moderate-growth scenarios, highlighting the untapped potential in emerging markets. In parallel, CSP capacity is forecasted to increase by 11–74 times by 2030, potentially reaching up to 474 GW, particularly in regions with high direct normal irradiance and supportive policy environments (IRENA, 2023).

From an investment standpoint, solar energy continues to attract unprecedented capital flows, underpinned by declining LCOE and enhanced investor confidence. Global investments in solar energy surpassed USD 251 billion in 2021 and are showing a continuous increase to USD 503 billion in 2024 (Figure 15.3), with capital allocations heavily concentrated in the APAC, European, and North American regions (IEA, 2022). Collectively, these three regions accounted for approximately 92% of global solar capacity, with countries such as China, the United States, India, Germany, Japan, Vietnam, and Brazil leading the charge. However, this investment trajectory reveals stark regional imbalances, developing economies in Latin America, the Middle East, Africa, and the Caribbean remain underrepresented in solar deployment and financing, largely due to structural barriers such as inadequate grid infrastructure, policy uncertainty, and limited access to concessional finance (Schmela et al., 2023). Bridging this disparity will require targeted international cooperation, de-risking mechanisms, and innovative financial instruments to unlock latent solar potential in the Global South.

In summary, solar energy's ascendancy as a cornerstone of the global renewable energy transition underscores the importance of niche-level technological and financial innovation. The accelerated deployment of solar PV, underpinned by supportive regulatory environments and evolving investment ecosystems, exemplifies the transformative capacity of renewable technologies when synergized with enabling market and policy conditions. As the world progresses toward net-zero

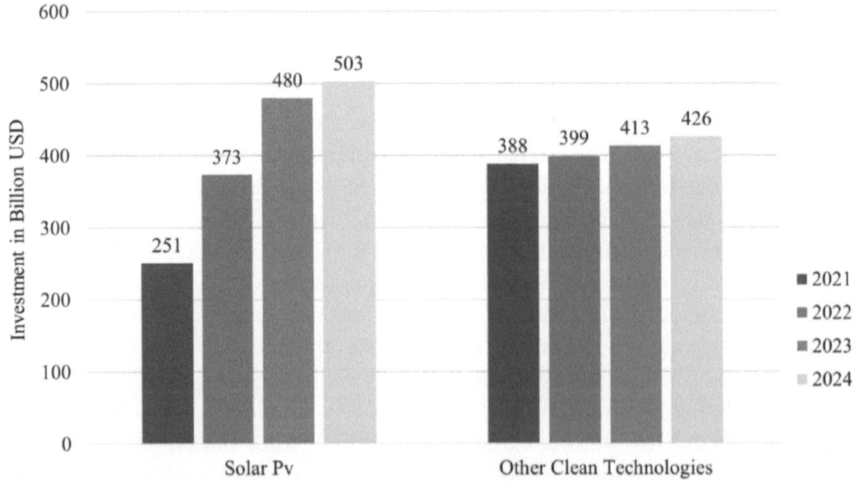

FIGURE 15.3 Global annual investment in solar PV and other generation technologies, 2021–2024.

objectives, solar power will remain a critical enabler, not only in reducing carbon intensity but also in democratizing energy access and enhancing energy sovereignty across both developed and developing economies.

Wind energy, on the other hand, represents a mature yet continually evolving innovation, propelling the global renewable energy transition. Over the past two decades, wind power has progressed through significant technological improvements in turbine architecture, aerodynamic design, and digital control systems, enabling it to become one of the most scalable and cost-effective renewable energy technologies (IRENA, 2023; GWEC, 2023). The global installed wind capacity exceeded 940 GW by the end of 2022, generating over 2,300 TWh of electricity and contributing significantly to the decarbonization of power systems (IEA, 2023). Despite the challenges posed by the pandemic, the global wind energy market continues to show strong potential, with an anticipated annual growth rate of 4%, driven largely by supportive governmental policies such as China's FiT and the United States' Production Tax Credit (PTC) (Lee and Zhao, 2021; Lee, 2022). Innovations such as floating offshore wind turbines, advanced composite materials, and predictive analytics for operations and maintenance have further improved the reliability and efficiency of wind systems, especially in offshore and high-resource zones (IRENA, 2022), targeting both performance optimization and cost reduction. Parallel to structural upgrades, breakthroughs in power electronics, especially in the application of doubly-fed induction generators (DFIG) and variable-speed generator systems have facilitated more stable grid integration, reduced mechanical stress, and optimized energy yield across variable wind conditions (Tan et al., 2022).

Moreover, the LCOE for onshore wind has decreased by approximately 70% since 2010, making it competitive or cheaper than fossil fuels in many regions (IRENA, 2023). Countries like Denmark, Germany, and the UK have pioneered wind-centric energy systems, while markets in Latin America, South-East Asia, and Africa are increasingly integrating wind to diversify their power generation portfolios (GWEC, 2023). However, the sector faces systemic constraints, including land-use conflicts, infrastructure bottlenecks, and supply chain vulnerabilities tied to rare earth materials such as neodymium and dysprosium—critical for permanent magnet generators (IEA, 2022). These supply risks underscore the growing importance of technological sovereignty and resilient procurement strategies to secure wind energy's long-term viability. As these niche-level innovations scale and align with broader landscape pressures like climate imperatives and energy security concerns, wind energy is playing a transformative role in restructuring global energy geopolitics. Despite recent headwinds stemming from inflationary pressures and elevated interest rates,

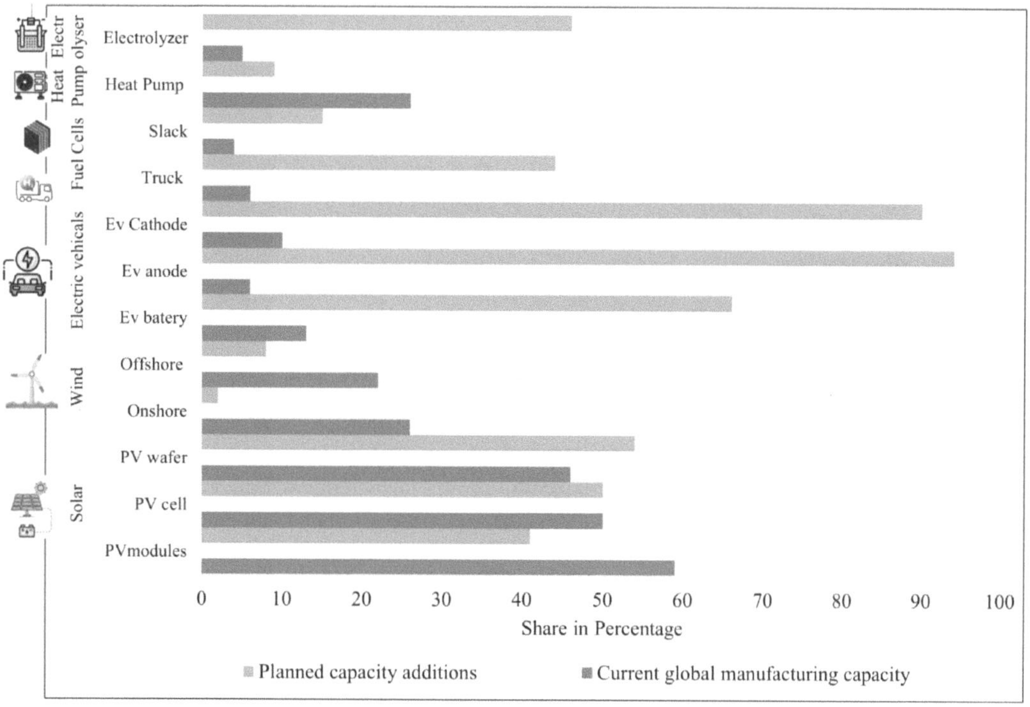

FIGURE 15.4 Renewable energy capacity.

which have constrained investment and project deployment, the wind energy sector is positioned for a robust medium-term recovery. This resurgence will be supported by adaptive policy frameworks tailored to new macroeconomic realities, as well as efforts to streamline permitting processes and improve the financial viability of wind projects. Projections indicate that, relative to the 2017–2023 period, global onshore wind capacity additions could nearly double by 2030, while offshore wind installations are expected to experience a nearly fourfold increase (IEA 2024). Global installed renewable power capacity continued its upward trajectory in 2023, led predominantly by wind and solar technologies. This expansion was underpinned by robust policy support and record-breaking investment levels, an 8.1% increase compared to 2022. Solar PV, wind power, EV, electrolyzers, and other clean technologies continue to grow in planned capacity, highlighting the deepening global commitment to decarbonizing power systems and accelerating the energy transition (Figure 15.4).

Similarly, hydrogen at the niche level has emerged as a promising and highly versatile energy vector, instrumental in advancing the renewable energy transition. Its multifaceted applications ranging from industrial feedstocks in chemical and refining sectors (Gonzalez-Garay et al., 2022; Nyangon & Darekar, 2024) to large-scale energy storage solutions for heating and electricity generation (Avargani et al., 2022; Nyangon & Darekar, 2024), as well as its potential as a clean fuel alternative for hard-to-decarbonize sectors, such as aviation, shipping, and heavy road transport (Ansell, 2022; Rissman et al., 2020) position hydrogen as a strategic enabler in decarbonization pathways. Green hydrogen, produced through electrolysis using renewable energy, enables better integration of solar and wind into the grid while reducing emissions (Wang et al., 2018; Uyar & Beşikci, 2017). However, the upscaling of hydrogen technologies has been constrained by cost and infrastructure-related challenges (Egeland-Eriksen et al., 2021).

Recent policy interventions, particularly in advanced economies, signal a significant shift in supporting the hydrogen sector's maturation. A notable example is the United States' IRA of 2022, which introduced one of the world's most comprehensive clean hydrogen incentive schemes (The White House, 2023). The IRA provides a clean hydrogen PTC of up to \$3/kg based on the carbon intensity of production, alongside options for investment tax credits measures that are expected to significantly reduce production costs and enhance investor confidence. Complementing these incentives is the development of seven federally funded hydrogen hubs, which aim to catalyze regional innovation ecosystems around clean hydrogen. These initiatives are expected to accelerate commercialization, improve economies of scale, and drive cost reductions in production and distribution technologies (Yarmuth, 2022). Furthermore, projections by Glenk and Reichelstein (2019) suggest hydrogen's increasing cost competitiveness in niche applications, particularly in power-to-gas (P2G) systems, reinforcing its potential as a cornerstone of a future low-carbon energy system. Collectively, these advancements demonstrate the growing institutional and technological momentum within the hydrogen niche, enabling it to challenge the dominant fossil fuel regime and contribute meaningfully to the global energy transition.

However, since 2010, the global share of renewables in the electricity mix has risen substantially, reaching 30% in 2023, an increase of 10% points over the past decade. This growth has been driven by ambitious renewable energy policies and sharp cost declines in solar and wind technologies. Countries endowed with significant hydropower resources, such as Brazil, Colombia, Canada, New Zealand, Sweden, and Norway, continue to exhibit the highest shares of renewables in their power mix, with renewables accounting for over two-thirds of electricity generation. Nevertheless, the combined effects of climate change and El Niño led to a 3% reduction in global hydropower output, stagnating the renewable share in countries like the United States and China, and even causing declines in India, Canada, and Mexico. Conversely, the EU saw a notable increase in its renewable share, rising by nearly 6% points to 45%, supported by both hydropower recovery and continued solar and wind expansion. Similarly, Australia and Chile experienced significant gains, with renewable shares increasing by 2.8% and 7% points, respectively, reaching approximately 34% and 64% of total electricity generation (Figure 15.5).

Decentralized Energy Systems

DES represent a transformative approach at the niche level in the renewable energy transition, offering localized and sustainable energy solutions tailored to community-specific needs. Unlike conventional centralized systems, DES focuses on the effective utilization of locally available

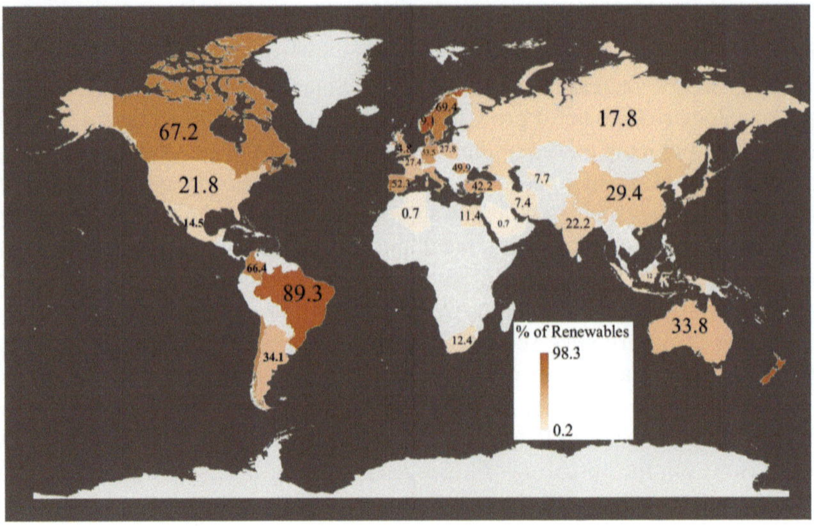

FIGURE 15.5 Share of renewables in electricity production. (Source: Prepared by author based on Enterdata. World Energy & Climate Statistics—Yearbook 2024 https://yearbook.enerdata.net/renewables/renewable-in-electricity-production-share.html.)

renewable resources such as wind, hydro, biomass, and solar energy to generate power for small towns, rural villages, and off-grid areas. As highlighted by Surya et al. (2021), the integration of renewable sources has the potential to meet regional energy demands while minimizing environmental impact. DES not only promotes environmental sustainability but also addresses energy inequality by empowering marginalized and geographically isolated populations, particularly in rural settings (Hiremath et al., 2009). Moreover, the flexibility of DES is enhanced through Hybrid Renewable Energy Systems (HRES), which combine multiple energy sources such as solar–wind or biomass–solar paired with suitable energy storage, ensuring a reliable and uninterrupted supply. This hybrid approach mitigates the intermittency issues common to single-source systems. Tools such as hybrid optimization of multiple energy resources (HOMER), long-term energy alternative planning system (LEAP), and network planner have been instrumental in assessing the feasibility and designing customized energy solutions for remote areas, especially in regions like Africa (Kemausuor et al., 2018). Ultimately, decentralized systems not only reduce carbon emissions and support climate goals but also foster inclusivity, resilience, and energy independence at the community level.

In addition to their environmental and social benefits, DES play a crucial role in reshaping the geopolitical dimensions of energy access and governance. By shifting energy production closer to the point of consumption, DES reduces dependency on large-scale fossil fuel infrastructure and international energy markets, thus insulating communities from global energy price volatility and geopolitical supply disruptions. This localized model empowers communities to take control of their energy future, fostering participatory planning, local economic development, and job creation in renewable energy sectors. Furthermore, DES provides a critical foundation for climate-resilient infrastructure. The integration of distributed energy resources (DERs), such as rooftop solar panels, small wind turbines, and biomass generators, supported by battery storage, microgrids, and smart metering technologies, enhances grid flexibility and energy security. Hybrid configurations—like solar-diesel or wind-biomass systems—are particularly vital in areas with variable renewable resource availability. These systems ensure that energy needs are met reliably even under fluctuating weather conditions, thereby supporting health services, education, and livelihoods in off-grid and underserved regions. However, the successful deployment of DES is contingent upon supportive policy environments, financing mechanisms, and institutional capacity. Challenges such as inadequate regulatory frameworks, lack of technical expertise, and limited initial investment remain barriers to scaling decentralized solutions. As Kemausuor et al. (2018) emphasize, the use of tools like HOMER and LEAP in energy planning can aid in designing cost-effective, context-specific systems that align with national energy access and climate mitigation targets. In summary, DES at the niche level not only complements national and global decarbonization goals but also contributes to democratizing energy access, fostering local innovation, and building resilient, inclusive energy systems. As such, they are a key pillar in the broader renewable energy transition and offer a pathway toward more equitable and geopolitically stable energy futures.

DES encompassing microgrids, community solar initiatives, and off-grid solutions are increasingly recognized as transformative forces in the global energy transition, driving significant geopolitical shifts. As highlighted by Adil and Ko (2016), DES are evolving from isolated stand-alone units to interconnected smart grids, with socio-technical dimensions shaping this transformation. Technologically, grid-connected DES enhances power reliability by balancing the variability of local renewable generation through integration with national or regional electricity networks. This reduces dependence on centralized fossil-fuel-based power, thereby weakening the geopolitical leverage historically held by major fossil fuel–exporting nations. Socially, DES empowers consumers as "prosumers," fostering energy democratization and local agency in electricity production. Community-based power planning facilitates clustered microgrid developments, enhancing local energy sovereignty, and reducing reliance on imported energy. Furthermore, the integration of Information and Communication Technologies (ICT) such as smart meters and sensors enables real-time data exchange and efficient system operation, transforming passive users into informed

TABLE 15.1
Technology Evolution Pathways: Drivers and Outcomes

Stage of Development	Dimension	Drivers/Enablers	Outcomes/Impacts
Initial phase	**Engineering focus**	Declining costs due to innovations in Renewable Energy Technologies (RETs)	Encourages grid-integrated setups via shared access points; drives demand for advanced energy management
	Behavioral & market trends	Increased affordability and attractiveness of solar PV and thermal systems	Greater adoption; initiates community-based influence patterns (e.g., neighbor adoption effects)
			Facilitates better communication between consumers and utilities via smart technology at the distribution level
Transitional phase	**Technology architecture**	Introduction of distributed generation (DG) systems and MicroGrid frameworks	Evolves toward Smart MicroGrid systems with enhanced coordination and intelligence
Expansion phase	**Societal engagement**	Growth of self-generating users (prosumers)	Emergence of local energy communities and new collaborative economic models
	System optimization	Challenges due to variable renewable output and bi-directional energy flow	Necessitates smart grid enhancements (e.g., advanced meters, ICT); enables scalable MicroGrid integration

Source: Modified by author based on Adil and Ko (2016).

participants in energy governance. These advances shift geopolitical dynamics by decentralizing control over energy systems, promoting energy independence for communities and countries, and disrupting traditional energy trade routes. As countries adopt these localized, resilient systems, especially in regions underserved by national grids, DES pave the way for a multipolar energy landscape, diminishing the global dominance of fossil energy powers and redistributing geopolitical influence in favor of energy innovators and technology leaders. Table 15.1 illustrates the evolving socio-technical dynamics of DES across various stages of technological intervention, highlighting how both technological advancements and social transformations contribute to the transition from distributed generation (DG) to smart microgrids.

The growing adoption of DES, particularly DG, is reshaping not only local energy landscapes but also the broader geopolitical structure of energy governance. As individual DG units increasingly coalesce into neighborhood-scale microgrids, the traditional centralized model of utility service provision is being fundamentally challenged. This transition is strongly influenced by social dynamics—especially peer effects and community-level interactions—which have been shown to significantly enhance the uptake of renewable technologies such as solar PV systems. These community-driven initiatives often exert a greater impact on adoption than conventional informational or financial incentives, highlighting the transformative role of collective agency in accelerating the energy transition. From a geopolitical perspective, this decentralization process redefines control over energy production and consumption, enabling communities to act as both producers and consumers, or "prosumers." This shift dilutes the authority of vertically integrated utilities and introduces a variety of new governance arrangements, ranging from community-owned models to partnerships with public or private entities. Such localized control mechanisms contribute to national and regional energy resilience, reduce dependency on imported fossil fuels, and support a more distributed and participatory energy infrastructure—an evolution with significant implications for global energy power dynamics.

Technological Sovereignty and Rare Earth Competition

At the niche level of the global energy transition, where DES, community microgrids, and hybrid renewable solutions are gaining traction, technological sovereignty has become a pivotal concern. Technological sovereignty, broadly understood as the capacity of states or localized systems to independently develop, manage, and sustain essential energy technologies, is increasingly constrained by geopolitical dependencies on critical materials such as REEs (Kiggins, 2015; Scholten et al., 2020). The transition to decentralized and clean energy systems is heavily reliant on technologies such as wind turbines, EVs, and energy storage systems, which in turn depend on materials like neodymium, dysprosium, and lithium (Ali, 2014). As niche-level initiatives scale up especially in off-grid or rural regions through community solar, hydrogen microgrids, and battery storage local actors often rely on imported technologies embedded with these critical minerals. This creates a paradox in the energy transition: while decentralization aims to increase local resilience and autonomy, the technological backbone may still be subject to volatile global supply chains and concentrated market power, particularly from China, which controls over 80% of REE processing (Gulley et al., 2018; Humphries, 2010).

Renewable energy systems such as solar PV arrays, onshore wind farms, and EVs are significantly more mineral-intensive than their hydrocarbon-based counterparts. For instance, the construction of a typical EV requires approximately six times the mineral inputs of a conventional internal combustion engine vehicle, while an onshore wind power plant demands nearly nine times more mineral resources than a natural gas-fired power station. As the global energy transition accelerates, this material intensity has grown substantially; since 2010, the average volume of minerals required per megawatt of newly installed power generation capacity has increased by around 50%, in direct correlation with the rising share of renewables in energy investments. This growing dependency on critical minerals such as lithium, cobalt, nickel, REEs, and copper raises concerns about the long-term strategic autonomy of nations pursuing rapid energy transitions (Figure 15.6).

Countries that lack domestic access to these resources may find their technological sovereignty compromised, increasingly reliant on concentrated and geopolitically sensitive supply chains dominated by a few producing nations. As such, ensuring secure, sustainable, and diversified access to

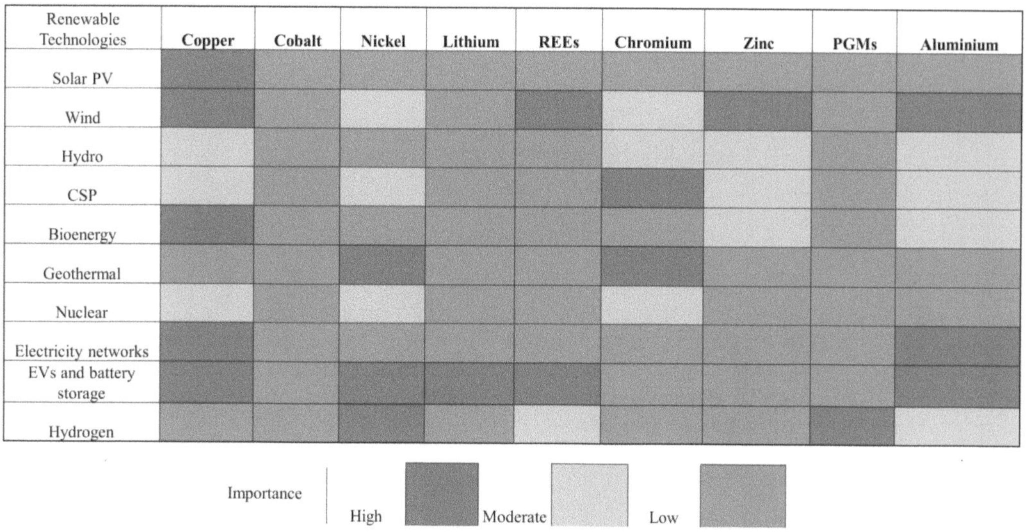

FIGURE 15.6 Essential materials for renewable energy technology. (Source: Prepared by author based on IEA-2021 https://www.iea.org/reports/the-role-of-critical-minerals-in-clean-energy-transitions/mineral-requirements-for-clean-energy-transitions.)

energy-critical minerals becomes a prerequisite for preserving national control over energy technologies and maintaining resilience in the face of global supply disruptions. The mineral composition of clean energy technologies varies significantly by application, each relying on a distinct set of critical elements. For example, lithium, nickel, cobalt, manganese, and graphite play indispensable roles in the performance, durability, and energy density of batteries, which underpin the functionality of EVs and stationary energy storage systems. Simultaneously, REEs such as neodymium, dysprosium, and praseodymium are integral to the manufacture of permanent magnets, which are essential components in wind turbine generators and EV traction motors. Moreover, the expansion and modernization of electricity networks central to decarbonization strategies require substantial volumes of conductive metals, particularly copper and aluminum, with copper serving as a foundational material across nearly all electricity-related technologies (IEA, 2021).

This growing dependence on mineral-intensive technologies highlights a critical aspect of technological sovereignty: control over the raw materials that underpin strategic energy infrastructure. Countries that lack domestic reserves or the capacity to process these critical minerals risk becoming structurally dependent on a small number of resource-rich nations, many of which exert geopolitical leverage through their market dominance. As such, the geopolitical contest for rare earths and other critical minerals is not only a question of supply security, but also of maintaining autonomy over national energy transitions and reducing vulnerability to global market shocks and political instability (Herrington, 2021; Sovacool et al., 2022). Historically, the energy sector accounted for only a minor share of total mineral consumption; however, as low-carbon technologies scale, clean energy applications are rapidly becoming the fastest-growing segment in the global minerals economy. According to the IEA's Sustainable Development Scenario (SDS), which aligns with the Paris Agreement objectives, clean energy technologies are projected to account for over 40% of total demand for copper and REEs, 60%–70% for nickel and cobalt, and nearly 90% for lithium by 2040 (IEA, 2021). EVs and battery storage systems have already surpassed consumer electronics to become the dominant source of lithium demand and are expected to overtake stainless steel as the leading consumer of nickel within the next two decades. This intensification of mineral dependency has profound implications for technological sovereignty, particularly for countries that lack access to these critical resources. As strategic control over lithium, cobalt, and rare earth supply chains become more concentrated among a few geopolitical actors, energy-importing nations may face vulnerabilities analogous to past fossil fuel dependencies (Figure 15.7). These evolving

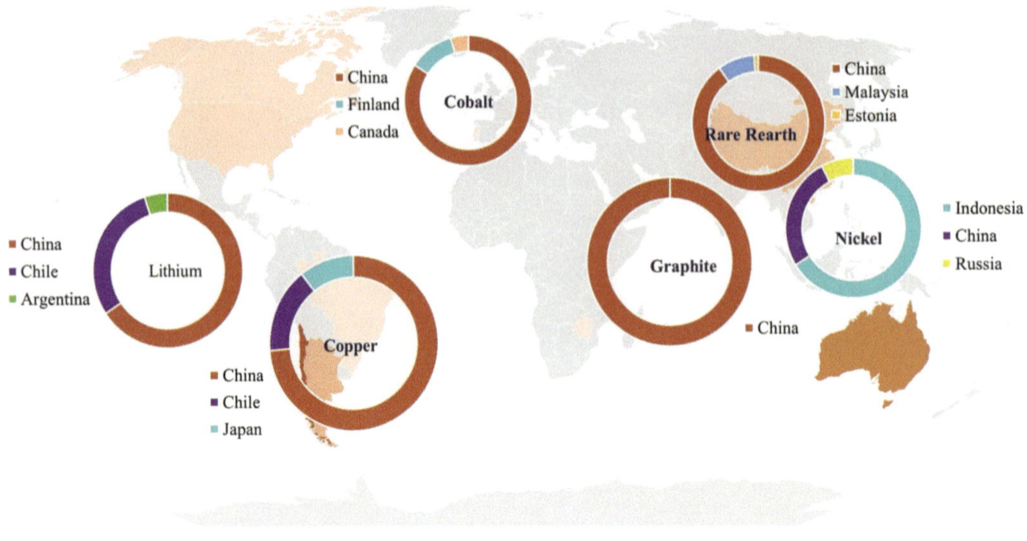

FIGURE 15.7 Share of top three producing countries in processing of selected minerals (Source: Prepared by author based on data IEA02023.)

dynamics underscore the need for diversified sourcing, investment in recycling technologies, and international cooperation to avoid resource bottlenecks and ensure resilient energy transitions (Sovacool et al., 2022).

THE GEOPOLITICS OF MINERAL SUPPLY FOR CLEAN ENERGY TRANSITION

The accelerating shift toward clean energy technologies is fundamentally reshaping the geopolitical landscape of mineral supply chains. Over the next two decades, the global demand for critical minerals will flow as countries increase climate ambition and deploy technologies such as EVs, wind turbines, and battery storage at unprecedented scales. According to projections by the IEA, mineral demand from clean energy technologies is expected to double by 2040 in the Stated Policies Scenario (STEPS) and quadruple in the SDS. EVs and battery storage alone will account for nearly half of this growth, with battery-related mineral demand increasing over 30 times in the SDS (IEA, 2021). Demand for lithium is set to increase by more than 40-fold, while significant growth is also projected for graphite, copper, nickel, and REEs—the latter being vital for wind turbine magnets and EV motors.

These shifting dynamics place mineral-rich nations, especially those with dominant market shares in cobalt (e.g., the Democratic Republic of Congo), rare earths (e.g., China), and lithium (e.g., Chile, Argentina, and Australia), at the center of a new global geopolitical contest. As clean energy transitions advance, access to and control over these critical minerals are becoming strategic levers of influence, potentially replacing oil and gas as the basis of geopolitical power in the 21st century (Herrington, 2021). The geographic concentration of mineral production and processing capabilities raises concerns around supply security, trade restrictions, and market manipulation. Consequently, countries with high mineral import dependency, particularly in Europe, North America, and parts of Asia, are increasingly pursuing policies aimed at resource diversification, domestic production, recycling, and international partnerships to safeguard technological sovereignty. While clean energy transitions promise to decarbonize economies, they simultaneously introduce new geopolitical fault lines that must be addressed through resilient, transparent, and sustainable mineral supply chains (Vakulchuk et al., 2020).

To reconfigure geopolitical powers around green energy, political implications of activities across the mineral value chain and/or trade of the natural resource are significant, because they are increasingly politically sensitive as witnessed by the rise of critical raw material (CRM) designations worldwide (Andersson, 2020; Ghorbani et al., 2024). In fact, all known critical energy transition minerals (ETMs) are also dual use, supporting both military and civilian uses, which implies that they are of high political priority, especially where conflict could result in their deprivation (European Commission, 2020a). In particular, a key aspect of their political sensitivity pertains to the risk of supply disruption (e.g., European Commission, 202b), because regardless of the supply rate of any material, its mass delivered across the supply chain is conserved. Other clear lines of evidence in support of condition, include an increasingly high-level policy driving actions to secure such materials (e.g., European Commission, 2023). Foreign policy is also a line of evidence, such as multiple nations that have expressed the intent to establish friendly trading blocs (Ghorbani et al., 2024). As such, strong political involvement is demonstrably occurring in the trade and productivity of CRMs.

The geopolitics of mineral supply in the energy transition is increasingly defined by the asymmetric distribution of geologically essential minerals and materials (GEMMs), and the strategic behavior of nations seeking to secure or leverage these resources. As demand for lithium, cobalt, graphite, and REEs surges, potentially increasing by more than 50 times by 2040, depending on technological pathways (IEA, 2021), countries with large reserves are gaining newfound geopolitical relevance. For example, Chile's nationalization of its lithium sector mirrors its earlier experience with copper, suggesting that state-led resource control may again play a central role in shaping global markets (Villegas & Scheyder, 2023; Singh, 2010). At the same time, major players like

China, which dominates the refining and processing stages of critical minerals, are embedding their geopolitical influence through expansive foreign investments and initiatives like the Belt and Road (Kalantzakos, 2020). Such dynamics introduce new vulnerabilities for import-dependent regions such as the EU, which has responded by forming the CRMs club and enacting the CRMs Act to foster resource partnerships and reduce dependency (European Commission, 2023a). Meanwhile, producer countries like Bolivia, Mongolia, and the Democratic Republic of Congo (DRC) may strengthen their bargaining power by forming cooperative blocs, potentially reshaping the supply-side geopolitics of energy transitions. This evolving global landscape illustrates how mineral supply chains, once seen as purely economic, are now central to international strategic competition, cooperation, and technological sovereignty (Overland et al., 2022; Klinger, 2018).

The geopolitical landscape of the clean energy transition is being profoundly reshaped by the geographic distribution of CRMs and the strategic formation of resource trade blocs. Based on the geographic distribution of critical ETMs and the global dominance of certain nations in mining, refining, and reserve capacity, this study envisions the formation of future economic trade blocs centered around critical mineral value chains. These blocs could emerge either through geopolitical alignment, industrial strategy, or shared resource endowments. Here's a projection of five potential trade blocs (Figure 15.8) that could shape future clean energy geopolitics and trade:

- Critical Mineral Refining & Technology Bloc (China-Centric)
- Lithium Triangle and Latin American Raw Material Bloc
- Africa's Resource-Rich Position
- Indo-Pacific Mining & Processing Bloc or Nickel Bloc
- Western Alliance for Ethical and Transparent Mineral Supply Chains

Strategic resource distribution of the critical ETMs creates potential conflicts over resource access (Overland et al., 2022; European Commission, 2023b). Consequently, access to ETMs supply could depend on a purchaser's international relations with a sovereign country, among other things. As the global demand for lithium, cobalt, nickel, copper, graphite, and REEs intensifies, countries endowed with these resources or those dominating their processing are consolidating their influence. For instance, China, already responsible for refining 65% of global lithium, 74% of cobalt, and

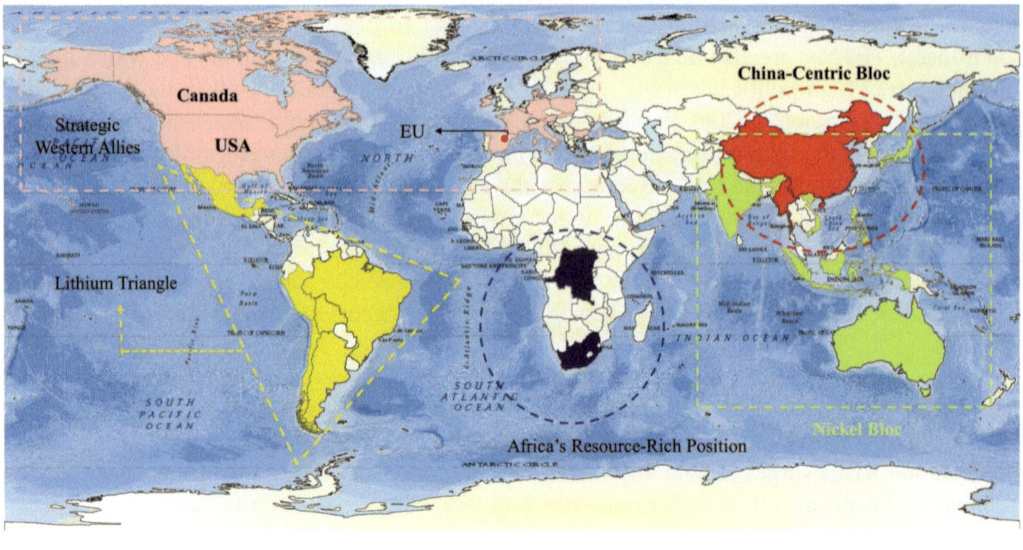

FIGURE 15.8 Strategic resource alliances and the emergence of critical mineral trade blocs. (Source: Mapped by author.)

TABLE 15.2
Critical Minerals and the New Geopolitics of Energy Transition

Issue	Simplified Tension/Trend	References
Access vs. sovereignty	Countries rich in minerals (e.g., DRC, Chile) want more control over how their minerals are processed and exported.	IEA (2021), and Sovacool et al. (2020b)
Tech sovereignty	Countries like China, the US, and the EU aim to control clean tech (e.g., battery and solar manufacturing).	IRENA (2023), IEA (2023), and Zhang and Chen (2022)
Mineral alliances	Groups like the IEA, G7, and new UN frameworks are forming rules for fair and stable mineral trade.	UN Framework on Critical Minerals (2024) and G7 Minerals Security Partnership (2023)
Geopolitical shift	Future global power struggles may focus more on control of mineral routes than oil regions.	Overland (2019), Vakulchuk et al. (2020), and Klare (2012)
ESG and trust	Western countries want minerals to come with transparency, ethical mining, and local community benefits.	World Bank (2020), and IPCC (2021)

90% of REEs, continues to maintain dominance through its vertically integrated mineral supply chains. Similarly, Australia, Chile, and Argentina, key players in the lithium "triangle," are leveraging their reserves to forge deeper trade partnerships, particularly with consumer markets in East Asia and Europe. Indonesia, the world's largest producer of nickel and home to over 40% of global reserves, is emerging as the epicenter of a Southeast Asian battery material corridor, along with the Philippines and New Caledonia. This region is increasingly linked to refining centers in China and Japan, enabling the formation of a critical mineral belt with high geopolitical stakes. Meanwhile, Africa, particularly the Democratic Republic of Congo (DRC)—which supplies over 70% of the world's cobalt—is being courted by both Western and Chinese stakeholders, raising questions around resource sovereignty, developmental equity, and environmental standards. These regional resource clusters are giving rise to new economic alliances focused on mineral security, resembling the historical importance of OPEC for oil. Examples include the EU's CRMs Act, US-led Minerals Security Partnership (MSP), and collaborative efforts by UNEP and the World Bank to foster just and sustainable supply chains. The emergence of these blocs underscores the geopolitical reordering driven not by fossil fuels but by mineral dependencies, where trade dynamics, processing capacities, and investment flows are pivotal to securing strategic autonomy in the energy transition (Table 15.2).

POLICY RECOMMENDATIONS AND FUTURE OUTLOOK

Based on the research findings, a set of comprehensive policy recommendations is essential to guide a just and geopolitically stable energy transition.

- First, it is critical to support a just and inclusive transition by ensuring social equity and fostering public–private partnerships, especially in resource-rich but economically vulnerable regions such as sub-Saharan Africa and Latin America. Transparent and ethical supply chains must be established through international certification schemes and robust ESG standards to avoid exploitation, human rights abuses, and environmental degradation. Diversifying and securing critical mineral supply chains is also vital, reducing overreliance on dominant processing nations like China by encouraging strategic stockpiling, new exploration in geopolitically stable countries, and mutually beneficial partnerships with the Global South.

- Strengthening global governance through multilateral alliances, such as the G7 Critical Minerals Alliance and the UN Panel on Critical Energy Transition Minerals, can help build enforceable agreements that ensure equitable access and trade stability. Additionally, promoting circular economy principles through recycling, material substitution, and improved efficiency can reduce pressure on primary extraction and contribute to long-term sustainability. To support industrial development in mineral-rich nations, regional policies must prioritize local processing and value addition, empowering countries to move up the supply chain and avoid remaining raw material exporters. At the same time, countries with technological leadership should maintain a balance between national security and international collaboration in clean energy innovation.
- Finally, it is crucial to integrate geopolitical risk assessments into national energy and climate strategies. As the world shifts from oil chokepoints to "mineral corridors," proactive planning, scenario modeling, and strategic diplomacy will be necessary to prevent new forms of resource-related conflict. These coordinated efforts, anchored in fairness, sustainability, and resilience, will help ensure that the energy transition is not only green but also globally inclusive and geopolitically balanced.

CONCLUSION

This study emphasizes the multifaceted nature of the global energy transition, highlighting not only its technological and environmental dimensions but also its profound geopolitical implications. By applying the MLP framework from a top-down lens, the research reveals how national policies, global regimes, and strategic innovations interact to drive the shift from fossil fuel dependence to SES. The transition, while critical to achieving climate goals and long-term energy security, is reshaping global power dynamics, diminishing the influence of traditional hydrocarbon-rich states and elevating actors with dominance in renewable technologies and critical mineral supply chains. As the demand for strategic materials such as lithium, cobalt, and REEs surges, a new geopolitical landscape is emerging centered not on oil fields and shipping lanes, but on mineral corridors and technological sovereignty. However, this shift is not without challenges. Environmental degradation, social inequities, and new dependencies on politically volatile regions threaten to undermine the sustainability and justice of the transition. As such, the future of the energy order will depend heavily on effective governance, international collaboration, and equitable policy frameworks that prioritize responsible mining, circularity, and inclusive development. Ultimately, a successful and just energy transition must reconcile the urgent need for decarbonization with the imperative of geopolitical stability and socioeconomic fairness, ensuring that the green transformation benefits all, rather than deepening existing global divides.

REFERENCES

Adil, A. M., & Ko, Y. (2016). Socio-technical evolution of Decentralized Energy Systems: A critical review and implications for urban planning and policy. *Renewable and Sustainable Energy Reviews*, 57, 1025–1037.

Ali, S. H. (2014). Social and environmental impact of the rare earth industries. *Resources*, 3(1), 123–134.

Ali, S. H., Giurco, D., Arndt, N., Nickless, E., Brown, G., Demetriades, A., & Yakovleva, N. (2017). Mineral supply for sustainable development requires resource governance. *Nature*, 543(7645), 367–372.

Ali, S. H., Kalantzakos, S., Eggert, R., Gauss, R., Karayannopoulos, C., Klinger, J., Pu, X., Vekasi, K., & Perrons, R. K. (2022). Closing the infrastructure gap for decarbonization: The case for an integrated mineral supply agreement. *Environmental Science & Technology*, 56(22), 15280–15289. https://doi.org/10.1021/acs.est.2c05413.

Andersson, P. (2020). Chinese assessments of "critical" and "strategic" raw materials: Concepts, categories, policies, and implications. *The Extractive Industries and Society*, 7(1), 127–137. https://doi.org/10.1016/j.exis.2020.01.008.

Ansell, P. J. (2022). Hydrogen-electric aircraft technologies and integration: Enabling an environmentally sustainable aviation future. *IEEE Electrification Magazine*, 10(2), 6–16.

Avargani, V. M., Zendehboudi, S., Cata Saady, N. M., & Dusseault, M. B. (2022). A comprehensive review on hydrogen production and utilization in North America: Prospects and challenges. *Energy Conversion and Management*, 269, 115927.

Berkeley, N., Bailey, D., Jones, A., & Jarvis, D. (2017). Assessing the transition towards battery electric vehicles: A multi-level perspective on drivers of, and barriers to, take up. *Transportation Research part A: Policy and Practice*, 106, 320–332.

Bloomberg. (2023). Renewable investments. https://www.bloomberg.com/graphics/climate-change-data-green/investment.html.

Bloomberg New Energy Finance. (2023). Global low-carbon energy technology investment surges past $1 trillion for the first time. https://about.bnef.com/blog/global-low-carbon-energy-technology-investment-surges-past-1-trillion-for-the-first-time.

Bouzarovski, S., & Konieczny, M. (2010). Landscapes of paradox: Public discourses and policies in Poland's relationship with the Nord Stream pipeline. *Geopolitics*, 15(1), 1–21.

Bradshaw, M. J. (2009). The geopolitics of global energy security. *Geography Compass*, 3(5), 1920–1937.

Brulle, R. J. (2018). The climate lobby: A sectoral analysis of lobbying spending on climate change in the USA, 2000 to 2016. *Climatic Change*, 149(3), 289–303.

Brunet, C., Savadogo, O., Baptiste, P., Bouchard, M. A., Cholez, C., Gendron, C., & Merveille, N. (2021). The three paradoxes of the energy transition-assessing sustainability of large-scale solar photovoltaic through multi-level and multi-scalar perspective in Rwanda. *Journal of Cleaner Production*, 288, 125519.

Chishti, M. Z., Sinha, A., Zaman, U., & Shahzad, U. (2023). Exploring the dynamic connectedness among energy transition and its drivers: Understanding the moderating role of global geopolitical risk. *Energy Economics*, 119, 106570 https://doi.org/10.1016/j.eneco.2023.106570.

Costa, E., Wells, P., Wang, L., & Costa, G. (2022). The electric vehicles and renewable energy: Changes in boundary conditions that enhance business model innovations. *Journal of Cleaner Production*, 333, 130034. https://doi.org/10.1016/j.jclepro.2021.130034.

Dall-Orsoletta, A. D., Ferreira, P., & Dranka, G. G. (2022). Low-carbon technologies and just energy transition: Prospects for electric vehicles. *Journal of Energy Conversion and Management*, 16, 100271. https://doi.org/10.1016/j.ecmx.2022.100271.

DivestInvest. (2023). Fossil Free Finance: Global Divestment Commitments https://www.divestinvest.org

Egeland-Eriksen, T., Hajizadeh, A., & Sartori, S. (2021). Hydrogen-based systems for integration of renewable energy in power systems: Achievements and perspectives. *International Journal of Hydrogen Energy*, 46(63), 31963–31983.

European Commission. (2020a). National energy and climate plans. Accessed from https://commission.europa.eu/energy-climate-change-environment/implementation-eu-countries/energy-and-climate-governance-and-reporting/national-energy-and-climate-plans_en.

European Commission. (2020b). GECO 2017: Global Energy and Climate Outlook 2017: How climate policies improve air quality. Accessed from https://joint-research-centre.ec.europa.eu/projects-and-activities/global-energy-and-climate-outlook/past-geco-editions/geco-2017_en#:~:text=GECO2017%20examines%20the%20evolution%20of%20the%20world%20energy,the%20preparations%20for%20the%20future%20global%20stock-taking%20exercises.

European Commission. (2023). Corporate Sustainability Reporting Directive (CSRD). Accessed from https://finance.ec.europa.eu/capital-markets-union-and-financial-markets/company-reporting-and-auditing/company-reporting/corporate-sustainability-reporting_en.

Fatemi, A. M., &Fooladi, I. J. (2013). Sustainable finance: A new paradigm. *Global Finance Journal*, 24(2), 101–113.

Flynn, D., Rather, Z., Årdal, A. R., D'Arco, S., Hansen, A. D., Cutululis, N. A., ... & Wang, Y. (2019). Technical impacts of high penetration levels of wind power on power system stability. In P. D. Lund, J. Byrne, R. Haas, & D. Flynn (Eds.), *Advances in energy systems: The large-scale renewable energy integration challenge* (pp. 47–65). https://doi.org/10.1002/9781119508311.ch3.

G7 Minerals Security Partnership. (2023). G7 cooperation to de-risk minerals investments in the global south. https://www.csis.org/analysis/g7-cooperation-de-risk-minerals-investments-global-south#:~:text=In%20the%20G7%20Climate,%20Energy%20and%20Environment%20Ministers%E2%80%99,of%20diversification%20of%20existing%20suppliers%20of%20critical%20minerals.%E2%80%9D.

Geels, F. W. (2002). Technological transitions as evolutionary reconfiguration processes: a multi-level perspective and a case-study. *Research Policy*, 31(8–9), 1257–1274.

Geels, F. W., & Schot, J. (2007). Typology of sociotechnical transition pathways. *Research Policy*, 36(3), 399–417.

Geels, F. W., Sovacool, B. K., Schwanen, T., & Sorrell, S. (2017). Sociotechnical transitions for deep decarbonization. *Science*, 357(6357), 1242–1244.

Ghorbani, Y., Zhang, S. E., Bourdeau, J. E., Chipangamate, N. S., Rose, D. H., Valodia, I., & Nwaila, G. T. (2024). The strategic role of lithium in the green energy transition: Towards an OPEC-style framework for green energy-mineral exporting countries (GEMEC). *Resources Policy*, 90, 104737.
Gkillas, K., Gupta, R., Pierdzioch, C., & Yoon, S. M. (2021). OPEC news and jumps in the oil market. *Energy Economics*, 96, 105096 https://doi.org/10.1016/j.eneco.2021.105096.
Gleick, P. H., Cooley, H., Famiglietti, J. S., Lettenmaier, D. P., Oki, T., Vörösmarty, C. J., & Wood, E. F. (2013). Improving understanding of the global hydrologic cycle: observation and analysis of the climate system: The global water cycle. In G. R. Asrar & J. W. Hurrell (Eds.), *Climate Science for Serving Society: Research, Modeling and Prediction Priorities*, pp. 151–184.
Glenk, G., & Reichelstein, S. (2019). Economics of converting renewable power to hydrogen. *Nature Energy*, 4(3), 216–222.
Global Wind Energy Council (GWEC). (2023). Global Wind Report 2023. https://gwec.net/global-wind-report-2023/
Goldthau, A., & Sitter, N. (2021). Horses for courses. The roles of IPE and global public policy in global energy research. *Policy and Society*, 40(4), 467–483.
Gonzalez-Garay, A., Bui, M., Freire Ordóñez, D., High, M., Oxley, A., Moustafa, N.,... & Shah, N. (2022). Hydrogen production and its applications to mobility. *Annual Review of Chemical and Biomolecular Engineering*, 13(1), 501–528.
Gulley, A. L., Nassar, N. T., & Xun, S. (2018). China, the United States, and competition for resources that enable emerging technologies. *Proceedings of the National Academy of Sciences*, 115(16), 4111–4115.
Gunther, S. J., & Bernell, D. (2019). Challenging the system: The role of community choice aggregation in California's transition to a renewable energy future. *The Electricity Journal,* 32(10), 106679.
Hao, F., & Shao, W. (2021). What really drives the deployment of renewable energy? A global assessment of 118 countries. *Energy Research & Social Science*, 72, 101880.
Herrington, R. (2021). Mining our green future. *Nature Reviews Materials*, 6(6), 456–458.
Hiremath, R. B., Kumar, B., Deepak, P., Balachandra, P., Ravindranath, N. H., & Raghunandan, B. N. (2009). Decentralized energy planning through a case study of a typical village in India. *Journal of Renewable and Sustainable Energy*, 1(4), 43–50. https://doi.org/10.1063/1.3167282.
Hirsch, R. L., Bezdek, R., & Wendling, R. (2005). Peaking of world oil production: Impacts, mitigation, & risk management (No. DOE/NETL-IR-2005–093; NETL-TPR-2319). National Energy Technology Laboratory (NETL), Pittsburgh, PA, Morgantown, WV, and Albany.
Hsiang, S. M., Burke, M., & Miguel, E. (2013). Quantifying the influence of climate on human conflict. *Science*, 341(6151), 1235367. https://finance.ec.europa.eu/capital-markets-union-and-financial-markets/company-reporting-and-auditing/company-reporting/corporate-sustainability-reporting_en
Humphries, M. (2010). *Rare Earth Elements: The Global Supply Chain.* Diane Publishing.
IEA. (2020b). World Energy Outlook 2020. International Energy Agency, pp. 1–464.
IEA. (2021a). Global Energy Review 2021. International Energy Agency, pp. 1–36.
IEA. (2021b). Share of Top Producing Countries in Total Processing of Selected Minerals and Fossil Fuels. IEA, Paris. https://www.iea.org/data-and-statistics/charts/share-of-top-producing-countries-in-total-processing-of-selected-minerals-and-fossil-fuels-2019, Licence: CC BY 4.0
IEA. (2022). The Role of Critical Minerals in Clean Energy Transitions. International Energy Agency.
IEA. (2023). Energy Technology Perspectives 2023. International Energy Agency, pp. 1–464.
IEA. (2023a). Global Energy and Climate Model. International Energy Agency, pp. 1–127.
IEA. (2023b). Tracking Clean Energy Progress 2023. International Energy Agency. https://www.iea.org/reports/tracking-clean-energy-progress-2023#overview.
IMF. (2023). World Economic Outlook https://www.imf.org/en/Publications/WEO/Issues/2023/10/10/world-economic-outlook-october-2023
InfluenceMap. (2019). Big Oil's Real Agenda on Climate Change. Londres, Influence Map. https://influencemap.org/report/Climate-Lobbying-by-the-Fossil-Fuel-Sector
IPCC (2007). *Weather and climate extreme events in a changing climate*. Accessed from https://www.ipcc.ch/report/ar6/wg1/chapter/chapter-11/.
IPCC. (2021). *Climate Change 2021: The Physical Science Basis. Contribution of Working Group I to the Sixth Assessment Report of the Intergovernmental Panel on Climate Change* [Masson-Delmotte, V., P. Zhai, A. Pirani, S. L. Connors, C. Péan, S. Berger, N. Caud, Y. Chen, L. Goldfarb, M. I. Gomis, M. Huang, K. Leitzell, E. Lonnoy, J. B. R. Matthews, T. K. Maycock, T. Waterfield, O. Yelekçi, R. Yu, and B. Zhou (eds.)]. Cambridge University Press, Cambridge, and New York, In press, doi:10.1017/9781009157896.
IRENA. (2019b). Climate Change and Renewable Energy: National Policies and the Role of Communities, Cities and Regions. A Report to the G20 Climate and Sustainability Working Group (CSWG).

IRENA. (2019a). Global Energy Transformation: A Roadmap to 2050. International Renewable Energy Agency, pp. 1–52.
IRENA. (2021). Global Renewables Outlook: Energy Transformation 2050. International Renewable Energy Agency, pp. 1–291.
IRENA (2022). Renewable Energy Statistics. Accessed from https://www.irena.org/publications/2022/Jul/Renewable-Energy-Statistics-2022.
IRENA. (2023a). Global Renewables Outlook 2023. International Renewable Energy Agency.
IRENA. (2023b). World Energy Transitions Outlook 2023: 1.5°C Pathway, vol. 1. International Renewable Energy Agency, pp. 1–176.
Jacobsen, G. (2019). The Impact of Energy Booms on Local Workers, vol. 468. IZA World of Labor. https://doi.org/10.15185/izawol.468.
Jia-Hai, Y. (2014). Developing super smart grids in China: perspective of socio-technical systems transition. In D. Mah, P. Hills, V. O. K. Li, R. Balme (Eds.), *Smart Grid Applications and Developments*, pp. 279–301. London: Springer London. https://link.springer.com/chapter/10.1007/978-1-4471-6281-0_14.
Johnson, C. (2017). Competing para-sovereignties in the borderlands of Europe. *Geopolitics*, 22(4), 772–793.
Kalantzakos, S. (2020). The race for critical minerals in an era of geopolitical realignments. *Internal Spectrometry,* 55 (3), 1–16. https://doi.org/10.1080/ 03932729.2020.1786926.
Kemausuor, F., Sedzro, M. D., & Osei, I. (2018). Decentralised energy systems in Africa: Coordination and integration of off-grid and grid power systems—review of planning tools to identify renewable energy deployment options for rural electrification in Africa. *Current Sustainable/Renewable Energy Reports*, 5, 214–223.
Kiggins, R. D. (2015). *The Political Economy of Rare Earth Elements: Rising Powers and Technological Change*. Springer.
Klare, M. (2012). *The Race for What's Left: The Global Scramble for the World's Last Resources*. Macmillan.
Klinger, J. M. (2018). Rare earth elements: Development, sustainability and policy issues. *The Extractive Industries and Society*, 5(1), 1–7. https://doi.org/10.1016/j.exis.2017.12.016.
Lantushenko, V., & Schellhorn, C. (2023). The rising risks of fossil fuel lobbying. *Global Finance Journal*, 56, 100829.
Lee, J. (2022). The evolving role of tax incentives in U.S. renewable energy policy: Implications for wind power deployment. *Energy Policy*, 165, 112938. https://doi.org/10.1016/j.enpol.2022.112938
Lee, J., & Zhao, X. (2021). Policy support and market dynamics in global wind energy: A comparative study of China and the United States. *Renewable and Sustainable Energy Reviews*, 146, 111136. https://doi.org/10.1016/j.rser.2021.111136
Li, K., Tan, X., Yan, Y., Jiang, D., & Qi, S. (2022). Directing energy transition toward decarbonization: The China story. *Energy*, 261, 124934.
Luft, G., & Korin, A. (Eds.). (2009). *Energy Security Challenges for the 21st Century: A Reference Handbook*. Bloomsbury Publishing USA.
Mahmood, N., Zhao, Y., Lou, Q., & Geng, J. (2022). Role of environmental regulations and eco-innovation in energy structure transition for green growth: Evidence from OECD. *Technological Forecasting and Social Change*, 183, 121890.
Ministry of New and Renewable Energy (MNRE), Government of India. (2023). India's Renewable Energy Roadmap 2030: Policy Framework and Deployment Strategy. https://mnre.gov.in
National Development and Reform Commission (NDRC). (2021). The 14th Five-Year Plan for National Economic and Social Development of the People's Republic of China and the Outline of the Long-Range Objectives Through the Year 2035. Central Committee of the Communist Party of China & State Council, Beijing. https://www.ndrc.gov.cn/fggz/fzzlgh/gjjgh/202203/P020220311591338505192.pdf
National Renewable Energy Laboratory (NREL). (2023). Perovskite-Silicon Tandem Solar Cells Surpass 30% Efficiency. U.S. Department of Energy. https://www.nrel.gov/news/program/2023/perovskite-silicon-tandem-surpass-30-percent.html
Net Zero Tracker. (2023). Net Zero Stocktake 2023. Energy & Climate Intelligence Unit, Data-Driven EnviroLab, NewClimate Institute, and Oxford Net Zero. https://zerotracker.net/analysis/net-zero-stocktake-2023
NRC (National Research Council). (2002) *Review of USGCRP Plan for a New Science Initiative on the Global Water Cycle*. National Academy Press, Washington, DC.
Nyangon, J., & Darekar, A. (2024). Advancements in hydrogen energy systems: A review of levelized costs, financial incentives and technological innovations. *Innovation and Green Development*, 3(3), 100149.
OPEC. (2023). Maximum energy, minimum emissions: OPEC commentary for the G7 Hiroshima Summit. Accessed from https://www.opec.org/pn-detail/13-17-may-2023.html.
OpenSecrets. (2023). OpenSecrets.org lobbying (https://www.opensecrets.org/federal-lobbying/) and revolving door (https://www.opensecrets.org/revolving/) datasets in compliance with its Creative Commons license.

Overland, I. (2015). Future petroleum geopolitics: Consequences of climate policy and unconventional oil and gas. In *Handbook of Clean Energy Systems*, pp. 1–29. https://doi.org/10.1002/9781118991978.hces203.
Overland, I. (2019). The geopolitics of renewable energy: Debunking four emerging myths. *Energy Research & Social Science*, 49, 36–40.
Overland, I., Juraev, J., & Vakulchuk, R. (2022). Are renewable energy sources more evenly distributed than fossil fuels? Renew. *Energy*, 200, 379–386. https://doi.org/10.1016/j.renene.2022.09.046.
Owen, J. R., Kemp, D., Schuele, W., & Loginova, J. (2023). Misalignment between national resource inventories and policy actions drives unevenness in the energy transition. Communications Earth & Environment, 4, 454. https://doi.org/10.1038/s43247-023-01134-4.
PwC. (2023). Global Consumer Insights Survey 2023. PricewaterhouseCoopers. https://www.pwc.com/gx/en/industries/consumer-markets/consumer-insights-survey.html
Rissman, J., Bataille, C., Masanet, E., Aden, N., Morrow III, W. R., Zhou, N., Elliott, N., et al. (2020). Technologies and policies to decarbonize global industry: Review and assessment of mitigation drivers through 2070. *Applied Energy*, 266, 114848.
Rubin, S., Veloz, T., & Maldonado, P. (2021). Beyond planetary-scale feedback self- regulation: Gaia as an autopoietic system. *Biosystems* 199, 104314. https://doi.org/10.1016/j.biosystems.2020.104314.
Santos, T. (2020). The prize and the quest of energy beyond the oil paradigm: Renewable energy deployment under the sustainable development agenda. *Sul Global*, 1(1), 60–83.
Sattich, T., Freeman, D., Scholten, D., & Yan, S. (2021). Renewable energy in EU-China relations: Policy interdependence and its geopolitical implications. *Energy Policy*, 156, 112456. https://doi.org/10.1016/j.enpol.2021.112456.
SBTi. (2023). Science-Based Targets initiative: Companies Taking Action. https://sciencebasedtargets.org/companies-taking-action
Schmela, M., Rossi, R., Lits, C., Chunduri, S. K., Shah, A., Muthyal, R.,... &Saratchandra, P. (2023). Advancements in solar technology, markets, and investments–A summary of the 2022 ISA World Solar Reports. *Solar Compass*, 6, 100045
Scholten, D., Bazilian, M., Overland, I., & Westphal, K. (2020). The geopolitics of renewables: New board, new game. *Energy Policy*, 138, 111059.
Singh, J. T. N. (2010). Governing the extractive sector: The politics of globalisation and copper policy in Chile. *Journal of Critical Globalisation Studies*, 3, 60–80.
Sovacool, B. K., Ali, S. H., Bazilian, M., Radley, B., Nemery, B., Okatz, J., & Mulvaney, D. (2020a). Sustainable minerals and metals for a low-carbon future. *Science*, 367(6473), 30–33.
Sovacool, B. K., Hook, A., Martiskainen, M., & Baker, L. (2019). The whole systems energy injustice of four European low-carbon transitions. *Global Environmental Change*, 58, 101958.
Sovacool, B. K., Ryan, S. E., Stern, P. C., Janda, K., & Rochlin, G. (2020b). Integrating behavior, energy, and climate research. *Nature Climate Change*, 10, 109–115. (not found in google scholar and not in form of APA format)
Sovacool, B. K., Newell, P., Carley, S., & Fanzo, J. (2022). Equity, technological innovation and sustainable behaviour in a low-carbon future. *Nature Human Behaviour*, 6(3), 326–337.
Surya, B., Muhibuddin, A., Suriani, S., Rasyidi, E. S., Baharuddin, B., Fitriyah, A. T., & Abubakar, H. (2021). Economic evaluation, use of renewable energy, and sustainable urban development mamminasata metropolitan, Indonesia. *Sustainability*, 13(3), 1165.
Tan, J. D., Chang, C. C. W., Bhuiyan, M. A. S., Nisa'Minhad, K., & Ali, K. (2022). Advancements of wind energy conversion systems for low-wind urban environments: A review. *Energy Reports*, 8, 3406–3414.
The White House. (2023). Building a Clean Energy Economy: A Guidebook to the Inflation Reduction Act's Investments in Clean Energy and Climate Action. https://www.whitehouse.gov/cleanenergy/inflation-reduction-act-guidebook/
Trout, K., Muttitt, G., Lafleur, D., Van de Graaf, T., Mendelevitch, R., Mei, L., & Meinshausen, M. (2022). Existing fossil fuel extraction would warm the world beyond 1.5C. *Environmental Research Letters*, 17(6), 064010.
UN. (2023). UN Climate Change 2023 Highlights. United Nations. Accessed from https://unfccc.int/about-us/2023-highlights.
U.S. Securities and Exchange Commission (SEC). (2023). Proposed Rule: The Enhancement and Standardization of Climate-Related Disclosures for Investors. https://www.sec.gov/news/press-release/2022-46
UNFCCC. (2023). Glasgow Financial Alliance for Net Zero (GFANZ): Progress Report 2023. United Nations Framework Convention on Climate Change. https://unfccc.int/news/gfanz-2023-progress-report

UN Framework on Critical Minerals. (2024). UN guidance for action on critical energy transition minerals: A document of the UN secretary-general's working group on transforming the extractive industries for sustainable development. Accessed from https://wedocs.unep.org/handle/20.500.11822/48298#:~:text=In%20December%202024,%20the%20United%20Nations%20Secretary-General%E2%80%99s%20Panel,proposing%20seven%20Guiding%20Principles%20and%20five%20Actionable%20Recommendations.

Uyar, T. S., & Beşikci, D. (2017). Integration of hydrogen energy systems into renewable energy systems for better design of 100% renewable energy communities. *International Journal of Hydrogen Energy*, 42(4), 2453–2456.

Vakulchuk, R., Overland, I., & Scholten, D. (2020). Renewable energy and geopolitics: A review. *Renewable and Sustainable Energy Reviews*, 122, 109547.

Villegas, A., & Scheyder, E. (2023). Chile plans to nationalize its vast lithium industry. *Reuters*. https://www.reuters.com/markets/commodities/chiles-boric-announces -plan-nationalize-lithium-industry-2023–04–21/.

Wainstein, M. E., & Bumpus, A. G. (2016). Business models as drivers of the low carbon power system transition: A multi-level perspective. *Journal of Cleaner Production*, 126, 572–585.

Wang, D., Muratori, M., Eichman, J., Wei, M., Saxena, S., & Zhang, C. (2018). Quantifying the flexibility of hydrogen production systems to support large-scale renewable energy integration. *Journal of Power Sources*, 399, 383–391.

Wang, S., Wang, J., & Wang, W. (2023). Do geopolitical risks facilitate the global energy transition? Evidence from 39 countries in the world. *Resources Policy*, 85, 103952.

World Bank. (2020). World bank releases guide for sovereign issuers to engage with investors on Environmental, Social and Governance (ESG) issues. Accessed from https://www.worldbank.org/en/news/press-release/2020/11/08/world-bank-releases-guide-for-sovereign-issuers-to-engage-with-investors-on-environmental-social-and-governance-esg-issues.

Xlinks. (2023). Xlinks Morocco–UK Power Project. Retrieved from: https://www.xlinks.co

Yarmuth, J. (2022). Explanatory Statement for Division D – Energy and Water Development and Related Agencies Appropriations Act, 2022. U.S. House of Representatives, Committee on Appropriations. https://appropriations.house.gov/

Yu, H., Wen, B., Zahidi, I., Chow, M. F., Liang, D., & Madsen, D. Ø. (2024). The critical role of energy transition in addressing climate change at COP28. *Results in Engineering*, 22, 102324. https://doi.org/10.1016/j.rineng.2024.102324.

Zhang, S., & Chen, W. (2022). Assessing the energy transition in China towards carbon neutrality with a probabilistic framework. *Nature Communications*, 13(1), 1–15.

16 Impact of Climate Change on Glacier Ice Volume

A Comparative Study between Two Major Glaciers of the Himalayan Region

Mohit Joshi, Ajay Dashora, and Rukhsar Anjum

INTRODUCTION

The ice at mountain peaks that moves under gravity is termed glacier ice. Glacier ice tends to slide on the ground under the influence of gravity. Growth of ice in glaciers is termed as snow accumulation, whereas loss of ice due to melting and breaking off is termed as glacial retreat. In polar glaciers, glacial retreat is caused due to melting as well as breaking, whereas in Himalayan glaciers main reason for glacial retreat is melting of ice. The movement of snow from locations receiving heavy snow accumulation to locations undergoing a large amount of mass loss through either melting or breaking of ice blocks is represented by the flowing ice visible at the surface of glaciers. By estimating the rate of flow of this flowing ice, we can predict the effect of global warming on glaciers. Hence, we can say that in order to have an improved knowledge of glaciers and hazards related to them, the measurement of glacier flow rates is of utmost importance.

Fresh water demand is growing manyfold since the last century. The main reasons behind this rapidly growing demand are the exponential growth of population, intensifying agriculture, and constant growth of industries and cities. Glaciers are the key source of getting fresh water through rivers, and hence they are vital for understanding the disruptions in the earth's environment with growing population and water demands (Abdelouhed et al., 2021). More precisely, glaciers are the best indicator to assess and predict climate change scenarios. The most important consequences of growing industrialization and urbanization on climate that were detected during the last few decades are the shrinking and thinning of glaciers (Meier et al., 2004). Any change in glaciers' area and volume affects the hydrology of downstream areas. Hence, it is very important to study glacier mass balance, study climate changes, and make future policies for sustainable development with a limited effect of urbanization and industrialization on the environment.

About 40,900 km^2 area of the Karakoram and Himalayan mountains is glaciated (Bolch et al., 2012). The significant resources of the Himalayan Rivers for satisfying water requirements and economic actions of the population living in downstream areas depend on melted water obtained from annual snowfall in higher terrains and glaciers. It means that the seasonality of water and its amount released from the glaciers is affected by the varied landscape and climate of the Himalaya. The current and future climatic scenarios will affect the availability of water. These changes will negatively impact the plants, animals, and climatic conditions of the region. Thus, it is mandatory to evaluate the future health of a glacier so that policies are developed keeping future climate change scenarios in mind, for sustainable growth.

For evaluating a glacier's melting rate, its volume change is the most significant parameter to be assessed, as a volume study of a glacier gives a clear indication about the balance between

DOI: 10.1201/9781003620747-19

accumulation and retreat rate of the glacier. With the help of glacier volume estimates, we can also predict future ice mass balance scenarios. This can prove to be of immense importance for moving toward sustainable development by making policies that can work toward fulfilling human needs with minimal damage to natural resources. Glacier volume studies are comparatively easier in Polar Regions as compared to the harsh and difficult terrain of the Himalayas. Himalaya, as we all know, is having very uneven topography with severe climatic scenarios. This limits the quantity of on-site studies on Himalayan glaciers, as going to the Himalayas for conducting field studies requires large equipment to be transported to hilly terrain and a need to provide essentials of livelihood to scientists and personnel working in harsh and tough climatic conditions of the Himalaya.

In the last few decades, the glaciers in the Himalayan and Karakoram region have been constantly retreating and getting thin due to global warming and climatic change (Bolch et al., 2012; Kulkarni & Karyakarte, 2014). Based on the type of mountainous terrain, melting glaciers may cause the development of supraglacial lakes. If the glaciers retreat rate is faster than the accumulation rate, it will lead to widening of these supraglacial lakes, which increases the chances of flash floods that can occur due to sudden breaking of boundaries and melting of ice in lakes. Since the advancement of studies related to glaciers, many such flash floods have been reported during the last century. Such a glacial lake outburst flood in India was first encountered in Jammu and Kashmir in the Shyok glacier in the year 1926. This flood destroyed the Abudan village (Mason, 1929). Two such events of glacial lake outburst floods were reported in the Himachal Pradesh in the Kinnaur valley in 1981 and 1988. The latest such reported incident in India is of a flash flood in Uttarakhand in Kedarnath in 2013, in which the glacial lake outburst flood of Chorabari lake was accelerated because of heavy rainfall in the nearby regions (Dobhal et al., 2013). To predict future flash floods due to glacial lake outbursts, proper assessment of the volume of newly formed glacier lakes and studies related to the expansion of existing supraglacial lakes are essential (Worni et al., 2013).

Various image registration algorithms have claimed glacier displacement finding over the years, but detecting geometric shifts between two images precisely is still a challenging task (Le Moigne, J. et al., 2011). So, there is a need for an image co-registration technique which is fast and applicable for all kinds of vast amounts of remotely sensed data accessible through various government and commercial satellites. The most difficult task is to design an algorithm that can handle the data obtained across the world through different satellites at diverse periods with varying sensors.

With various technological reforms, more reliable and accurate remote sensing techniques were invented in order to measure glacier velocities. Some of the most common techniques are INSAR (Interferometric Synthetic Aperture Radar), feature tracking with the help of GPS, DINSAR (differential interferometric synthetic aperture radar), offset tracking, etc. Each of the abovementioned techniques includes its own pros and cons. As they are used to extract data of different time scales, they can be used as complementary to each other to validate the results. This chapter aims to develop a Python-based methodology to measure surface displacement of glaciers, measure surface velocity of glaciers, measure ice thickness of glaciers, and estimate glacier volume using an open-source Python module AROSICS.

STUDY AREA

The Gangotri glacier is situated in India's Uttarakhand's Uttarkashi District. This glacier is one of the main sources of the river Ganga. Gangotri glacier is one of the biggest glaciers in the Himalaya, measuring 30.2 km in length and 1.5 km in mean width. It is situated between 30°43'22″N–30°55'49″N latitude and 79°4'41″E–79°16'34″E longitude (Figure 16.1). The surface elevation of Gangotri glacier ranges from 4,000 m to 7,000 m above sea level (Jain, 2008) and the surface area is 140 km^2. Bara Shigri glacier is located in Lahaul district, Himachal Pradesh, India. This glacier is one of the key sources of the river Chenab. Bara Shigri glacier is 28 km long and has a surface area of 130 km^2. It is situated between 32°05'00″N-32°20'00″N latitude and 77°28'31″E–77°77'44″E longitude.

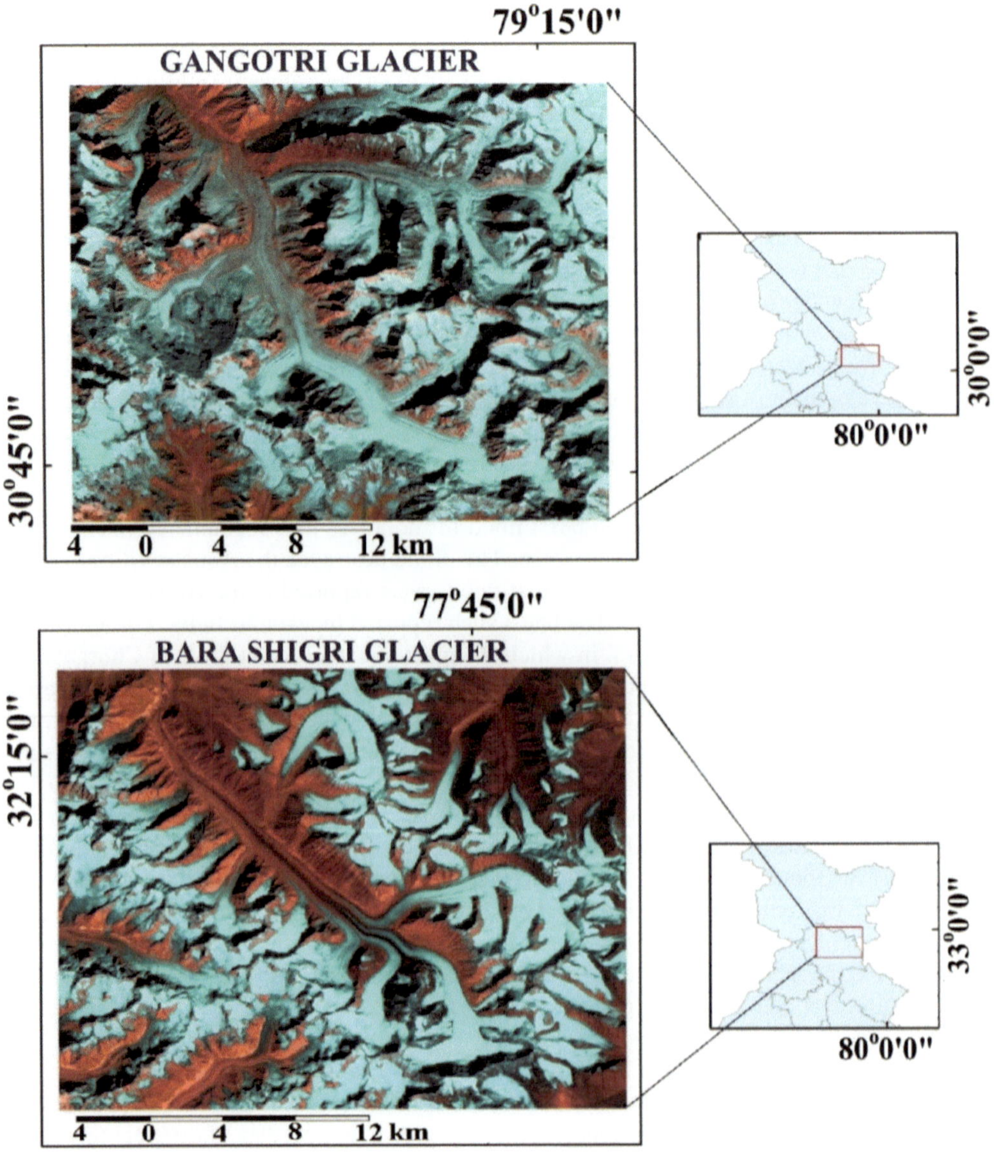

FIGURE 16.1 Location map: Gangotri glacier, and Bara Shigri glacier in the Indian Himalayan region.

Since a significant portion of the work on the "line of perpetual snows" was written in the historical time span of 1840–1850 by two distinguished writers, Jack in 1844 and Humboldt in 1845, the Himalayas, also known as the "Abode of Snow," have drawn the attention of many observers. At the time, Pindari glacier, which is located in Uttarakhand's lower Shiwalik hill, was the most visited glacier. Geologists and cartographers observed glaciers in the 19th century in order to create maps using the triangulation method. Godwin Austin, a well-known renounced geologist who mapped and investigated K2, sometimes called Mount Godwin-Austen, published extensively on the upper Indus River basin in 1862. After visiting the Kabru and Pandim glaciers in Sikkim State in 1891, P.N. Bose became among the first Indian geologists to write on the Indian Himalayan glaciers. The Indian Geological Survey also took the initiative to start an international program in 1907 to monitor and investigate glacier changes. The program lasted for three years, ending in 1910.

People made significant contributions to this discipline during the period of independence as well. Kanwar Sain was the first scientist to study how ice, snow, and glaciers affect the Himalaya's surface water supplies (1946). Numerous independent studies were conducted between 1910 and 1957 by various scientists.

Another attempt to watch glaciers was initiated in 1957–1958, which was regarded as the International Geophysical Year. During this program, a great deal of data was gathered about the history of glaciers and their changes. Later, during 1964–1974, a decadal program was launched as an international effort to research numerous glaciers. The first scientist to estimate the total amount of ice cover in the Himalayas was Wissman (1959), who noted that roughly 17% of the region was covered in ice, spanning 33,200 km². The entire ice cover was later assessed and written down to have increased to 35,110 km² by Qin and Qi (1999). The 1800s saw a great deal of small-scale labor. However, a lot of effort was made in the previous three decades due to technological advancements, the need for sources and resources, and an increase in understanding.

DATABASE

In this chapter, we have used an open-source Python module, AROSICS, to perform image co-registration and detect ground displacement. Daniel Scheffler (German Research Centre of Geosciences) developed AROSICS as part of the GeoMultiSens project, which is financed by the German Federal Ministry of Education and Research (https://git.gfz-potsdam.de/danschef/arosics). For evaluating a glacier's melting rate, its volume change is the most significant parameter to be assessed, as the volume study of a glacier gives a clear indication about the balance between accumulation and retreat rate of the glacier. With the help of glacier volume estimates, we can also predict future ice mass balance scenarios. Hence, in order to accomplish the objectives, a Python code is developed to carry out the following measurements: (i) Measure surface displacement of glaciers, (ii) measure surface velocity of glaciers, (iii) measure ice thickness of glaciers, and (iv) glacier volume estimate. In order to calculate the measurements listed above, the study was decided to be carried out in the sequence of activities. The steps followed are summarized in the form of a flow chart in Figure 16.2.

To analyze glacier changes, a time series analysis must be generated by incorporating satellite images from various sensors over the area during months with no rainfall or cloud cover, because

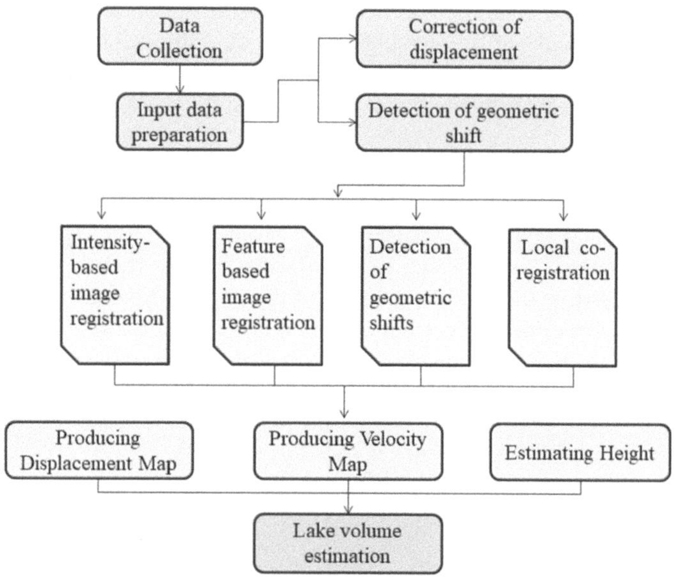

FIGURE 16.2 Flow chart of methodologies employed.

glacier changes are not uniform in place and time. September to November are the most suitable for carrying out these studies.

The datasets obtained from different sensors need to be processed before using them for further calculations (Table 16.1). This study utilized freely available Landsat 8 images (https://earthexplorer.usgs.gov/) of Gangotri glacier and Bara Shigri glacier from 2014 to 2019 for calculating time-dependent factors like glacier displacement vectors and glacier surface velocity. Freely available ASTER Global Digital Elevation Model for creating contour maps was also used in order to calculate ice thickness and make glacier ice volume estimates.

METHODOLOGY

In order to execute the Python code, the following steps were performed.

Intensity-Based Image Registration Technique

According to Goshtasby (2005), in the case of intensity-based methods, correlation matrices were used to compare the intensity patterns of photographs. Intensity-based approaches register whole pictures or sub-images. After registering images, the centers of the associated images are considered important feature points. In the intensity-based image recognition technique, similar gray value patterns of the reference image and the target image are compared, and based on this similarity, the displacements are quantified.

Feature-Based Image Registration Technique

According to Goshtasby (2005), in the case of feature-based methods, the focus is on the relation between constant features of a landform concerning time and space, such as points, lines, and contours. Feature-based methods establish a relationship between all spatially located distinct locations in the reference image and the target image. After knowing the relation between several spatially located distinct points in the reference image and the target image, it can superimpose the target image on the reference image, thus finding the geometrical transformation. This geometrical transformation is then used in establishing a point-by-point relation between the reference and the target image. According to Chen H. et al. (2003), most of the feature-based techniques are computationally cheap, and they can calculate displacement vectors in the sub-pixel range if a sufficient number of distinct features are available on the ground. Intensity-based image recognition techniques examine the sub-pixel displacement vectors without depending on the presence of salient image features.

Detection of Geometric Shifts

Geometric shifts among the reference image and the target image were calculated by converting both images into the frequency field using FFTW (fastest Fourier transform in the west). FFTW is an openly accessible and is the fastest execution of the discrete Fourier transform (DFT) (Frigo and Johnson, 2005). A cross-power spectrum of both images was generated using FFTW, which is checked for geometric shifts. Using inverse FFTW, the cross-power spectrum was changed back into the spatial domain. Detection of geometric shifts using FFTW was done using a local co-registration approach.

Local Co-Registration Approach

In the local co-registration approach, correlation was done by moving a window of fixed size over the whole image. Hence, a number of tie points or ground control points (GCPs) were obtained in local co-registration approach on the basis of which target image was shifted (Jawak et al., 2019).

TABLE 16.1
Description of Landsat 8 and ASTER Datasets Used for the Study

Gangotri Glacier				Bara Shigri Glacier				ASTER GDEM	
Data	Date of Acquisition	Band	Resolution (m)	Data	Date of Acquisition	Band	Resolution (m)	Data	Glacier
LC08_L1TP_146039_20141023_20170418_01_T1_B4	23-Oct-14	4	30	LC08_L1TP_147038_20140928_20170419_01_T1_B4	28-Sep-14	4	30	ASTGTMV003_N30E079_dem	Gangotri
LC08_L1TP_145039_20141117_20170417_01_T1_B4	17-Nov-14	4	30	LC08_L1TP_147038_20141030_20170418_01_T1_B4	30-Oct-14	4	30	ASTGTMV003_N32E077_dem	Bara Shigri
LC08_L1TP_146039_20150924_20170403_01_T1_B4	24-Sep-15	4	30	LC08_L1TP_147038_20150915_20170404_01_T1_B4	15-Sep-15	4	30		
LC08_L1TP_145039_20151003_20170403_01_T1_B4	03-Oct-15	4	30	LC08_L1TP_147038_20151001_20170403_01_T1_B4	01-Oct-15	4	30		
LC08_L1TP_145039_20160919_20170321_01_T1_B4	19-Sep-16	4	30	LC08_L1TP_147038_20161003_20170319_01_T1_B4	03-Oct-16	4	30		
LC08_L1TP_146039_20160926_20170320_01_T1_B4	26-Sep-16	4	30	LC08_L1TP_147038_20161019_20170319_01_T1_B4	19-Oct-16	4	30		
LC08_L1TP_146039_20170929_20171013_01_T1_B4	29-Sep-17	4	30	LC08_L1TP_147038_20170904_20170916_01_T1_B4	04-Sep-17	4	30		
LC08_L1TP_145039_20171008_20171023_01_T1_B4	08-Oct-17	4	30	LC08_L1TP_147038_20170920_20170930_01_T1_B4	20-Sep-17	4	30		
LC08_L1TP_146039_20181002_20181010_01_T1_B4	02-Oct-18	4	30	LC08_L1TP_147038_20181025_20181031_01_T1_B4	25-Oct-18	4	30		
LC08_L1TP_145039_20181027_20181115_01_T1_B4	27-Oct-18	4	30	LC08_L1TP_147038_20181110_20181127_01_T1_B4	10-Nov-18	4	30		
LC08_L1TP_145039_20191014_20191018_01_T1_B4	14-Oct-19	4	30	LC08_L1TP_147038_20191012_20191018_01_T1_B4	12-Oct-19	4	30		
LC08_L1TP_146039_20191021_20191030_01_T1_B4	21-Oct-19	4	30	LC08_L1TP_147038_20191028_20191030_01_T1_B4	28-Oct-19	4	30		

$$C_{(u,v)} = \frac{\sum_{x,y}\left(f(x,y)-\bar{f}\right)\times\left(g(x+u,y+v)-\bar{g}(x+u,y+v)\right)}{\sqrt{\sum_{x,y}\left(\left(f(x,y)-\bar{f}\right)\right)^2}\times\sqrt{\sum_{x,y}\left(\left(g(x+u,y+v)-\bar{g}(x+u,y+v)\right)\right)^2}} \quad (16.1)$$

Here,

$f(x, y)$ and $g(x, y)$ are pixel values in windows W and W' of image 1 and image 2;
u and v=offsets between W and W';
$\bar{f}$ and $\bar{g}$ (u, v)=average pixel values of W and W'.

Correction of Displacements

The correction of displacements was done to minimize the effect of geometric misregistration. The correction of displacements in the target image was carried out by warping the target image by using many tie points obtained in a local co-registration approach.

Producing Displacement Map

After following all the above steps, the target image was superimposed over the reference image based on the GCPs obtained through the local co-registration approach, and the displacement intensity was obtained at various coordinates. These displacements were plotted as an intensity map.

Producing Surface Velocity Map

The methodology of producing glacier surface velocity is very basic, and it was evaluated from the fact that velocity at a given point is the amount of displacement per unit time. The resulting velocity was calculated using Eq. (16.2):

$$V = \frac{\text{Displacement intensity}}{\left(\text{Difference in acquisition dates between image pairs}\right)} \quad (16.2)$$

Estimating Ice Thickness

With the help of the surface velocity obtained, the height of the glacier surface from bedrock at various points was estimated using Glen's flow equation for glaciers. The Glen's flow equation gives a relation between shear strain rate and stress, which can be further equated for finding the ice thickness of a glacier at a particular point if the surface glacier velocity, ice surface slope, friction coefficient is known to us. Thus, ice thickness (H) at a particular point where surface velocity is known is given by Glen (1955).

$$H = \left(\frac{1.5Us}{A\left(f\rho g\sin\alpha\right)^3}\right)^{\frac{1}{4}} \quad (16.3)$$

where

Us=surface velocity
A (creep parameter)=3.24 * 10^{-24} $P/a^3/s^1$ (Cuffey and Paterson, 2010)
H=ice thickness
f is friction coefficient=0.8 (Haeberli and Hoelzle, 1995)
ρ (Density of ice)=900 kg/m^3 (Farinotti et al., 2009)
g=9.8 m/s^2
α = Ice surface slope at a contour interval of 100 m calculated from ASTER DEM

Glacier Volume Estimates

On estimating ice thickness, a digital elevation model (DEM) was generated for the whole glacier area. With the help of this DEM, the ice volume in the glacier was estimated. On comparing the volume change of glaciers over time, we can estimate the effect of climate change on glaciers' health. For evaluating a glacier's melting rate, its volume change is the most significant parameter, which can assessed as a volume study of a glacier that gives a clear indication about the balance between accumulation and retreat rate of the glacier.

RESULTS

In this probe, an effort was made to develop a Python code to estimate the glacier ice volume. An understanding of glaciers' volume change is critical for understanding the long-range repercussions of climate change due to temperature rise and its effect on glacier flow velocity and mass balance. The results show that the glacier volume changes in inaccessible areas can be monitored by developing a Python code, which can analyze satellite images using an intensity-based phase correlation technique.

The newly developed Python code enables the extraction of glacier shapes from pictures and the creation of vector files (Table 16.2). The criteria for defining glacier melt in filtered images are indeed complicated, involving the length and gradient level onsets for an individual or a group of pixels. The segments are extracted and transformed to a "shp." format, which is then exported to ArcGIS, a geospatial analysis and processing program. On the other hand, in terms of logic and competence, most of the images extracted from the satellite panchromatic band seem to be very significant and better correspond to geographical processes like ice melting or surface velocity. Upon analysis, the ice volume was 98.34 km^3 in 2018, which reduced to 70.77 km^3 in Gangotri glacier, whereas in Bara Shigri glacier, it was 54.73 km^3 in 2018, which reduced to 49.38 km^3 in 2019, respectively. Surface velocity at various locations of Gangotri glacier varies between 43.56 m/year to 314.26 m/year (Figure 16.3). The results obtained from the Python code show that for Gangotri glacier, ice thickness is more toward the lower elevations and less on higher mountains (Figure 16.4). In Gangotri glacier, results of the years 2014 and 2018, and in Bara Shigri glacier, results of the year 2014 are distorted and vary largely from other years. This is mainly due to the presence of cloud cover over Gangotri glacier in the year 2014 during September to November, and heavy precipitation in the year 2018 during September to November.

For Gangotri glacier and Bara Shigri glacier, surface displacements are more toward the lower elevations and less in the mountainous terrain (Figures 16.5 and 16.6). The average surface velocity of Gangotri glacier is 212.467 m/year (Figure 16.7). The average ice volume in Gangotri glacier is 79.983 km^3 (Figure 16.8). Ice thickness at various locations of Gangotri glacier varies between 136.43 and 414.24 m. Keeping data obtained from the years 2014 and 2018 aside, it is visible that the volume of ice in the glacier decreased from 87.8 km^3 in 2015 to 70.77 km^3 in 2019. For Bara Shigri glacier, ice thickness is more toward the lower elevations and less on higher mountains (Figure 16.9). The average surface velocity of Bara Shigri glacier is 154.26 m/year (Figures 16.10 and 16.11). Ice thickness at various locations of Bara Shigri glacier varies between 200.46 and 371.155 m (Figure 16.12). Total ice volume in Bara Shigri glacier is 49.38 km^3. Keeping data obtained from the year 2014 aside, it is visible that the volume of ice in the Bara Shigri glacier decreased from 50.77 km^3 in 2015 to 44.631 km^3 in 2019, and in the Gangotri glacier, the ice volume reduced from 87.8 km^3 to 79.98 km^3. In order to know future volume projections of Gangotri glacier, datasets containing a wide range of years are needed so that distorted datasets with false values due to cloud cover and heavy precipitation can be neglected.

TABLE 16.2
Glacier Properties Yearly Variation (2014–2019)

Gangotri Glacier								Bara Shigri Glacier					
	Glacier Volume		Surface Velocity		Ice Thickness			Glacier Volume		Surface Velocity		Ice Thickness	
Year	V_{avg} (m/year)	Volume (km³)	V_{min} (m/year)	V_{max} (m/year)	H_{min} (m)	H_{max} (m)	Year	$V_{avg.}$ (m/year)	Volume (km³)	v_{min} (m/year)	v_{max} (m/year)	H_{min} (m)	H_{max} (m)
2014	131.27	61.46	22.43	313.3	140.64	370.79	2014	95.36	40.7	30.73	310.46	176.63	320.22
2015	252.76	87.8	24.53	315.13	142.26	443.69	2015	196.31	50.77	62.35	312.23	204.72	380.6
2016	256.16	86.33	99.7	315.34	139.89	436.29	2016	151.12	50.27	17.67	311.82	201.58	374.5
2017	250.32	75.2	49.58	314.94	121.85	380.016	2017	183.59	55.22	30.98	314.05	221.42	411.36
2018	133.68	98.34	9.32	313.1	159.35	496.99	2018	192.707	54.73	65.12	310.03	219.48	407.76
2019	250.49	70.77	55.8	313.8	114.6	357.66	2019	106.52	44.631	23.85	305.5	178.97	332.49
Avg.	212.467	79.983	43.56	314.268	136.43	414.24	Avg.	154.26	49.38	38.45	310.68	200.46	371.155

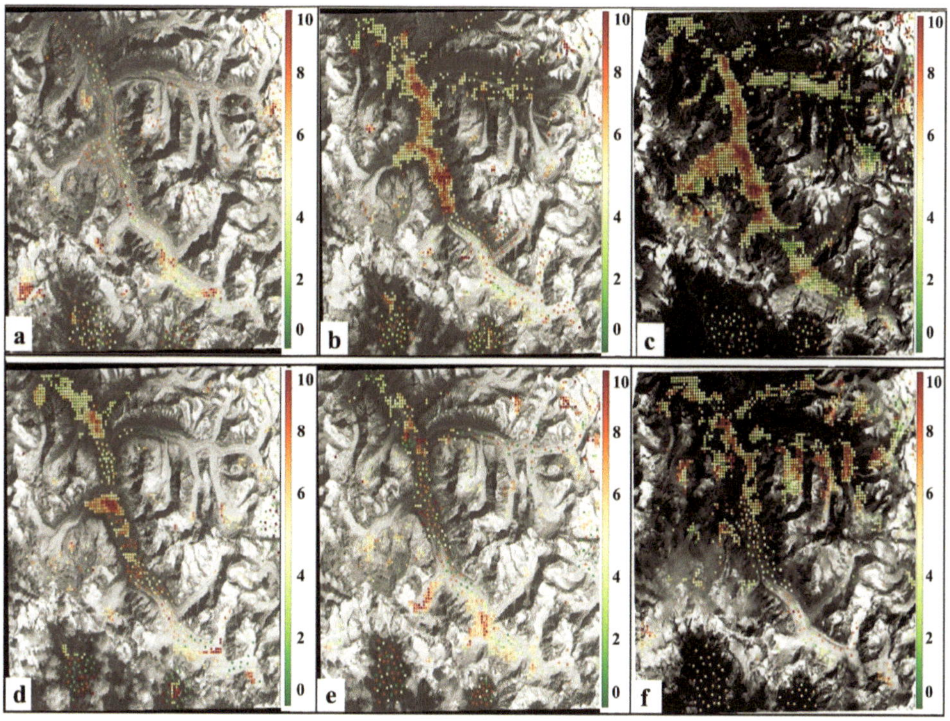

FIGURE 16.3 Gangotri glacier surface velocity intensity map ((a) 2014, (b) 2015, (c) 2016, (d) 2017, (e) 2018, (f) 2019).

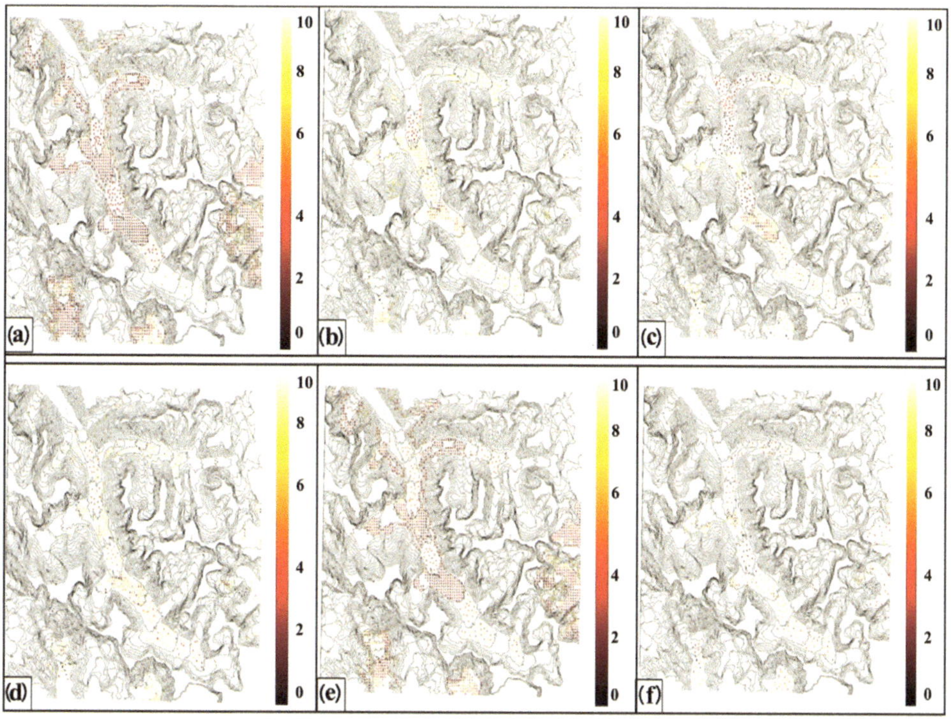

FIGURE 16.4 Gangotri glacier ice thickness intensity map ((a) 2014, (b) 2015, (c) 2016, (d) 2017, (e) 2018, (f) 2019).

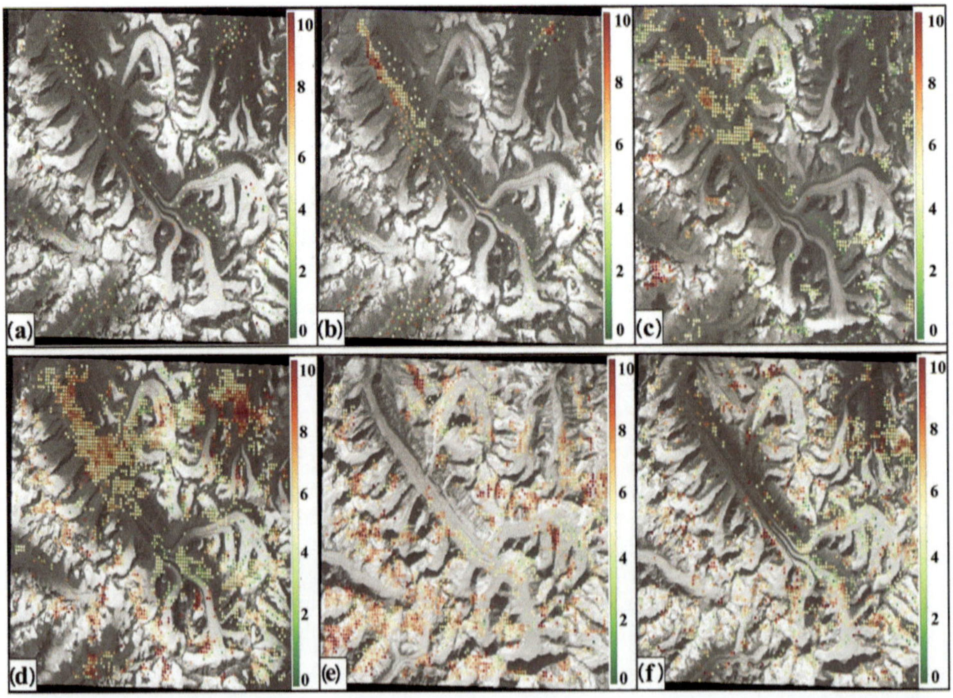

FIGURE 16.6 Bara Shigri glacier displacement intensity map ((a) 2014, (b) 2015, (c) 2016, (d) 2017, (e) 2018, (f) 2019).

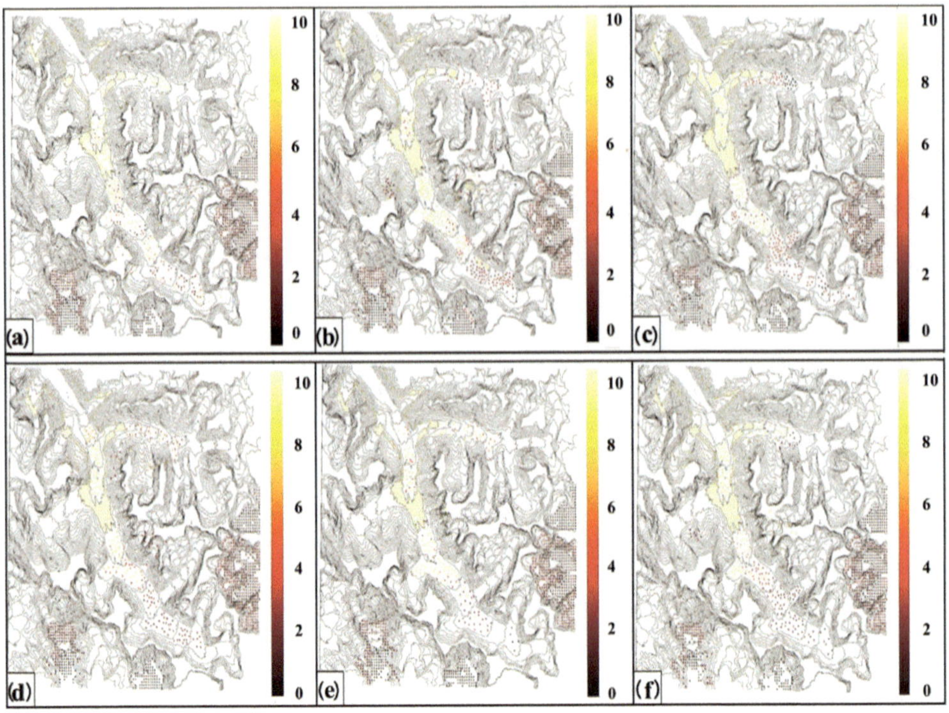

FIGURE 16.5 Gangotri glacier displacement intensity map ((a) 2014, (b) 2015, (c) 2016, (d) 2017, (e) 2018, (f) 2019).

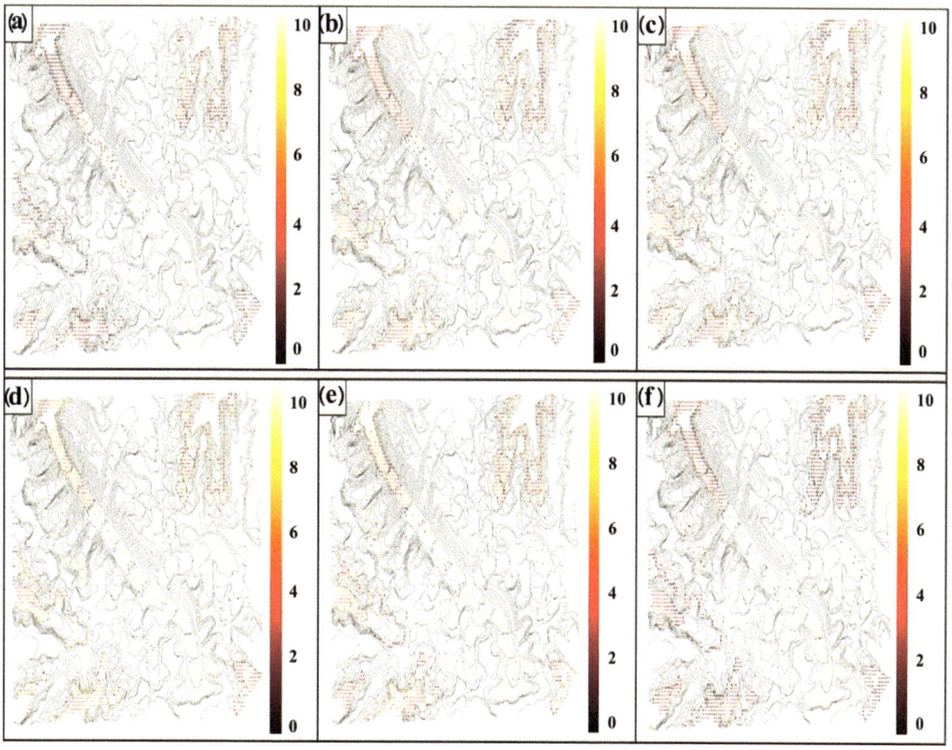

FIGURE 16.7 Gangotri glacier surface velocity variations (2014–2019).

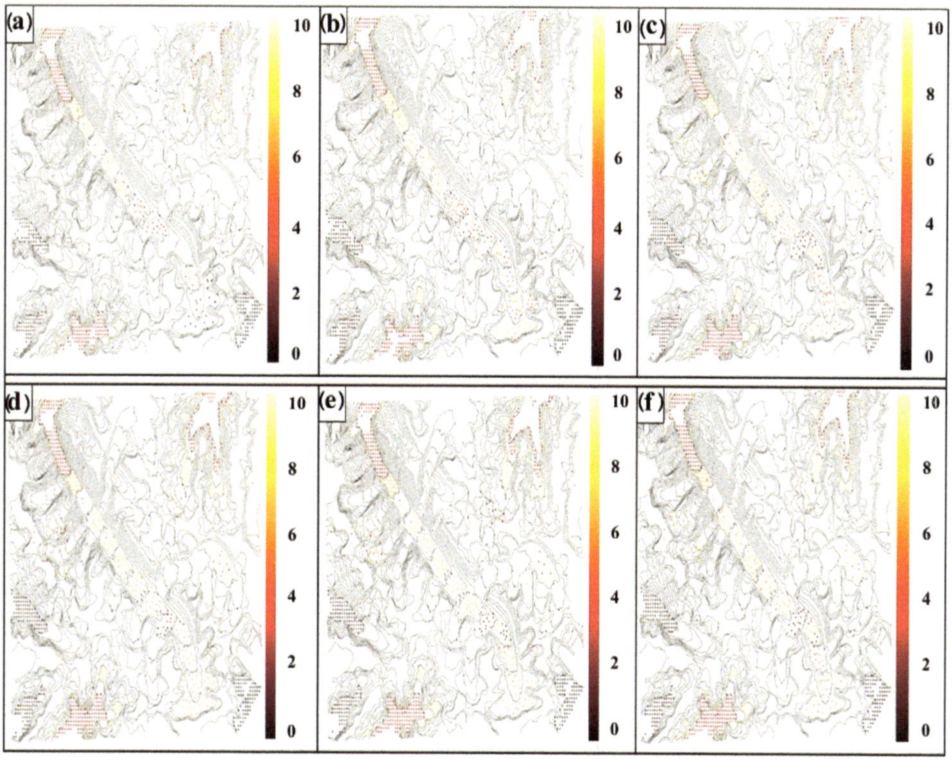

FIGURE 16.8 Gangotri glacier ice thickness variations (2014–2019).

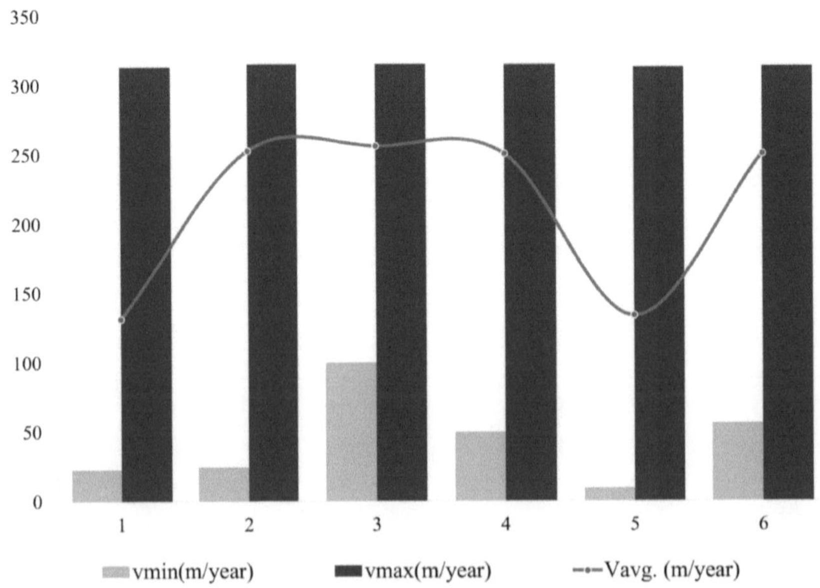

FIGURE 16.9 Bara Shigri glacier ice thickness intensity ((a) 2014, (b) 2015, (c) 2016, (d) 2017, (e) 2018, (f) 2019).

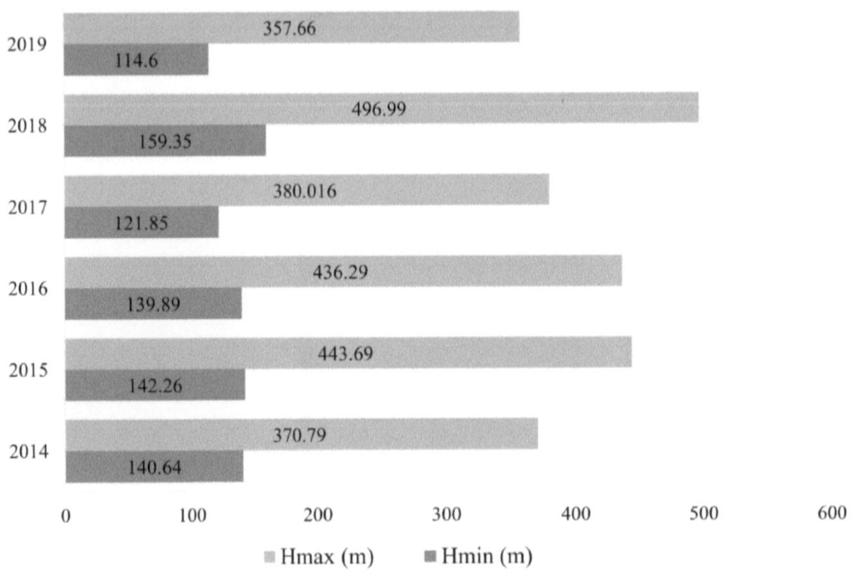

FIGURE 16.10 Bara Shigri glacier surface velocity variations (2014–2019).

DISCUSSION

Our comparative analysis of Gangotri and Bara Shigri glaciers (2014–2019) reveals accelerated mass loss rates that exceed previous decadal averages, consistent with broader Himalayan amplification of climate change impacts (Wester et al., 2019). The combined thermal and pollution pressures that occurred between 2014 and 2019, which mark a shift to faster, differentiated glacier retreat in the Himalayas, considerably increase Gangotri's susceptibility. This study found rapid ice

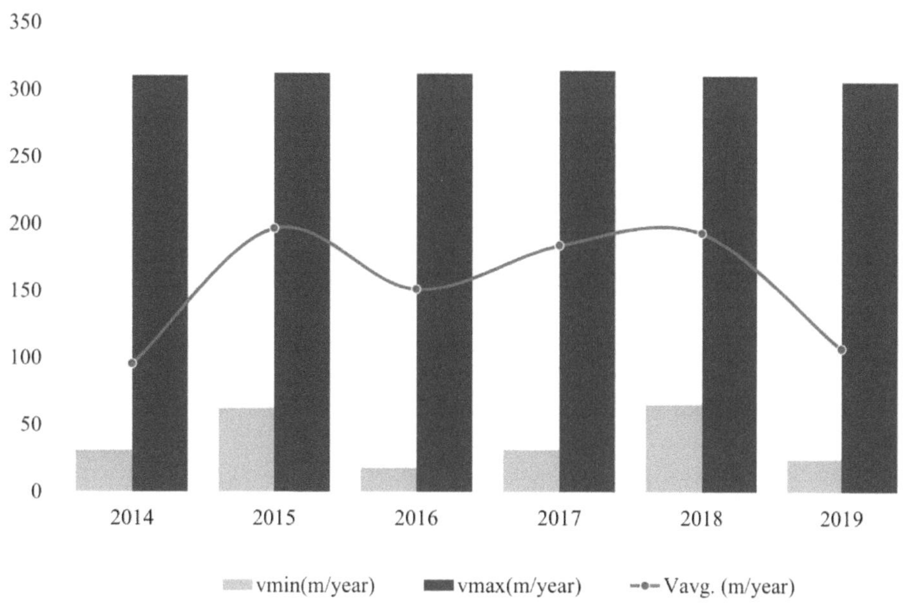

FIGURE 16.11 Bara Shigri glacier surface velocity intensity ((a) 2014, (b) 2015, (c) 2016, (d) 2017, (e) 2018, (f) 2019).

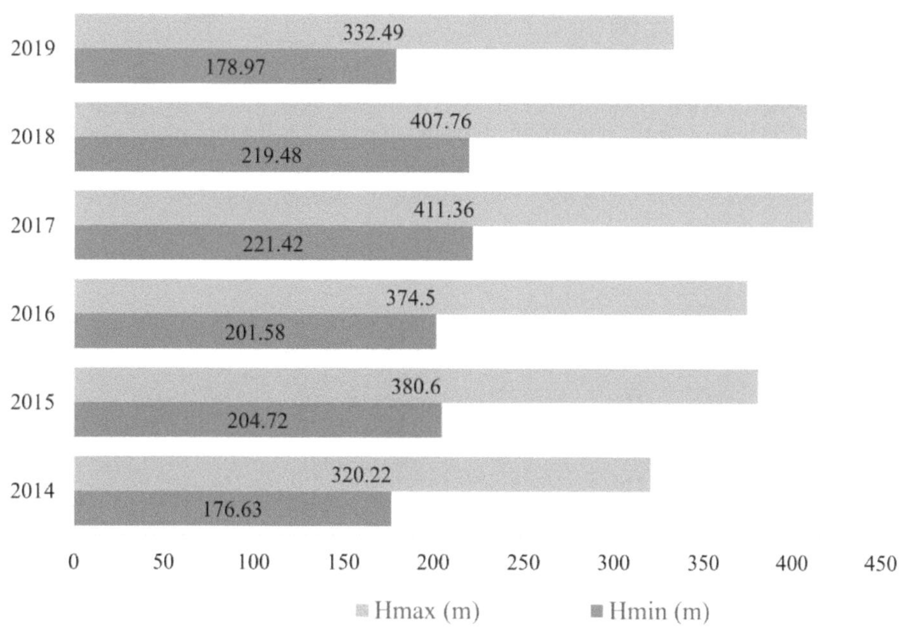

FIGURE 16.12 Bara Shigri glacier ice thickness variations (2014–2019).

volume loss in Gangotri glacier and Bara Shigri glacier occurred during the 2014–2019 period of record-breaking warming in the Himalayas (Sharma et al., 2019). Similar outcomes were found in recent studies. Thanveer et al. (2025) observed negative mass balance for Gangotri glacier from the year 2000 to 2023, suggesting that the aftermath of climate change is accelerating. The differences among these glaciers highlight the localized cryospheric and climatic conditions that influence retreat patterns. The degree of frontal retreat was progressively accelerated from 2017 to 2022, paralleling the years from 2011 to 2017, respectively (Thanveer et al., 2025). Thakur et al. (2023) showed that the creation and amalgamation of supraglacial lakes and proglacial lakes led to the loss of ice volume. Das et al. (2022) compared four-decade data and found a sharp increase in mass loss in Bara Shigri glacier, resulting in the upward movement of equilibrium line.

The study showed a linkage between variability of albedo, fluctuation of temperature, and glacier ice volume loss. The study period coincided with 2015–2019, the warmest five-year span in Himalayan data (Lalande et al., 2021). Because of its lower elevation (3,900 m), the Gangotri glacier was especially vulnerable, with temperatures during the summer at the equilibrium line altitude (ELA) above 0°C for 25% more days than between 2000 and 2013 (Mandal et al., 2022).

Santra et al. (2019), Zhang et al. (2020), and Williamson and Menounos (2021) concluded that climate change intensifies the rate of glacier melt by the reduction of Albedo caused by high temperature fluctuation, light-absorbing pollutants, and black carbon (BC). Due to wind patterns, Gangotri was disproportionately affected by the 30% increase in black carbon accumulation from the Indo-Gangetic plains between 2014 and 2019 (Thind et al., 2019). Both glaciers experienced a significant acceleration in ice loss between 2014 and 2019, with warming and pollution driving Gangotri's retreat and snowfall buffering ensuring Bara Shigri's stability. Hudda et al. (2020) and Gurjar et al. (2016) revealed that there is a recent surge of aerosol concentration, which might be associated with a rise in anthropogenic activities like street traffic, biomass burning, and industrialization.

CONCLUSION

Himalayan glaciers are constantly undergoing volume loss due to an increase in temperature. For instance, in our study, we found that Gangotri glacier's volume decreased from 87.8 km^3 in 2015 to 70.77 km^3 in 2019. Similarly, Bara Shigri glacier's volume decreased from 50.77 km^3 in 2015 to 44.631 km^3 in 2019 (Figure 16.13). This is an alarming situation, as the melting of glaciers can bring flash floods downstream and can ultimately lead to water scarcity in the future. The average volume of ice in Gangotri glacier basin came out to be 79.983 km^3. According to Gantayat et al. (2014), the

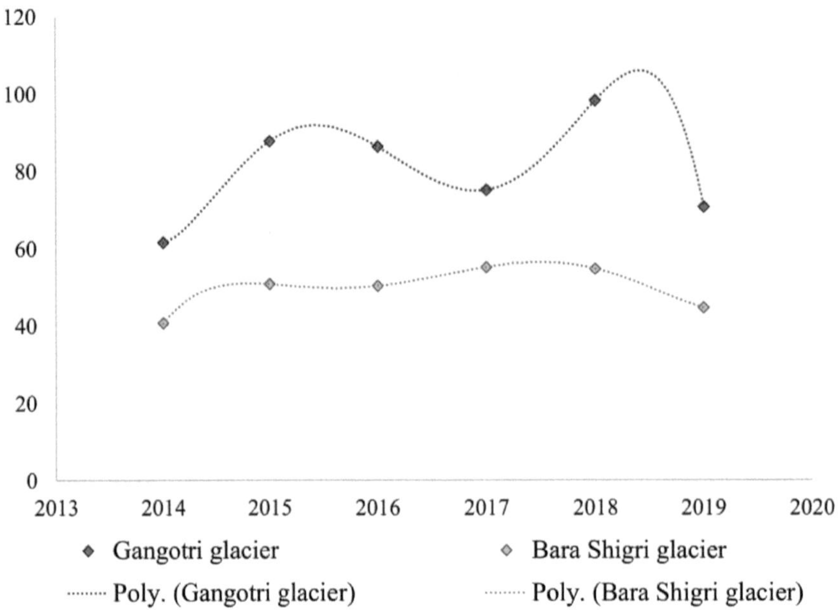

FIGURE 16.13 Glacier volume variations from 2014 to 2019 (Volume in km^3).

average ice volume of Gangotri glacier is 27.4 km^3. This difference is mainly because a rectangular area of Gangotri glacier was considered for the analysis, hence, nearby rocks and mountains also got included in our final resultant output. Moreover, this study also has a limitation of not including annual precipitation data over the glacier. Glaciers have more ice volume concentrated in the regions of low elevations, that is, in the valley region of a glacier, as compared to the higher mountainous terrains. Yearly precipitation data from the India Meteorological Department needs to be analyzed so that future predictions can be made and a complete picture of glacier mass loss over time can be obtained. For instance, in the case of Gangotri glacier the data of the year 2018 is completely different from other years because of heavy precipitation over the Gangotri basin in the year 2018.

This work attempted to know about the present condition and rate of changes occurring in two of the Himalayan glaciers as an impact of global rise in temperature and other climate change phenomena. However, the estimation of yearly water yield from the glacier, keeping glacier volume loss and yearly precipitation in account needs to be done in the future. The development of a Python code to predict future mass balance, volume loss, future volume, area loss, future area, and response time by considering annual precipitation and glacier melting rate along with flood modeling of downstream area in case of a flash flood, due to sudden melting of the glacier ice is needed to be done. The present work will soon be advantageous for the investigation studies of glacier ice melting and as well as in other sectors. Moreover, glacier-specific data must be urgently incorporated into regional climate adaptation programs in light of these findings, and further investigation is needed for a better understanding of the role of contributing factors that cause glacier ice melt.

ACKNOWLEDGMENTS

The authors extend their appreciation to anonymous reviewers for their valuable suggestions.

CREDIT AUTHORSHIP CONTRIBUTION STATEMENT

MJ Base concept, writing—original draft, Methodology; ***AD*** Review, Methodology, Validation; ***RA*** Writing—final draft, Methodology.

FUNDING

This manuscript is self-financed, and no external funds were received from anywhere.

DATA AVAILABILITY

The data will be provided on request to the corresponding author.

DECLARATIONS

Ethics Approval: Not applicable.

Consent to Participate: All authors reviewed and approved the final manuscript.

Competing Interests: The authors declare no competing interests.

REFERENCES

Abdelouhed, F., Ahmed, A., Abdellah, A., Mohammed, I., & Zouhair, O. (2021). Extraction and analysis of geological lineaments by combining ASTER-GDEM and Landsat 8 image data in the central high atlas of Morocco. *Natural Hazards*, 111(2), 1907–1929.

Bolch, T., Kulkarni, A., Kääb, A., Huggel, C., Paul, F., Cogley, J. G.,... & Bajracharya, S. (2012). The state and fate of Himalayan glaciers. *Science*, 336(6079), 310–314.

Chen, H. M., Arora, M. K., & Varshney, P. K. (2003). Mutual information-based image registration for remote sensing data. *International Journal of Remote Sensing*, 24(18), 3701–3706.

Cuffey, K. M., & Paterson, W. S. B. (2010). *The Physics of Glaciers*, 4th edn. Butterworth-Heinemann, Oxford.

Das, S., Taloor, A. K., Singh, A. K., & Kothyari, G. C. (2022). Monitoring the dynamics of Bara Shigri glacier using Synthetic Aperture Radar data of Sentinel-1 satellite and effect of climate on mass balances. *Advances in Space Research*, 70(12), 3948–3958.

Dobhal, D. P., Gupta, A. K., Mehta, M., & Khandelwal, D. D. (2013). Kedarnath disaster: Facts and plausible causes. *Current Science*, 105, 171–174.

Farinotti, D., Huss, M., Bauder, A., Funk, M., & Truffer, M. (2009). A method to estimate ice volume and ice-thickness distribution of alpine glaciers. *Journal of Glaciology*, 55(191), 422–430.

Frigo, M., & Johnson, S. G. (2005). The design and implementation of FFTW3. *Proceedings of the IEEE*, 93(2), 216–231.

Gantayat, P., Kulkarni, A. V., & Srinivasan, J. (2014). Estimation of ice thickness using surface velocities and slope: Case study at Gangotri Glacier, India. *Journal of Glaciology,* 60(220), 277–282.

Glen, J. W. (1955). The creep of polycrystalline ice. *Proceedings of the Royal Society of London. Series A, Mathematical and Physical Sciences (1934–1990)*, 228 (1175), 519–538.

Goshtasby, A. A. (2005). *2-D and 3-D Image Registration: For Medical, Remote Sensing, and Industrial Applications*. John Wiley & Sons.

Gurjar, B. R., Ravindra, K., & Nagpure, A. S. (2016). Air pollution trends over Indian megacities and their local-to-global implications. *Atmospheric Environment*, 142, 475–495.

Haeberli, W., & Hoelzle, M. (1995). Application of inventory data for estimating characteristics of and regional climate-change effects on mountain glaciers: A pilot study with the European Alps. *Annals of Glaciology*, 21, 206–212.

Hudda, N., Simon, M. C., Patton, A. P., & Durant, J. L. (2020). Reductions in traffic-related black carbon and ultrafine particle number concentrations in an urban neighborhood during the COVID-19 pandemic. *Science of the Total Environment*, 742, 140931.

Jain, S. K. (2008). Impact of retreat of Gangotri glacier on the flow of Ganga river. *Current Science*, 95(8), 1012–1014.

Jawak, S. D., Joshi, M., Luis, A. J., Pandit, P. H., Kumar, S., Wankhede, S. F., & Somadas, A. T. (2019). Mapping velocity of the potsdam Glacier, East Antarctica using Landsat-8 data. *International Archives of the Photogrammetry, Remote Sensing & Spatial Information Sciences*, 42, 1753–1757.

Kulkarni, A. V., & Karyakarte, Y. (2014). Observed changes in Himalayan glaciers. *Current Science*, 106, 237–244.

Lalande, M., Ménégoz, M., Krinner, G., Naegeli, K., & Wunderle, S. (2021). Climate change in the High Mountain Asia in CMIP6. *Earth System Dynamics*, 12(4), 1061–1098.

Le Moigne, J., Netanyahu, N. S., & Eastman, R. D. (Eds.). (2011). *Image Registration for Remote Sensing*. Cambridge University Press.

Mandal, A., Angchuk, T., Azam, M. F., Ramanathan, A., Wagnon, P., Soheb, M., & Singh, C. (2022). 11-year record of wintertime snow surface energy balance and sublimation at 4863 m asl on Chhota Shigri Glacier moraine (western Himalaya, India). *The Cryosphere Discussions*, 2022, 1–41.

Mason, K. (1929). Indus floods and Shyok glaciers. *Himalayan Journal*, 1, 10–29.

Meier, H. M., Döscher, R., & Halkka, A. (2004). Simulated distributions of Baltic Sea-ice in warming climate and consequences for the winter habitat of the Baltic ringed seal. *AMBIO: A Journal of the Human Environment*, 33(4), 249–257.

Qin, X., & Qin, D. (1999). Chemical characteristics of four kinds of water in the rongbuk glacier basin, Mount Qomolangma. *Chinese Geographical Science*, 9(3), 274–278.

Santra, S., Verma, S., Fujita, K., Chakraborty, I., Boucher, O., Takemura, T.,... & Sharma, M. (2019). Simulations of black carbon (BC) aerosol impact over Hindu Kush Himalayan sites: validation, sources, and implications on glacier runoff. *Atmospheric Chemistry and Physics*, 19(4), 2441–2460.

Sharma, P., Patel, L. K., Singh, A. T., Meloth, T., & Ravindra, R. (2019). Glacier response to climate in Arctic and Himalaya during last seventeen years: A case study of Svalbard, Arctic and Chandra Basin, Himalaya. In P. S. Goel, R. Ravindra, S. Chattopadhyay (Eds.), *Climate change and the white world* (pp. 139–156). Cham: Springer International Publishing. https://link.springer.com/chapter/10.1007/978-3-030-21679-5_10.

Thakur, P. K., Pandey, P., & Chauhan, P. (2023). Assessment of dynamics of frontal part of Gangotri glacier, India, from 2017 to 2021 using remote sensing data. *Journal of the Indian Society of Remote Sensing*, 51(4), 691–698.

Thanveer, J., Ramiz, M., Siddiqui, M. A., Pulpadan, Y. A., & Remya, S. N. (2025). Dynamics of Gangotri Glacier, India: Unravelling the influence of climatic and anthropogenic factors. *Environmental Science and Pollution Research*, 32(3), 1498–1524.

Thind, P. S., Chandel, K. K., Sharma, S. K., Mandal, T. K., & John, S. (2019). Light-absorbing impurities in snow of the Indian Western Himalayas: Impact on snow albedo, radiative forcing, and enhanced melting. *Environmental Science and Pollution Research*, 26, 7566–7578.

Wester, P., Mishra, A., Mukherji, A., & Shrestha, A. B. (2019). *The Hindu Kush Himalaya assessment: mountains, climate change, sustainability and people* (p. 627). Springer Nature. https://link.springer.com/book/10.1007/978-3-319-92288-1.

Williamson, S. N., & Menounos, B. (2021). The influence of forest fire aerosol and air temperature on glacier albedo, western North America. *Remote Sensing of Environment*, 267, 112732.

Worni, R., Huggel, C., & Stoffel, M. (2013). Glacial lakes in the Indian Himalayas from an area-wide glacial lake inventory to on-site and modeling-based risk assessment of critical glacial lakes. *Science of the Total Environment*, 468, S71–S84.

Zhang, Y., Gao, T., Kang, S., Sprenger, M., Tao, S., Du, W.,... & Meng, W. (2020). Effects of black carbon and mineral dust on glacial melting on the Muz Taw glacier, Central Asia. *Science of the Total Environment*, 740, 140056.

17 Assessing the Household-Level Landslide Adaptation and Mitigation Strategies in Shimla District of Himachal Pradesh, India

Aastha Sharma, Haroon Sajjad, Rayees Ali, Daawar Bashir Ganaie, and Nirsobha Bhuyan

INTRODUCTION

Landslide is a devastating natural phenomenon that poses a threat to life and socioeconomic development (Ahmed, 2021; Sujatha and Sridhar, 2021). Intense and erratic rainstorms have escalated the global occurrence of climate change-induced landslides (Chai and Wu, 2023; Jakob, 2022). It is anticipated that the South Asian countries will progressively experience climate variability and its consequent monsoon rainfall regimes by the 2050s (IPCC, 2022). These changes will make these countries highly vulnerable to climate-related disasters. Philippines, Indonesia, India, Colombia and Mexico are the most disaster-affected countries due to their exposure to geophysical events (World Risk Report, 2024). Among all natural disasters, landslides accounted for higher annual losses and damages (Chen et al., 2017). This disaster alone has made India more vulnerable to fatal occurrences (Martha et al., 2021). Nearly 12.6% of India's total landmass is vulnerable to landslides (NRSC, 2023). The Indian Himalayas, in particular, experienced high susceptibility due to rugged topography, active geotectonic movement and concentrated monsoonal rainfall (Dikshit et al., 2020). Developmental activities and unplanned human settlement have further catapulted the vulnerability to disaster (Du et al., 2020; Martha et al., 2021; Singh et al., 2021; Gupta et al., 2022; Roy et al., 2023; Sana et al., 2024). About 30% of the global landslide-induced economic losses were registered in the Himalayas (Haigh and Rawat, 2011; Hasegawa et al., 2009; Meena et al., 2022).

Adaptation strategies to landslides stem from the severity of the disaster and its long-term detrimental effects on communities (Sudmeier-Rieux et al., 2012; Maikhuri et al., 2017). These strategies exhibit significant variation across different regions and communities. However, it is predominantly the poor and marginalized people bearing the brunt of ambient climate change-induced disasters (Sharma et al., 2023). The Adaptation Gap Report (UNEP, 2023) defines adaptation as the process of adjustment to expected or actual climate and its effects. Adaptation seeks to moderate or avoid harm or exploit beneficial opportunities. It has gained momentum as a critical strategy for managing and adjusting to the barriers and challenges posed by extreme conditions. Adaptive capacity has become a critical policy concern in landslide susceptible areas. For instance, Gupta et al. (2020) observed that indigenous communities residing in high-elevation areas (1000–2000 m) in the Indian Himalayas are vulnerable. They further emphasized the need for adopting proactive approaches to mitigate disasters. Structural measures such as building retaining walls, structural

 DOI: 10.1201/9781003620747-20

reinforcement and improved housing conditions have been widely recognized for landslide mitigation (for example, see Fotopoulou and Pitilakis, 2017; Shukla et al., 2011; Nasiri and Hajiazizi, 2021; Chai and Wu, 2023; Sharma et al., 2024). However, Sultana and Tan (2021) argued that non-structural measures (i.e., community-based programs, early warning dissemination and emergency preparedness plans) are cost-effective and instrumental in mitigating landslides. Yuniawan et al. (2022) emphasized the effective early warning system for exposure reduction and prevention from climate change-induced landslides in Indonesia. Wang et al. (2024) investigated diverse agricultural and non-agricultural adaptation measures adopted by smallholder farmers for landslides in Northwest China. Their study revealed that farmers' adaptation choices are influenced by a complex interplay of external (e.g., household assets, resources and social networks) and internal (e.g., awareness, knowledge and skills) factors. A study by Raška et al. (2023) analyzed how communities in Central Europe adopted varied post-disaster responses such as relocation, abandonment, marginalization, resilient recovery and halting development to cope with landslide impacts. They also emphasized the need for context-specific adaptation strategies.

Inadequate preparedness, limited access to resources and reduced adaptive capacity have intensified challenges for communities in the Himalayas (Bera et al., 2020a; Gupta et al., 2020; Rehman and Azhoni, 2023). These conditions put their traditional lifestyles at considerable risk, complicating efforts to address landslide-related issues. This study embryonically investigated community-based adaptation strategies, emergency preparedness and early warning systems in the Shimla district of northwest India. The impact of climate change-induced rainfall, alongside rapid development, has significantly accelerated the incidence of landslides in the area. The encroachment of settlements on unstable slopes has diminished the natural capacity of hills and mountains to stabilize land and manage water flow, leading to increased soil erosion. The increased vulnerability is evident through displacement, disruption and damage to infrastructure. The fragile and dynamic nature of the mountain ecosystem necessitates effective adaptation and mitigation strategies to reduce the impacts of landslides. Consequently, this research focused on examining household-level adaptation strategies in response to landslide threats in Shimla district. The study asserted that its findings may contribute to the formulation of community-oriented initiatives aimed at reducing vulnerability and enhancing adaptive capacity.

STUDY AREA

Shimla district is located between 30° 45′–31° 44′ N latitudes and 77° 0′– 78° 19′ E longitudes (Figure 17.1). It is divided into 10 developmental blocks (administrative division of the state). Approximately 0.81 million people inhabit the district with a density of 159 persons per km^2. The majority of the population resides in rural areas (75%). The district has a sex ratio of 915 females per 1,000 males. The literacy rate in the district is 75.3%, with male literacy (80.7%) being higher than female literacy (69.4%). Employment in the district is largely driven by agriculture, horticulture and tourism sectors (Census of India, 2011). The district exhibits distinctive climate characteristics and experiences a subtropical highland type of climate. Humidity ranges from 80% to 90% during the monsoon season (June and September) and declines by about 36% during the summer season.

The soil consists of Entisols and Inceptisols according to the soil taxonomic classification. It is marked by high nutrient content and suitability for supporting diverse vegetation and agricultural activities. Pabbar, Giri and Sutlej are the major rivers that drain the study area. The predominant tree species in the district are pine, deodar, oak and rhododendron. Geologically, the study area is characterized by the lesser Himalayan sequence rocks and Greater Himalayan crystalline formations with abundant faults and thrusts, interlocking spurs, and narrow and steep-sided valleys. Landslides and flash floods are evident in the mountainous landscape, causing significant challenges to soil stability and having profound impacts on the socioeconomic conditions of the local communities (Panchal and Shrivastava, 2021; Sharma et al., 2024).

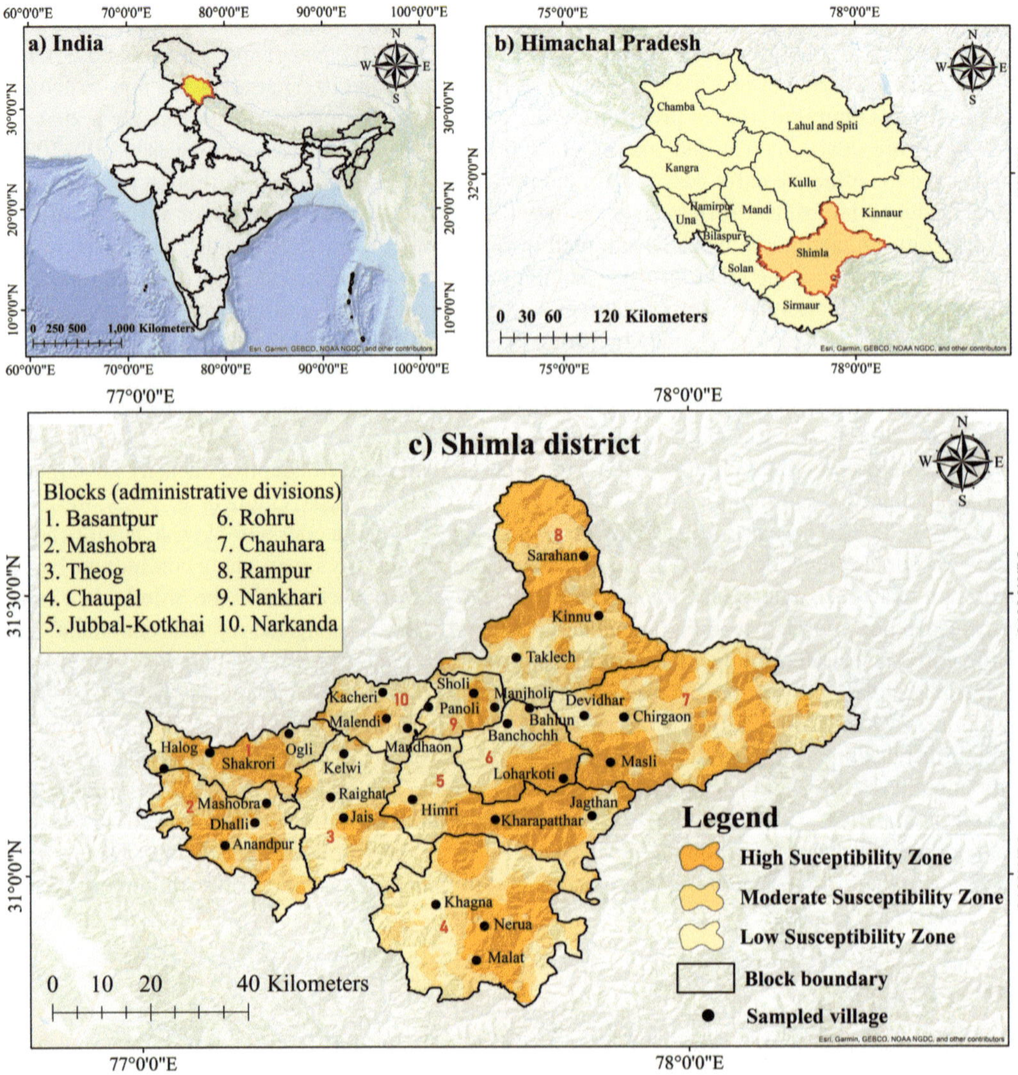

FIGURE 17.1 Location of the surveyed villages: (a) Himachal Pradesh in India. (b) Shimla district and (c) surveyed villages in different susceptible zones.

DATABASE AND METHODOLOGY

A survey was conducted at the household level that aimed to assess vulnerability to landslides in the Shimla district. Data was gathered through a structured questionnaire during the intense storm periods of July and August 2023. This period was chosen for the survey since these monsoonal months reflected the conditions contributing to landslide occurrences. A stratified random sampling method was employed to select villages and households across the district, which consisted of 10 administrative blocks. In the first stage, the villages were categorized into three strata, namely high, moderate and low, based on their susceptibility to landslides. One village was then randomly chosen from each stratum. Therefore, three villages were selected from each block. This way, a total of 30 villages were selected for undertaking the survey. In the second stage, the number of households was selected. The Census of India has categorized the working population into four categories, that is, cultivators, agricultural laborers, household industry workers and other workers. Five households were randomly selected from each category of workers. In this way, 600 households were sampled from the district. Every household included in the survey responded positively to the questionnaire, which helped achieve a response rate of 100% (Figure 17.2).

During the field visit, the socioeconomic status of households was analyzed to assess the climate change-induced landslide and its impacts (Table 17.1).

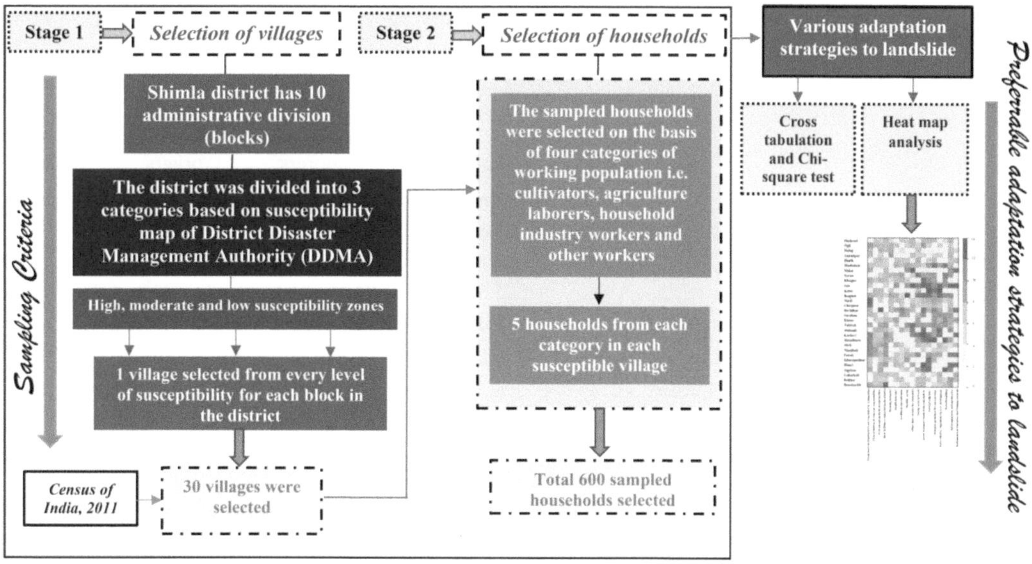

FIGURE 17.2 Methodological flowchart.

TABLE 17.1
Summary Statistics of Housing Characteristics among the Sampled Households

Housing Characteristics		Percentage	Mean	First Quantile	Third Quantile	Standard Deviation
Housing type	Muddy house	58.1	33.33	19.77	49.4	17.63
	Semi-cemented house	18.6				
	Fully cemented house	23.3				
Housing facilities	Electricity	88.7	82.9	78.1	85.57	5.14
	Drinking water	76.2				
	Latrine	83.8				
Roof material	Light weight (wood, mud, grass, thatch, sheets and asbestos)	61.1	33.33	14.73	53.05	21.1
	Heavy weight (tiles and slate)	28.9				
	Flat roof (brick, stone, concrete)	10.0				
Combustion material	Firewood	53.6	25	1.65	48.4	23.7
	LPG	43.1				
	Kerosene	3.1				
	Dung cake	0.2				

CROSS-TABULATION

Cross-tabulation and chi-square test were conducted to examine the relationship between different occupational sectors and various adaptation strategies (see Table 17.2). The Pearson's chi-square test is a non-parametric statistical method used to ascertain a significant relationship between

TABLE 17.2
Various Adaptation Strategies Employed across Different Occupations

Adaptation Strategies		Occupational Structure							
		Cultivators		Agricultural Laborers		Household Industries Workers		Other Workers	
		Yes	No	Yes	No	Yes	No	Yes	No
Early warning	Immediate neighborhood and relatives	58.8	41.2	51.3	48.7	66.3	33.7	66.7	33.3
	Local authority	40.1	59.8	30.1	69.9	65.1	34.9	71.3	28.7
System	News (TV, Radio, Print)	5.6	94.3	3.4	96.6	60.2	39.8	85.1	14.9
	Detailed Inventory map	7.8	92.1	5.2	94.8	45.7	54.3	65.5	34.6
	Basic hazard map	8.5	91.4	6.8	93.2	50.1	49.9	60.2	39.8
Emergency response system	Permanent coordination between responders in communities with specialized equipment and well-trained rescue services	6.2	93.7	4.6	95.4	60.1	39.9	75.3	24.7
	Clear definition of roles & responsibilities at local level and proportionate allocation of resources	38.2	61.7	30.6	69.4	59.1	40.9	74.1	25.9
	First aid and health facilities availed during the disaster	50.1	49.8	45.5	54.6	65.1	34.9	78.2	21.8
	Basic facilities availed during disaster (food, water, transport & emergency shelter)	5.3	94.7	5.7	94.3	10.3	89.7	25.1	74.9
Adaptation by communities	Housing insurance	3.1	96.8	3.5	96.6	20.1	79.9	30.2	69.8
	Health insurance	4.5	95.4	5.2	94.8	30.1	69.9	40.5	59.6
	Livelihood diversification	20.3	79.6	15.1	84.9	60.1	39.9	75.3	24.7
	Monthly savings	35.4	64.5	25.7	74.3	60.2	39.8	73.2	26.8
	Other home outside the community	10.2	89.7	8.5	91.6	25.1	74.9	35.5	64.6
	Strong social network or intracommunity cooperation	65.1	34.8	58.1	41.9	70.5	29.6	85.1	14.9
	Access to disaster funds from stakeholders	15.2	84.7	12.5	87.6	45.2	54.8	50.3	49.7
	Saving food grains during a different time of year	80.1	19.8	75.2	24.8	70.1	29.9	65.2	34.8
	Stringent guidelines for constructions activities	8.3	91.7	2.2	97.8	13.2	86.8	35.1	64.9
	Fairly effective regulations for new development, however, potential problems with older constructions	30.1	69.8	28.5	71.6	45.2	54.8	60.1	39.9
	Retrofit buildings	1.4	98.6	0.2	99.8	15.2	84.8	30.5	69.6
	Soil management and agroforestry	70.1	29.8	85.5	14.6	30.2	69.8	45.1	54.9
	Participation in awareness and training program	6.23	93.7	4.6	95.4	60.1	39.9	75.3	24.7

two categorical variables. This is achieved by comparing the observed frequencies in the cross-tabulation with the expected frequencies under the null hypothesis of independence. A statistically significant chi-square result shows that the choice of adaptation strategies is influential and well associated with the categories of workers. Further, a heatmap was created to visualize the preferred and least preferred existing early warning systems, emergency response systems and adaptation strategies across the surveyed villages. These analyses will provide valuable insights into the interplay between occupational sectors and diverse adaptation strategies for understanding the socioeconomic factors determining adaptation choices.

RESULTS

Profile of the Sampled Households

Among the total respondents, 52% were male, while the remaining 48% identified as female. The average age of the participants was reported to be 58.33 years. Approximately 78% of the households adhered to Hinduism, followed by Islam at 18% and Sikhism at 4%.

The majority of households surveyed belonged to the general caste (59%), followed by other backward castes (19%), scheduled castes (16%) and scheduled tribes (7%). Most households were led by males, who represented 93% of the heads, while only 7% were female-led. Approximately 41% of these heads of households had completed higher secondary education, with 23% having finished secondary education and 18% having completed primary education. Only 2% of the respondents were graduates. About 67% of the households were joint families, whereas 33% were nuclear families. Most respondents (37%) reported a monthly income of less than 15,000 Indian rupees (1 US $=82.90 Indian rupees), followed by incomes between ₹15,001 and ₹30,000 (29%), ₹30,001–₹45,000 (20%) and over ₹45,000 (13%). In terms of occupation, 46% were involved in agriculture, 22% were engaged in small businesses, 18% worked as laborers, 7% raised livestock, 4% were in services and 3% were involved in fishing. The largest portion of the workforce (61%) was classified as main workers, while 39% were marginal workers. A significant majority of respondents (81%) had lived in the district since birth, but nearly 19% had started migrating to larger cities in search of improved economic opportunities and relief from harsh weather conditions. Among the households sampled, around 89% reported having sufficient access to electricity; however, 11% experienced issues with unreliable supply and limited access. Most households relied on firewood as their primary cooking fuel, accounting for 54%, followed by liquefied petroleum gas at 43% and kerosene at 3%.

PERCEPTION ON CLIMATE CHANGE AND LANDSLIDE OCCURRENCE

The majority of the sampled respondents (88%) have reported a significant increase in climate change and landslide occurrence (Figure 17.3). Nearly 63% have strong agreement with the changes in the annual temperature. Around 80% of respondents have perceived changes in summer temperature, while more than half of the sampled respondents asserted that winters had become less severe. An analysis of the seasonal frequency of rainfall revealed that heavy rainfall was the most common phenomenon during the monsoon season (June–September), followed by the pre-monsoon (March–May) and post-monsoon (October–November). An increase in rainfall intensity was observed by a large number of respondents. However, the study indicated neutral perception regarding the delayed onset of the monsoon. Many individuals perceived shifts in the timing of monsoon rain, while others did not observe such changes in natural variability. Nearly 51% of the sampled respondents reported that rainfall-induced landslide frequency had increased during the last 40 years.

Impact on Socioeconomic and Environmental Conditions due to Landslide

The sampled households were asked to disclose their experiences regarding the implications of climate change-induced landslides. Social well-being, food security, property damage and local

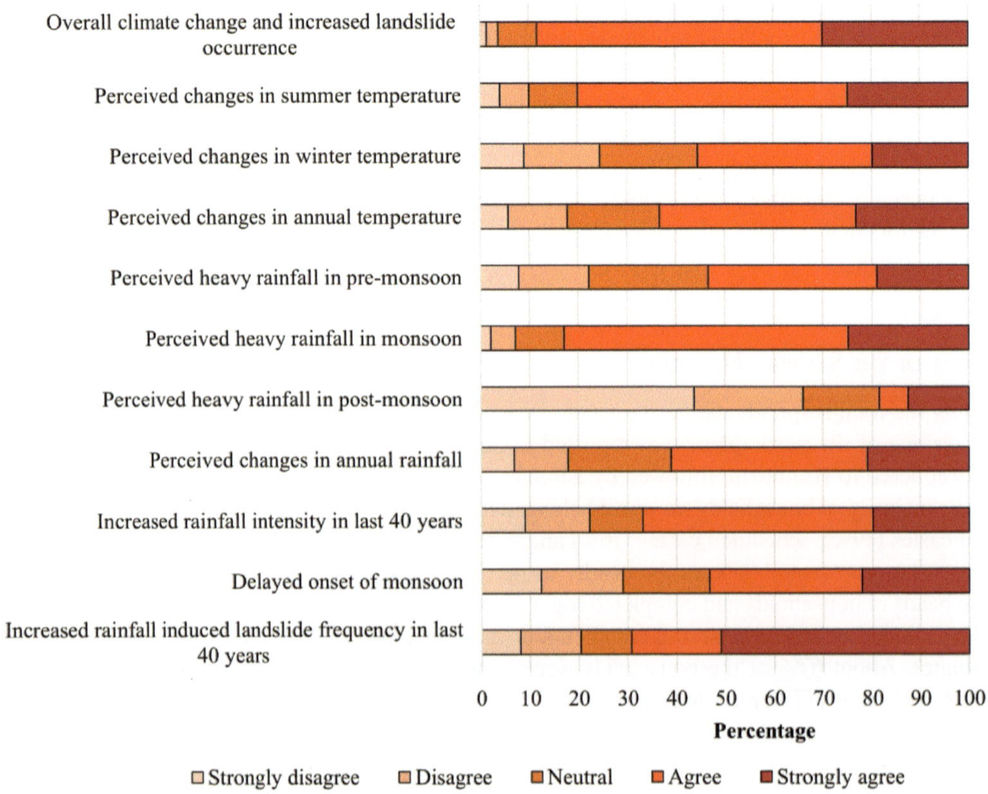

FIGURE 17.3 Perception of the sampled respondents on climate change and landslide.

infrastructure were analyzed among the households (Figure 17.4). Nearly 73% of respondents reported severe deformation in the house structure due to the intense landslide. The force had directly impacted buildings and caused immediate physical damage such as cracks, buckling and destruction. Respondents have also disclosed instances of land subsidence and wide fissures in Sarahan and Kinnu villages of Rampur block. Further, these villages are situated in seismic zone IV, which renders the house structure susceptible to geotectonic movement. Since 1970, the area has been subjected to severe earthquakes of great magnitude. Significant damage to material assets was observed by 53% of respondents, followed by not suffered (22%), moderately suffered (19%) and least suffered (5%). The survey data revealed food scarcity, with 41% of participants experiencing severe distress due to a shortage of food availability. Respondents from non-agricultural sectors reported that landslides disrupted the supply chain for fruits, vegetables and grocery items. The prevalence of road blockages has severely affected 69% of households and contributed to significant hindrance and restricted access to markets. The sampled respondents stated severe soil erosion and contamination (54%) followed by moderately suffered (20%), least suffered (10%) and not suffered (15%) soil erosion. Changes in the cropping pattern were another concern driven by climate change-induced landslides by 52% of the sampled households, followed by not suffering (30%), moderately suffering (13%) and least suffering (6%). A substantial proportion of respondents (81%) expressed concern regarding the severe deterioration of surface water quality while only 1% reported no adverse effects.

SOURCES OF EARLY WARNING SYSTEMS

The efficacy of early warning systems was assessed among the sampled households (Figure 17.5). Among the various sources of early warning systems, immediate neighborhood and relatives (57%) were found highly effective followed by local authorities (49%), warning signs/message alerts from

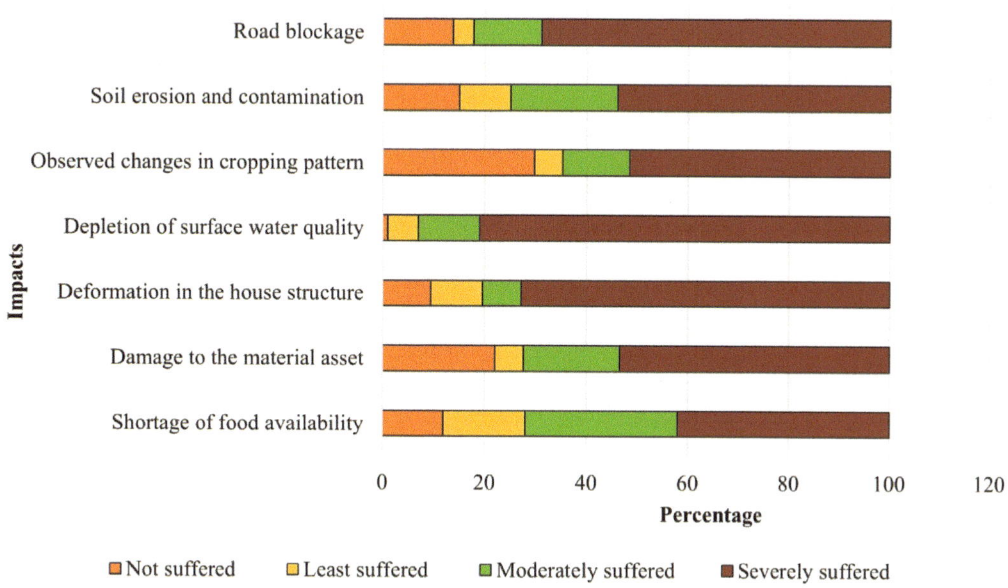

FIGURE 17.4 Socioeconomic and environmental implications of landslide.

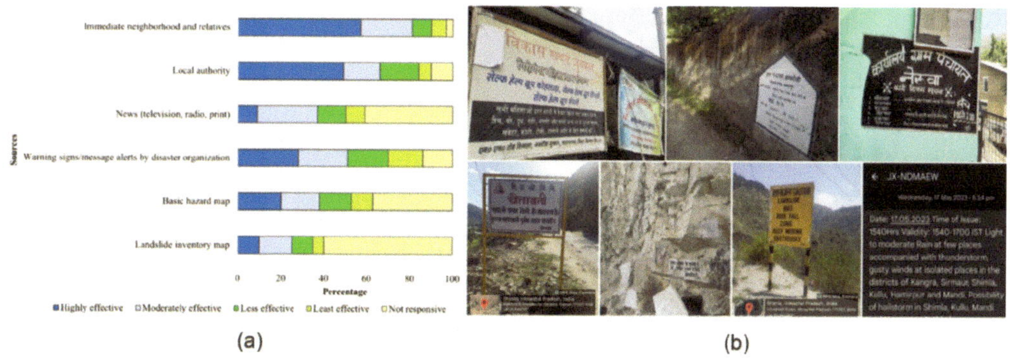

FIGURE 17.5 (a) Sources of the early warning system in Shimla district. (b) Sources of the early warning system available to the sampled household. (Source: Field photographs captured during survey, 2024.)

disaster organization (28%), basic hazard map (20%), detailed inventory map (10%) and newspapers or electronic media (9%). Households reported that gram panchayats (village assembly), police and self-help groups were reliable sources for disseminating information through public announcements. However, the limited number of respondents received information from other sources. Some households did not receive any forewarning information that significantly hindered their ability to prepare for landslides. Announcements by local authorities were identified as highly effective according to 49% of sampled respondents, while nearly 18% stated that local authorities were less a responsive during landslides, followed by moderately effective (17%), not responsive (10%) and least effective (6%). Nearly 28% found the hazard warning signs and message alert as highly effective, followed by moderately effective (23%), less effective (19%), least effective (16%) and not responsive (14%).

In the case of the basic hazard map, nearly 37% of sampled households were not responsive, while 20% were highly responsive, followed by moderately (18%), less (15%) and least (10%) effective. Nearly 60% of the sampled respondents were not responsive to the detailed inventory map, followed by moderately effective (15%), highly effective (10%), less effective (10%) and least effective (5%) in the early warning system. Nearly 41% of respondents found news (television, radio, print)

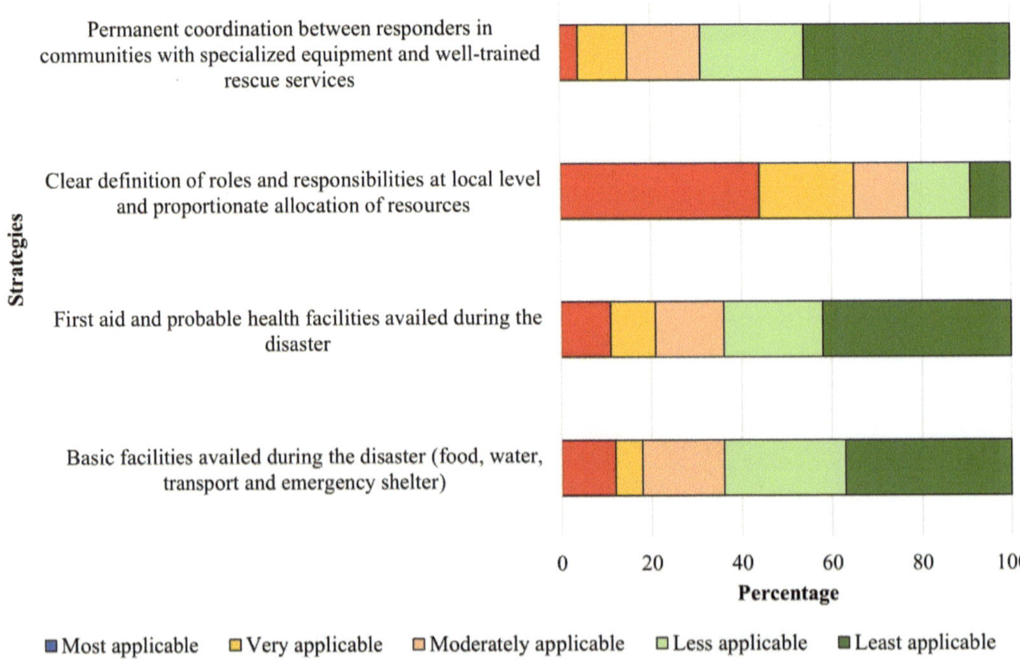

FIGURE 17.6 Strategies implemented by sampled households during the landslide.

not responsive, while a meager respondents (9%) disclosed it as highly effective, moderately effective (28%), less effective (13%) and least effective (9%).

EFFECTIVENESS OF THE EMERGENCY RESPONSE SYSTEM

The disaster response system involves timely actions to reduce suffering during the disaster (Figure 17.6). It involves permanent coordination with well-trained rescue services, clearly defining local roles and resource allocation, providing first aid and health facilities and availing basic disaster facilities. The sampled households stated that a clear definition of roles and responsibilities at the local level (44%) was mostly adopted by the respondents. There was a lack of agreement over the notion that villages had established permanent coordination between responders in communities with specialized equipment and well-trained rescue services during the disaster. Lack of training, mock drills and effective communication were substantial barriers to disaster prevention and response.

The perception of the sampled households regarding access to essential amenities such as food, water, transportation and emergency shelter showed significant variation. Permanent coordination between responders (46%) and basic facilities availed during the disaster (37%) were observed to be least applicable among the surveyed respondents.

LANDSLIDE ADAPTATION STRATEGIES ADOPTED BY LOCAL COMMUNITIES

Households were asked to disclose their adaptation strategies adopted before and during the landslide (Figure 17.7). The most preferred strategy adopted by the sampled households was soil management and agroforestry (77%), followed by diversification in livelihood practices (64%), saving food grains (62%) and monthly savings (51%). Strong social network (39%), owning homes outside the community (37%) and access to disaster funds from stakeholders were identified as moderately

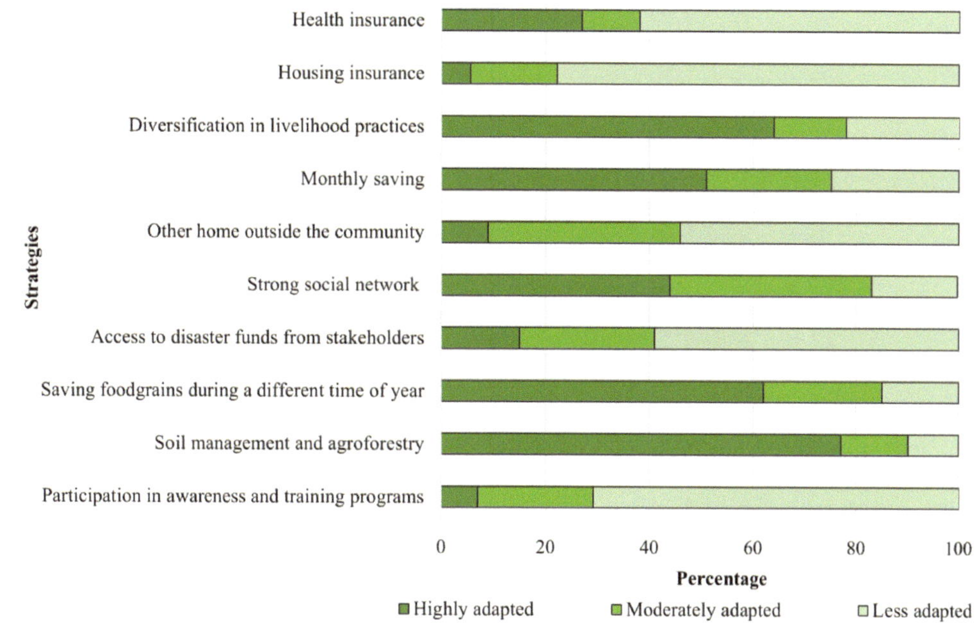

FIGURE 17.7 Landslide adaptation strategies adopted by communities.

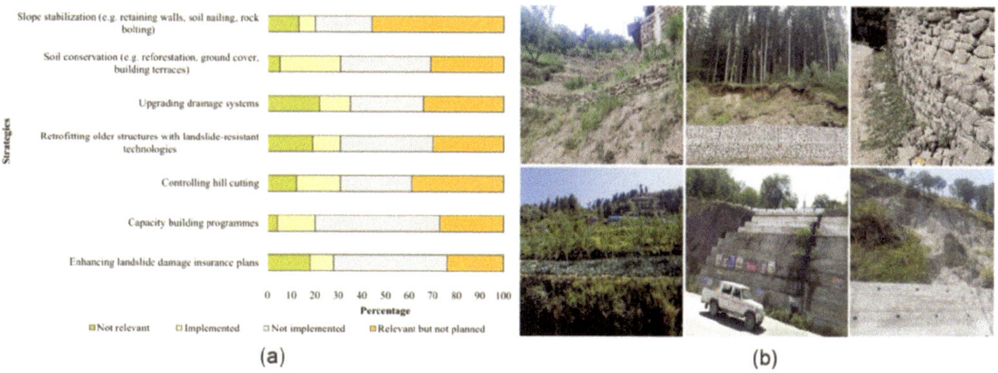

FIGURE 17.8 (a) Landslide mitigation strategies. (Source: Field photographs captured during survey, 2024.) (b) Landslide mitigation strategies adopted in Shimla district.

adapted strategies. Housing insurance (78%), participation in awareness and training programs (71%) and health insurance (62%) were less adopted by the sampled respondents.

LANDSLIDE MITIGATION STRATEGIES

Landslide mitigation strategies involved slope stabilization (e.g., retaining walls, soil nail, rock bolt), bioengineering (e.g. reforestation, ground cover, building terraces), upgrading drainage systems (e.g. surface drainage, subsurface drainage), retrofitting older structures with landslide-resistant technologies, stringent guidelines for constructions and land-use activities, capacity building programs and enhancing landslide damage insurance plans (Figure 17.8).

Slope stabilization measures (56%) were found relevant but not planned in the study area, followed by stringent guidelines for constructions and land-use activities (39%), upgradation of the

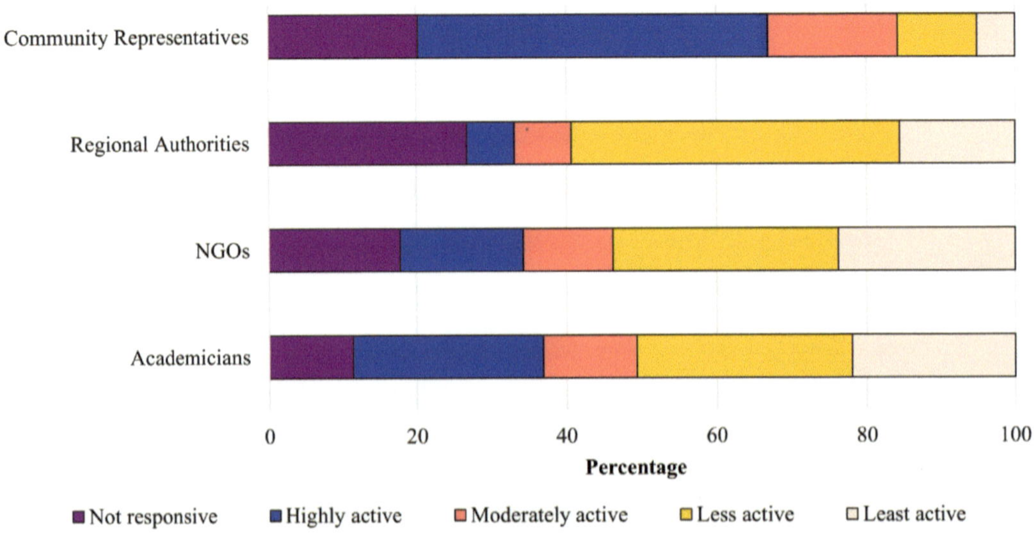

FIGURE 17.9 Perception of sampled households on the role of various stakeholders.

drainage system (34%) and soil conservation (31%). Respondents disclosed that capacity building programs (53%), landslide insurance schemes (48%) and retrofitting older structures (39%) were not implemented. While soil conservation practices (26%) and controlling hill cutting (19%) emerged leading implemented strategies. Nearly 22% of the sampled respondents considered the upgradation of drainage systems not relevant to landslide mitigation strategies.

STAKEHOLDER INVOLVEMENT IN LANDSLIDE RISK REDUCTION

The role of community representatives, regional authorities, non-governmental organizations and academicians was examined among the sampled households (Figure 17.9).

The involvement of stakeholder analysis revealed that community representatives (64%) demonstrated significant involvement and were classified as highly active and moderately active, while 20% of the sampled households regarded them as not responsive. This indicates a robust level of community engagement, which is essential for effective disaster response. However, the presence of 11% less active and 5% least active representatives suggests there is still potential for enhanced participation. Non-governmental organizations (56%) were considered least responsive in landslide risk reduction, followed by regional authorities (34%) and academicians (29%). The distribution across various stakeholders shows that while some groups were highly engaged, there is a need to increase the activity levels of the less engaged groups to ensure an effective response in landslide management.

PREFERRED LANDSLIDE ADAPTATION STRATEGIES

Soil management, saving food grain, strong social network, clear definition of roles and responsibilities at the local level, monthly savings, immediate neighborhoods and relatives for dissemination information and fairly effective regulations for new developments; however, potential problems with older constructions were preferred by all occupations in the study area (Table 17.2). However, the least preferred strategies among respondents were basic facilities availed during disasters, housing and health insurance, retrofitting buildings, access to disaster funds from stakeholders and low participation in awareness and training programs. In the case of village-wise analysis, Ogli, Raighat, Sholi, Manjholi, Panoli, Kacheri and Mandhaon villages performed relatively well

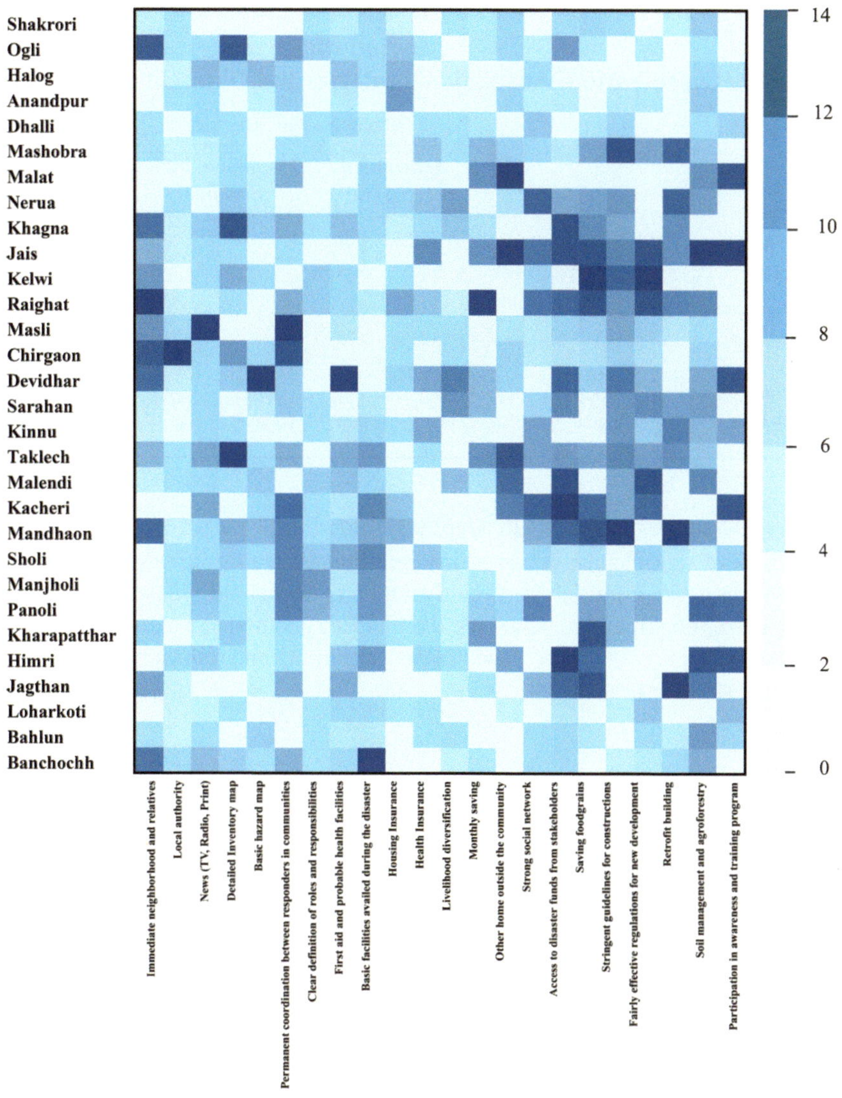

FIGURE 17.10 Heatmap illustrating landslide adaptation strategies in the sampled villages.

in adaptation strategies except in participation in awareness and training programs and insurance schemes. Effective social ties and livelihood diversification have reduced vulnerability in these villages (Figure 17.10). Further, the villages, namely Ogli and Raighat, are dominated by the agrarian sector where practices such as contour farming, mulching, crop rotation and intercropping have contributed to soil conservation and the maintenance of soil fertility.

Results of the chi-square test revealed a significant relationship between adaptation strategies and occupational groups (Table 17.3). The analysis for the early warning system revealed a strong and statistically significant relationship for other workers and cultivators at the 0.01 level of significance. For household industry workers, a moderate relationship was found at the 0.05 level of significance, while no significant relationship was observed among agricultural laborers. The chi-square test also revealed that a positive and statistically significant relationship was observed for the emergency response system in other workers at the 0.01 level of significance. Conversely, no significant relationship was identified for household industry workers. In terms of local adaptation strategies, the results were not significantly associated with cultivators, agricultural laborers and household

TABLE 17.3
Chi-Square Test between Various Adaptation Strategies and Different Occupations

	Occupational Structure							
	Cultivators		Agricultural Laborers		Household Industry Workers		Other Workers	
Adaptation Strategies	χ^2 (df)	*p*	χ^2 (df)	*p*	χ^2 (df)	*p*	χ^2 (df)	*p*
Early warning system	61.96 (3)	0.0000***	5.94 (3)	0.2312	9.67 (3)	0.0215**	16.61 (3)	0.0008***
Emergency response system	8.82 (3)	0.0042**	6.78 (3)	0.0024*	3.44 (3)	0.3284	16.03 (3)	0.0011***
Local adaptation strategies	15.71 (12)	0.6540	11.13 (12)	0.7320	18.89 (12)	0.9150	166.79 (12)	0.0000***

df, degree of freedom.
*** Significant at 0.01 level of significance, **0.05 level of significance, *0.10 level of significance.

industry workers. However, a statistically significant relationship was observed with other workers at the 0.01 level of significance. It can be inferred that adaptation strategies are less effectively implemented among agricultural laborers, cultivators and household industry workers. Thus, it is essential to prioritize investment in enhancing their capacity to manage and respond to challenges. This involves expanding access to emergency response training and reinforcing the implementation of local adaptation strategies.

DISCUSSION

The study intended to understand the perception of the sampled households regarding the impact of climate change and landslides in Shimla district of Himachal Pradesh. With a particular emphasis on early warning, emergency response systems and local adaptation strategies, these measures tried to address the increasing casualties and socioeconomic impacts. Effective pre- and post-disaster adaptive responses are instrumental in coping with landslide events and may significantly mitigate long-term implications (Rehman and Azhoni, 2023). Climate change-induced landslides and their adaptation have emphasized selecting site-specific indicators and addressing the social, economic, environmental and infrastructural challenges in affected areas. The adaptive strategies relied on integrating traditional knowledge, community preparedness and proper understanding of the disaster (Sahana et al., 2023; Sharma et al., 2024). A paradigm shift has occurred from descriptive analyses to empirical studies in analyzing the effects of different adaptation strategies (Wang et al., 2024). The severe implications have drawn the attention of researchers and policymakers to both the short-term and long-term consequences of these events.

Landslide adaptation assessment requires dynamic indicators to enhance the ability of an individual or community to resist, recover and manage financial strain over time. In the Garhwal Himalayas, Mirdda et al. (2022) identified livelihood diversity, higher income status, housing insurance and access to disaster awareness programs as the prominent landslide adaptation measures employed by rural communities. Measures such as reinforced concrete housing structures, mass prediction and investment in disaster prevention measures were highlighted as effective landslide coping strategies for farmer communities in the Three Gorges Reservoir Area of China (Peng et al., 2019). On the other hand, regular participation in organizations such as women's unions, farmer interest groups, farmers' unions and agricultural cooperatives, along with diversified income sources, was considered an ideal strategy for reducing vulnerability (Thanh Thi Pham et al., 2020). There was broad consensus among households regarding the influence of climate change on the increasing frequency and magnitude of landslides (Ahmed, 2021; Jakob, 2022; Chai and Wu, 2023). Factors such as inadequate structural resistance of houses, economic disparities, social marginalization and limited access to natural resources presented great challenges to the coping capacity. These findings align with existing literature on the Himalayan state (Bera et al., 2020a; Martha et al., 2021; Meena et al., 2022; Mirdda et al., 2022; Singh et al., 2021). The present analysis revealed that several villages located in the highly susceptible zones experienced economic hardship, inevitable loss of material assets and restricted access to basic necessities during the disaster. A significant proportion of marginalized communities and low-income groups residing there were affected due to limited social networks for support and poor housing conditions. The lack of education and awareness about disaster risk and preparedness measures also leaves them ill-equipped to cope with and adapt to such events.

The primary source of livelihood for the sampled households was agriculture and horticulture, which relied on small landholdings for subsistence. These households were disproportionately affected by landslides, with changes in crop yield and cropping patterns having a direct impact on their livelihood. The findings are in tune with Nahalomo et al. (2023), who also observed that the cascading effects of landslides reduced agricultural productivity, which not only threatened food security but also deepened financial vulnerabilities, making recovery even more challenging for marginalized groups. Results also indicated that the existing early warning and emergency response systems were quite poor in the study area. Immediate neighborhood and relatives were the major sources for disseminating information and others had poor access to early warning systems. The findings are in tune with Sahana et al. (2023). Respondents engaged in non-agricultural activities exhibited a preference for out-migration as a coping mechanism. The affected respondents received financial aid from the stakeholders; however, the distribution process was inefficient, and limited monetary help may not cover the full extent of damages to homes and property. Most respondents were not compensated for the loss and damages from the aftermath landslide. Low coverage of housing and health insurance schemes, less participation in awareness programs and absence of retrofitting in the house structures further intensified vulnerability. Similar studies have been reported by (Bera et al., 2020b; Thanh Thi Pham et al., 2020; Mirdda et al., 2022). The low performance of Halog, Taklech, Masli, Malendi, Bahlun, Jais, Nerua, Anandpur, Mashobra, Chirgaon, Shakrori, Loharkoti and Malat villages also indicates a less adaptive capacity and a high degree of landslide implications.

Devising effective adaptation and mitigation strategies and their execution are essential for minimizing the consequences of the losses in an era of increasing climate change. It is imperative to concentrate on localized, community-driven initiatives that directly address the needs of vulnerable households. Strengthening disaster preparedness, improving access to resources for recovery and enhancing local knowledge-sharing networks will be essential for building resilience at the grassroots level. This analysis is not only limited to examining the perception of climate change-induced landslide occurrence but provides practical insights into household-level adaptation strategies across different occupations.

CONCLUSION

The present study examined the perception analysis on climate change-induced landslide and diverse adaptation strategies adopted by rural communities in Shimla district of Himachal Pradesh. Findings revealed that saving food grains during different times of the year, a clear definition of roles and responsibilities at the local level, strong social network, soil management, monthly savings and fairly effective regulations for new developments; however, potential problems with older constructions were most preferred by all occupations. Housing insurance scheme, adoption of retrofitting buildings and low participation in awareness and training were observed to be least preferable. Village-wise analysis revealed that Malendi, Bahlun, Jais and Nerua villages toward the low performance, whereas Ogli, Raighat, Sholi, Manjholi, Panoli, Kacheri and Mandhaon villages performed relatively well in adaptation strategies due to strong social networks and a clear definition of roles and responsibilities at the local level. The study calls for timely intervention for promoting eco-engineering solutions, bio-reinforced slope stabilization and designing flexible evacuation infrastructure to mitigate the implications of landslides. Access to reliable early warning systems, provision of constructing emergency shelter, raising awareness and modernizing indigenous landslide adaptation techniques may effectively enhance coping mechanisms among local communities. The methodology provides a practical and theoretical reference for assessing climate change-induced landslide adaptation and mitigation strategies and can be utilized in other mountainous ecosystems.

REFERENCES

Ahmed, B., 2021. The root causes of landslide vulnerability in Bangladesh. *Landslides* 18, 1707–1720. https://doi.org/10.1007/s10346-020-01606-0

Bera, S., Guru, B., Chatterjee, R., Shaw, R., 2020a. Geographic variation of resilience to landslide hazard: A household-based comparative studies in Kalimpong hilly region, India. *Int. J. Disaster Risk Reduct.* 46. https://doi.org/10.1016/j.ijdrr.2019.101456

Bera, S., Guru, B., Oommen, T., 2020b. Indicator-based approach for assigning physical vulnerability of the houses to landslide hazard in the Himalayan region of India. *Int. J. Disaster Risk Reduct.* 50. https://doi.org/10.1016/j.ijdrr.2020.101891

Census of India, 2011. *District Census Handbook.* https://censusindia.gov.in/nada/index.php/catalog/483/download/1580/DH_2011_0211_PART_B_DCHB_SHIMLA.pdfT_B_DCHB_SHIMLA.pdf. Accessed on 23 Nov 2024

Chai, J., Wu, H.-Z., 2023. Prevention/mitigation of natural disasters in urban areas. *Smart Constr. Sustain. Cities* 1. https://doi.org/10.1007/s44268-023-00002-6

Chen, W., Pourghasemi, H.R., Panahi, M., Kornejady, A., Wang, J., Xie, X., Cao, S., 2017. Spatial prediction of landslide susceptibility using an adaptive neuro-fuzzy inference system combined with frequency ratio, generalized additive model, and support vector machine techniques. *Geomorphology* 297, 69–85. https://doi.org/10.1016/j.geomorph.2017.09.007

Dikshit, A., Sarkar, R., Pradhan, B., Segoni, S., Alamri, A.M., 2020. Rainfall induced landslide studies in indian himalayan region: A critical review. *Appl. Sci.* 10. https://doi.org/10.3390/app10072466

Du, J., Glade, T., Woldai, T., Chai, B., Zeng, B., 2020. Landslide susceptibility assessment based on an incomplete landslide inventory in the Jilong Valley, Tibet, Chinese Himalayas. *Eng. Geol.* 270. https://doi.org/10.1016/j.enggeo.2020.105572

Fotopoulou, S.D., Pitilakis, K.D., 2017. Probabilistic assessment of the vulnerability of reinforced concrete buildings subjected to earthquake induced landslides. *Bull. Earthq. Eng.* 15, 5191–5215. https://doi.org/10.1007/s10518-017-0175-x

Gupta, A.K., Negi, M., Nandy, S., Kumar, M., Singh, V., Valente, D., Petrosillo, I., Pandey, R., 2020. Mapping socio-environmental vulnerability to climate change in different altitude zones in the Indian Himalayas. *Ecol. Indic.* 109. https://doi.org/10.1016/j.ecolind.2019.105787

Gupta, V., Kumar, S., Kaur, R., Tandon, R.S., 2022. Regional-scale landslide susceptibility assessment for the hilly state of Uttarakhand, NW Himalaya, India. *J. Earth Syst. Sci.* 131. https://doi.org/10.1007/s12040-021-01746-4

Haigh, M., Rawat, J.S., 2011. Landslide causes: Human impacts on a Himalayan landslide swarm. *Belgeo* 201–219. https://doi.org/10.4000/belgeo.6311

Hasegawa, S., Dahal, R.K., Yamanaka, M., Bhandary, N.P., Yatabe, R., Inagaki, H., 2009. Causes of large-scale landslides in the Lesser Himalaya of central Nepal. *Environ. Geol.* 57, 1423–1434. https://doi.org/10.1007/s00254-008-1420-z

IPCC, 2022. *Climate Change 2022: Impacts, Adaptation and Vulnerability.* https://www.ipcc.ch/report/sixth-assessment-report-working-group-ii/. Accessed on 21 Nov 2024

Jakob, M., 2022. Landslides in a changing climate. In Davies, T., Rosser N., Shroder, J.F. (eds.). *Landslide Hazards, Risks, and Disasters*, 505–579. https://doi.org/10.1016/b978-0-12-818464-6.00003-2

Maikhuri, R.K., Nautiyal, A., Jha, N.K., Rawat, L.S., Maletha, A., Phondani, P.C., Bahuguna, Y.M., Bhatt, G.C., 2017. Socio-ecological vulnerability: Assessment and coping strategy to environmental disaster in Kedarnath valley, Uttarakhand, Indian Himalayan Region. *Int. J. Disaster Risk Reduct.* 25, 111–124. https://doi.org/10.1016/j.ijdrr.2017.09.002

Martha, T.R., Roy, P., Jain, N., Khanna, K., Mrinalni, K., Kumar, K.V., Rao, P.V.N., 2021. Geospatial landslide inventory of India—an insight into occurrence and exposure on a national scale. *Landslides* 18, 2125–2141. https://doi.org/10.1007/s10346-021-01645-1

Meena, S.R., Soares, L.P., Grohmann, C.H., van Westen, C., Bhuyan, K., Singh, R.P., Floris, M., Catani, F., 2022. Landslide detection in the Himalayas using machine learning algorithms and U-Net. *Landslides* 19, 1209–1229. https://doi.org/10.1007/s10346-022-01861-3

Mirdda, H.A., Bera, S., Chatterjee, R., 2022. Vulnerability assessment of mountainous households to landslides: A multidimensional study in the rural Himalayas. *Int. J. Disaster Risk Reduct.* 71. https://doi.org/10.1016/j.ijdrr.2022.102809

Nahalomo, A., Iversen, P.O., Andreassen, B.A., Kaaya, A., Rukooko, A.B., Rukundo, P.M., 2023. Food insecurity, dietary diversity and the right to adequate food among households in landslide-prone communities in Eastern Uganda: A cohort study. *PLoS One* 18. https://doi.org/10.1371/journal.pone.0283078

Nasiri, M., Hajiazizi, M., 2021. An experimental and numerical investigation of reinforced slope using geotextile encased stone column. *Int. J. Geotech. Eng.* 15, 543–552. https://doi.org/10.1080/19386362.2019.1651029

NRSC, 2023. Landslide atlas of India (Mapping, Monitoring and R&D studies using Remote Sensing data). https://www.nrsc.gov.in. Accessed on 13 Nov 2024

Panchal, S., Shrivastava, A.K., 2021. A comparative study of frequency ratio, shannon's entropy and analytic hierarchy process (Ahp) models for landslide susceptibility assessment. *ISPRS Int. J. Geo-Information* 10. https://doi.org/10.3390/ijgi10090603

Peng, L., Xu, D., Wang, X., 2019. Vulnerability of rural household livelihood to climate variability and adaptive strategies in landslide-threatened western mountainous regions of the Three Gorges Reservoir Area, China. *Clim. Dev.* 11, 469–484. https://doi.org/10.1080/17565529.2018.1445613

Raška, P., Riezner, J., Bíl, M., Klimeš, J., 2023. Long-term landslide impacts and adaptive responses in rural communities: Using historical cases to validate the cumulative causation approach. *Int. J. Disaster Risk Reduct.* 93. https://doi.org/10.1016/j.ijdrr.2023.103748

Rehman, S., Azhoni, A., 2023. Analyzing landslide susceptibility, health vulnerability and risk using multi-criteria decision-making analysis in Arunachal Pradesh, India. *Acta Geophys.* 71, 101–128. https://doi.org/10.1007/s11600-022-00943-z

Roy, D., Sarkar, A., Kundu, P., Paul, S., Chandra Sarkar, B., 2023. An ensemble of evidence belief function (EBF) with frequency ratio (FR) using geospatial data for landslide prediction in Darjeeling Himalayan region of India. *Quat. Sci. Adv.* 11. https://doi.org/10.1016/j.qsa.2023.100092

Sahana, M., Patel, P.P., Rehman, S., Rahaman, M.H., Masroor, M., Imdad, K., Sajjad, H., 2023. Assessing the effectiveness of existing early warning systems and emergency preparedness towards reducing cyclone-induced losses in the Sundarban Biosphere Region, India. *Int. J. Disaster Risk Reduct.* 90. https://doi.org/10.1016/j.ijdrr.2023.103645

Sana, E., Kumar, A., Robson, E., Prasanna, R., Kala, U., Toll, D.G., 2024. Preliminary assessment of series of landslides and related damage by heavy rainfall in Himachal Pradesh, India, during July 2023. *Landslides* 21, 919–931. https://doi.org/10.1007/s10346-023-02209-1

Sharma, A., Sajjad, H., Rahaman, M.H., Saha, T.K., Bhuyan, N., 2024. Effectiveness of hybrid ensemble machine learning models for landslide susceptibility analysis: Evidence from Shimla district of North-west Indian Himalayan region. *J. Mt. Sci.* 21, 2368–2393. https://doi.org/10.1007/s11629-024-8651-7

Sharma, A., Sajjad, H., Roshani, Rahaman, M.H., 2023. A systematic review for assessing the impact of climate change on landslides: Research gaps and directions for future research. *Spat. Inf. Res.* https://doi.org/10.1007/s41324-023-00551-z

Shukla, S.K., Sivakugan, N., Das, B.M., 2011. A state-of-the-art review of geosynthetic-reinforced slopes. *Int. J. Geotech. Eng.* 5, 17–32. https://doi.org/10.3328/IJGE.2011.05.01.17-32

Singh, A., Pal, S., Kanungo, D.P., 2021. An integrated approach for landslide susceptibility–vulnerability–risk assessment of building infrastructures in hilly regions of India. *Environ. Dev. Sustain.* 23, 5058–5095. https://doi.org/10.1007/s10668-020-00804-z

Sudmeier-Rieux, K., Jaquet, S., Derron, M.H., Jaboyedoff, M., Devkota, S., 2012. A case study of coping strategies and landslides in two villages of Central-Eastern Nepal. *Appl. Geogr.* 32, 680–690. https://doi.org/10.1016/j.apgeog.2011.07.005

Sujatha, E.R., Sridhar, V., 2021. Landslide susceptibility analysis: A logistic regression model case study in Coonoor, India. *Hydrology* 8. https://doi.org/10.3390/hydrology8010041

Sultana, N., Tan, S., 2021. Landslide mitigation strategies in southeast Bangladesh: Lessons learned from the institutional responses. *Int. J. Disaster Risk Reduct.* 62. https://doi.org/10.1016/j.ijdrr.2021.102402

Thanh Thi Pham, N., Nong, D., Raghavan Sathyan, A., Garschagen, M., 2020. Vulnerability assessment of households to flash floods and landslides in the poor upland regions of Vietnam. *Clim. Risk Manag.* 28. https://doi.org/10.1016/j.crm.2020.100215

UNEP, 2023. Underfinanced. Underprepared. https://doi.org/https://doi.org/10.59117/20.500.11822/43796. Accessed on 3 Nov 2024

Wang, J., Wang, H., Wu, Z., 2024. Factors influencing farmers' adaptation willingness under landslide risks: Using an extended sustainable livelihood framework. *Int. J. Disaster Risk Reduct.* 107. https://doi.org/10.1016/j.ijdrr.2024.104512

World Risk Report, 2024. Berlin: Bündnis Entwicklung Hilft. https://reliefweb.int/report/world/worldriskreport-2024-focus-multiple-crises. Accessed on 8 Nov 2024

Yuniawan, R.A., Rifa'i, A., Faris, F., Subiyantoro, A., Satyaningsih, R., Hidayah, A.N., Hidayat, R., Mushthofa, A., Ridwan, B.W., Priangga, E., Muntohar, A.S., Jetten, V.G., van Westen, C.J., Den Bout, B. V., Sutanto, S.J., 2022. Revised rainfall threshold in the Indonesian landslide early warning system. *Geosciences* 12. https://doi.org/10.3390/geosciences12030129

18 Climate Change and Phytoplankton Habitat Suitability in Nainital Lake

Insights from Species Distribution Models

Zainab Khan, Ateeque Ahmad, and Amit Kumar

INTRODUCTION

The Kumaun Himalayas, located in Uttarakhand, India, are renowned for their diverse and scenic freshwater lakes formed through tectonic activities approximately 4,000 years ago (Thomas, 1952). Among these, Nainital Lake, situated in the lesser Himalayas, is one of the most significant. The hydrology of the lake is sustained by both rainfall and perennial springs, making it a perennial water body (Khan and Ahmad, 2021; Khan et al., 2020). As a critical freshwater resource for the surrounding urban population, Nainital Lake also supports a diverse range of aquatic species, including phytoplankton that form the base of the trophic level. However, the enclosed nature of the lake, combined with its relatively small size and high altitude, amplifies its sensitivity to climatic variations and anthropogenic pressures, such as nutrient loading and thermal stress (Catalan et al., 2006). These factors significantly influence the physicochemical properties of the lake, thereby impacting its biological communities, particularly phytoplankton, which play an essential role in biogeochemical cycles, primary production, and food web dynamics (Litchman et al., 2015; Ferreira et al., 2024).

Phytoplankton distributions are highly sensitive to variations in temperature, light availability, and nutrient levels, making them reliable indicators of climate-induced changes in aquatic environments (Cloern, 1999; Hubert et al., 2025; Hutchins & Tagliabue, 2024). The physicochemical parameters of aquatic ecosystems, including temperature, pH, dissolved oxygen (DO), electrical conductivity (EC), total dissolved solids (TDS), and Secchi depth, play a pivotal role in determining the distribution and abundance of phytoplankton (Gogoi et al., 2019; Machate et al., 2023). Temperature directly influences metabolic rates, nutrient cycling, and stratification patterns within lakes, thereby impacting the growth and spatial distribution of phytoplankton communities (Ben Ezra et al., 2024; Dory et al., 2024; Fernández et al., 2024). Changes in pH can alter nutrient availability and competitive dynamics among phytoplankton species, while DO levels influence respiratory activities and redox conditions critical for phytoplankton survival (Jakobsen et al., 2015). EC and TDS reflect the ionic composition and dissolved substances in the water, which are crucial for osmoregulation in phytoplankton cells (Wetzel, 2002). Secchi depth serves as an indirect measure of turbidity and light penetration, affecting the photosynthetic efficiency of phytoplankton (Williamson et al., 2009).

Nainital Lake's high altitude and relatively small size make it particularly susceptible to temperature fluctuations, which can significantly impact its thermal stratification, DO levels, and overall ecosystem dynamics (Choudhary et al., 2013). Additionally, the influx of tourists during summer months exerts substantial pressure on the lake's resources, exacerbating nutrient enrichment

DOI: 10.1201/9781003620717-21

and reducing water quality (Raghava, 2020; Dash et al., 2020). Rising temperatures and altered hydrological cycles in the Himalayan region, as highlighted by previous studies, underscore the need for focused research on climate impacts at a localized scale (Sharma, 2014). However, limited research has been conducted to quantitatively assess the impact of projected climate change scenarios on phytoplankton dynamics in Nainital Lake.

This chapter aims to predict the distribution patterns of *Cladophora dalmatica* Kützing, the dominant phytoplankton species in Nainital Lake, under future temperature scenarios using random forest-assisted species distribution modeling (SDM) techniques. The analysis is based on geocoded physicochemical data, including temperature, pH, DO, EC, TDS, and Secchi depth, collected from 41 locations across the lake. Phytoplankton samples from 26 locations were analyzed using light microscopy to determine species composition and abundance. To simulate the lake surface temperature for future scenarios, statistically downscaled Coupled Model Intercomparison Project Phase 5 (CMIP5) climate projections were employed, providing a robust basis for assessing temperature-driven changes in phytoplankton distribution. The chapter focuses on understanding how future temperature variations alone might influence phytoplankton patterns, assuming other physicochemical parameters remain constant.

By isolating the impact of warming, this research provides valuable insights into the resilience of the trophic structure of the lake under future climate scenarios. The findings are expected to contribute to targeted conservation strategies for lentic ecosystems facing climate-induced stress, emphasizing the need for adaptive management practices to mitigate the impacts of rising temperatures.

CANDIDATE LAKE

The Kumaun Himalayas, located in the Indian state of Uttarakhand, are known for their diverse and scenic freshwater lakes, which were formed through tectonic activities approximately 4,000 years ago (Thomas, 1952). These lakes are primarily replenished by rainfall and perennial springs, making them perennial in nature (Khan and Ahmad, 2021). Among these, Nainital Lake stands out due to its high altitude of 1,938 m and an area of 0.435 km^2, making it one of the largest and most significant lakes in the region. As a rainfed and spring-fed lake, Nainital plays a crucial role in providing water resources to the local population and supporting a diverse range of aquatic species, including phytoplankton that form the base of the trophic level. However, the enclosed nature of Nainital Lake also makes it particularly vulnerable to climatic variations and anthropogenic pressures, such as nutrient loading and thermal stress. The high altitude and relatively small size of the lake amplify its sensitivity to temperature fluctuations, which can significantly impact its thermal stratification, DO levels, and overall ecosystem dynamics (Choudhary et al., 2013). Additionally, the influx of tourists during the summer months exerts substantial pressure on the lake's resources, further aggravating issues of nutrient enrichment and reduced water quality (Raghava, 2020; Dash et al., 2020). The selection of Nainital Lake (Figure 18.1) for this study is based on its significant surface area and altitude, which provide a unique environment for examining the impacts of climate-induced temperature changes on phytoplankton distribution.

DATA AND METHODS

DATA COLLECTION

Geocoded physicochemical data, including water temperature, pH, DO, EC, TDS, and Secchi depth, were measured in the field from selected lake at 41 different locations using a combination of a GPS device, Lutron's WA-2015, and a black and white Secchi disk. The calibration of the pH probe was done using buffer solutions, while the DO probe was calibrated in the air with atmospheric oxygen. Each sampling point was georeferenced to ensure spatial accuracy. The measurements were taken from a constant depth of 20 cm. Phytoplankton presence and absence data were obtained from

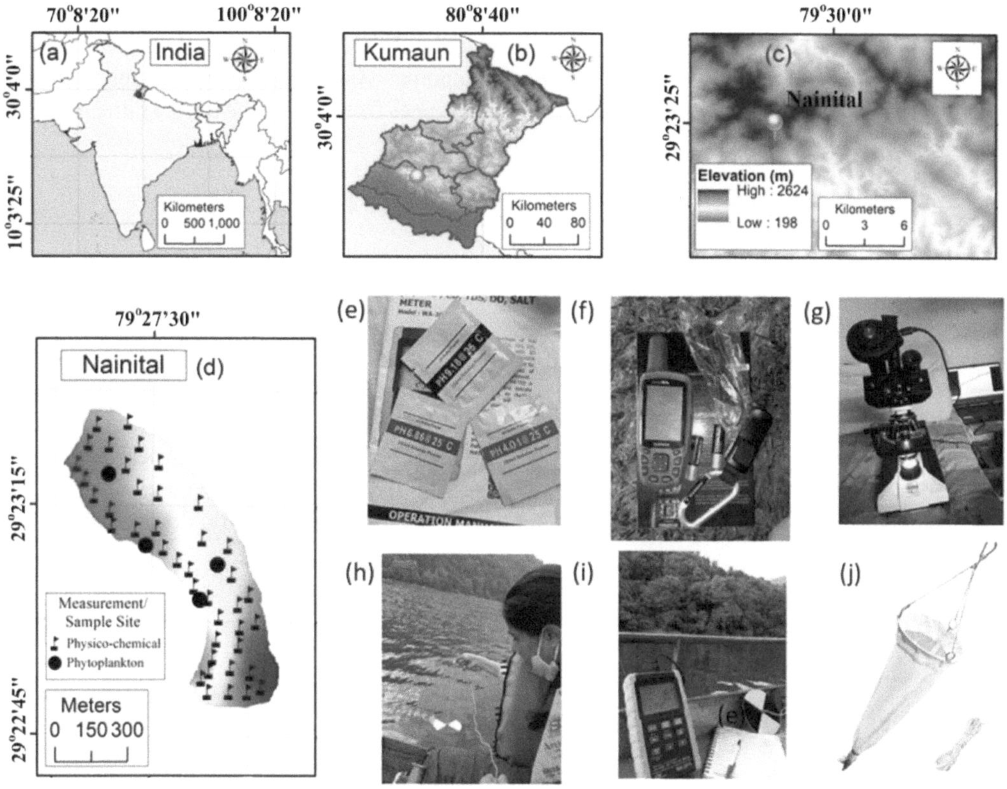

FIGURE 18.1 Location of candidate lake (d), in Kumaun (b) within India (a), locate and high elevation (c), buffer solution sachets (e), GPS (f), optical microscope (g), researcher in the field (h), Lutron's WA-2015 water multiparameter (i), and plankton net (j). (Source: Prepared by researchers.)

26 locations through field sampling using a plankton net, ensuring comprehensive spatial coverage across the lakes. The instrumentation and data collection phase spanned from March 22nd, 2023, to March 27th, 2023. Microscopic identification was performed using light microscopy with the help of ESAW trinocular microscope (used by Sen and Mandal, 2018; Dhige and Singh, 2025) at magnifications ranging from 400× to 1,000× to classify phytoplankton species, with samples prepared within six hours of collection to maintain naturalistic observation without staining (Karlson et al., 2010) following the standards of American Public Health Association (2017), which provides comprehensive guidelines for classifying and enumerating phytoplankton species. Phytoplankton enumeration was carried out using a Sedgwick-Rafter counting chamber, allowing precise quantification of species abundance.

GIS-BASED INTERPOLATION

To convert discrete physicochemical measurements into continuous surface data representations, a GIS-based interpolation technique was employed. Surface interpolation, a key GIS technique, estimates values across a geographic area from discrete data points (Andrews et al., 2002; Blachowski and Ellefmo, 2023). Among the common methods, cubic spline interpolation stands out, using cubic polynomials to connect adjacent data points smoothly, ensuring continuity in both value and slope (Guo et al., 2020; Rabbath and Corriveau, 2019; Gülüm et al., 2019; Haibing et al., 2021). This method, used in studies such as Nagalakshmi et al. (2016) and Al Mousawi et al. (2023), facilitates the creation of 2D continuous surfaces by interpolating geocoded samples from survey measurements of physicochemical parameters, enabling a detailed representation of these variations across the spatial extent of the lake. The formula for cubic spline interpolation involves using a set of cubic polynomials to interpolate between data points while maintaining continuity in value and slope. Given a set of data points (*x*[*i*], *y*[*i*]), where *i* ranges from 0 to *n*, the cubic spline interpolation formula for the interval [*x*[*i*], *x*[*i*+1]] is as follows:

$$Si(x) = ai + bi * (x - xi) + ci * (x - xi)^\wedge 2 + di * (x - xi)^\wedge 3 \tag{18.1}$$

where

Si(*x*) is the cubic spline function for the interval [*x*[*i*], *x*[*i*+1]].
ai, bi, ci, and *di* are coefficients specific to the interval [*x*[*i*], *x*[*i*+1]].
x is the input value within the interval [*x*[*i*], *x*[*i*+1]].

The cubic spline interpolation technique within the GIS environment creates 2D continuous surfaces using geocoded samples from precise survey measurements of physicochemical parameters. This method ensures a smooth, continuous surface, effectively representing the variations in these parameters across the spatial extent of the candidate lakes, thus aiding in the detailed analysis of the study area.

SPECIES DISTRIBUTION MODELING

SDM was conducted to assess the relationship between the spatial distribution of dominant phytoplankton species and the physicochemical characteristics of the lakes. SDM is a powerful predictive tool used to estimate the potential distribution of species by integrating environmental variables with species occurrence data (Srivastava et al., 2019). This method helps in identifying suitable habitats, understanding species–environment relationships, and predicting potential shifts in distribution under various environmental scenarios (Bruneel et al., 2018). In this chapter, the geocoded physicochemical data, including temperature, pH, DO, EC, TDS, and Secchi depth, were used as conditioning variables. The presence and absence data of phytoplankton served as the inventory for the SDM. The modeling was performed in RStudio (RStudio Team, 2023) using a machine regression-based random forest algorithm (Eq. 2) to predict the current distribution patterns of phytoplankton where the prediction is the average of the outputs from all decision trees

$$\hat{y}_i = \frac{1}{k}\sum_{i=1}^{k}\hat{y}_i \tag{18.2}$$

where

$\hat{y}_i$ = predicted value from the *i*th tree.
k = total number of trees.

To evaluate the performance and accuracy of the models, the receiver operating characteristic (ROC) curve and the area under the curve (AUC) metrics were employed. The ROC curve was utilized to assess the ability of the model to distinguish between the presence and absence of phytoplankton, with the AUC providing a quantitative measure of the model's discriminatory power. An AUC value close to 1 indicated excellent model performance, while a value near 0.5 suggested a model with no better accuracy than random guessing. The model with the highest AUC score was selected as the most suitable for predicting phytoplankton distribution.

To assess the potential impacts of climate change on lake ecosystems, future lake surface temperature was simulated using air temperature data from the CMIP5. After Rodhe's thermodynamic equation (Eq. 18.3), the scenario air temperature was translated into lake surface temperatures.

$$T_{w,t} = (1 - e^{-k\Delta t})\, T_{a,t} + e^{-k\Delta t} T_{w,t-\Delta t} \tag{18.3}$$

where

$T_{w,t}$ is the water temperature (LSTs) at a given time t;
Δt is the time interval, that is, 17 and 20 years;
k is a constant;
$T_{a,t}$ is the temperature of the air at time t.

The CMIP5 data were statistically downscaled using local weather station data taken from the Aryabhatta Research Institute of Observational Sciences (ARIES), Devasthal, Nainital, to calibrate the scenario data with observed data, ensuring higher accuracy with quantile mapping (Eq. 5).

$$X_{\text{corrected}} = F_{\text{obs}}^{-1}\left(F_{\text{sim}}\left(X_{\text{sim}}\right)\right) \tag{18.4}$$

where

F_{sim} = simulated value from the climate model.
X_{sim} = cumulative distribution function (CDF) of the simulated data.
F_{obs}^{-1} = inverse CDF (or quantile function) of the observed data.
$X_{\text{corrected}}$ = bias-corrected or adjusted value.

For the future scenarios, only the lake temperature was replaced with simulated data corresponding to CMIP5 projections for the years 2040, 2060, 2080, and 2100, while other conditioning variables, such as pH, DO, EC, TDS, and Secchi depth, were assumed to remain constant. The SDM was rerun for these scenarios.

ACCURACIES OF THE MODELS

The evaluation of model accuracy is a critical aspect of the present study, ensuring the reliability of predictions related to the distribution of *C. dalmatica* Kützing in Nainital Lake. A 70–30 data split of 26 points was adopted for training and testing the model, following methodologies outlined by Khan et al. (2023), Khan et al. (2024), Carrington et al. (2022). This approach aims to balance the model's ability to learn patterns effectively while maintaining an unbiased assessment of its predictive capabilities.

To evaluate the performance of the SDM, an ROC curve was employed. The ROC curve is a graphical representation that illustrates the performance of a classification model across different classification thresholds by plotting two essential metrics: true positive rate (TPR) and false positive rate (FPR). This curve plots two key factors, TPR and FPR. TPR, also known as recall, is defined by Eq. (18.5):

$$\text{TPR} = \frac{\text{TP}}{\text{TP} + \text{FN}} \tag{18.5}$$

where TP represents true positives, and FN represents false negatives.

Similarly, FPR is defined by Eq. (18.6):

$$\text{FPR} = \frac{\text{FP}}{\text{FP} + \text{TN}} \tag{18.6}$$

where FP represents false positives, and TN represents true negatives.

RESULTS

The physicochemical characteristics of Nainital Lake (Table 18.1; Figure 18.2a–f) reveal substantial spatial and temporal variations that influence phytoplankton distribution. The temperature of the lake ranged from 20.2°C to 30.1°C, with a mean of 25.53°C, showing notable spatial heterogeneity (Figure 18.2a). Higher temperatures were observed in the northern parts of the lake, which could be

TABLE 18.1
Physico-Chemical Parameters of Nainital Lake

	Temperature (°C)	pH	DO (mg/L)	EC (μS)	TDS (ppm)	Secchi Depth (m)
Min	20.2	8.01	4.2	555	355	1.13
Max	30.1	8.78	6.6	595	425	1.7
Mean	25.53	8.4	5.37	585	397.44	1.37

Note: 41 samples studied (*Source:* Estimated by researchers).

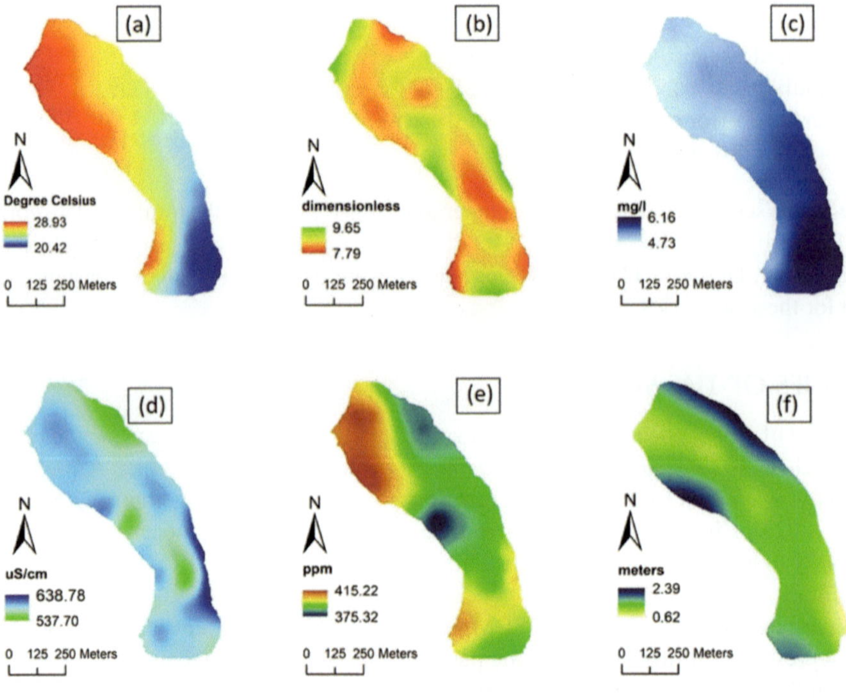

FIGURE 18.2 Physicochemical parameters of Nainital, temperature (a), pH (b), DO (c), EC (d), TDS (e), and Secchi depth (f). (Source: Prepared by researchers.)

attributed to shallower depths and greater exposure to solar radiation, potentially enhancing phytoplankton growth in these areas.

The pH levels varied between 8.01 and 8.78, with an average of 8.4, indicating slightly alkaline conditions conducive to phytoplankton proliferation (Figure 18.2b). Elevated pH near the shorelines may result from increased biological activities and nutrient inflow, suggesting localized eutrophication effects. DO ranged from 4.2 mg/L to 6.6 mg/L, with a mean of 5.37 mg/L, which falls within acceptable limits for supporting aquatic life. However, lower DO levels in the southern regions of the lake (Figure 18.2c) may indicate localized hypoxia, potentially limiting phytoplankton diversity in these areas.

EC values spanned from 555 μS/cm to 595 μS/cm, averaging 585 μS/cm, reflecting moderate ionic concentrations (Figure 18.2d). Elevated EC in the central parts of the lake suggests potential input of dissolved ions, which could influence osmoregulation processes in phytoplankton. Similarly, TDS varied between 355 and 425 ppm, with a mean of 397.44 ppm (Figure 18.2e), displaying a spatial distribution pattern similar to that of EC, further corroborating the presence of moderate ionic content.

TABLE 18.2
Diversity and Composition of Phytoplankton Species Identified in Nainital Lake

S.N.	Species	Genus	Phylum	Kingdom	Abundance (cells/mL)	Food Synthesis
1	*C. dalmatica* Kützing	*Cladophora*	Chlorophyta	Plantae	**11**	Photosynthetic
2	*C. oleofaciens* Trainor & Bold	*Chlorococcum*	Chlorophyta	Plantae	2	Photosynthetic
3	*C. sorokiniana* Shihira & R.W. Krauss	*Chlorella*	Chlorophyta	Plantae	1	Photosynthetic
4	*G. graciledictum* Reichardt	*Gomphonem*	Ochrophyta	Chromista	9	Photosynthetic
5	*Nitzchia* sp.	*Nitzchia*	Ochrophyta	Chromista	1	Photosynthetic
6	*Navicula* sp.	*Navicula*	Ochrophyta	Chromista	3	Photosynthetic
7	*Roholtiella* sp.	*Roholtiella*	Cyanobacteria	Bacteria	1	Photosynthetic

Source: Identified and counted by researchers.

Secchi depth, an indicator of water clarity, ranged from 1.13 to 1.7 m, with an average of 1.37 m (Figure 18.2f). Reduced Secchi depth in the northern parts of the lake suggests higher turbidity, likely due to suspended particulate matter or phytoplankton blooms, which could limit light penetration and photosynthetic efficiency for phytoplankton below the surface.

The phytoplankton community of Nainital Lake consists of seven species from three phyla (Table 18.2; Figure 18.3a–h), that is, *Chlorophyta*, *Ochrophyta*, and *Cyanobacteria. Chlorophyta* was the most diverse, represented by *C. dalmatica* Kützing, *Chlorococcum oleofaciens*, and *Chlorella sorokiniana. C. dalmatica* Kützing emerged as the dominant species with a concentration of 11 cells/mL, indicating that the slightly alkaline pH and moderate ionic content of the lake are conducive to its growth. *Gomphonema graciledictum* was the second most abundant species with 9 cells/mL, suggesting that moderate light penetration and DO levels support its presence.

In contrast, species like *C. sorokiniana*, *Nitzchia* sp., and *Roholtiella* sp. were present in much lower concentrations (1 cell/mL each), likely due to localized variations in physicochemical conditions. The dominance of *Cladophora* suggests that it plays a significant role in the lake's primary production and trophic dynamics, highlighting the need for effective management strategies to prevent potential algal blooms under changing climate conditions.

SIMULATION OF LAKE SURFACE TEMPERATURE

To evaluate the influence of climate change on phytoplankton habitat dynamics, lake surface temperatures were simulated for the years 2040, 2060, 2080, and 2100 using statistically downscaled CMIP5 air temperature data and Rodhe's thermodynamic equation. Table 18.3 and Figure 18.4 present the projected surface temperatures for Bhimtal, Nainital, and Sadiyatal lakes. Notably, Nainital Lake is projected to experience a consistent warming trend, with temperatures rising from 24.45°C in 2040 to 29.12°C by 2100. This increase of nearly 5°C underscores the potential thermal stress that could significantly alter the ecological balance within the lake. By integrating these simulated temperatures into SDM, the study captures how projected warming scenarios may reshape habitat suitability for phytoplankton in Nainital Lake. The model outcomes reflect spatial and temporal variations in phytoplankton occurrence probabilities, highlighting the need for continuous monitoring and adaptive lake management strategies under changing climatic conditions.

The results indicate a progressive decline in the probability of occurrence for *C. dalmatica* Kützing in Nainital Lake from 2023 to 2100, highlighting the potential impacts of rising temperatures on its distribution. As shown in Figure 18.5, the probability of occurrence in 2023 ranged from 0.055 to 0.980, with higher suitability observed in the northern and central regions of the lake. This distribution suggests that the existing physicochemical conditions in these areas are

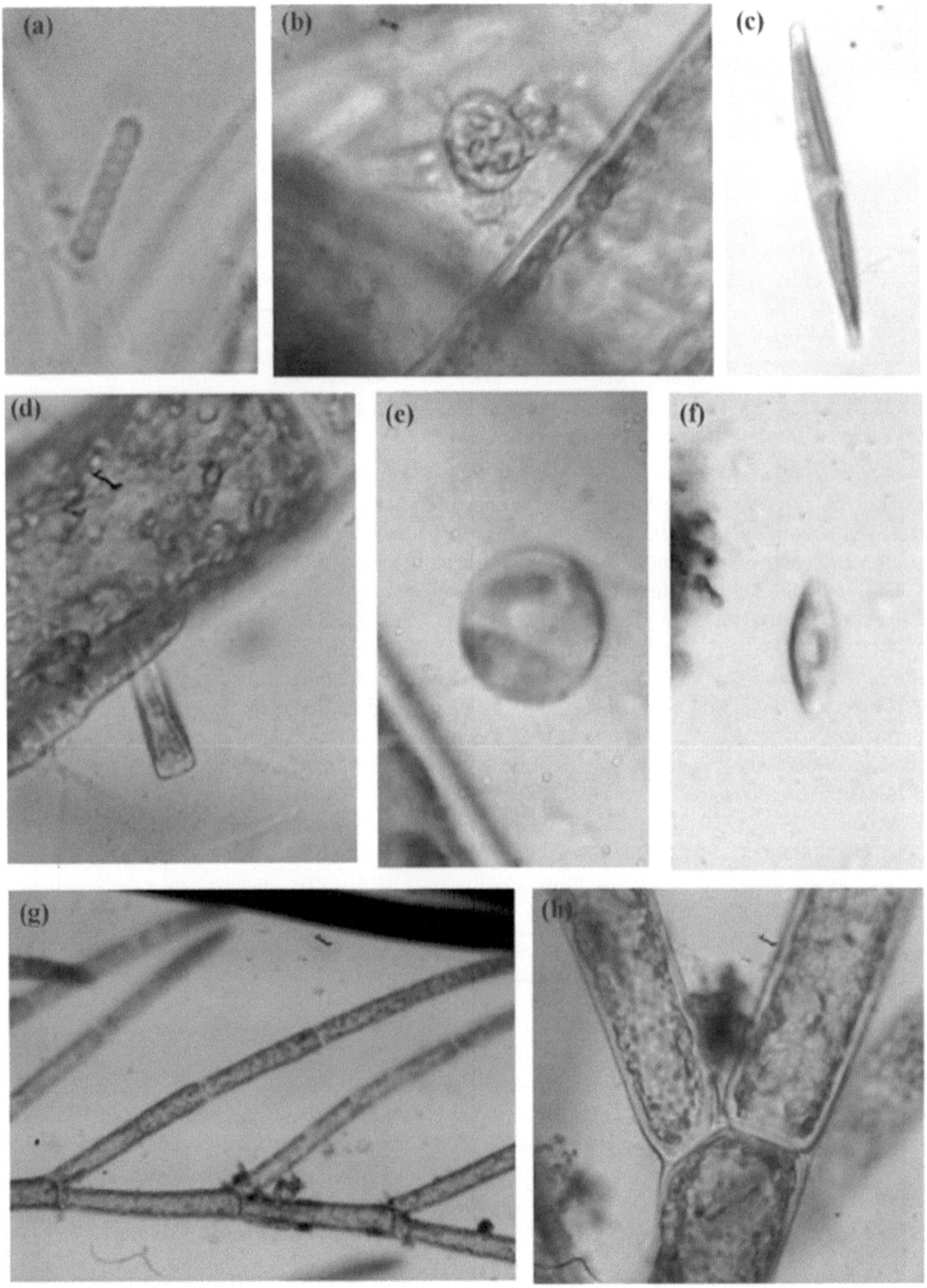

FIGURE 18.3 Phytoplankton of Nainital *Roholtiella* sp. (a), *Chlorococcum oleofaciens* Trainor & Bold (b), *Nitzchia* sp. (c), *Gomphonema graciledictum* Reichardt (d), *Chlorella sorokiniana* Shihira & R.W. Krauss (e), *Navicula* sp. (f), and *Cladophora dalmatica* Kützing (g, h). (Source: Captured by researchers.)

TABLE 18.3
Simulated Lake Surface Temperatures (in °C)

Year	Bhimtal	Nainital	Sadiyatal
2040	29.189	24.447	25.604
2060	28.943	24.733	25.728
2080	29.674	26.861	26.350
2100	31.235	29.117	27.261

Source: Simulated by researchers.

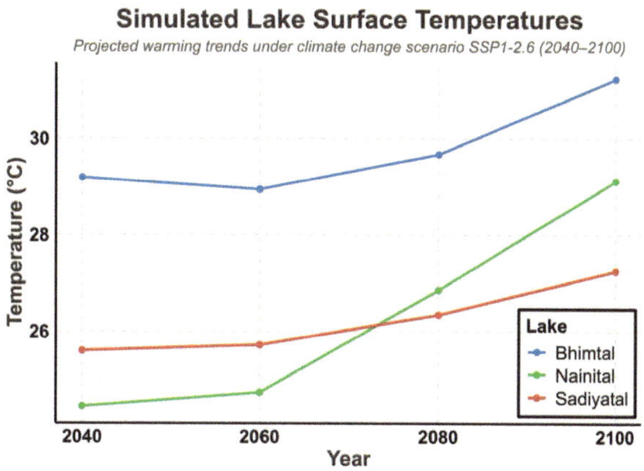

FIGURE 18.4 Simulated lake surface temperatures (in °C).

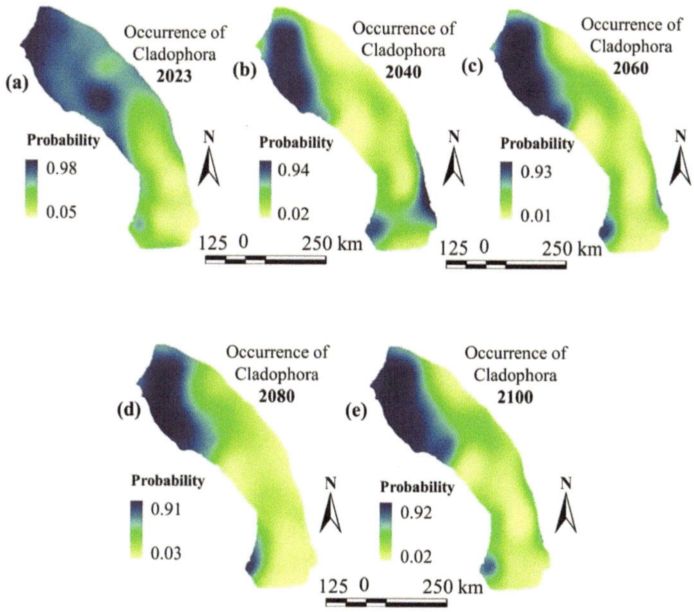

FIGURE 18.5 Probability of occurrence for *Cladophora* in Nainital in 2023 (a), 2040 (b), 2060 (c), 2080 (d), and 2100 (e). (Source: Prepared by researchers.)

TABLE 18.4
Mean Probability of Occurrence for Cladophora

Lentic Ecosystem	2023	2040	2060	2080	2100	Change (2020–2100)
Nainital	0.431	0.421	0.460	0.471	0.309	–0.122

Source: Estimated by researchers.

favorable for *Cladophora*. However, by 2040, the range narrowed to 0.002–0.940, indicating an initial decline in high-suitability zones, particularly in the southern parts of the lake. This declining trend continued into 2060, with the probability range further contracting to 0.011–0.933, and high-suitability areas becoming increasingly confined to the northern parts of the lake. By 2080, the range shifted to 0.038–0.915, and the spatial extent of areas with high probability (>0.8) continued to diminish. The most significant reduction was observed in 2100, where the probability range declined to 0.020–0.918, indicating a substantial loss of optimal habitats for *C. dalmatica* Kützing due to climate-induced warming.

The changes in mean probability values presented in Table 18.4 corroborate this trend, with a decrease from 0.431 in 2023 to 0.421 in 2040, suggesting the onset of thermal stress. Interestingly, a slight increase to 0.460 in 2060 and further to 0.471 in 2080 might reflect short-term adaptation or local environmental variability that temporarily favored *Cladophora* growth. However, this trend was not sustained, as the mean probability significantly dropped to 0.309 by 2100. The overall change of –0.122 in mean probability from 2023 to 2100 underscores the potential vulnerability of *Cladophora* to sustained temperature increases.

The observed spatial contraction in high-suitability areas, particularly in the southern parts of the lake, suggests that elevated temperatures could exceed the optimal range for *C. dalmatica* Kützing, leading to habitat loss and a decline in its population. This reduction in the probability of occurrence highlights the risk to the lake's primary production and trophic stability, emphasizing the need for adaptive management strategies to mitigate the impacts of rising temperatures on Nainital Lake's ecosystem.

Figure 18.6a–e illustrates the ROC curves and the corresponding AUC values for the probability of occurrence of *C. dalmatica* Kützing in Nainital Lake for the years 2023, 2040, 2060, 2080, and 2100. The AUC values indicate the predictive accuracy of the SDM based on random forest, with higher AUC values representing better model performance. In 2023, the model achieved an AUC of 0.91, suggesting excellent predictive accuracy, while in 2040, the AUC decreased slightly to 0.87, still indicating good model performance. A further decline in AUC to 0.83 in 2060 reflects a gradual reduction in accuracy, potentially due to increased environmental variability. Interestingly, the AUC value for 2080 showed a minor improvement to 0.85, which might be attributed to short-term local adaptations or temporarily favorable environmental conditions for *Cladophora*. However, by 2100, the AUC dropped significantly to 0.74, indicating reduced model reliability and a higher level of uncertainty in predicting the occurrence of *Cladophora* under long-term climate scenarios. The ROC curves further support these findings, as they demonstrate a shift from high sensitivity and specificity in 2023 and 2040 to a less distinct separation between true positives and false positives by 2100. The progressively declining AUC values suggest that long-term climate variability and cumulative environmental stressors may complicate the accurate prediction of *Cladophora* distribution, underscoring the need for incorporating additional environmental variables and higher-resolution data in future models to enhance predictive robustness.

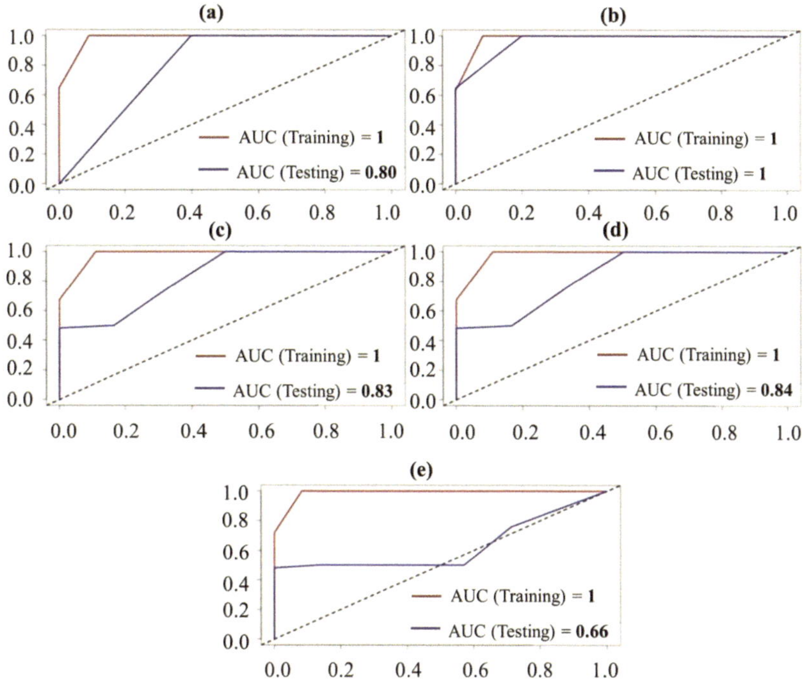

FIGURE 18.6 Accuracies of probability of occurrence for *Cladophora* in Nainital in 2023 (a), 2040 (b), 2060 (c), 2080 (d), and 2100 (e). (Source: Prepared by researchers.)

DISCUSSION

The findings of this study underscore the significant influence of climate-induced temperature changes on the distribution patterns of *C. dalmatica* Kützing in Nainital Lake, highlighting the potential risks to the lake's trophic dynamics and primary production. The observed spatial and temporal variations in the probability of occurrence for *Cladophora* from 2023 to 2100 reflect a progressive decline in suitable habitats, particularly in response to rising water temperatures projected under CMIP5 climate scenarios. The reduction in mean probability from 0.431 in 2023 to 0.309 in 2100, as indicated in Table 18.3, suggests a substantial loss of optimal habitats, aligning with previous studies that emphasize the vulnerability of chlorophytic algae to thermal stress (Paerl and Huisman, 2008; Reynolds, 2006).

The spatial distribution maps in Figure 8.5 reveal a northward contraction of high-suitability zones, with the southern and central parts of the lake experiencing the most pronounced declines. This pattern can be attributed to elevated temperatures exceeding the optimal range for *C. dalmatica* Kützing, leading to reduced photosynthetic efficiency and competitive exclusion by other phytoplankton species better adapted to warmer conditions. The highest probability values observed in 2023 (0.980) were progressively reduced to 0.918 by 2100, indicating a declining trend in habitat suitability that could potentially alter the lake's primary production and nutrient cycling processes. These findings align with the hypothesis that temperature is a critical determinant for phytoplankton distribution in lentic systems (Paerl et al., 2011). Interestingly, the temporary increase in mean probability to 0.460 in 2060 and 0.471 in 2080 might suggest short-term local adaptations or environmental variability that favored *Cladophora* growth. Such fluctuations could result from transient changes in water column stratification or nutrient availability, which may have temporarily offset the negative impacts of warming. However, the subsequent drop to 0.309 by 2100 indicates that these short-term adaptations are unlikely to sustain under prolonged thermal stress, underscoring the need for continuous monitoring and adaptive management strategies.

The observed spatial patterns in physicochemical parameters, including temperature, pH, DO, EC, TDS, and Secchi depth, further elucidate the mechanisms underlying the distribution of *Cladophora*. The slightly alkaline pH (8.01–8.78) and moderate ionic content, indicated by EC (555–595 μS/cm) and TDS (355–425 ppm), appear conducive to the proliferation of *Cladophora*, as suggested by its dominance among the identified phytoplankton species. However, the decreasing Secchi depth, ranging from 1.13 to 1.7 m, implies increased turbidity possibly due to suspended particulates or phytoplankton blooms, which could further limit light availability and exacerbate the impacts of rising temperatures on photosynthetic efficiency.

The findings highlight the potential for significant ecological shifts in Nainital Lake's trophic structure due to climate change, particularly the risk of reduced primary production driven by the decline of *C. dalmatica* Kützing. Given that *Cladophora* plays a key role in the lake's primary production and nutrient dynamics, its decline could cascade through the food web, affecting higher trophic levels, including zooplankton and fish communities. The observed patterns emphasize the need for comprehensive management strategies focusing on maintaining water quality and mitigating thermal stress through measures such as riparian vegetation restoration, nutrient input control, and enhanced monitoring of physicochemical parameters.

Furthermore, the reliance on a single-pixel approach due to resource limitations, as outlined in the limitations section, suggests that the actual spatial heterogeneity of *Cladophora* distribution might be underrepresented. Future research should consider employing dynamic downscaling and higher spatial resolution data to capture fine-scale variations more accurately. Additionally, integrating other environmental factors such as nutrient dynamics, light penetration, and hydrodynamic processes could provide a more comprehensive understanding of phytoplankton responses to climate change.

In conclusion, the study underscores the critical role of temperature in shaping phytoplankton distribution in Nainital Lake and calls for targeted conservation efforts to mitigate the impacts of climate change on its aquatic ecosystems. The observed trends not only highlight the potential risks to primary production and trophic dynamics but also serve as a basis for formulating adaptive management strategies aimed at enhancing the resilience of lentic systems under projected climate scenarios.

CONCLUSION

This study highlights the significant impact of climate-induced temperature changes on the distribution of *C. dalmatica* Kützing in Nainital Lake, revealing a clear trend of declining habitat suitability from 2023 to 2100. The observed reduction in mean probability of occurrence, from 0.431 in 2023 to 0.309 in 2100, underscores the vulnerability of this dominant phytoplankton species to sustained thermal stress. The spatial distribution patterns indicate a contraction of high-suitability areas, particularly in the southern and central parts of the lake, suggesting that rising temperatures could surpass the optimal range for *Cladophora*, leading to significant habitat loss.

The findings emphasize the critical role of temperature as a determinant of phytoplankton distribution, with potential implications for primary production and trophic stability in Nainital Lake. The slight increases in mean probability observed in 2060 and 2080 likely reflect short-term local adaptations or environmental variability; however, the subsequent decline to 0.309 by 2100 indicates that such adaptations are insufficient to counteract prolonged warming.

The physicochemical conditions of the lake, including moderate ionic content, slightly alkaline pH, and decreasing Secchi depth, appear to have initially favored *Cladophora* proliferation. However, the projected warming trends threaten to disrupt these conditions, potentially triggering shifts in phytoplankton community composition and cascading effects on higher trophic levels.

Overall, this study underscores the urgent need for adaptive management strategies focused on mitigating the impacts of rising temperatures on Nainital Lake's primary production. Conservation efforts should prioritize maintaining water quality, controlling nutrient inputs, and enhancing monitoring of physicochemical parameters to safeguard the resilience of this high-altitude lentic system against climate-induced stress. The outcomes also call for further research incorporating higher-resolution data and additional environmental variables to refine predictions of phytoplankton responses to future climate scenarios.

LIMITATIONS

This study has certain limitations that may have influenced the results and interpretations. Due to limited resources, dynamic downscaling could not be performed for the CMIP 5 data; consequently, the analysis was restricted to a single air temperature data point, with one pixel representing the entire lake's surface. This may have overlooked spatial heterogeneity within the lake.

The use of unanchored floating platforms, such as boats, introduced potential measurement errors. Wind-induced swaying of the boats might have caused slight variations in sensor readings for parameters like temperature, pH, DO, EC, TDS, and Secchi depth.

The data collection process, which extended over several hours, could have also introduced temporal variability, particularly in water temperature. Changes in solar radiation and sun angle during this period may have influenced the visibility of the Secchi disk, thereby affecting the accuracy of Secchi depth measurements. These factors should be considered when interpreting the results of this study.

ACKNOWLEDGMENTS

The authors express their sincere gratitude to the Atmospheric Science Division, ARIES, Devasthal, Nainital, for providing valuable data and support essential for this study. The assistance and resources offered by the team significantly contributed to the successful completion of this research.

DECLARATION OF FAIR USE OF AI ASSISTANCE

The AI assistance was employed ethically using ChatGPT 3.5 and following the book's guidelines to enhance readability and coherence. All intellectual contributions, data analysis, and interpretations remain the sole work of the author. No part of the research design, methodology, results, or conclusions was generated or influenced by any AI tools.

FUNDING

This research was supported by the Maulana Azad National Fellowship (MANF) provided by the Government of India (Candidate ID: MANF-2018–19-UTT-94191). The funding was provided to Zainab Khan in the form of a doctoral fellowship.

REFERENCES

Al Mousawi, E., Jahad, U. A., Chabuk, A., Al-Ansari, N., Majdi, A., & Laue, J. (2023). Applying different water quality indices and GIS to assess the water quality, case study: Euphrates river in Qadisiyah Province. *Polish Journal of Environmental Studies*, *32*(5), 4201–4217.

Andrews, B., Gares, P. A., & Colby, J. D. (2002). Techniques for GIS modeling of coastal dunes. *Geomorphology*, *48*(1—3), 289–308.

Ben Ezra, T., Blachinsky, A., Gozali, S., Tsemel, A., Fadida, Y., Tchernov, D., Lehahn, Y., Tsagaraki, T. M., Berman-Frank, I., & Krom, M. (2024). Interannual changes in nutrient and phytoplankton dynamics in the Eastern Mediterranean Sea (EMS) predict the consequences of climate change; results from the Sdot-Yam Time-series station 2018—2022. *Deep-Sea Research Part I: Oceanographic Research Papers*, *213*. https://doi.org/10.1016/j.dsr.2024.104403

Blachowski, J., & Ellefmo, S. L. (2023). Mining ground deformation estimation based on pre-processed InSAR open data—A Norwegian case study. *Minerals*, *13*(3), 328.

Bruneel, S., Gobeyn, S., Verhelst, P., Reubens, J., Moens, T., & Goethals, P. (2018). Implications of movement for species distribution models-rethinking environmental data tools. *Science of the Total Environment*, *628*, 893–905.

Carrington, A. M., Manuel, D. G., Fieguth, P. W., Ramsay, T., Osmani, V., Wernly, B.,... & Holzinger, A. (2022). Deep ROC analysis and AUC as balanced average accuracy, for improved classifier selection, audit and explanation. *IEEE Transactions on Pattern Analysis and Machine Intelligence*, *45*(1), 329–341.

Catalan, J., Camarero, L., Felip, M., Pla, S., Ventura, M., Buchaca, T.,... & Quijano, D. D. D. (2006). High mountain lakes: Extreme habitats and witnesses of environmental changes. *Limnetica*, *25*(1—2), 551–584.

Choudhary, P., Routh, J., & Chakrapani, G. J. (2013). A 100-year record of changes in organic matter characteristics and productivity in Lake Bhimtal in the Kumaon Himalaya, NW India. *Journal of paleolimnology*, *49*(2), 129–143.

Cloern, J. E. (1999). The relative importance of light and nutrient limitation of phytoplankton growth: A simple index of coastal ecosystem sensitivity to nutrient enrichment. *Aquatic Ecology*, *33*, 3–15.

Dash, P., Tandel, R., Baruah, D., & Sarma, D. (2020). Mahseer sanctuaries of Meghalaya: A conservation and recreational perspective. *Aquaculture Asia Magazine*, *24*(1), 3–7.

Dhige, B., & Singh, M. R. (2025). Harnessing the past: Vibration analysis of organic additives in ancient plasters for sustainable building solutions. *Vibrational Spectroscopy*, *136*, 103751.

Dory, F., Nava, V., Spreafico, M., Orlandi, V., Soler, V., & Leoni, B. (2024). Interaction between temperature and nutrients: How does the phytoplankton community cope with climate change? *Science of the Total Environment*, *906*, 167566.

Fernández, R., Alcocer, J., Oseguera, L. A., Zuñiga-Ramos, C. A., & Vilaclara, G. (2024). Phytoplankton communities' seasonal fluctuation in two neighboring tropical high-mountain lakes. *Plants*, *13*(21). https://doi.org/10.3390/plants13213021

Ferreira, A., Mendes, C. R. B., Costa, R. R., Brotas, V., Tavano, V. M., Guerreiro, C. V., Secchi, E. R., & Brito, A. C. (2024). Climate change is associated with higher phytoplankton biomass and longer blooms in the West Antarctic Peninsula. *Nature Communications*, *15*(1). https://doi.org/10.1038/s41467-024-50381-2

Gogoi, P., Sinha, A., Das Sarkar, S., Chanu, T. N., Yadav, A. K., Koushlesh, S. K.,... & Das, B. K. (2019). Seasonal influence of physicochemical parameters on phytoplankton diversity and assemblage pattern in Kailash Khal, a tropical wetland, Sundarbans, India. *Applied Water Science*, *9*, 1–13.

Gülüm, M., Yesilyurt, M. K., & Bilgin, A. (2019). The performance assessment of cubic spline interpolation and response surface methodology in the mathematical modeling to optimize biodiesel production from waste cooking oil. *Fuel*, *255*, 115778.

Guo, B., Zhang, J., Meng, X., Xu, T., & Song, Y. (2020). Long-term spatio-temporal precipitation variations in China with precipitation surface interpolated by ANUSPLIN. *Scientific Reports*, *10*(1), 81.

Haibing, H., Zheng, X., Yin, J., & Wang, Y. (2021). Research on O-ring dimension measurement algorithm based on cubic spline interpolation. *Applied Sciences*, *11*(8), 3716.

Hubert, Z., Louchart, A. P., Robache, K., Epinoux, A., Gallot, C., Cornille, V., Crouvoisier, M., Monchy, S., & Artigas, L. F. (2025). Decadal changes in phytoplankton functional composition in the Eastern English Channel: Possible upcoming major effects of climate change. *Ocean Science*, *21*(2), 679–700. https://doi.org/10.5194/os-21-679-2025

Hutchins, D. A., & Tagliabue, A. (2024). Feedbacks between phytoplankton and nutrient cycles in a warming ocean. *Nature Geoscience, 17*(6), 495–502. Nature Research. https://doi.org/10.1038/s41561-024-01454-w

Jakobsen, H. H., Blanda, E., Staehr, P. A., Højgård, J. K., Rayner, T. A., Pedersen, M. F.,... & Hansen, B. W. (2015). Development of phytoplankton communities: Implications of nutrient injections on phytoplankton composition, pH and ecosystem production. *Journal of Experimental Marine Biology and Ecology*, *473*, 81–89.

Karlson, B., Godhe, A., Cusack, C., & Bresnan, E. (2010). Introduction to methods for quantitative phytoplankton analysis. In B. Karlson, C. Cusack and E. Bresnan (Eds.). *Microscopic and Molecular Methods for Quantitative Phytoplankton Analysis* (pp. 5–20). UNESCO.

Khan, Z., & Ahmad, A. (2021). Retrieved from https://www.researchgate.net/publication/349692352_Modelling_Physicochemical_Properties_of_Bhimtal_Lake_using_Geospatial_Techniques

Khan, Z., Ali, S. A., Parvin, F., Mohsin, M., Shamim, S. K., & Ahmad, A. (2023). Predicting the effects of climate change on prospective Banj oak (Quercus leucotrichophora) dispersal in Kumaun region of Uttarakhand using machine learning algorithms. *Modeling Earth Systems and Environment*, *9*(1), 145–156.

Khan, Z., Bano, N., Shamim, S. K., & Ahamd, A. (2020). *Assessing the Possible Causes of Water Quantity Depletion and Assessment of Water Quality of Naini Lake, Nainital.* Old Delhi, Delhi: Keystone Publishing House.

Khan, Z., Mohsin, M., Ali, S. A., Vashishtha, D., Husain, M., Parveen, A.,... & Ahmad, A. (2024). Comparing the performance of machine learning algorithms for groundwater mapping in Delhi. *Journal of the Indian Society of Remote Sensing*, *52*(1), 17–39.

Litchman, E., de Tezanos Pinto, P., Edwards, K. F., Klausmeier, C. A., Kremer, C. T., & Thomas, M. K. (2015). Global biogeochemical impacts of phytoplankton: A trait-based perspective. *Journal of Ecology*, *103*(6), 1384–1396.

Machate, O., Schmeller, D. S., Schulze, T., & Brack, W. (2023). Review: Mountain lakes as freshwater resources at risk from chemical pollution. In *Environmental Sciences Europe* (Vol. 35, Issue 1). Springer. https://doi.org/10.1186/s12302-022-00710-3

Nagalakshmi, R., Prasanna, K., & Chandar, S. P. (2016). Water quality analysis using gis interpolation method in serthalaikadu Lagoon, east coast of India. *Rasayan Journal of Chemistry*, *9*(4), 634–640.

Paerl, H. W., Hall, N. S., & Calandrino, E. S. (2011). Controlling harmful cyanobacterial blooms in a world experiencing anthropogenic and climatic-induced change. *Science of the Total Environment*, *409*(10), 1739–1745.

Paerl, H. W., & Huisman, J. (2008). Blooms like it hot. *Science*, *320*(5872), 57–58.

Rabbath, C. A., & Corriveau, D. (2019). A comparison of piecewise cubic Hermite interpolating polynomials, cubic splines and piecewise linear functions for the approximation of projectile aerodynamics. *Defence Technology*, *15*(5), 741–757.

Raghava, P. (2020). Impact of urbanisation on water resources & opportunity of RWH structure in Nainital. Major Project Dissertation. Accessed from https://www.cedarhimalaya.org/pdf/Palash-converted.pdf.

Reynolds, C. S. (2006). *The ecology of phytoplankton*. Cambridge University Press.

RStudio Team. (2023). *RStudio: Integrated Development Environment for R*. Posit Software, PBC. https://posit.co

Sen, K., & Mandal, R. (2018). Fresh-water fish diseases in west Bengal, India. *International Journal of Fisheries and Aquatic Studies*, *6*(5), 356–362.

Sharma, M. (2014). Water quality assessment of the central Himalayan Lake, Nainital. *Advances in Environmental Chemistry*, *2014*(1), 473074.

Srivastava, V., Lafond, V., & Griess, V. C. (2019). Species distribution models (SDM): Applications, benefits and challenges in invasive species management. *CABI Reviews*, *2019*, 1–13.

Thomas, A. N. (1952). On the origins of the Kumaon lakes in the Himalayas. *Geological Magazine*, *89*(6), 385–400.

Wetzel, R. G. (2002). *Lake and River Ecosystems*. Limnology; San Diego, CA: Academic Press.

Williamson, C. E., Saros, J. E., Vincent, W. F., & Smol, J. P. (2009). Lakes and reservoirs as sentinels, integrators, and regulators of climate change. *Limnology and Oceanography*, *54*(6part2), 2273–2282.

19 Mapping Climate-Driven Persistent Flooding Using Google Earth Engine and Its Impacts in the Indo-Gangetic Basin

Mujahid Husain, Rohit Kumar Azad, Deepika Vashishtha, Amit Kumar, and Zainab Khan

INTRODUCTION

Climate change has significantly altered global hydrological patterns, leading to an increase in the frequency, intensity, and duration of flood events. Rising temperatures intensify the hydrological cycle by increasing evaporation and atmospheric moisture, resulting in heavier and more erratic rainfall. Additionally, glacial melt and sea-level rise contribute to riverine and coastal flooding, particularly in low-lying and densely populated regions. These climate-driven changes exacerbate existing vulnerabilities, causing widespread damage to agriculture, infrastructure, and human settlements. Understanding the connection between climate change and flooding is critical for developing effective adaptation strategies, enhancing disaster preparedness, and building climate-resilient communities.

In Southeast Asia, hydro-meteorological disasters are persistent phenomena that have widespread impacts on both natural ecosystems and built environments (Dutta and Herath, 2004; Ahamed and Bolten, 2017). India, with its dense riverine network, is particularly prone to flooding, making contiguous human settlements—such as villages, towns, and cities—highly vulnerable (Mohanty et al., 2020). The Indian subcontinent receives the majority of its annual precipitation during the monsoon months of July to September (Dhar and Nandargi, 2003), with more than 75% of rainfall occurring in this period. This concentrated rainfall frequently results in flooding across various parts of the country (Mishra and Shah, 2018; Guhathakurta et al., 2011).

However, identifying flood-prone areas using traditional survey-based methods is challenging due to the dynamic nature of surface water bodies (Kiage et al., 2005). Remote sensing has emerged as an efficient alternative, capable of accurately capturing changes in surface water dynamics and effectively delineating flood-vulnerable zones (Rahman and Thakur, 2018). Among all land-use types, croplands are the most vulnerable during flood events, as they are typically located on fertile floodplains (Brémond et al., 2013). In India, nearly all floodplain regions are used for agriculture and are adversely affected each time a flood occurs (Adnan et al., 2020; Shrestha et al., 2021). This vulnerability is often compounded by unsustainable practices and inadequate preparedness measures (Merz et al., 2010). The growing scale of crop losses has made damage assessment a critical component of large-scale flood risk management (Hendrawan and Komori, 2021). Accurate estimation of flood-affected cropland areas and the resulting economic losses is essential for understanding recurring damages (Tapia-Silva et al., 2011; Kourgialas and Karatzas, 2016). Such estimations inform financial appraisals, crop yield predictions, pricing, risk analysis, and damage compensation,

DOI: 10.1201/9781003620747-22

ultimately contributing to the country's sectoral development (Del Ninno et al., 2003; Klaus et al., 2016). Moreover, these assessments are crucial for planning strategies to combat future flood hazards (Rahman et al., 2021).

Monitoring natural disasters plays a vital role in reducing associated risks and impacts. It also supports other phases of disaster management, including preparedness and response (Van Westen, 2013; Ghaffarian et al., 2018). The use of geospatial technologies in disaster monitoring, especially for hydro-meteorological events like floods, is increasing due to its cost-effectiveness and ability to provide high-temporal-resolution data over large areas (Sarti et al., 2001; Kabenge et al., 2017). During the response phase, geospatial tools help identify flooded areas in near real time, aiding in resource mobilization and prioritizing relief distribution (Patel and Srivastava, 2013). In the preparedness and recovery phases, these tools support post-event activities such as effective land-use planning and the formulation of disaster mitigation policies (Opolot, 2013; Rahman and Di, 2017). Due to the rapidly changing extent of surface water during floods, traditional methods struggle to provide timely mapping. In contrast, the revisit capabilities of satellites offer a reliable alternative for real-time flood monitoring (Bagheri & Liu, 2024).

In long-term flood risk management, mapping inundated areas and assessing event-related uncertainties are of critical importance, especially for planning emergency response activities (Hostache et al., 2009; Pierdicca et al., 2013). Synthetic aperture radar (SAR) data offer unique advantages, including all-weather, day-and-night imaging capabilities, and the ability to distinguish flooded from non-flooded areas by analyzing different scattering behaviors (Stephens et al., 2012; Schumann et al., 2014). These properties make SAR a valuable tool for flood mapping, with applications in hydraulic model calibration and validation (Di Baldassarre et al., 2009).

Optical datasets are often less effective during the monsoon season due to cloud cover (Qadir and Mondal, 2020). SAR datasets, by contrast, can be acquired under any atmospheric conditions, making them especially suitable for flood monitoring (Lu et al., 2014; Cao et al., 2019). Over time, numerous methodologies have been developed for flood mapping using SAR data, including automatic classification algorithms (Hess et al., 1995; Bonn and Dixon, 2005), change detection methods (Xiong et al., 2012), visual interpretation (Dellepiane et al., 2010), image fusion (Bovolo and Bruzzone, 2005), post-event comparisons (Liu et al., 2010), multi-resolution analysis (Celik and Ma, 2010), and thresholding techniques (Chen et al., 2021).

Flood duration estimation typically involves generating inundation maps through time series analysis using binary flood masks (Schumann and Moller, 2015; Shen et al., 2019). These maps show the number of times each pixel has been flooded within a given period, and their accuracy depends on the quality of the input data (Agnihotri et al., 2019). Automated flood mapping has significantly advanced this field, replacing time-consuming manual methods with more efficient processes (Voigt et al., 2007; Dell'Acqua et al., 2011). Modern techniques enable simultaneous processing of multiple flood events and rapid generation of outputs (Giustarini et al., 2012). Thus, this study aims to analyze climate-driven persistently flooded areas in the Indian state of Uttar Pradesh during the period 2017 to 2021, providing insights into flood dynamics and their impacts on croplands and infrastructure.

STUDY AREA

Uttar Pradesh is located in the north-central part of India, covering an area of 2,40,928 km^2, with a population density of 830 persons per km^2 (Figure 19.1). It is the fourth-largest state in India by area and has the highest share of the inhabited population among all Indian states. Uttar Pradesh shares an international border with Nepal to the north and state borders with Madhya Pradesh to the south, Bihar to the east, and Rajasthan and Haryana to the west. The state is traversed by two major rivers, the Ganga and the Yamuna, along with 32 other small and large rivers. Administratively, the state comprises 75 districts, with Lucknow serving as the capital and Kanpur as its financial and industrial hub. The climate of Uttar Pradesh is classified as Humid Subtropical with Dry Winters

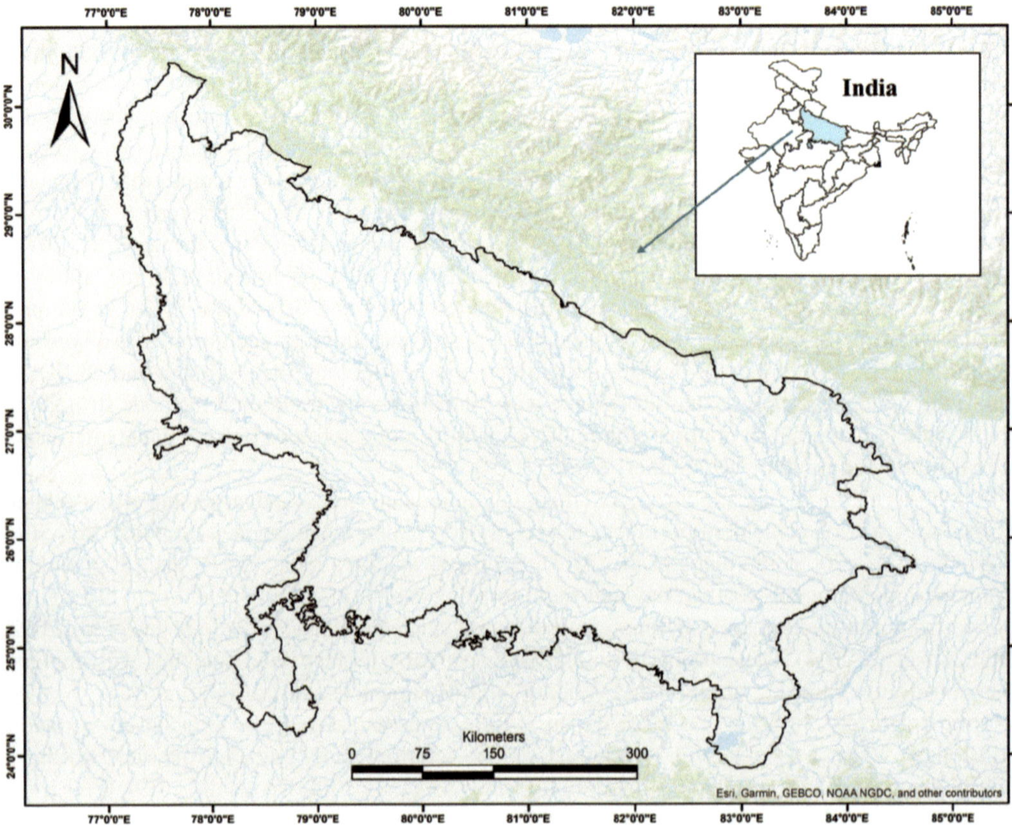

FIGURE 19.1 Location map of study area.

(CWa) under the Köppen climate classification system, with an average seasonal rainfall of 840 mm. Flooding in the state primarily occurs during the monsoon season due to the overflowing of major rivers such as the Ganga, Yamuna, Ghagra, and Gomti.

Floods in Uttar Pradesh are increasingly being linked to the broader impacts of climate change, which have altered regional weather patterns and intensified the hydrological cycle. Rising global temperatures have led to increased evaporation and higher atmospheric moisture content, resulting in more intense and erratic monsoon rainfall across the Indo-Gangetic Plain. This shift has contributed to the frequent overflow of major rivers such as the Ganga, Yamuna, Ghagra, and Gomti, triggering widespread and recurrent flooding in the state. Additionally, the melting of Himalayan glaciers—another consequence of climate change—has increased river discharge during the monsoon season, further exacerbating flood risks. These climate-induced changes, coupled with unplanned urban expansion and deforestation, have significantly reduced the natural capacity of the region to absorb excess water. As a result, flood events in Uttar Pradesh have become more severe and unpredictable, posing growing challenges to agriculture, infrastructure, and human settlements.

DATABASE AND METHODOLOGY

Database

This study has been initiated at the gram panchayat level of Uttar Pradesh, which is the lowest administrative unit. The gram panchayat boundary map was obtained from the web portal https://bhuvan-panchayat3.nrsc.gov.in/. Mapping of flood events has been done with the help of C-Band

SAR data acquired by the Sentinel 1 satellite. The generation of historical flood layers for the period 2017–2020 was done using the Java script on Google Earth Engine (GEE). Other input data such as land use and land cover, and a drainage map were also obtained from SIS-DP. The details of the dataset used and their sources are mentioned in Table 19.1.

METHODOLOGY

The following study aims at identifying persistently flooded low-lying areas of Uttar Pradesh at the gram panchayat level. The generation of the flood layers has been performed using code scripted solely for Sentinel-1A data using a thresholding method. The whole process of generating flood layers has been performed in GEE using JavaScript. The study could not be conducted for the period before 2017 due to non-availability of Sentinel 1. A threshold value of 1.25 was used for all years (2017–2021) to segregate water features and other classes. After the extraction of flood layers, permanent water bodies and false pixels have been erased using drainage network of the study area. Therefore, a total of 86 historical flood layers have been generated from 2017 to 2021 (Table 19.2).

Moreover, to find out persistently flooded gram panchayats of Uttar Pradesh, a vulnerability map has been prepared that has been categorized into three categories, namely, low, medium, and high. Subsequently, to find out the persistency of flood events in an area, the attributes of the overall historical flood layers obtained have been computed in MS Excel with the help of "group by" function. The converted attribute tables from ArcGIS have been appended into a single worksheet for estimating the persistency of gram panchayats. Thus, the classification of the persistent areas has been

TABLE 19.1
Datasets and Ancillary Data Utilized for Mapping Flood Events and Delineating Persistently Flooded Areas

S. no.	Data	Spatial Resolution	Temporal Resolution	Source
1.	Gram panchayat boundary of Uttar Pradesh	Vector format	–	https://bhuvan-panchayat3.nrsc.gov.in/
2.	Sentinel-1 (C-Band)	5 m	6 days	https://earthengine.google.com/
3.	Land use and Landcover (Crops)	10k	–	https://www.nrsc.gov.in/
4.	Drainage Network	30 m	16 days	https://earthexplorer.usgs.gov
5.	Road Network	Vector format	–	https://www.openstreetmap.org/
6	Rail Network	Vector format	–	https://www.openstreetmap.org/

TABLE 19.2
Overall Historical Flood Layers Generated with the Help of Google Earth Engine Using Sentinel-1 Microwave Data

S. No.	Period	Flood Layers Generated	GEE Assess URL
1	July–September 2017	19	https://earthengine.google.com/
2	July–September 2018	15	https://earthengine.google.com/
3	July–September 2019	13	https://earthengine.google.com/
4	July–September 2020	19	https://earthengine.google.com/
5	July–September 2021	20	https://earthengine.google.com/

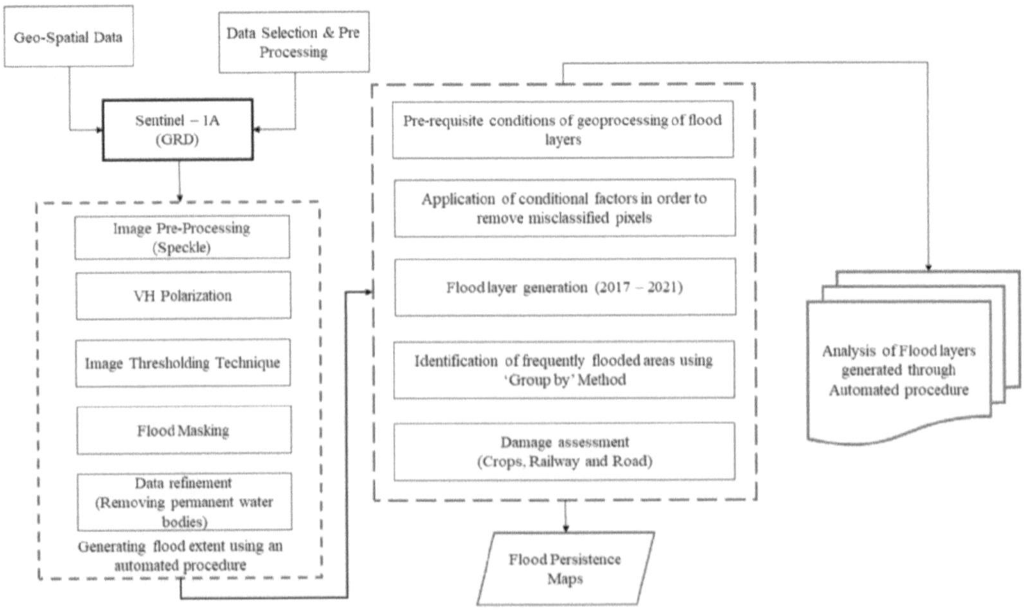

FIGURE 19.2 Procedure followed for generating historical flood layers and finding out persistently flooded areas.

done based on the persistency of flood events in five years, for example, 1–5 (low), 5–10 (medium), and >10 (high). The final output of the vulnerability map was obtained by spatially joining the Excel sheet and shapefile of the study area in order to join the attributes of both datasets. After joining the data, a vulnerability map of gram panchayat level is obtained. Moreover, damage assessment of flood is an essential part of any flood-related study, and for estimating the total submerged areas, intersect tools from the ArcGIS toolbar have been used. The intersect tool has been used for calculating the total crop area submerged in a particular period. Ultimately, the total number of gram panchayats affected, as well as the total cropland affected, has been estimated in the following study in the past 5 years (Figure 19.2).

RESULTS

Impacts of Floods on Cultivation and Infrastructure (2017–2021)

Uttar Pradesh witnesses most of the rainfall during the south-west monsoon season (July–Sep). Heavy and prolonged local rainfall in the monsoon period causes flooding triggers in this part of the Indo-Gangetic Basin (IGB), which causes damage to property and loss of human life every year. The amount and concentration of rainfall, as well as the existing river water levels, are the main determinants of flood intensity and penetration.

GEE-based flood inundation analysis has been done in the identification of worst-affected gram panchayat, in terms submerged area of crop land and public services, viz., road network and railway facilities. In Uttar Pradesh, the majority of people depend mostly on agriculture for their living. More than 70% of the population is dependent on agriculture and related industries directly or indirectly. In the year 2017, 29,116.11 km^2 agricultural land of Shahjahanpur, Chitrakoot, Band, Sonbhadra, Farrukhabad, Budaun, and some parts of Mahoba, Jhansi, Kannauj, Kanpur Nagar, Lucknow, and Jalaun district was submerged. Big flood events of 2017 have been recorded during the second week of July (Figure 19.3). Whereas, in 2018 affected agricultural land substantially decreased to 24,524.83 km², which included districts such as Budaun, Etawah, Farrukhabad, Jalaun, Shahjahanpur, and some parts of St. Kabir Nagar, Raebareilly, Mirzapur, Jhansi, and Gorakhpur. Flood big events of 2018 have been recorded during second week of July (Figure 19.4). Again, in 2019 a downfall in the affected area was noticed, and only 1,173.63 km² area was submerged, including parts of Ballia, Bareilly, Gorakhpur, Gazipur, Kanpur Nagar, Pilibhit, and Mainpuri district of Uttar Pradesh. Flood big events of 2019 have also been recorded during the second week of July (Figure 19.5). In 2020, the affected area has shown an increase and it increased to 21,374.3 km². Flood big events of 2020 have been recorded during the second week of July (Figure 19.6). Apparently in 2021, the affected area has shown a significant increase to 16,954.68 km². Flood big events of 2021

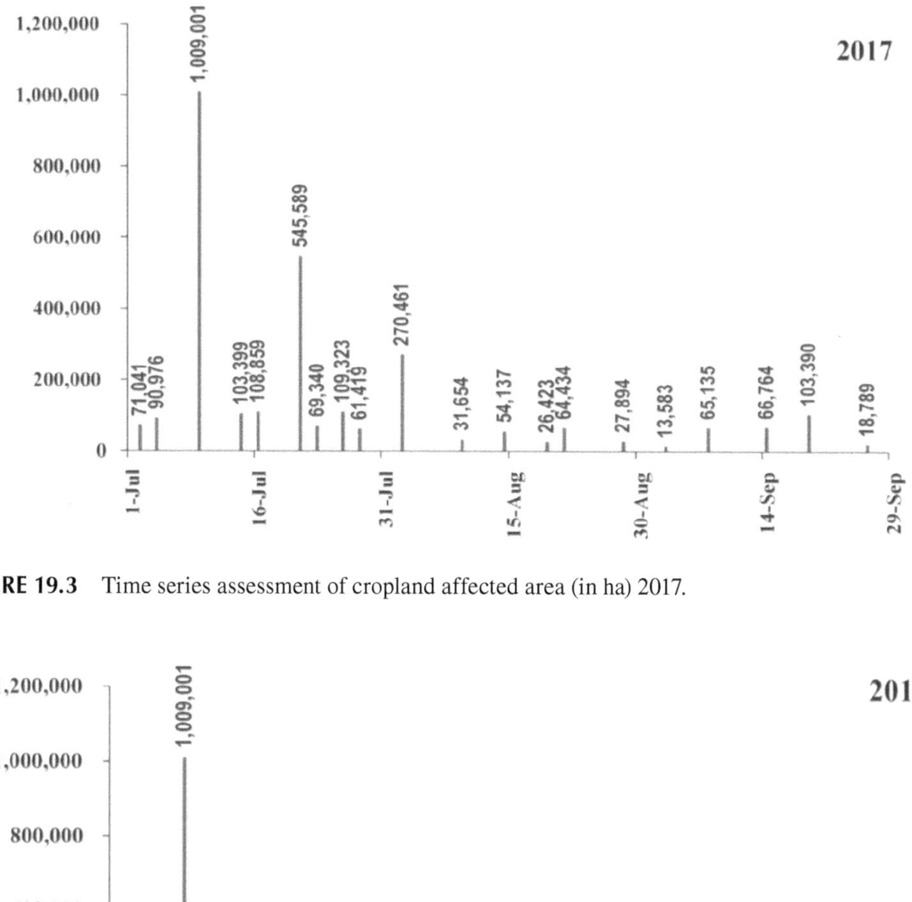

FIGURE 19.3 Time series assessment of cropland affected area (in ha) 2017.

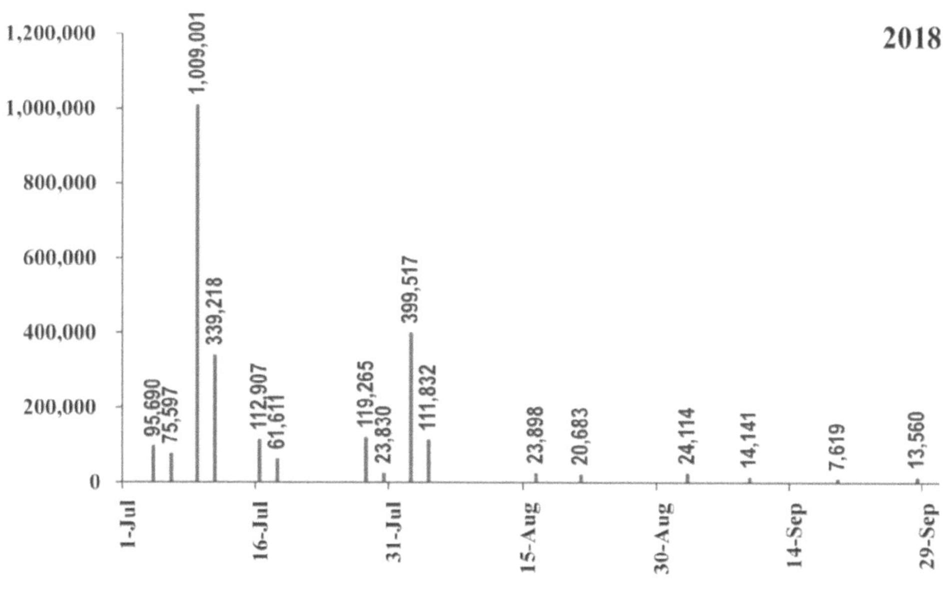

FIGURE 19.4 Time series assessment of cropland affected area (in ha) 2018.

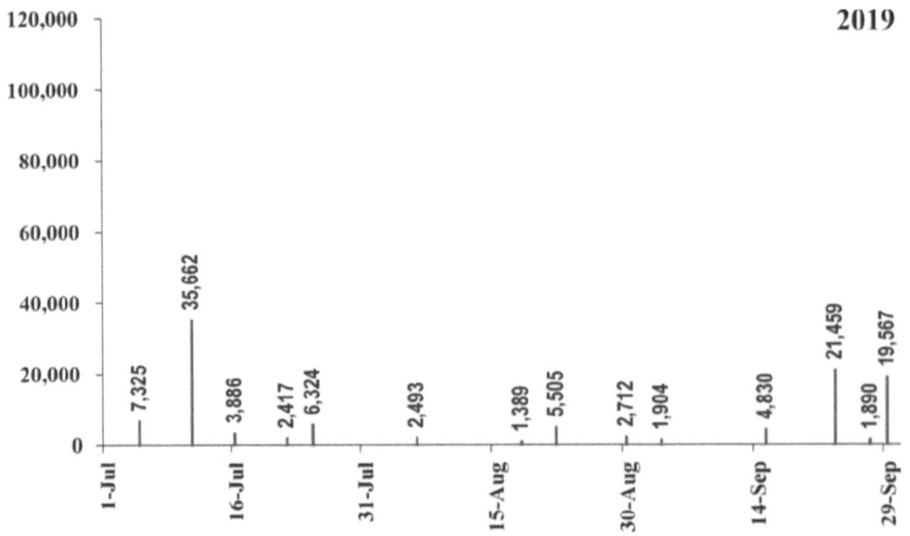

FIGURE 19.5 Time series assessment of cropland affected area (in ha) 2019.

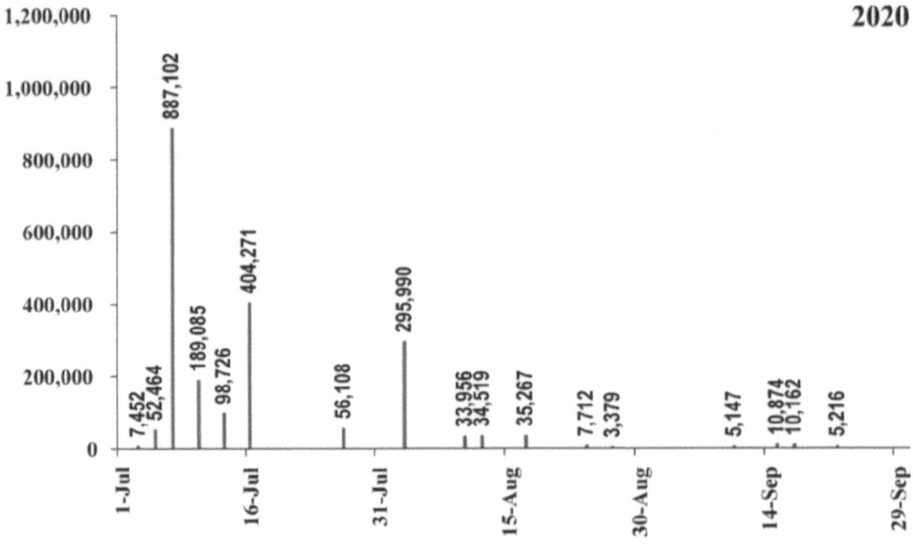

FIGURE 19.6 Time series assessment of cropland affected area (in ha) 2020.

have been recorded during the second week of July (Figure 19.7 and Table 19.3). The annual flood region over the Uttar Pradesh (UP) from 2017 to 2021 is also presented in Figures 19.8 and 19.9.

This study has been done with the objective to assess the annual and cumulative flood extent, and its effect on agricultural land, road, and railway lines in recent years' floods (i.e., 2017–2021) in Uttar Pradesh (Figures 19.10–19.12). The analysis of the results showed that about 49,16,616 ha area of the state have been submerged during the study period in different phases of flood events. About 55% districts were observed to have been directly impacted by flooding (Figure 19.13).

Districts like Ballia, Gazipur, Deoria, and Gorakhpur, situated in eastern UP and along major rivers (Ghagra and Ganga) confluence, experienced flooding every year in the monsoon season. In the southern part, the districts of Jhansi, Jalaun, Hamirpur, Banda, and Chitrakoot has experienced devastation of crops due to flooding (Figure 19.10). Infrastructure networks, particularly

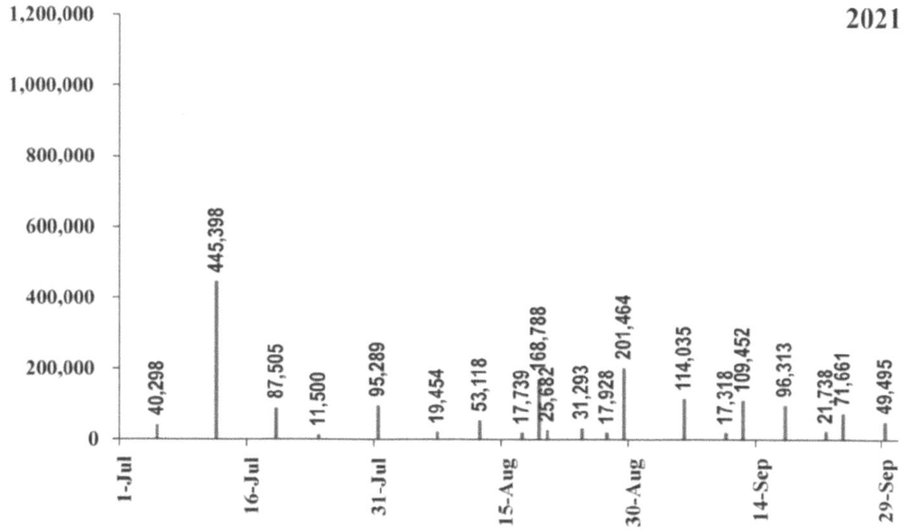

FIGURE 19.7 Time series assessment of cropland affected area (in ha) 2021.

TABLE 19.3
Crop Submerged in Flood Events 2017–2021

Year	Crops Affected (in km^2)
2017	29,116.11
2018	24,524.83
2019	1,173.63
2020	21,374.3
2021	16,954.68

transportation systems, are considered the backbone of cities, ensuring their livability and prosperity. In the context of a global economy reliant on the movement of goods, information, and people, transportation networks are vital for community safety and economic stability. Floods can severely damage roads, worsening hazardous conditions. Flooded roads have two primary effects: direct damage, such as injuries and vehicle destruction, and indirect consequences like traffic disruptions that negatively impact the economy, society, and security (Tacnet et al., 2012). Additionally, road construction can exacerbate natural flood impacts by disrupting habitats and blocking water, sediment, and nutrient flow (Douven et al., 2009). Uttar Pradesh's Road network, managed by public works department, includes 7,550 km of national highways, 7,530 km of state highways, 5,761 km of major district roads, 3,254 km of other district roads, and 1,38,702 km of village roads. However, around 60% of state highways, 62% of major district roads, and 83% of other district roads are narrower than 7 m (Uttar Pradesh Core Road Network Development Project).

Data from 2017 to 2021 show significant variations in road submergence. In 2017, 3,761 km of major roads were submerged, particularly in districts like Ballia, Chandauli, and Gorakhpur. In 2018, 2,970 km of roads were submerged during 15 flood events. However, in 2019, road submergence dropped sharply to only 123 km, attributed to a weaker monsoon. In 2020, submergence rose again to 1,107 km, with roads most affected in districts like Lalitpur, Jhansi, and Kanpur Nagar.

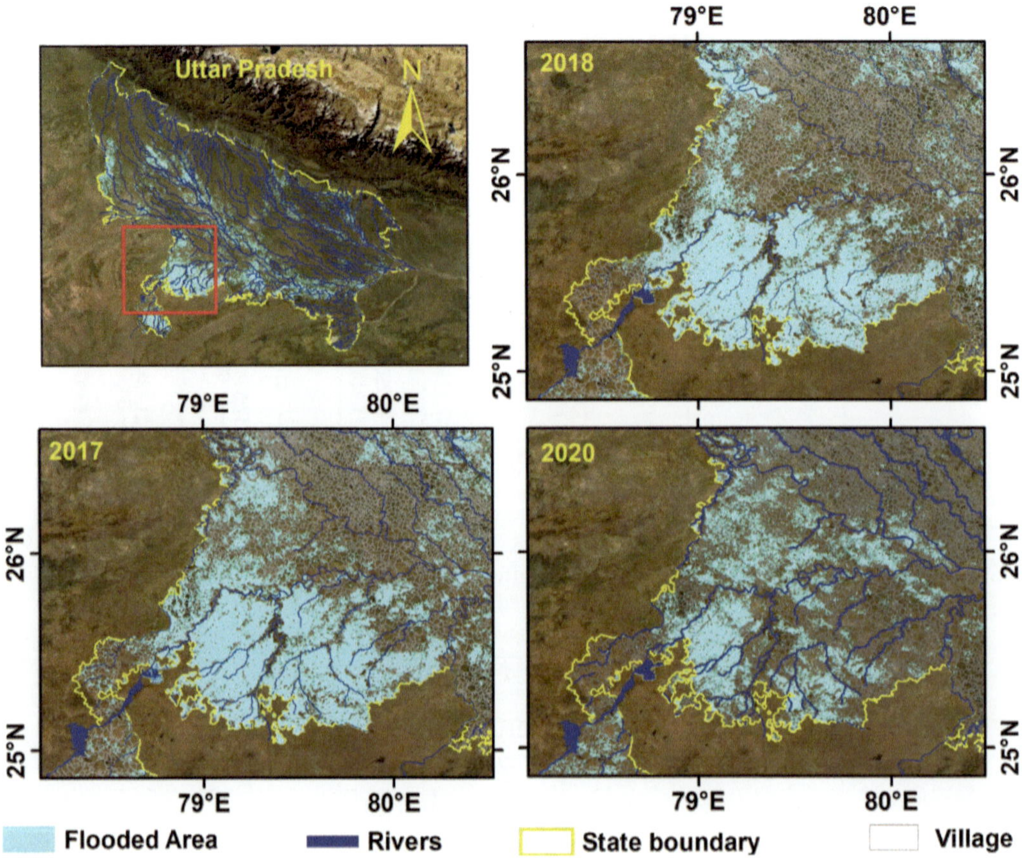

FIGURE 19.8 Annual flood regions over Uttar Pradesh state, 2017, 2018, and 2020.

In 2021, the submerged network increased further to 1,943 km, affecting districts such as Agra, Aligarh, and Mathura.

In recent decades, extreme precipitation, storms, and flooding have become more frequent and severe. Many sections of railway stations are constructed on level terrain and low-lying areas, and railway tunnels are constructed where the terrain is lower than the surrounding areas. There is limited drainage capacity in low-lying areas that can be overwhelmed by a heavy rainfall event. Trains need to slowdown in order to avoid damaging themselves when floodwaters rise above the railroad rails. Flooding can result in a short circuit, a power outage, or both if the railway has a live conductor rail. As a result, traveling may be hampered not only during the flood but also afterward. Thus, a heavy rainfall event can cause serious disruptions to the railways.

Overall, there are 1,134 railway stations in Uttar Pradesh state. The five main railway stations of Uttar Pradesh are Allahabad Junction, Jhansi Junction, Pt. Deen Dayal Upadhaya Junction, Ghaziabad, and Kanpur Central. In almost every monsoon season, the railway network of Uttar Pradesh gets affected due to heavy rainfall events. Tables 19.4 and 19.5 illustrate total railway tracks submerged due to flood events occurring between 2017 and 2021.

IMPACTS OF FLOODS ON DISADVANTAGED OR MARGINALIZED COMMUNITIES

Floods in Uttar Pradesh have a disproportionately severe impact on disadvantaged and marginalized communities, including scheduled castes (dalits), landless laborers, tribal groups, and women. These groups often inhabit the most vulnerable areas—such as riverbanks, embankments, and

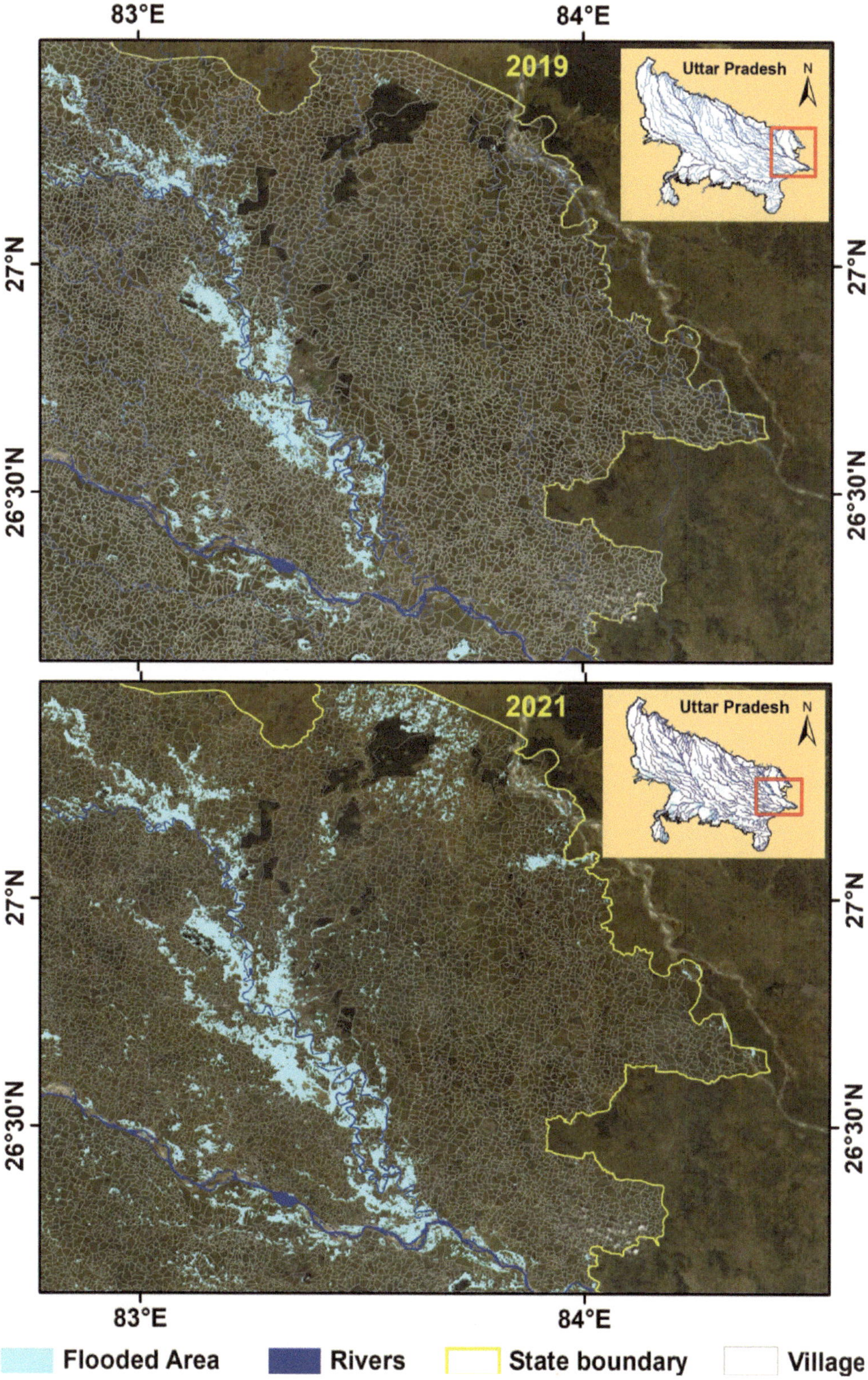

FIGURE 19.9 Annual flood region over Uttar Pradesh state, 2019 and 2021.

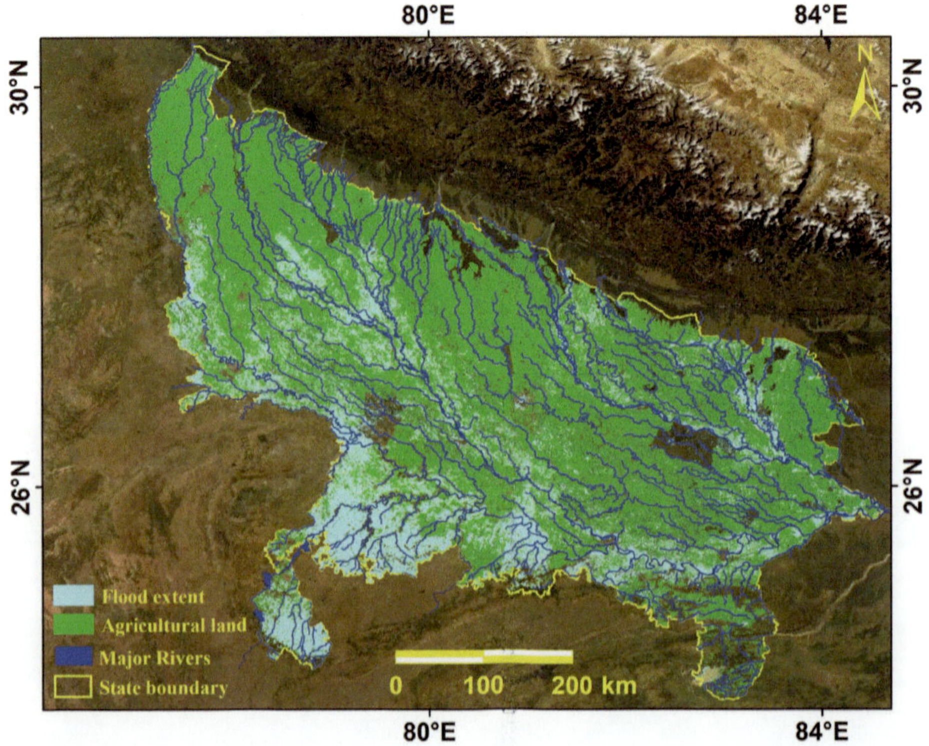

FIGURE 19.10 Crop submerged in flood during 2017–2021.

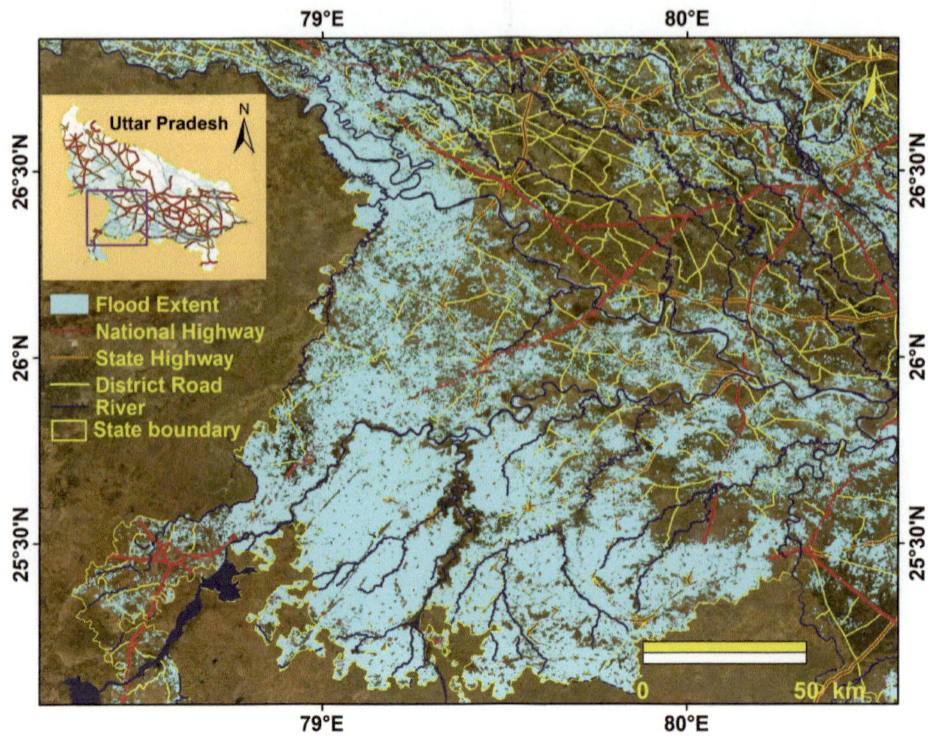

FIGURE 19.11 Different categories of road network submerged in floods during the period 2017–2021.

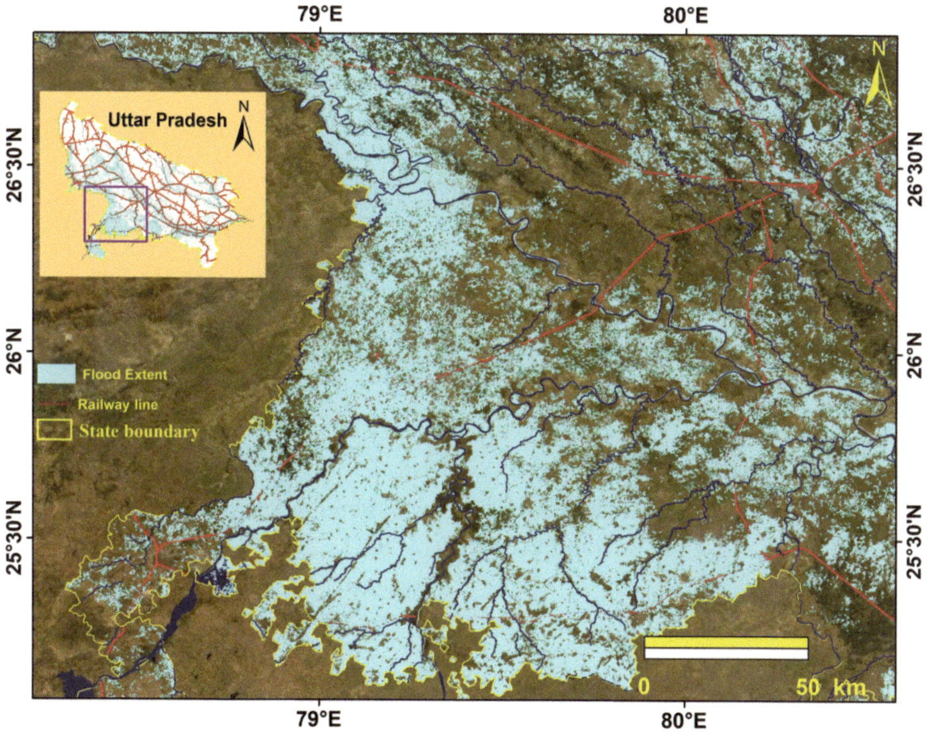

FIGURE 19.12 Railways network submerged in flood during the period 2017–2021.

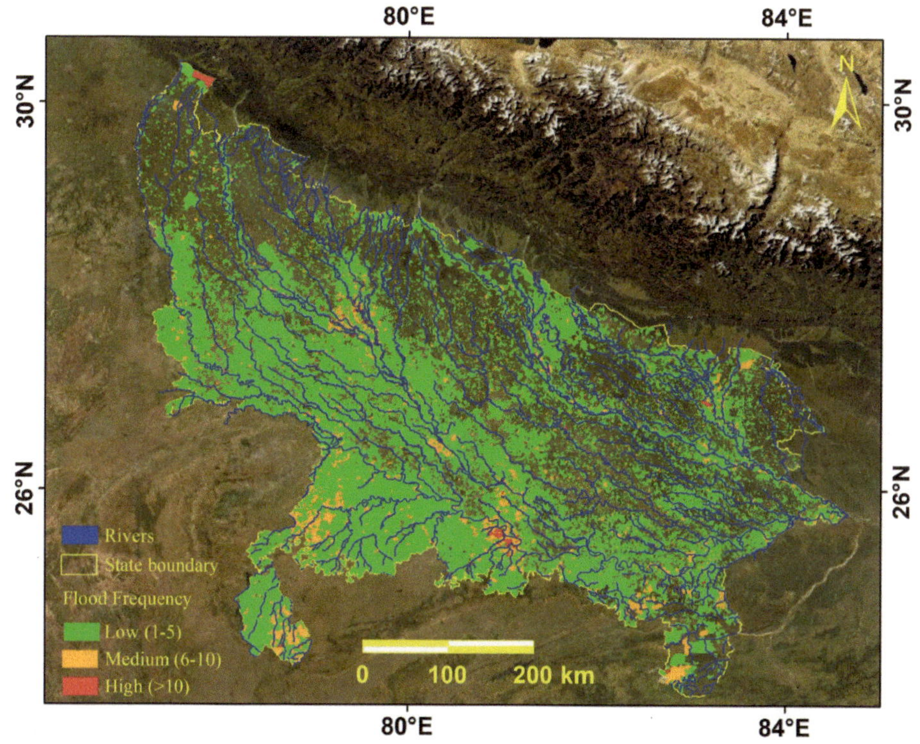

FIGURE 19.13 Flood frequency in Gram Panchayats (GP) of Uttar Pradesh 2017–2021.

TABLE 19.4
Different Categories of Roads Submerged in Flood Events during Period 2017–2021

Year	National Highway	State Highway	District road	Total Roads Submerged (km)
2017	428	683	2,650	3,761
2018	349	451	2,170	2,970
2019	18	12	93	123
2020	96	159	851	1,107
2021	224	194	1,525	1,943

TABLE 19.5
Total Railway Tracks Submerged in Flood Events 2017–2021

Year	Railway Tracks Submerged (in km)
2017	489
2018	394
2019	20
2020	97
2021	160

low-lying floodplains—due to historical socioeconomic exclusion and lack of access to secure land tenure (Chowdhury et al., 2017; Singh and Pandey, 2021). As a result, their houses, typically made of mud, thatch, or other nondurable materials, are easily destroyed during flood events. The loss of shelter is compounded by limited access to early warning systems and evacuation infrastructure, often due to geographic isolation and lack of digital or linguistic access to alerts (Sengupta et al., 2020). These communities rely heavily on agriculture and daily wage labor, sectors that are among the first to be disrupted by flooding, leading to long-term income loss and food insecurity. Furthermore, in the aftermath of floods, bureaucratic procedures for compensation and aid distribution often require proof of residence or land ownership, documents that many marginalized families do not possess, thereby excluding them from formal relief and rehabilitation programs (Government of Uttar Pradesh, 2022).

The post-flood recovery phase highlights and exacerbates existing inequalities. Relief and aid distribution mechanisms are frequently influenced by local power dynamics, wherein marginalized groups are deprioritized due to a lack of political voice and representation (Kumar and Mishra, 2020). Women from these communities face unique and heightened risks, including exposure to sexual and gender-based violence in overcrowded and poorly managed relief shelters, as well as loss of access to menstrual hygiene products and reproductive healthcare services (Dasgupta et al., 2019). The disruption of schooling during and after floods disproportionately affects children from disadvantaged backgrounds, who often lack access to digital education tools and face higher dropout rates due to economic pressures (UNICEF, 2021). The cumulative effects of repeated flood exposure—asset loss, displacement, health crises, and educational disruptions—contribute to an ongoing cycle of poverty and marginalization. Addressing these challenges requires a more inclusive and equity-focused approach to disaster risk reduction and climate adaptation planning in

TABLE 19.6
Total Gram Panchayat Affected during 2017–2021

Year	Gram Panchayat Affected
2017	19,342
2018	15,684
2019	3,703
2020	18,881
2021	24,830

TABLE 19.7
Gram Panchayat Flood Vulnerability

Vulnerability	Total Gram Panchayats
Low (0–5)	34,032
Medium (5–10)	3,179
High (<10)	101

Uttar Pradesh, one that actively involves marginalized voices in decision-making processes and ensures equitable access to resources and support (World Bank, 2020).

FLOOD VULNERABILITY MAP 2017–2021

Villages of Uttar Pradesh experience floods in every monsoon season, in which many families and the poor section of society get affected, and many of them even lose their source of livelihood. Some people even have to migrate to other places, leaving behind their permanent residence. To calculate the vulnerability of gram panchayat during 2017–2021, the flood events (Table 19.6) and gram panchayat have been classified into three categories based on flood events. Gram Panchayats (GP) that experienced flood events in the past five years (Figure 19.13) were classified as highly vulnerable. For example, Rampura Khurd GP in Budaun district, Amlokhar GP in Banda district, Khoobpur Anchitpur GP in Shahjahanpur district, and Chilli Mal Banger GP in Chitrakoot district each recorded as many as 15 flood events (Tables 19.7 and 19.8).

DISCUSSION

The IGB, one of the most densely populated and agriculturally productive regions in the world, is increasingly vulnerable to persistent and recurrent flooding due to climate change. Shifting monsoon patterns, glacial melt from the Himalayas, and rising extreme weather events have intensified the frequency and severity of floods in the basin (Mall et al., 2018). These floods not only disrupt livelihoods and agricultural systems but also exacerbate socioeconomic disparities, particularly among marginalized rural communities. Traditional flood monitoring approaches are often limited by data scarcity, slow processing, and a lack of spatial detail. Hence, there is a pressing need for scalable, real-time, and high-resolution tools that can systematically map and monitor flood events across large and diverse terrains like the IGB.

TABLE 19.8
Top 20 Most Affected Gram Panchayats of Uttar Pradesh during 2017–2021

S. No	Gram Panchayat	Block Panchayat	District	Frequency	Category
1	Rampura Khurd	Dataganj	Budaun	16	High
2	Amlokhar	Kamasin	Banda	15	High
3	Khoobpur Anchitpur	Jaitipur	Shahjahanpur	15	High
4	Chilli Mal Banger	Pahari	Chitrakoot	15	High
5	Asafpur	Mion	Budaun	12	High
6	Bachheli Daranagar	Usawan	Budaun	12	High
7	Kilapur Kalaan	Mirzapur	Shahjahanpur	12	High
8	Labhari	Mion	Budaun	12	High
9	Thatharmai	Jalabad	Shahjahanpur	12	High
10	Bhundi	Usawan	Budaun	12	High
11	Arhaiya	Dhata	Fatehpur	12	High
12	Bardwara	Pahari	Chitrakoot	12	High
13	Beera	Kamasin	Banda	12	High
14	Bena Mau	Kamasin	Banda	12	High
15	Bilas Banger	Pahari	Chitrakoot	12	High
16	Bir Ghumai Surki	Pahari	Chitrakoot	12	High
17	Chak Malik Bhiti	Dalmau	Raebareilly	12	High
18	Ganeshpur	Bhitaura	Fatehpur	12	High
19	Hasta Bangar	Pahari	Chitrakoot	12	High
20	Jamaunapur Kachhar	Sarsawan	Kaushambi	12	High

GEE, a cloud-based geospatial processing platform, offers powerful capabilities for mapping climate-driven flood events using multi-temporal satellite data and advanced algorithms. This research theme is significant because it leverages GEE to generate spatially explicit and time-sensitive flood maps that can help identify flood persistence, duration, and extent with greater accuracy and speed (Gorelick et al., 2017; Kumar et al., 2021). Such data-driven insights are crucial for policymakers, disaster management authorities, and development planners to prioritize intervention zones, assess vulnerability, and design climate-resilient infrastructure. Additionally, integrating the spatial analysis of floods with socioeconomic data can enhance our understanding of flood impacts on vulnerable communities, paving the way for more inclusive climate adaptation strategies in the IGB (World Bank, 2020).

The research theme titled "Mapping Climate-Driven Persistent Flooding Using Google Earth Engine and its Impacts in the Indo-Gangetic Basin" distinguishes itself from existing literature by integrating high-resolution, long-term satellite data analysis with a socio-environmental lens to assess both the spatial patterns and human impacts of persistent flooding. While prior studies have largely focused on short-term flood events or localized case studies using traditional GIS methods or field-based assessments (Chatterjee et al., 2019; Roy and Ahmad, 2021), this research leverages the computational power of GEE to conduct large-scale, temporal flood mapping over decades, thereby capturing the evolving nature of climate-driven flood persistence across the IGB. Additionally, unlike many technical flood mapping studies that overlook socioeconomic dimensions, this approach explicitly links geospatial flood trends with the vulnerabilities of marginalized populations, offering a more holistic view of flood risk under changing climate regimes (Sharma et al., 2022). The integration of cloud computing, remote sensing, and social vulnerability analysis positions this research at the intersection of climate science, geospatial technology, and disaster justice.

This research also helps in highlighting significant potential for benefiting *char* inhabitants—communities living on the riverine sandbars and islands of the IGB who are among the most vulnerable to flood hazards. These populations often face chronic displacement, loss of livelihoods, and lack of access to infrastructure due to the dynamic and unstable nature of the chars, which are frequently submerged or reshaped by floodwaters (Islam and Rashid, 2011). By leveraging the capabilities of GEE for high-resolution, near-real-time flood mapping, this research can help identify patterns of persistent inundation and highlight the spatial extent and frequency of flood-prone zones. Such information is crucial for targeted early warning systems, adaptive land-use planning, and the prioritization of relief and rehabilitation efforts for char dwellers, who are often overlooked in mainstream disaster response frameworks (Rahman et al., 2019). Moreover, the integration of satellite-based flood analytics can support policymakers in designing climate-resilient development interventions that address the specific vulnerabilities of char communities in the IGB (Shahriar et al., 2020).

CONCLUSION

In this research work, a web-based JavaScript code has been developed to process and extract huge datasets of Sentinel-1 SAR on a cloud-computing platform GEE. The damage assessment of agricultural land, road, and railways using all-weather datasets of SAR on large-scale study has been done. The study evaluated flood events of UP during 2017–2021, and the results showed that about 49,166 km^2 area has been flooded, and the cropped area of about 41,035 km^2, road of 70,522 km, and railway track of 686 km length were submerged. The study showed that 43 districts were badly affected every year. But some districts, namely Jalaun, Hamirpur, Mahoda, Lalitpur, Jhansi, Etawah, Gorakhpur, Chitrakoot, and Banda districts, are more vulnerable. Further, the results show that more than 37,000 gram panchayats were affected by floods, moderate to severe (Table 19.8). Flood persistency has been calculated for flood events, and its result depicts that more than 3,100 gram panchayats experienced 5–10 flood events during the study period (Figure 19.13).

The outputs of the study in the form of estimated flood-affected crop fields, road, and railway lines and their statistics may be helpful for decision-makers and policy makers to take preventive measures for areas which experience flood persistently. Frequent flood-affected areas mitigation strategies, as well as response and recovery plans, should be formulated, which can be implemented to reduce the risk of flood disaster. Such plans should be formulated at all levels of hierarchy, that is, state, district, tehsil, and gram panchayat. It is important to develop and promote a village-level plan for disaster mitigation that involves the entire village population. The focus should be on the villages' own preparation and equipping to handle any eventualities in the larger catastrophe area. Sustainable and effective strategies should be formulated for high-risk areas, which should aim toward reducing the impact of flood disaster on the lives and property of people, also reducing socioeconomic impacts as well. Therefore, the creation of historical flood events can aid in pre- and post-disaster management efforts, which can significantly raise public awareness of disaster risk and lead to a demand for safer communities in which to live and work and significantly lower the risks of loss of life, injury, financial burden, and destruction of natural and cultural resources as a result of flood disasters.

LIMITATIONS OF THIS STUDY

One limitation of this research lies in the dependence on remote sensing data and cloud-based geospatial platforms, which, while powerful, have inherent constraints in spatial and temporal resolution. Satellite imagery often struggles with cloud cover during peak monsoon periods, potentially leading to underestimation of flood extent or duration. Additionally, while GEE allows for large-scale analysis, it may not capture localized variations in flood impact, especially in densely populated or heterogeneous landscapes like the IGB. Ground-truthing data is often limited or unavailable, which can affect the accuracy of flood mapping and impact assessments. Moreover, socioeconomic data integrated into the study may be outdated or inconsistent across regions, limiting the robustness of conclusions regarding vulnerable populations. These limitations highlight the need for complementary field data, higher-resolution datasets, and multidisciplinary approaches to ensure a more comprehensive understanding of climate-driven flooding in the region.

DECLARATION

Funding for research: No funding was received for this research work.

Disclosure of Potential Conflict of Interests: All authors declare that they have no conflict of interest.

Data availability: Data will be available on reasonable request.

REFERENCES

Adnan, M. S. G., Abdullah, A. Y. M., Dewan, A., & Hall, J. W. (2020). The effects of changing land use and flood hazard on poverty in coastal Bangladesh. *Land Use Policy*, 99, 104868.

Agnihotri, A. K., Ohri, A., Gaur, S., Das, N., & Mishra, S. (2019). Flood inundation mapping and monitoring using SAR data and its impact on Ramganga River in Ganga basin. *Environmental Monitoring and Assessment*, 191(12), 1–16.

Ahamed, A., & Bolten, J. D. (2017). A MODIS-based automated flood monitoring system for southeast asia. *International Journal of Applied Earth Observation and Geoinformation*, 61, 104–117.

Bagheri, A., & Liu, G. J. (2024). Climate change and urban flooding: assessing remote sensing data and flood modeling techniques: a comprehensive review. *Environmental Reviews*, 33, 1–14.

Bonn, F., & Dixon, R. (2005). Monitoring flood extent and forecasting excess runoff risk with RADARSAT-1 data. *Natural Hazards*, 35(3), 377–393.

Bovolo, F., & Bruzzone, L. (2005). A detail-preserving scale-driven approach to change detection in multitemporal SAR images. *IEEE Transactions on Geoscience and Remote Sensing*, 43(12), 2963–2972.

Brémond, P., Grelot, F., & Agenais, A. L. (2013). Flood damage assessment on agricultural areas: review and analysis of existing methods. *Natural Hazards and Earth System Sciences,* 13, 2493–2512. https://doi.org/10.5194/nhess-13-2493-2013.

Cao, H., Zhang, H., Wang, C., & Zhang, B. (2019). Operational flood detection using Sentinel-1 SAR data over large areas. *Water*, 11(4), 786.

Celik, T., & Ma, K. K. (2010). Multitemporal image change detection using undecimated discrete wavelet transform and active contours. *IEEE Transactions on Geoscience and Remote Sensing*, 49(2), 706–716.

Chatterjee, S., Singh, R., & Jha, S. K. (2019). Flood risk assessment and mapping in the Ganga River Basin using GIS and remote sensing. *Natural Hazards,* 96(2), 661–682.

Chen, S., Huang, W., Chen, Y., & Feng, M. (2021). An adaptive thresholding approach toward rapid flood coverage extraction from Sentinel-1 SAR imagery. *Remote Sensing*, 13(23), 4899.

Chowdhury, A., Jha, R., & Dutta, A. (2017). Flood risk and social vulnerability in India: A case study of Uttar Pradesh. *Environmental Hazards,* 16(4), 321–336.

Dasgupta, R., Sahoo, S., & Sinha, A. (2019). Gendered impacts of floods in northern India. *Gender & Development,* 27(1), 69–85.

Del Ninno, C., Dorosh, P. A., & Smith, L. C. (2003). Public policy, markets and household coping strategies in Bangladesh: Avoiding a food security crisis following the 1998 floods. *World Development*, 31(7), 1221–1238.

Dell'Acqua, F., Bignami, C., Chini, M., Lisini, G., Polli, D. A., & Stramondo, S. (2011). Earthquake damages rapid mapping by satellite remote sensing data: L'Aquila april 6th, 2009 event. *IEEE Journal of Selected Topics in Applied Earth Observations and Remote Sensing*, 4(4), 935–943.

Dellepiane, S., Angiati, E., & Vernazza, G. (2010, July). Processing and segmentation of COSMO-SkyMed images for flood monitoring. In *2010 IEEE International Geoscience and Remote Sensing Symposium*, Honolulu, Hawaii, U.S.A., pp. 4807–4810.

Dhar, O. N., & Nandargi, S. (2003). Hydrometeorological aspects of floods in India. *Natural Hazards*, 28(1), 1–33.

Di Baldassarre, G., Schumann, G., & Bates, P. D. (2009). A technique for the calibration of hydraulic models using uncertain satellite observations of flood extent. *Journal of Hydrology*, 367(3–4), 276–282.

Douven, W. J. A. M., Goichot, M., & Verheij, H. J. (2009). Roads and floods: Best practice guidelines for the integrated planning and design of economically sound and environmentally friendly roads in the Mekong floodplains of Cambodia and Vietnam. https://resolver.tudelft.nl/uuid:b71f58d5-eef4-4d77-b4e8-b9819c599451

Dutta, D., & Herath, S. (2004). Trend of floods in Asia and flood risk management with integrated river basin approach. In *Proceedings of the 2nd International Conference of Asia-Pacific Hydrology and Water Resources Association,* Singapore, pp. 55–63.

Ghaffarian, S., Kerle, N., & Filatova, T. (2018). Remote sensing-based proxies for urban disaster risk management and resilience: A review. *Remote Sensing*, 10(11), 1760.

Giustarini, L., Hostache, R., Matgen, P., Schumann, G. J. P., Bates, P. D., & Mason, D. C. (2012). A change detection approach to flood mapping in urban areas using TerraSAR-X. *IEEE Transactions on Geoscience and Remote Sensing*, 51(4), 2417–2430.

Gorelick, N., Hancher, M., Dixon, M., Ilyushchenko, S., Thau, D., & Moore, R. (2017). Google Earth Engine: Planetary-scale geospatial analysis for everyone. *Remote Sensing of Environment, 202*, 18–27.

Government of Uttar Pradesh. (2022). *Disaster management policy report.* Department of Revenue and Disaster Management, Government of Uttar Pradesh.

Guhathakurta, P., Sreejith, O. P., & Menon, P. A. (2011). Impact of climate change on extreme rainfall events and flood risk in India. *Journal of Earth System Science*, 120(3), 359–373.

Hendrawan, V. S. A., & Komori, D. (2021). Developing flood vulnerability curve for rice crop using remote sensing and hydrodynamic modeling. *International Journal of Disaster Risk Reduction*, 54, 102058.

Hess, L. L., Melack, J. M., Filoso, S., & Wang, Y. (1995). Delineation of inundated area and vegetation along the Amazon floodplain with the SIR-C synthetic aperture radar. *IEEE Transactions on Geoscience and Remote Sensing*, 33(4), 896–904.

Hostache, R., Matgen, P., Schumann, G., Puech, C., Hoffmann, L., & Pfister, L. (2009). Water level estimation and reduction of hydraulic model calibration uncertainties using satellite SAR images of floods. *IEEE Transactions on Geoscience and Remote Sensing*, 47(2), 431–441.

Islam, M. R., & Rashid, S. M. A. (2011). Vulnerability and coping strategies of char dwellers in Bangladesh: A case study of char Kalira in Kurigram District. *Bangladesh Journal of Public Administration,* 20(1), 87–100.

Kabenge, M., Elaru, J., Wang, H., & Li, F. (2017). Characterizing flood hazard risk in data-scarce areas, using a remote sensing and GIS-based flood hazard index. *Natural Hazards*, 89(3), 1369–1387.

Kiage, L. M., Walker, N. D., Balasubramanian, S., Babin, A., & Barras, J. (2005). Applications of Radarsat-1 synthetic aperture radar imagery to assess hurricane-related flooding of coastal Louisiana. *International Journal of Remote Sensing*, 26(24), 5359–5380.

Klaus, S., Kreibich, H., Merz, B., Kuhlmann, B., & Schröter, K. (2016). Large-scale, seasonal flood risk analysis for agricultural crops in Germany. *Environmental Earth Sciences*, 75(18), 1–13.

Kourgialas, N. N., & Karatzas, G. P. (2016). A flood risk decision making approach for Mediterranean tree crops using GIS; climate change effects and flood-tolerant species. *Environmental Science & Policy*, 63, 132–142.

Kumar, A., Jain, S., & Singh, R. (2021). Flood inundation mapping using Google Earth Engine and Sentinel-1 SAR data: A case study in Eastern Uttar Pradesh. *Journal of Hydrology: Regional Studies,* 35, 100824.

Kumar, N., & Mishra, V. (2020). Politics of relief distribution in post-flood Uttar Pradesh. *Indian Journal of Social Work,* 81(1), 89–103.

Liu, K., Ai, B., & Wang, S. (2010). Fish-pond change detection based on short term time series of RADARSAT images and object-oriented method. In *2010 3rd International Congress on Image and Signal Processing*, pp. 2175–2179. IEE Explore. https://ieeexplore.ieee.org/xpl/conhome/5623876/proceeding.

Lu, J., Giustarini, L., Xiong, B., Zhao, L., Jiang, Y., & Kuang, G. (2014). Automated flood detection with improved robustness and efficiency using multi-temporal SAR data. *Remote Sensing Letters*, 5(3), 240–248.

Mall, R. K., Srivastava, R. K., Banerjee, T., Mishra, O. P., Bhatt, D., & Sonkar, G. (2018). Disaster risk reduction including climate change adaptation over South Asia: Challenges and ways forward. *International Journal of Disaster Risk Science,* 9(2), 105–116.

Merz, B., Disse, M., Günther, K., & Schumann, A. (2010). Preface Outcomes of the RIMAX programme: Risk management of extreme flood events. *Natural Hazards and Earth System Sciences (NHESS)*, (Special Issue), 1–3.

Mishra, V., & Shah, H. L. (2018). Hydro climatological perspective of the Kerala flood of 2018. *Journal of the Geological Society of India*, 92(5), 645–650.

Mohanty, M. P., Mudgil, S., & Karmakar, S. (2020). Flood management in India: A focussed review on the current status and future challenges. *International Journal of Disaster Risk Reduction*, 49, 101660.

Opolot, E. (2013). Application of remote sensing and geographical information systems in flood management: A review. *Research Journal of Applied Sciences Engineering and Technology*, 6(10), 1884–1894.

Patel, D. P., & Srivastava, P. K. (2013). Flood hazards mitigation analysis using remote sensing and GIS: Correspondence with town planning scheme. *Water Resources Management*, 27(7), 2353–2368.

Pierdicca, N., Pulvirenti, L., Chini, M., Guerriero, L., & Candela, L. (2013). Observing floods from space: Experience gained from COSMO-SkyMed observations. *Acta Astronautica*, 84, 122–133.

Qadir, A., & Mondal, P. (2020). Synergistic use of radar and optical satellite data for improved monsoon cropland mapping in India. *Remote Sensing*, 12(3), 522.

Rahman, M., & Di, L. (2017). The state of the art of spaceborne remote sensing in flood management. *Natural Hazards*, 85(2), 1223–1248.

Rahman, M., Di, L., Yu, E., Lin, L., & Yu, Z. (2021). Remote sensing based rapid assessment of flood crop damage using novel disaster vegetation damage index (DVDI). *International Journal of Disaster Risk Science*, 12(1), 90–110.

Rahman, M. R., & Thakur, P. K. (2018). Detecting, mapping and analysing of flood water propagation using synthetic aperture radar (SAR) satellite data and GIS: A case study from the Kendrapara District of Orissa State of India. *The Egyptian Journal of Remote Sensing and Space Science*, 21, S37–S41.

Rahman, M. M., Di Baldassarre, G., & Gain, A. K. (2019). The dynamics of poverty and risk in the char lands of Bangladesh. *Natural Hazards,* 99(1), 337–354.

Roy, N., & Ahmad, M. (2021). Evaluating flood hazards in Bihar, India: A remote sensing and GIS-based approach. *International Journal of Disaster Risk Reduction,* 53, 102019.

Sarti, F., Inglada, J., Landry, R., & Pultz, T. (2001). Risk management using remote sensing data: Moving from scientific to operational applications. *Present. X SBSR* April, 23, 27.

Schumann, G. J. P., & Moller, D. K. (2015). Microwave remote sensing of flood inundation. *Physics and Chemistry of the Earth, Parts A/B/C*, 83, 84–95.

Schumann, G. P., Vernieuwe, H., De Baets, B., & Verhoest, N. E. C. (2014). ROC-based calibration of flood inundation models. *Hydrological Processes*, 28(22), 5495–5502.

Sengupta, R., Roy, P., & Chattopadhyay, S. (2020). Access to climate information and early warning systems in marginalized communities. *Climate Policy Brief,* 8(3), 14–19.

Shah, T., & Mishra, A. (2020). Climate change and groundwater: India's hidden water crisis. *Water Policy Research Highlight*. International Water Management Institute (IWMI).

Shahriar, M., Mahmud, M. S., & Hossain, M. T. (2020). Satellite-based flood monitoring and risk mapping for sustainable development in flood-prone areas of Bangladesh. *Remote Sensing Applications: Society and Environment,* 18, 100291.

Sharma, P., Bansal, A., & Mehrotra, R. (2022). Climate change, floods, and social vulnerability: A review of emerging challenges in South Asia. *Environmental Research Letters,* 17(4), 045002.

Singh, R., & Pandey, A. (2021). Spatial inequality and disaster vulnerability in the Indo-Gangetic Plain. *Journal of Rural Development Studies,* 37(2), 45–60.

Shen, X., Wang, D., Mao, K., Anagnostou, E., & Hong, Y. (2019). Inundation extent mapping by synthetic aperture radar: A review. *Remote Sensing*, 11(7), 879.

Shrestha, B. B., Kawasaki, A., & Zin, W. W. (2021). Development of flood damage functions for agricultural crops and their applicability in regions of Asia. *Journal of Hydrology: Regional Studies*, 36, 100872.

Stephens, E. M., Bates, P. D., Freer, J. E., & Mason, D. C. (2012). The impact of uncertainty in satellite data on the assessment of flood inundation models. *Journal of Hydrology*, 414, 162–173.

Tacnet, J. M., Mermet, E., & Maneerat, S. (2012). Analysis of importance of road network exposed to natural hazards. In *AGILE'2012 International Conference on Geographic Information System*, Avignon.

Tapia-Silva, F. O., Itzerott, S., Foerster, S., Kuhlmann, B., & Kreibich, H. (2011). Estimation of flood losses to agricultural crops using remote sensing. *Physics and Chemistry of the Earth, Parts A/B/C*, 36(7–8), 253–265.

UNICEF. (2021). *Impact of floods on child education in rural India.* United Nations Children's Fund India. https://www.unicef.org/india

Van Westen, C. J. (2013). Remote sensing and GIS for natural hazards assessment and disaster risk management. *Treatise on Geomorphology*, 3, 259–298.

Voigt, S., Kemper, T., Riedlinger, T., Kiefl, R., Scholte, K., & Mehl, H. (2007). Satellite image analysis for disaster and crisis-management support. *IEEE Transactions on Geoscience and Remote Sensing*, 45(6), 1520–1528.

World Bank. (2020). *Building resilient communities in South Asia: Equity and inclusion in disaster risk management.* The World Bank Group. https://www.worldbank.org

Xiong, B., Chen, J. M., & Kuang, G. (2012). A change detection measure based on a likelihood ratio and statistical properties of SAR intensity images. *Remote Sensing Letters*, 3(3), 267–275.

20 Unraveling the Vulnerability of Dry Tropical Grassland to Climate Change

A Case Study from an Asymmetrical Warming Experiment

Mohammad Sirazuddin Ansari, R. Sagar, Aakansha Pandey, Deepankar Yadav, Anshul Jaiswal, and Richa Rawat

INTRODUCTION

Tropical grasslands are one of the most widely distributed terrestrial biomes, approximately 1.5×10^9 hm^2 land area on the Earth (Cao et al., 2024). It provides several ecological services and economic benefits for humans (Ru et al., 2023; Scholtz and Twidwell, 2022). As per AR6 of the Intergovernmental Panel on Climate Change (IPCC), anthropogenic activities caused the global earth surface temperature to increase by 0.84°C–1.10°C between 2001 and 2020 compared to 1850 and 1900 (IPCC, 2023). This increase is expected to further increase by 2°C–6°C by the end of this century (IPCC, 2022). The average surface temperature of the Earth is currently 1.45°C higher than it was during pre-industrialization (WMO, 2024). More specifically, the rate of nighttime temperature increase was approximately twice that of the daytime over land between 1950 and 1993 (Du et al., 2019). The last 50 years have witnessed an asymmetric warming pattern as a result of a rise in daily minimum soil temperature over the daily maximum soil temperature (Phillips et al., 2011; Stone and Weaver, 2003). Studies have suggested that it will continue more severely in upcoming times than in the past (Peng et al., 2013; Stone and Weaver, 2003). Xia et al. (2010) reported that in Northern China, nighttime warming (NW) caused more rise in mean and daily minimum temperatures (0.74°C and 0.99°C, respectively) during NW than during daytime warming (DW) (0.60°C and 0.66°C, respectively) in a temperate steppe ecosystem. However, there were no significant changes observed in the increase in mean and daily maximum temperature between day (0.81°C and 1.13°C, respectively) and night (0.81°C and 1.10°C, respectively) during DW. Conversely, studies in India attributed an increasing trend of daily maximum temperature and insignificant change in daily minimum temperature (Alexander et al., 2006; Joshi et al., 2020). Few studies reported a sharp decrease in both daily minimum and maximum temperatures (Kothawale et al., 2012; Yadav et al., 2004). Thus, this uncertain change in daily minimum and daily maximum temperatures may harm soil and vegetation and human well-being (Vinnarasi et al., 2017).

Besides, the latest studies have focused attention on asymmetric warming, that is, DW and NW on species diversity, phenology, and vegetation (Cao et al., 2018; Shen et al., 2016). In China's Songnen Plain, marsh vegetation is asymmetrically affected by daytime maximum and nighttime minimum temperatures, according to the study by Wang et al. (2020). Due to warming-induced water shortages, rising maximum temperatures may hinder the growth of marsh vegetation, but compensatory stimulated photosynthesis in the Songnen Plain may favor rising nighttime minimum

DOI: 10.1201/9781003620747-23

temperatures (Wang et al., 2020). Other studies have shown that the DW reduces the temporal stability of an ecosystem by influencing bud growth (Rossi and Isabel, 2017) due to more photosynthetic activities during the daytime and decreasing the abundance of dominant plant species in semiarid grassland ecosystems (Yang et al., 2017). Additionally, NW suggests that if daily minimum temperatures increase more than maximum temperatures under continued climate change, the potential to sequester carbon can be enhanced in grasslands. In temperate steppe, continually increasing soil temperature by 0.64°C during nighttime might led to more carbon sequestration, whereas a 0.36°C increase in daytime temperature caused more carbon efflux (Wang et al., 2020). Since most of the studies have been largely explored in alpine marshes, tropical semiarid and temperate ecosystems, focused on aboveground vegetation and processes.

Soil microorganisms, often referred to as the "tip of the iceberg" in the ecosystem functioning (Brown et al. 2024), drive organic matter breakdown, nutrient cycling, carbon sequestration, plant diversity, productivity, and overall ecosystem resilience (Bardgett and Van Der Putten, 2014; Schimel, 2016). The responses of aboveground vegetation to asymmetrical warming patterns are a reflection of what happens belowground in soil. Soil microbes play a crucial role in determining ecosystem functioning and thereby ecosystem resilience (Wagg et al., 2014). Therefore, delimited studies of asymmetrical warming on belowground processes need to be explored, in addition to aboveground processes. In a study, NW rather than DW creates more favorable conditions for fungal growth, and fungi to fungi-to-bacteria ratio increased, thus inducing the mineralization of carbon and nitrogen (Zhao et al., 2024a). Long-term DW rather than NW increases the bacterial population, decreases the fungal population, decreases the fungi-to-bacteria ratio, and thus increases soil respiration (Feng et al., 2020). These observations demonstrated conflicting scenarios in responses of belowground nutrient cycling to asymmetrical warming in grassland ecosystems.

Majority of single and multiple climate change driving factor experiments were done in temperate Northern Hemisphere ecosystems (USA, Europe, and China) to learn more about how asymmetrical warming affects ecosystem processes. However, none were done in tropical, specifically dry tropical grasslands (Song et al., 2019). Moreover, the majority of studies were concentrated on understanding the effects of DW and NW on the aboveground processes (Peng et al., 2013). Few studies have looked at how asymmetrical warming affects the variety of microbes in the soil (Zhao et al., 2024b); therefore, it is important to compare the short-term response of DW and NW on microbial communities in a changing climate to gain a better understanding of how things work below ground. In addition, there are several critical questions to address: Does the short-term effect of DW and NW have profoundly affected the belowground processes? How does the soil microbe respond to asymmetrical warming patterns? What are the factors and mechanisms involved in underlying such changes? To better understand the vulnerability of this ecosystem and fulfill the knowledge gaps, we designed a manipulative experiment in dry tropical grassland in the Botanical Garden, Banaras Hindu University, Varanasi. The objectives of this study are: (i) to check whether DW or NW treatments are differently regulating the soil microbes; (ii) to access the response of soil microbial communities and microbial biomass; (iii) to estimate the soil extracellular enzyme activities linked to carbon, nitrogen, and phosphorus cycling; (iv) to determine the relationship between soil variables, soil microbial abundance, and extracellular enzyme activities.

Based on the specific objectives we hypothesized that (i) NW would cause a more favorable environment for the microbes to adapt and regulate the belowground processes, (ii) Gram-positive (+) bacteria to Gram-negative (–) bacteria (G+/G–) and microbial biomass—carbon and nitrogen (MBC: MBN) ratios stoichiometry would shift differentially in DW and NW, respectively, and (iii) NW affects extracellular enzyme activities (EEAs) more pronounced than DW. The findings of this study will provide ecological insights into the impacts of asymmetrical warming on belowground processes in dry tropical grassland functioning and will emphasize sustainable grassland management practices and ecosystem modeling.

MATERIALS AND METHODS

Study Area

The study site was located at the Botanical Garden on the campus of Banaras Hindu University, which is located in Varanasi, Uttar Pradesh, India, at 25.28° N, 82.95° E, and 76 m above mean sea level (Figure 20.1). None of the studies have been conducted in dry tropical grassland, especially in India. This area has also witnessed asymmetrical warming patterns as per climate conditions included in Table 20.1. This site was chosen to carry out the manipulative experiment as it has rich biodiversity of forbs, sedges, and grasses that mimic the natural grassland. The research area

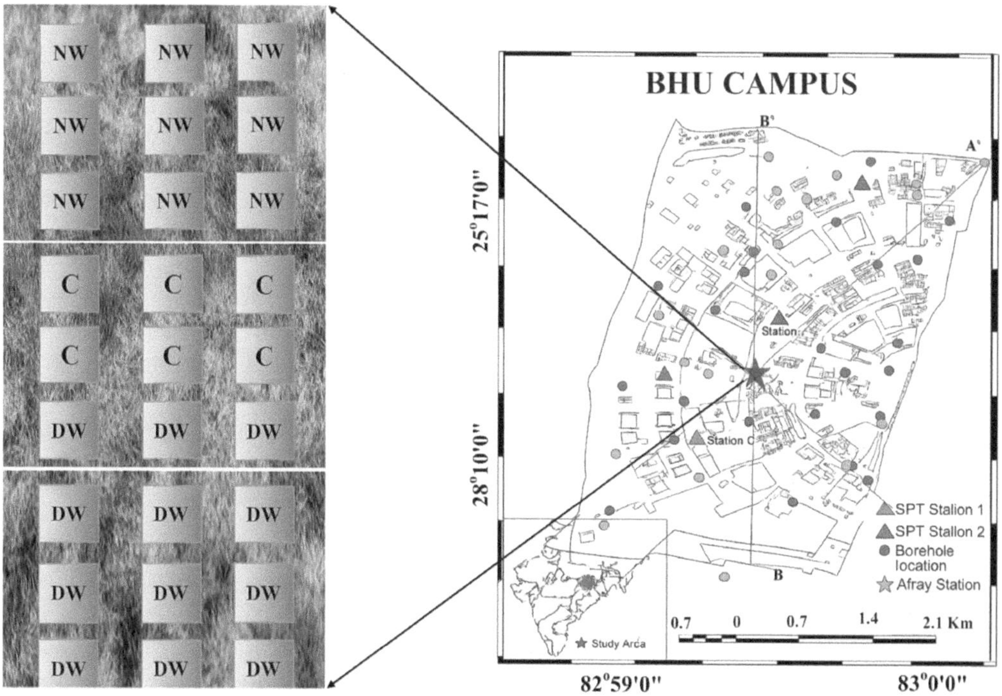

FIGURE 20.1 Spatial distribution of the study area in the Botanical Garden (left), located on the campus of Banaras Hindu University, Varanasi, India. (The map was generated by Sangeeta et al. 2022 (Department of Geophysics, BHU).)

TABLE 20.1
Variation of Daily Maximum Temperature, Daily Minimum Temperature and Daily Mean Temperature during Experimental Duration (2022–2024) in the Botanical Garden, Banaras Hindu University, Varanasi, India

Variables	2022 (°C)	2023 (°C)	2024 (°C)	Average Temperature (°C)	Rate (%)
DT_{max}	31.2	32.0	33.0	32.0	5.7
DT_{min}	17.7	18.5	19.9	18.7	12.4
DT_{mean}	24.4	25.2	26.4	25.3	8.1

Note: DT_{max}, DT_{min}, and DT_{mean} represent daily maximum temperature, daily minimum temperature, and daily mean temperature, respectively.

has a monsoonal climate categorized into three seasons: summer (April–June), rainy season (July–September), and winter (November–February). March marks the transition from winter to summer, while October marks the transition from rainy to winter. During the experimental period, the daily minimum temperature T_{min} and daily maximum temperature T_{max} were 18.7°C and 32.1°C, respectively, and mean annual precipitation was 1,100 m. The soil is described as an inceptisol, which has a silty loam texture, a neutral to alkaline pH, a pale brown color, and a low amount of nitrogen but substantial amounts of phosphorus and potassium. *Cynodon dactylon*, *Dicanthium annulatum*, and *Ageratum conyzoides* are the predominant species of the study area.

Experimental Design and Warming Treatments

An experimental plot was established in the grassland of the Botanical Garden of Banaras Hindu University. The experimental plot is divided into 27 subplots, each having a 2 m×2 m size. These subplots were further classified into three treatment levels: control (no warming), DW, and NW (Figure 20.1). Each treatment had nine replicates. The warming treatment was provided by installing the MSR-2420 infrared radiators at 2.25 m high from the ground (Kalglo Electronics Inc., Bethlehem, USA). It was continued for three consecutive years (2022–2024). The daily minimum temperature (T_{min}) was exceeded at a faster rate than the daily maximum temperature (T_{max}) over the experimental period. A detailed description is included in Table 20.1. In addition, to replicate the shading effects of the infrared radiator, a single "dummy" heater of the same size and form is used in each treatment. In order to simulate asymmetric warming, that is, $T_{min} > T_{max}$, for DW, the radiators remained switched on from 6 a.m. to 6 p.m. of each day with a temperature of 0.9°C higher than the ambient temperature, whereas for those for the NW, the radiators remained switched on from 6 p.m. to 6 a.m. with a temperature of 1.5°C higher than the nighttime ambient temperature (Xia et al., 2009).

FIELD SAMPLING

Three soil samples at 10-cm soil depth from the adjacent plots of each 2 m×2 m plot for each treatment were randomly collected. The randomly collected soil after removal of litter debris was stored at 4°C for the analysis of soil extracellular enzyme activities and at –80°C for lipid extraction and analysis and the assessment of soil bacterial and fungal communities. Also, just this analysis, the soil was allowed for the analysis of other soil attributes.

Measurement of Soil Properties and Microbial Community Composition

Soil temperature was measured using an ordinary thermometer; soil pH was determined from 5 g of soil mixed with 50 mL of deionized water; and soil moisture was calculated by oven-drying 5 g of soil (after recording its fresh weight) at – 80°C and then reweighing it (Verma and Sagar, 2020). Microbial biomass carbon (MBC) was estimated by the chloroform fumigation and incubation method (Jenkinson, 1988), and nitrogen (MBN) was estimated by the chloroform fumigation and extraction method (Vance et al., 1987).

Urease (Wu et al., 2024), β-1,4-N-acetylglucosaminidase (Tabatabai and Bremner, 1971), phenol oxidase (Iyyemperumal and Shi, 2008), acid phosphatase, and alkaline phosphatase (Tabatabai and Bremner, 1971; Eivazi and Tabatabai, 1977) were measured using spectrophotometric methods as per the standard protocol.

To analyze the soil microbial community, fatty acid assays were done. 0.5 g of sample was taken in the sample preparation tube, and then 2 mL BF_3, 1 mL toluene, and 1 mL methanol were added. The sample was then heated for 45 min in the water bath. The sample was then used for a fatty acid determination using the GC-FID test. The microbial community was then analyzed using the appropriate MIDI PLFAD1 software (MIDI, Inc., DE, USA), and biomarker fatty acids were used for categorizing and computing the occurrence and relative abundance of specific microbial groups (Zelles, 1999; Frostegård et al., 2011).

Statistical Analysis

The responses of DW and NW on soil microbial abundance, EEAs, and soil properties were analyzed using a one-way measure of analysis of variance with the help of SPSS software 12.0. Pearson bivariate correlation was performed to determine the degree of correlation among the soil microbial compositions, EEAs and soil variables. The relative change in response ratio (RR) to treatments was calculated using Eq. (20.1). For RR, the data were converted to natural logarithm [ln(*RR*)]. The natural logarithmic value of the mean ratio of the treatment group and the control group reflected the influence of treatment in the experiment (Hedges et al., 1999). The *RR* was calculated using the following equation:

$$RR = \ln\left(\frac{Xt}{Xc}\right) = \ln X_t - \ln X_c \tag{20.1}$$

where X_t and X_c are the means of each variable in the treatment groups and control groups, respectively. The *RR* was then transformed into percentage change.

RESULTS

Effects of DW and NW on Soil Microclimate

The soil temperature measured for DW and NW treatments was increased by 1.71% and 2.83%, respectively, as compared to the control. Compared to the control, the DW decreased the soil moisture by 10.45%, followed by a 7.25% reduction in NW, respectively. The soil pH varied in the NW treatment, increasing by 2.78%, whereas the DW treatment decreased it by 1.42%, as compared to the control, respectively (Figure 20.2). It seems that except for the soil moisture, the differences in temperature increase and pH change due to treatments were insignificant.

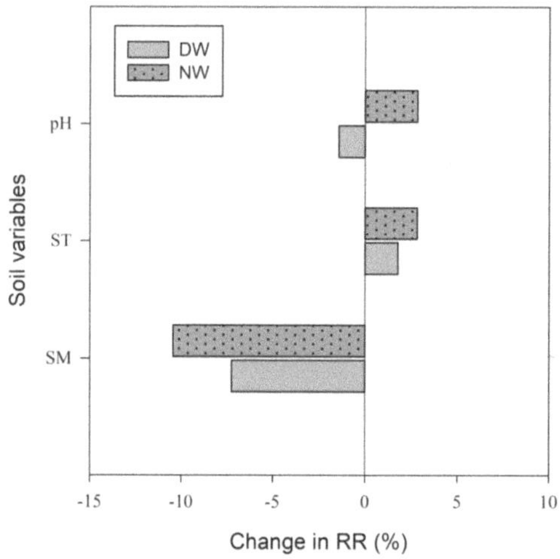

FIGURE 20.2 Response of soil variables to daytime and nighttime warming. (Note: ST and SM represent soil temperature and soil moisture, respectively.)

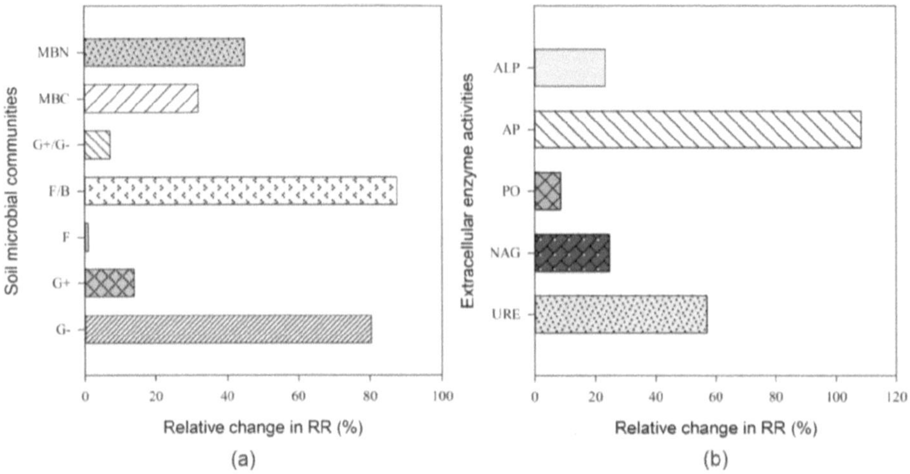

FIGURE 20.3 Relative change in response ratios (RR; %) of (a) soil microbial community composition and (b) extracellular enzyme activities under daytime and nighttime warming treatments in the tropical grassland.

Effects of DW and NW on Soil Microbial Community Composition

The MBC and MBN significantly varied during DW and NW, and their respective differences compared to the control were 31.94% and 45% ($p<0.05$) (Figure 20.3a), which could be caused by elevated soil temperature, reduced soil moisture, and soil pH. Compared to control warming, the NW treatment rather than DW treatment significantly decreased the relative abundance of fungi to bacteria (87.53%; $p<0.05$) and G+/G– (7.12%; $p<0.05$), respectively. The Gram (–) bacteria abundance has been observed more pronouncedly during NW as compared to DW, followed by the control (Figure 20.3a). Results showed that short-term NW and DW treatments significantly altered the microbial composition and their relative abundance.

Response of DW and NW Treatments on Soil EEAs

The EEAs showed differential responses to DW and NW. Among the N- and P-acquisition enzymes, the NAG increased by 24.76%, the URE by 57.07%, the AP by 108.35%, and the ALP by 23.37%. These increases were statistically significant ($p<0.01$) during NW, whereas PO's C-acquisition enzymatic activity significantly increased by 8.45% ($p<0.01$) during DW as compared to the control (Figure 20.3b).

Relationship between Soil Microbial Communities, EEAs and Soil Variables during DW and NW

Pearson's correlation analysis showed that soil G (+) bacteria and G (–) bacteria were significantly and positively correlated ($r=0.98$; $p<0.01$) and linked with the fungal community ($r=0.94$–0.98; $p<0.01$) and MBN ($r=0.99$; $p<0.01$). The NAG activity was significantly and negatively correlated with G (+) bacteria, G (–) bacterial and fungal community ($r=-0.85$ to -0.63; $p<0.05$), URE activities ($r=-0.81$; $p<0.01$), and AP activities ($r=-0.94$; $p<0.01$), respectively. SM and ST positively correlated with G (+) bacteria, G (–) bacteria, fungal community, MBC: MBN ratio, AP, and ALP activities ($r=0.77$–0.98; $p<0.01$) and negative correlation with NAG and PO ($r=0.93$; $p<0.01$). Soil pH showed significantly strong negative correlation with NAG and PO ($r=-0.87$; $p<0.01$), whereas significantly strong positive correlation with MBC: MBN and AP ($r=0.95$; $p<0.01$) (Table 20.2).

A summarized framework of all the observations under asymmetrical warming is depicted in Figure 20.4.

TABLE 20.2

Pearson Correlation Matrix Indicating the Relationship of Soil Physiochemical Properties to the Soil Microbial Community Composition and Soil Extracellular Enzyme Activities under DW and NW Treatments during the Experimental Duration in Dry Tropical Grassland Located in Botanical Garden, Banaras Hindu University, Varanasi, India

Variables	G+	G–	F	G+:G-	F:B	MBN	MBC	MBC:MBN	NAG	URE	PO	AP	ALP	ST	SM
G+															
G–	0.98**														
F	0.94**	0.98**													
G+:G–	0.99**	0.97**	0.92**												
F:B	0.12^{NS}	0.28^{NS}	0.44^{NS}	0.06^{NS}											
MBN	0.99**	0.99**	0.97**	0.98**	0.24^{NS}										
MBC	0.51^{NS}	0.64^{NS}	0.77*	0.46^{NS}	0.91**	0.62^{NS}									
MBC:MBN	0.68*	0.56^{NS}	0.41^{NS}	0.73*	–0.63^{NS}	0.59^{NS}	–0.26^{NS}								
NAG	–0.85**	–0.75*	–0.63^{NS}	–0.88**	0.41^{NS}	–0.77*	0.00^{NS}	–0.96**							
URE	–0.13^{NS}	0.02^{NS}	0.19^{NS}	–0.19^{NS}	0.96**	–0.01^{NS}	0.77*	–0.81**	0.63^{NS}						
PO	–0.85**	–0.75*	–0.63^{NS}	–0.87**	0.41^{NS}	–0.77*	0.00^{NS}	–0.96**	0.43^{NS}	0.63^{NS}					
AP	0.63^{NS}	0.50^{NS}	0.34^{NS}	0.67*	–0.68*	0.53^{NS}	–0.33^{NS}	0.99**	–0.94**	–0.85**	–0.94**				
ALP	0.99**	0.98**	0.96**	0.99**	0.17^{NS}	0.99**	0.55^{NS}	0.65^{NS}	–0.82**	–0.09^{NS}	–0.82**	0.59^{NS}			
ST	0.99**	0.99**	0.97**	0.98**	0.23^{NS}	0.99**	0.60^{NS}	0.60^{NS}	–0.78*	–0.03^{NS}	–0.78*	0.54^{NS}	0.99**		
SM	0.97**	0.93**	0.86**	0.98**	–0.06^{NS}	0.94**	0.34^{NS}	0.81**	–0.93**	–0.32^{NS}	–0.93**	0.77*	0.96**	–0.95**	
pH	0.52^{NS}	0.40^{NS}	0.22^{NS}	0.56^{NS}	–0.74*	0.41^{NS}	–0.40^{NS}	0.94**	–0.87**	–0.87**	–0.87**	0.95**	0.48^{NS}	0.43^{NS}	0.67*

Note: Values are Pearson's correlation coefficient (r), and asterisk (*) represents the level of significance. *Significant at $p<0.05$ and **significant at $p<0.01$, respectively.

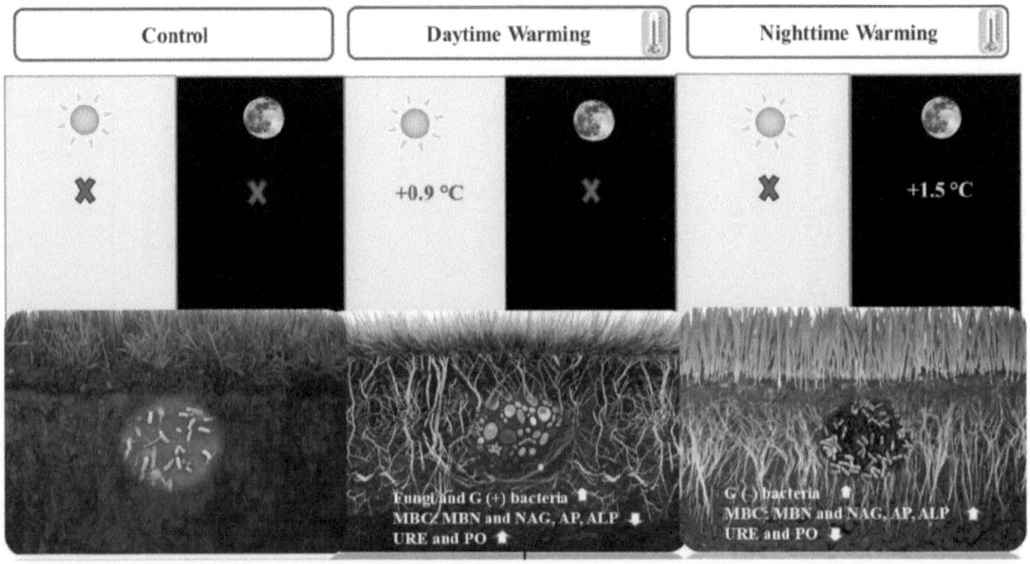

FIGURE 20.4 A conceptual framework of effects of daytime and nighttime warming on belowground processes, viz. soil microbial community composition, and extracellular enzyme activities in the tropical grassland. (Prepared by Mohammad Sirazuddin Ansari.)

DISCUSSION

The observed differences in microbial communities and their functional attributes due to asymmetric warming supported our first hypothesis, as the NW had a significantly greater influence on soil physiochemical properties than the DW, resulting in differences in microbial composition across the treatments. A global study by Guo et al. (2019), on the other hand, found that warming increased the number of microbes compared to when there was no warming. Also, Feng et al. (2023a) found that DW and NW had no effect on the microbial community in a semiarid grassland during a short-term study, but in a long-term study, DW decreased the number of fungi and the ratio of fungi to bacteria. A relative amount of total bacteria, Gram-positive bacteria, and fungi decreased significantly during the NW, but did not change significantly during the DW in a semiarid temperate steppe (Zhang et al., 2011). Similarly, other studies also reported changes in microbial community composition in response to DW and NW (Feng et al., 2023b; Zhao et al., 2024b).

In this study, the bacterial (both Gram-positive and Gram-negative) and fungal communities responded differently to the DW and NW treatments, which could be due to differences in EEAs responses to the patterns of warming led to changes in how the microbes functioned. This might have occurred due to changes in soil moisture and pH caused by warming-induced temperature increase. The increased soil temperature during the night leads to the greater proliferation of Gram-negative bacteria than Gram-positive bacteria (Zhang et al., 2011). This phenomenon may be attributed to the wide metabolic flexibility of Gram-negative bacteria and their ability to consume readily available labile substrates, both of which contribute to their increased abundance (Ali et al., 2018). In semiarid areas, evapotranspiration made the DW treated plots much drier than the NW treatment. In the same way, the different responses of the microbes in this study might be due to the lower soil moisture and pH in the DW treatment compared to the NW treatment. Also, there were more fungi than bacteria and more Gram-positive bacteria than Gram-negative bacteria. This could be because the Gram-positive bacteria released thick extracellular polymeric substances and endospores to protect themselves from water stress by keeping water and nutrients for a longer time or being more susceptible to dryness (Manzoni et al., 2012) and longer average path lengths and restrictions on dispersal (Zhong et al., 2023; Zhong et al., 2024), while the fungi formed large hyphal networks that helped the retaining of water in place over long distances (Kerfeld et al., 2018; Rodriguez-Caballero et al., 2018). A study by Wang and Kuzyako (2024) suggested a favorable impact of warming on Gram-negative bacteria, which were more responsive to fluctuating resource availability than fungi. A drop in the ratio of fungi to bacteria and the ratio of Gram-positive to Gram-negative bacteria under the NW suggested a shift toward r-strategist microbes that are fast-growing and can quickly use available nutrients (Gul and Whalen, 2022). This was shown by the strong negative correlation between soil pH and fungi-to-bacteria ratio, and the positive correlation between the Gram-positive to Gram-negative bacteria ratio and soil moisture. Thus, asymmetrical warming altered soil attributes, which in turn caused differential responses in microbial community composition and their functions in the tropical grassland.

Further, the NW treatment had greater MBN and slightly lower MBC, as compared to the DW treatment; this supported our second hypothesis. This could potentially be attributed to differences in warming patterns in soil dryness, as suggested by Xu and Yuan (2017), and the availability of more labile carbon and nitrogen substrates, as demonstrated by other studies (Cheng, 1999). Nevertheless, a global meta-analysis study has suggested that semiarid grassland ecosystems are more prone to soil dryness, so the MBN tends to decrease in warming magnitude below <1°C but increases beyond >1°C–2°C, whereas MBC tends to decrease in both warming magnitudes (Xu and Yuan, 2017).

The Preferential Substrate Utilization Hypothesis says that when the soil gets warmer and there is more labile carbon, microbes may choose to use the labile carbon instead of making hydrolytic enzymes like NAG, URE, AP, and ALP (Cheng, 1999; Qui et al., 2023). Interestingly, the reduction in the activities of these enzymes during the DW justified our third hypothesis. The substantial increase in nitrogen-(NAG and URE) and phosphorus-(AP and ALP) acquisition enzymes under NW treatment signifies that the microbial communities prioritize nutrient mobilization to compensate for potential resource limitations caused by higher microbial activities (Qiu et al., 2023). Because bacteria grow faster than fungi under NW treatment, this may be the reason why nitrogen and phosphorus acquisition enzymes are working harder. Bacteria need nutrients to grow quickly (Xu et al., 2020). The NAG and URE activities were more active during the NW treatment. This was likely because different microbes were competing for nutrients (Wang and Kuzyakov, 2024). For instance, fungi have a higher C:N ratio than Gram-positive bacteria (Paul, 2016; Wallenstein et al., 2006), which means they need different amounts of nutrients from the soil and are better at taking in carbon (Seaton et al., 2022). The NAG, primarily released by fungi, have a tendency to utilize the less energy-demanding substrates (labile carbon as substrate), which is also a competent nutrient source for Gram-positive bacteria. This is supported by a meta-analysis by Waring et al. (2013) that the Gram-positive bacteria utilize the chitin-derived labile carbon to fulfill the substrate requirement. Both are stable during the DW treatment, so during NW, they increase their NAG and URE enzyme activities to meet their nutritional needs. This finding is supported by occurrences of negative correlations of NAG and URE activities with the MBC:MBN ratio and a positive correlation of URE activity with the fungi-to-bacteria ratio, as well as a negative correlation of NAG activity with the Gram-positive to Gram-negative bacteria ratio.

In addition, the phosphorus-acquiring enzymes (AP and ALP) were more active during the NW treatment compared to the DW treatment. The acidification and drier conditions under DW might limit enzyme efficiency, while the slight alkalization under NW could enhance phosphorus availability due to high microbial turnover that causes an increase in AP and ALP activities (Feng et al., 2023b). A study by Margalef et al. (2017) found that AP activity can stay active in alkaline conditions. They also found that higher microbial turnover in semiarid regions boosts phosphatase activities, even when pH is high. This conclusion could be further justified by the positive correlation of AP and ALP activities with the Gram-positive to Gram-negative bacteria and a negative correlation between the AP and fungi-to-bacteria ratio. During the DW treatment, carbon substrates depleted more rapidly; this might drive expression of enzymes toward enzymes with higher turnover rate and affinities (Steinweg et al., 2013; Bradford, 2013). Fungal dominance during DW treatment supports recalcitrant compound degradation and physiological stress caused by evapotranspiration, leading to reliance on alternative energy sources (recalcitrant compounds). There might be a possibility of enhanced PO activity, as it is evident due to occurrences of a positive correlation between PO activity and the fungi-to-bacteria ratio and a negative correlation with the soil pH and soil moisture.

CONCLUSION AND ECOLOGICAL IMPLICATIONS

The DW dries and acidifies soil, while NW warms and dries and alkalizes it. NW promotes Gram-negative bacteria, while DW stabilizes the fungal population but lowers enzymatic function. NW enhances nitrogen and phosphorus acquisition enzyme activity and nutrient mobilization, while DW shifts microbial resource allocation through carbon decomposition. NW treatment may shift microbial composition more pronouncedly than DW treatment, making soil microbes more tolerant to intervention. This study shows how future asymmetrical warming and moisture fluctuation may alter soil microbial communities and nutrient cycling in dry tropical grassland. Dry tropical soil microbial communities and tropical "grasslands" climate change consequences are illuminated by short-term asymmetric warming. Future research on long-term consequences, global climate change causes, and aboveground vegetation-microbial interactions under asymmetric warming patterns may need a long-term study for a better feedback of microbial composition and their function to the climate warming.

Soil bacteria recycle plant growth nutrients. NW promotes grassland bacteria. NW may promote carbon-storing bacteria in grasslands to adapt to climate change. DW quickly dries soil, straining microorganisms and limiting grassland productivity. Inoculating tropical C_4 grass with *Azospirillum brasilense* and *Pseudomonas fluorescens* may increase photosynthesis, productivity, and feed quality for strengthening the tropical grasslands. This stabilizes fungi-to-bacteria ratios after asymmetrical temperature. Water-efficient C_4 grasses resist heat and drought. Compost, charcoal, and manure boost soil microbial activity and water retention, lowering heat and drying stress. Tillage, reseeding, artificial and semi-artificial grasslands, and mowing regenerate grasslands. Regular mowing and reseeding preserve plant diversity and avoid climate-induced dominance shifts in degraded grasslands. Further, legumes may be used to fix nitrogen and store carbon in tropical grasslands. Rotational grazing reduces overgrazing, maintains soil cover and root biomass, and strengthens grasslands against temperature and moisture variations. Mulching and contour farming moistens DW. Further, government afforestation, regulated burning, community grazing, replanting, and soil protection and native species restoration may be useful for the health of tropical grassland and human well-being. These strategies could be useful for restoring the tropical grasslands and their microbial composition under the warming scenario.

LIMITATIONS

Soil EEAs could be highly sensitive to microclimatic conditions, which might not be fully controlled in the field experiment. This could cause variability and affect reproducibility. Short-term studies may not capture long-term enzyme responses to environmental changes, limiting the extrapolation of findings to broader climate change scenarios. The susceptibility of dry tropical ecosystems to climate change will be better understood through long-term future research.

ACKNOWLEDGMENTS

MSA gratefully acknowledges financial support as DBT-SRF from the Regional Centre for Biotechnology, Department of Biotechnology, Government of India (DBTHRDPMU/DBT-JRF/2024-25/SRF/851). RS highly acknowledges Science and Engineering Research Board (SERB), New Delhi, for the financial support with file number: EEQ/2021/000356 and IOE, Institute of Science, Banaras Hindu University.

REFERENCES

Alexander, L. V., Zhang, X., Peterson, T. C., Caesar, J., Gleason, B., Klein Tank, A. M. G.,... & Vazquez-Aguirre, J. L. (2006). Global observed changes in daily climate extremes of temperature and precipitation. *Journal of Geophysical Research: Atmospheres*, *111*(D5), 1–22. https://doi.org/10.1029/2005JD006290.

Ali, R. S., Poll, C., & Kandeler, E. (2018). Dynamics of soil respiration and microbial communities: Interactive controls of temperature and substrate quality. *Soil Biology and Biochemistry*, *127*, 60–70.

Bardgett, R. D., & Van Der Putten, W. H. (2014). Belowground biodiversity and ecosystem functioning. *Nature*, *515*(7528), 505–511.

Bradford, M. A. (2013). Thermal adaptation of decomposer communities in warming soils. *Frontiers in Microbiology*, *4*, 333.

Brown, R. W., Reay, M. K., Centler, F., Chadwick, D. R., Bull, I. D., McDonald, J. E.,... & Jones, D. L. (2024). Soil metabolomics-current challenges and future perspectives. *Soil Biology and Biochemistry*, *193*, 109382.

Cao, R., Shen, M., Zhou, J., & Chen, J. (2018). Modeling vegetation green-up dates across the Tibetan Plateau by including both seasonal and daily temperature and precipitation. *Agricultural and Forest Meteorology*, *249*, 176–186.

Cao, X., Zhao, C., Jia, H., & Yang, J. (2024). The dynamic characteristics and influencing factors of soil respiration in different types of grasslands in the Barkol Lake Basin. *Agriculture;* Basel, *14*(12), 2182. https://doi.org/10.3390/agriculture14122182.

Cheng, W. (1999). Rhizosphere feedbacks in elevated CO_2. *Tree Physiology*, *19*(4–5), 313–320.

Du, Z., Zhao, J., Liu, X., Wu, Z., & Zhang, H. (2019). Recent asymmetric warming trends of daytime versus nighttime and their linkages with vegetation greenness in temperate China. *Environmental Science and Pollution Research*, *26*, 35717–35727.

Eivazi, F., & Tabatabai, M. A. (1977). Phosphatases in soils. *Soil Biology and Biochemistry*, *9*(3), 167–172.

Feng, J., Ru, J., Song, J., Qiu, X., & Wan, S. (2023a). Long-term daytime warming rather than nighttime warming alters soil microbial composition in a semi-arid grassland. *Biology*, *12*(5), 699.

Feng, L., Cao, B., & Wang, X. (2023b). Response of soil extracellular enzyme activity and stoichiometry to short-term warming and phosphorus addition in desert steppe. *PeerJ*, *11*, e16227.

Feng, S., Liu, J., Zhang, Q., Zhang, Y., Singh, V. P., Gu, X., & Sun, P. (2020). A global quantitation of factors affecting evapotranspiration variability. *Journal of Hydrology*, *584*, 124688.

Frostegård, Å., Tunlid, A., & Bååth, E. (2011). Use and misuse of PLFA measurements in soils. *Soil Biology and Biochemistry*, *43*(8), 1621–1625.

Gul, S., & Whalen, J. K. (2022). Perspectives and strategies to increase the microbial-derived soil organic matter that persists in agroecosystems. *Advances in Agronomy*, *175*, 347–401.

Guo, X., Zhou, X., Hale, L., Yuan, M., Ning, D., Feng, J.,... & Zhou, J. (2019). Climate warming accelerates temporal scaling of grassland soil microbial biodiversity. *Nature Ecology & Evolution*, *3*(4), 612–619.

Hedges, L. V., Gurevitch, J., & Curtis, P. S. (1999). The meta-analysis of response ratios in experimental ecology. *Ecology*, *80*(4), 1150–1156.

Iyyemperumal, K., & Shi, W. (2008). Soil enzyme activities in two forage systems following application of different rates of swine lagoon effluent or ammonium nitrate. *Applied Soil Ecology*, *38*(2), 128–136.

Jenkinson, D. S. (1988). Determination of microbial biomass carbon and nitrogen in soil. In *Advances in Nitrogen Cycling in Agricultural Systems.* Ed. J. R. Wilson. Wallingford: CABI, 368–386.

Joshi, M. K., Rai, A., Kulkarni, A., & Kucharski, F. (2020). Assessing changes in characteristics of hot extremes over India in a warming environment and their driving mechanisms. *Scientific Reports*, *10*(1), 2631.

Kerfeld, C. A., Aussignargues, C., Zarzycki, J., Cai, F., & Sutter, M. (2018). Bacterial microcompartments. *Nature Reviews Microbiology*, *16*(5), 277–290.

Kothawale, D. R., Kumar, K. K., & Srinivasan, G. (2012). Spatial asymmetry of temperature trends over India and possible role of aerosols. *Theoretical and Applied Climatology*, *110*, 263–280.

Manzoni, S., Schimel, J. P., & Porporato, A. (2012). Responses of soil microbial communities to water stress: results from a meta-analysis. *Ecology*, *93*(4), 930–938.

Margalef, O., Sardans, J., Fernández-Martínez, M., Molowny-Horas, R., Janssens, I. A., Ciais, P.,... & Penuelas, J. (2017). Global patterns of phosphatase activity in natural soils. *Scientific Reports*, *7*(1), 1337.

Paul, E. A. (2016). The nature and dynamics of soil organic matter: Plant inputs, microbial transformations, and organic matter stabilization. *Soil Biology and Biochemistry*, *98*, 109–126.

Peng, S., Piao, S., Ciais, P., Myneni, R. B., Chen, A., Chevallier, F.,... & Zeng, H. (2013). Asymmetric effects of daytime and night-time warming on Northern Hemisphere vegetation. *Nature*, *501*(7465), 88–92.

Phillips, C. L., Gregg, J. W., & Wilson, J. K. (2011). Reduced diurnal temperature range does not change warming impacts on ecosystem carbon balance of Mediterranean grassland mesocosms. *Global Change Biology*, *17*(11), 3263–3273.

Qiu, Y., Zhang, K., Zhao, Y., Zhao, Y., Wang, B., Wang, Y.,... & Hu, S. (2023). Climate warming suppresses abundant soil fungal taxa and reduces soil carbon efflux in a semi-arid grassland. *Mlife*, *2*(4), 389–400.

Rodriguez-Caballero, E., Belnap, J., Büdel, B., Crutzen, P. J., Andreae, M. O., Pöschl, U., & Weber, B. (2018). Dryland photoautotrophic soil surface communities endangered by global change. *Nature Geoscience*, *11*(3), 185–189.

Rossi, S., & Isabel, N. (2017). Bud break responds more strongly to daytime than night-time temperature under asymmetric experimental warming. *Global Change Biology*, *23*(1), 446–454.

Ru, J., Wan, S., Hui, D., & Song, J. (2023). Overcompensation of ecosystem productivity following sustained extreme drought in a semiarid grassland. *Ecology*, *104*(4), e3997.

Schimel, J. (2016). Microbial ecology: Linking omics to biogeochemistry. *Nature Microbiology*, *1*(2), 1–2.

Scholtz, R., & Twidwell, D. (2022). The last continuous grasslands on Earth: Identification and conservation importance. *Conservation Science and Practice*, *4*(3), e626.

Seaton, F. M., Reinsch, S., Goodall, T., White, N., Jones, D. L., Griffiths, R. I.,... & Robinson, D. A. (2022). Long-term drought and warming alter soil bacterial and fungal communities in an upland heathland. *Ecosystems*, *25*(6), 1279–1294.

Shen, M., Piao, S., Chen, X., An, S., Fu, Y. H., Wang, S.,... & Janssens, I. A. (2016). Strong impacts of daily minimum temperature on the green-up date and summer greenness of the Tibetan Plateau. *Global Change Biology*, *22*(9), 3057–3066.

Song, J., Wan, S., Piao, S., Knapp, A. K., Classen, A. T., Vicca, S.,... & Zheng, M. (2019). A meta-analysis of 1,119 manipulative experiments on terrestrial carbon-cycling responses to global change. *Nature Ecology & Evolution*, *3*(9), 1309–1320.

Steinweg, J. M., Dukes, J. S., Paul, E. A., & Wallenstein, M. D. (2013). Microbial responses to multi-factor climate change: Effects on soil enzymes. *Frontiers in Microbiology*, *4*, 146.

Stone, D., & Weaver, A. (2003). Factors contributing to diurnal temperature range trends in twentieth and twenty-first century simulations of the CCCma coupled model. *Climate Dynamics*, *20*(5), 435–445.

Tabatabai, M. A., & Bremner, J. M. (1971). Michaelis constants of soil enzymes. *Soil Biology and Biochemistry*, *3*(4), 317–323.

Thorne, P. W., Donat, M. G., Dunn, R. J. H., Williams, C. N., Alexander, L. V., Caesar, J.,... & Rennie, J. J. (2016). Reassessing changes in diurnal temperature range: Intercomparison and evaluation of existing global data set estimates. *Journal of Geophysical Research: Atmospheres*, *121*(10), 5138–5158.

Vance, E. D., Brookes, P. C., & Jenkinson, D. S. (1987). An extraction method for measuring soil microbial biomass C. *Soil Biology and Biochemistry*, *19*(6), 703–707.

Verma, P., & Sagar, R. (2020). Effect of nitrogen (N) deposition on soil-N processes: A holistic approach. *Scientific Reports*, *10*(1), 10470.

Vinnarasi, R., Dhanya, C. T., Chakravorty, A., & AghaKouchak, A. (2017). Unravelling diurnal asymmetry of surface temperature in different climate zones. *Scientific Reports, 7*(1), 7350.

Wagg, C., Bender, S. F., Widmer, F., & Van Der Heijden, M. G. (2014). Soil biodiversity and soil community composition determine ecosystem multifunctionality. *Proceedings of the National Academy of Sciences*, *111*(14), 5266–5270.

Wallenstein, M. D., McNulty, S., Fernandez, I. J., Boggs, J., & Schlesinger, W. H. (2006). Nitrogen fertilization decreases forest soil fungal and bacterial biomass in three long-term experiments. *Forest Ecology and Management*, *222*(1–3), 459–468.

Wang, C., & Kuzyakov, Y. (2024). Mechanisms and implications of bacterial–fungal competition for soil resources. *The ISME Journal*, *18*(1), wrae073.

Wang, Y., Shen, X., Jiang, M., & Lu, X. (2020). Vegetation change and its response to climate change between 2000 and 2016 in marshes of the Songnen Plain, Northeast China. *Sustainability*, *12*(9), 3569.

Waring, B. G., Averill, C., & Hawkes, C. V. (2013). Differences in fungal and bacterial physiology alter soil carbon and nitrogen cycling: Insights from meta-analysis and theoretical models. *Ecology Letters*, *16*(7), 887–894.

WMO. (2024). *WMO confirms 2024 as warmest year on record at about 1.55°C above pre-industrial level*. The World Meteorological Organization. Accessed from https://wmo.int/news/media-centre/wmo-confirms-2024-warmest-year-record-about-155degc-above-pre-industrial-level.

Wu, J., Wang, H., Li, G., & Chen, N. (2024). Effects of nitrogen deposition on soil nitrogen fractions and enzyme activities in wet meadow of the Qinghai-Tibet Plateau. *Scientific Reports*, *14*(1), 31848.

Xia, J., Chen, S., & Wan, S. (2010). Impacts of day versus night warming on soil microclimate: Results from a semiarid temperate steppe. *Science of the Total Environment*, *408*(14), 2807–2816.

Xia, J., Han, Y., Zhang, Z., Zhang, Z., & Wan, S. (2009). Effects of diurnal warming on soil respiration are not equal to the summed effects of day and night warming in a temperate steppe. *Biogeosciences*, *6*(8), 1361–1370.

Xu, W., & Yuan, W. (2017). Responses of microbial biomass carbon and nitrogen to experimental warming: A meta-analysis. *Soil Biology and Biochemistry*, *115*, 265–274.

Xu, Y., Zhao, Z., Tong, W., Ding, Y., Liu, B., Shi, Y., ... Zhao, G. (2020). An acid-tolerance response system protecting exponentially growing Escherichia coli. *Nature Communications*, *11*(1), 1496.

Yadav, R. R., Park, W. K., Singh, J., & Dubey, B. (2004). Do the western Himalayas defy global warming? *Geophysical Research Letters*, *31*(17), 1–5. https://doi.org/10.1029/2004GL020201.

Yang, Z., Zhang, Q., Su, F., Zhang, C., Pu, Z., Xia, J.,... & Jiang, L. (2017). Daytime warming lowers community temporal stability by reducing the abundance of dominant, stable species. *Global Change Biology*, *23*(1), 154–163.

Zelles, L. (1999). Fatty acid patterns of phospholipids and lipopolysaccharides in the characterisation of microbial communities in soil: A review. *Biology and Fertility of Soils*, *29*, 111–129.

Zhang, N., Xia, J., Yu, X., Ma, K., & Wan, S. (2011). Soil microbial community changes and their linkages with ecosystem carbon exchange under asymmetrically diurnal warming. *Soil Biology and Biochemistry*, *43*(10), 2053–2059.

Zhao, J., Yang, W., Tian, L., Qu, G., & Wu, G. L. (2024a). Warming differentially affects above-and below-ground ecosystem functioning of the semi-arid alpine grasslands. *Science of the Total Environment*, *914*, 170061.

Zhao, P., Huang, Y., Liu, B., Chen, J., Lei, Z., Zhang, Y.,... & Peng, S. (2024b). Effects of daytime and nighttime warming on soil microbial diversity. *Geoderma*, *447*, 116909.

Zhong, Z., Zhang, G., & Fu, G. (2023). Effect of experiment warming on soil fungi community of *Medicago sativa*, *Elymus nutans* and *Hordeum vulgare* in Tibet. *Journal of Fungi*, *9*(9), 885.

Zhong, Z., Zhang, G., & Fu, G. (2024). Response of soil bacteria community to experiment warming in three agroecosystems of the Tibet. *Global Ecology and Conservation*, *50*, e02837.

Index